Optical Properties of Condensed Matter and Applications

Wiley Series in Materials for Electronic and Optoelectronic Applications

Series Editors

Dr. Peter Capper, *SELEX Sensors and Airborne Systems Infrared Ltd., Southampton, UK*

Professor Safa Kasap, *University of Saskatchewan, Canada*

Professor Arthur Willoughby, *University of Southampton, Southampton, UK*

Published Titles

Bulk Crystal Growth of Electronic, Optical and Optoelectronic Materials, Edited by P. Capper

Properties of Group-IV, III–V and II–VI Semiconductors, Edited by S. Adachi

Charge Transport in Disordered Solids with Applications in Electronics, Edited by S. Baranovski

Forthcoming Titles

Thin Film Solar Cells: Fabrication, Characterization and Applications, Edited by J. Poortmans and V. Arkhipov

Liquid Phase Epitaxy of Electronic, Optical and Optoelectronic Materials, Edited by P. Capper and M. Mauk

Dielectric Films for Advanced Microelectronics, Edited by K. Maex, M.R. Baklanov and M. Green

Optical Properties of Condensed Matter and Applications

Edited by

Jai Singh
Charles Darwin University, Darwin, Australia

John Wiley & Sons, Ltd

Copyright © 2006   John Wiley & Sons Ltd, The Atrium, Southern Gate, Chichester,
West Sussex PO19 8SQ, England

Telephone (+44) 1243 779777

Email (for orders and customer service enquiries): cs-books@wiley.co.uk
Visit our Home Page on www.wileyeurope.com or www.wiley.com

Other Wiley Editorial Offices

John Wiley & Sons Inc., 111 River Street, Hoboken, NJ 07030, USA

Jossey-Bass, 989 Market Street, San Francisco, CA 94103-1741, USA

Wiley-VCH Verlag GmbH, Boschstr. 12, D-69469 Weinheim, Germany

John Wiley & Sons Australia Ltd, 42 McDougall Street, Milton, Queensland 4064, Australia

John Wiley & Sons (Asia) Pte Ltd, 2 Clementi Loop #02-01, Jin Xing Distripark, Singapore 129809

John Wiley & Sons Canada Ltd, 6045 Freemont Blvd, Mississauga, Ontario, L5R 4J3, Canada

Wiley also publishes its books in a variety of electronic formats. Some content that appears in print may not be available in electronic books.

Library of Congress Cataloging-in-Publication Data
Optical properties of condensed matter and applications / edited by Jai Singh.
    p. cm. – (Wiley series in materials for electronic and optoelectronic applications)
  Includes bibliographical references and index.
  ISBN-13: 978-0-470-02192-7 (cloth : alk. paper)
  ISBN-10: 0-470-02192-6 (cloth : alk. paper)
1. Condensed matter – Optical properties.   I. Singh, Jai.   II. Series.
    QC173.458.O66O68 2006
    530.4'12 – dc22
                                              2006014789

British Library Cataloguing in Publication Data

A catalogue record for this book is available from the British Library

ISBN-13 978-0-470-02192-7 (HB)
ISBN-10 0-470-02192-6 (HB)

Typeset in 10/12 pt Times by SNP Best-set Typesetter Ltd., Hong Kong
Printed and bound in Great Britain by Antony Rowe Ltd., Chippenham, Wiltshire
This book is printed on acid-free paper responsibly manufactured from sustainable forestry
in which at least two trees are planted for each one used for paper production.

Contents

Series Preface

WILEY SERIES IN MATERIALS FOR ELECTRONIC AND OPTOELECTRONIC APPLICATIONS

This book series is devoted to the rapidly developing class of materials used for electronic and optoelectronic applications. It is designed to provide much-needed information on the fundamental scientific principles of these materials, together with how these are employed in technological applications. The books are aimed at postgraduate students, researchers, and technologists engaged in research, development, and the study of materials in electronics and photonics, and at industrial scientists developing new materials, devices, and circuits for the electronic, optoelectronic, and communications industries. The development of new electronic and optoelectronic materials depends not only on materials engineering at a practical level, but also on a clear understanding of the properties of materials, and the fundamental science behind these properties. It is the properties of a material that eventually determine its usefulness in an application. The series therefore also includes such titles as electrical conduction in solids, optical properties, thermal properties, etc., all with applications and examples of materials in electronics and optoelectronics. The characterization of materials is also covered within the series in as much as it is impossible to develop new materials without the proper characterization of their structure and properties. Structure–property relationships have always been fundamentally and intrinsically important to materials science and engineering.

Materials science is well known for being one of the most interdisciplinary sciences. It is the interdisciplinary aspect of materials science that has led to many exciting discoveries, new materials, and new applications. It is not unusual to find scientists with a chemical engineering background working on materials projects with applications in electronics. In selecting titles for the series, we have tried to maintain the interdisciplinary aspect of the field, and hence its excitement to researchers in this field.

PETER CAPPER
SAFA KASAP
ARTHUR WILLOUGHBY

Preface

Optical properties and their experimental measurements represent one of the most important scientific endeavors in the history of materials research. Advances made to date in photonic devices that have enabled optical communications could not have been achieved without a proper understanding of the optical properties of materials and how these properties influence the overall device performance. Today, research on optical properties of materials draws on not only physicists, who used to be the usual traditional researchers in this field, but also scientists and engineers from widely different disciplines. Although there are several books in the market on optical properties of materials, they have tended to be either too theoretical or so general that they do not include some of the recent and exciting advances. In some cases, the books are too specialized. More significantly, most of these books do not present all the recent advances in the field in one accessible volume. Therefore, it is intended here to have a single volume covering from fundamentals to applications, with up-to-date advances in the field, and a book that is useful to practitioners.

Following a semiquantitative approach, this book summarizes the basic concepts, with examples and applications, and reviews some recent developments in the study of optical properties of condensed-matter systems. It covers examples and applications in the field of electronic and optoelectronic materials, including organic polymers, inorganic glasses, and photonic crystals. An attempt is made to cover both the experimental and theoretical developments in any field presented in this book, which consists of 16 chapters contributed by very experienced and well known scientists and groups on different aspects of optoelectronic properties of condensed matter. Most chapters are presented to be relatively independent with minimal cross-referencing, and chapters with complementary contents are arranged together to facilitate a reader with cross-referencing.

In chapters 1 and 2 by Kasap, coworkers and collaborators, the fundamental optical properties of materials are concisely reviewed and these chapters are expected to refresh the readers with basics and provide some useful optical relations for experimentalists. In chapter 3, Shimakawa et al. present an up to date review of the optical properties of disordered condensed matter, both the theory and experiments, and chapter 4 by Singh and Ruda covers the concept of excitons for crystalline and non-crystalline materials. In chapter 5, Aoki has presented a comprehensive review of experimental advances in the techniques of observing photoluminescence together with recent luminescence results for amorphous semiconductors, and chapter 6 by Singh deals with theoretical advances in the field of photoluminescence and photoinduced changes in non-crystalline condensed-matter systems. Thus, the contents of chapters 5 and 6 are complementary. In chapter 7 by Kugler et al., recent advances in the simulation and understanding of the light-induced volume changes in chalcogenide glasses are presented. Chapter 8 by Edgar covers an extensive discussion on the optical properties of glasses. In chapter 9, Ruda and Matsuura present a comprehensive review of properties and applications of photonic crystals. In chapter 10, Tanaka has presented an up to date review of the nonlinear optical properties of photonic glasses. Chapter 11 by Kobayashi and Naito discusses the fundamental optical properties of organic semiconductors, and in chapter 12 Zhu has presented a comprehensive review of the applications of organic semiconductors. These two chapters are also complementary. In chapter 13, Truong and Tanemura have presented the optical properties of thin films and their applications. Chapter 14 by Kielbasa et al. presents a detailed review of the optical properties of materials with negative refractive index and their applications. In chapter 15, Singh and Oh have discussed the excitonic processes in quantum wells, and in chapter 16, Murayama and Oka have presented the optical properties and spin dynamics of diluted magnetic semiconductor nanostructures.

The aim of the book is to present its readers with the recent developments in theoretical and experimental aspects of novel optical properties of condensed matter and applications. Accomplishments and technical challenges in device applications are also discussed. The readership of the book is expected to be senior undergraduate and postgraduate students, R&D staff, and teaching and research professionals.

JAI SINGH
Darwin, Australia

1 Fundamental Optical Properties of Materials I

W.C. Tan[1], K. Koughia[1], J. Singh[2], and S.O. Kasap[1]

[1]*Department of Electrical Engineering, University of Saskatchewan, Saskatoon, Canada*
[2]*Faculty of Technology, Charles Darwin University, Darwin, NT 0909, Australia*

1.1 INTRODUCTION

Optical properties of a material change or affect the characteristics of light passing through it by modifying its propagation vector or intensity. Two of the most important optical properties are the refractive index n and the extinction coefficient K, which are generically called *optical constants*; though some authors include other optical coefficients within this terminology. The latter is related to the attenuation or absorption coefficient α. In this chapter we present the complex refractive index, the frequency or wavelength dependence of n and K, so-called dispersion relations, how n and K are interrelated, and how n and K can be determined by studying the transmission as a function of wavelength through a thin film of the material. Physical insights into n and K are provided in Chapter 2.

Optical Properties of Condensed Matter and Applications Edited by J. Singh

The optical properties of various materials, with n and K being the most important, are available in the literature in one form or another, either published in journals, books and handbooks or posted on websites of various researchers, organizations (e.g., NIST) or companies (e.g., Schott Glass). Nonetheless, the reader is referred to the works of Wolfe [1.1], Klocek [1.2], Palik [1.3, 1.4], Ward [1.5], Efimov [1.6], Palik and Ghosh [1.7], Nikogosyan [1.8], and Weaver and Frederikse [1.9] for the optical properties of a wide range of materials. Adachi's books on the optical constants of semiconductors are highly recommended [1.10–1.12] along with Madelung's third edition of 'Semiconductors: Data Handbook' [1.13]. There are, of course, other books and handbooks that also contain optical constants in various chapters; see, for example, references [1.14–1.17].

There are available a number of experimental techniques for measuring n and K, some of which have been summarized by Simmons and Potter [1.18]. For example, ellipsometery measures changes in the polarization of light incident on a sample to sensitively characterize surfaces and thin films. The interaction of incident polarized light with the sample causes a polarization change in the light, which may then be measured by analysing the light reflected from the sample. Recently, Collins has provided an extensive in-depth review of ellipsometry for optical measurements [1.19]. One of the most popular and convenient optical measurements involves passing a monochromatic light through a thin sample, and measuring the transmitted intensity as a function of wavelength, $T(\lambda)$, using a simple spectrophotometer. For thin samples on a thick transparent substrate, the transmission spectrum shows oscillations in $T(\lambda)$ with the wavelength due to interferences within the thin film. Swanepoel's technique uses the $T(\lambda)$ measurement to determine n and K, as described in Section 1.4.

1.2 OPTICAL CONSTANTS

One of the most important optical constants of a material is its refractive index, which in general depends on the wavelength of the electromagnetic wave, through a relationship called *dispersion*. In materials where an electromagnetic wave can lose its energy during its propagation, the refractive index becomes complex. The real part is usually the refractive index, n, and the imaginary part is called the *extinction coefficient*, K. In this section, the refractive index and extinction coefficient will be presented in detail along with some common dispersion relations. A more practical and a semiquantitative approach is taken along the lines in [1.18, 1.20, 1.21] rather than a full dedication to rigour and mathematical derivations. More analytical approaches can be found in other texts, e.g. [1.22].

1.2.1 Refractive index and extinction coefficient

The refractive index of an optical or dielectric medium, n, is the ratio of the velocity of light c in vacuum to its velocity v in the medium; $n = c/v$. Using this and Maxwell's equations, one obtains the well known Maxwell's formula for the refractive index of a substance as $n = \sqrt{\varepsilon_r \mu_r}$, where ε_r is the static dielectric constant or relative permittivity and μ_r the relative permeability. As $\mu = 1$ for nonmagnetic substances, one gets $n = \sqrt{\varepsilon_r}$, which is very useful in relating the dielectric properties to optical properties of materials at any particular frequency of interest. As ε_r depends on the wavelength of light, the refractive index also depends on the wavelength of light, and this dependence is called *dispersion*. In addition to dispersion, an electromagnetic wave propagating through a lossy medium experiences attenuation, which means it loses its energy, due to various loss mechanisms such as the generation of phonons (lattice waves), photogeneration, free carrier absorption, scattering, etc. In such materials, the refractive index becomes a complex function of the frequency of the light wave. The complex refractive index, denoted usually by n^*, with real part n, and imaginary part K, called the extinction coefficient, is related to the complex relative permittivity, $\varepsilon_r = \varepsilon_r' - j\varepsilon_r''$, by:

$$n^* = n - jK = \sqrt{\varepsilon_r} = \sqrt{\varepsilon_r' - j\varepsilon_r''} \tag{1.1a}$$

where ε_r' and ε_r'' are, respectively, the real and imaginary parts of ε_r. Equation (1.1b) gives:

$$n^2 - K^2 = \varepsilon_r' \quad \text{and} \quad 2nK = \varepsilon_r'' \tag{1.1b}$$

In explicit terms, n and K can be obtained as:

$$n = \left(1/2^{1/2}\right)\left[\left(\varepsilon_r'^2 + \varepsilon_r''^2\right)^{1/2} + \varepsilon_r'\right]^{1/2} \tag{1.2a}$$

$$K = \left(1/2^{1/2}\right)\left[\left(\varepsilon_r'^2 + \varepsilon_r''^2\right)^{1/2} - \varepsilon_r'\right]^{1/2} \tag{1.2b}$$

The optical constants n and K can be determined by measuring the reflectance from the surface of a material as a function of polarization and the angle of incidence. For normal incidence, the reflection coefficient, r, is obtained as

$$r = \frac{1 - n^*}{1 + n^*} = \frac{1 - n + jK}{1 + n - jK} \tag{1.3}$$

The reflectance R is then defined by:

$$R = |r|^2 = \left|\frac{1 - n + jK}{1 + n - jK}\right|^2 = \frac{(1-n)^2 + K^2}{(1+n)^2 + K^2} \tag{1.4}$$

Notice that whenever K is large, for example over a range of wavelengths, the absorption is strong, and the reflectance is almost unity. The light is then reflected, and any light in the medium is highly attenuated. (Typical sample calculations and applications may be found in ref. [1.20].)

Optical properties of materials are typically presented by showing the frequency dependences (dispersion relations) of either n and K or ε_r' and ε_r''. An intuitive guide to explaining dispersion in insulators is based on a single-oscillator model in which the electric field in the light induces forced dipole oscillations in the material (displaces the electron shells to oscillate about the positive nucleus) with a single resonant frequency ω_o. The frequency dependences of ε_r' and ε_r'' are then obtained as:

$$\varepsilon_r' = 1 + \frac{N_{at}}{\varepsilon_o}\alpha_e' \quad \text{and} \quad \varepsilon_r'' = 1 + \frac{N_{at}}{\varepsilon_o}\alpha_e'' \tag{1.5}$$

where N_{at} is the number of atoms per unit volume, ε_o is the vacuum permittivity, and α_e' and α_e'' are the real and imaginary parts of the electronic polarizability, given respectively by:

$$\alpha_e' = \alpha_{eo} \frac{1-(\omega/\omega_o)^2}{\left[1-(\omega/\omega_o)^2\right]^2 + (\gamma/\omega_o)^2 (\omega/\omega_o)^2} \tag{1.6a}$$

and

$$\alpha_e'' = \alpha_{eo} \frac{(\gamma/\omega_o)(\omega/\omega_o)}{\left[1-(\omega/\omega_o)^2\right]^2 + (\gamma/\omega_o)^2 (\omega/\omega_o)^2} \tag{1.6b}$$

where α_{eo} is the DC polarizability corresponding to $\omega = 0$ and γ is the loss coefficient that characterizes the electromagnetic (EM) wave losses within the material system. Using Equations (1.1)–(1.2) and (1.5)–(1.6), the frequency dependence of n and K can be studied. Figure 1.1(a) shows the dependence of n and K on the normalized frequency ω/ω_o for a simple single electronic dipole oscillator of resonance frequency ω_o.

It is seen that n and K peak close to $\omega = \omega_o$. If a material has a $\varepsilon_r'' \gg \varepsilon_r'$, then $\varepsilon_r \approx -j\varepsilon_r''$ and $n = K \approx \sqrt{\varepsilon_r''/2}$ is obtained from Equation (1.1b). Figure 1.1(b) shows the dependence of the reflectance R on the frequency. It is observed that R reaches its maximum value at a frequency slightly above $\omega = \omega_o$, and then remains high until ω reaches nearly $3\omega_o$; thus the reflectance is substantial while absorption is strong. The normal dispersion region is the frequency range below ω_o where n falls as the frequency decreases, that is, n decreases as the wavelength λ increases. The anomalous dispersion region is the frequency range above ω_o where n decreases as ω increases. Below ω_o, K is small and if ε_{DC} is $\varepsilon_r(0)$, then n becomes:

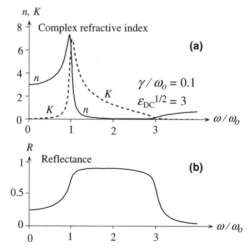

Figure 1.1   Refractive index, n and extinction coefficient K obtained from a single electronic dipole oscillator model. (a) n and K versus normalized frequency and (b) Reflectance versus normalized frequency [Reproduced from S.O. Kasap, *Principles of Electronic Materials and Devices*, 3rd Edition, McGraw-Hill, Boston, 2005]

$$n^2 \approx 1 + (\varepsilon_{DC} - 1)\frac{\omega_o^2}{\omega_o^2 - \omega^2}; \quad \omega < \omega_o \tag{1.7}$$

Since, $\lambda = 2\pi c/\omega$, defining $\lambda_o = 2\pi c/\omega_o$ as the resonance wavelength, one gets:

$$n^2 \approx 1 + (\varepsilon_{DC} - 1)\frac{\lambda^2}{\lambda^2 - \lambda_o^2}; \quad \lambda > \lambda_o \tag{1.8}$$

While intuitively useful, the dispersion relation in Equation (1.8) is far too simple. More rigorously, we have to consider the dipole oscillator quantum mechanically which means a photon excites the oscillator to a higher energy level, see, for example, Fox [1.21] or Simmons and Potter [1.18]. The result is that we would have a series of $\lambda^2/(\lambda^2 - \lambda_i^2)$ terms with various weighting factors A_i that add to unity, where λ_i represents different resonance wavelengths. The weighting factors A_i involve quantum mechanical matrix elements.

Figure 1.2 shows the complex relative permittivity and the complex refractive index of crystalline silicon in terms of photon energy $h\nu$. For photon energies below the bandgap energy (1.1 eV), both ε_r'' and K are negligible and n is close to 3.7. Both ε_r'' and K increase and change strongly as the photon energy becomes greater than 3 eV, far beyond the bandgap energy. Notice that both ε_r' and n peak at $h\nu \approx 3.5$ eV, which corresponds to a direct photoexcitation process, electrons excited from the valence band to the conduction band, as discussed later.

1.2.2 n and K, and Kramers–Kronig relations

If we know the frequency dependence of the real part, ε_r', of the relative permittivity of a material, then by using the Kramers–Kronig relations between the real and the imaginary

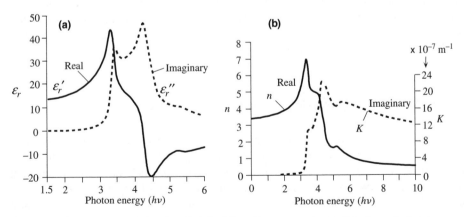

Figure 1.2 (a) Complex relative permittivity of a silicon crystal as a function of photon energy plotted in terms of real (ε_r') and imaginary (ε_r'') parts. (b) Optical properties of a silicon crystal versus photon energy in terms of real (n) and imaginary (K) parts of the complex refractive index [Data extracted from D.E. Aspnes and A.A. Studna, *Phys. Rev. B*, **27**, 985 (1983)].

parts we can determine the frequency dependence of the imaginary part ε_r'', and vice versa. The transform requires that we know the frequency dependence of either the real or imaginary part over as wide a range of frequencies as possible, ideally from zero (DC) to infinity, and that the material has linear behaviour, i.e., it has a relative permittivity that is independent of the applied field. The *Kramers–Kronig relations* for the relative permittivity $\varepsilon_r = \varepsilon_r' - j\varepsilon_r''$ are given by [1.23, 1.24] (see also Appendix 1C in [1.18])

$$\varepsilon_r'(\omega) = 1 + \frac{2}{\pi} P \int_0^\infty \frac{\omega' \varepsilon_r''(\omega')}{\omega'^2 - \omega^2} d\omega' \tag{1.9a}$$

and

$$\varepsilon_r''(\omega) = \frac{2\omega}{\pi} P \int_0^\infty \frac{\varepsilon_r'(\omega')}{\omega'^2 - \omega^2} d\omega' \tag{1.9b}$$

where ω' is the integration variable, P represents the Cauchy principal value of the integral, and the singularity at $\omega = \omega'$ is avoided. [A negative sign is inserted in Equation (1.9b) if we take $\varepsilon_r = \varepsilon_r' + j\varepsilon_r''$]

Similarly one can relate the real and imaginary parts of the polarizability, $\alpha'(\omega)$ and $\alpha''(\omega)$, and those of the complex refractive index, $n(\omega)$ and $K(\omega)$ as well. For $n^* = n(\omega) - jK(\omega)$,

$$n(\omega) = 1 + \frac{2}{\pi} P \int_0^\infty \frac{K(\omega')}{\omega' - \omega} d\omega' \quad \text{and} \quad K(\omega) = \frac{2}{\pi} P \int_0^\infty \frac{K(\omega')}{\omega' - \omega} d\omega' \tag{1.10}$$

A negative sign is needed in the second term if we use $n^* = n(\omega) + jK(\omega)$. It should be emphasized that the optical constants n and K have to obey what are called *f-sum rules* [1.25]. For example, the integration of $[n(\omega) - 1]$ over all frequencies must be zero, and the integration of $\omega K(\omega)$ over all frequencies gives $(\pi/2)\omega_p^2$, where $\omega_p = \hbar(4\pi NZe^2/m_e)^{1/2}$ is the free-electron plasma frequency in which N is the atomic concentration, Z is the total number of electrons per atom, and e and m_e are the charge and mass of the electron. The *f*-sum rules provide a consistency check and enable various constants to be interrelated.

1.3 REFRACTIVE INDEX AND DISPERSION

There are several popular models describing the spectral dependence of refractive index n in a material. Most of these are described below though some, such as the infrared refractive index, are covered under Reststrahlen absorption in Chapter 2 since it is closely related to the coupling of the EM wave to lattice vibrations. The most popular dispersion relation in optical materials is probably the Sellmeier relationship since one can sum any number of resonance-type terms to get as wide a range of wavelength dependence as possible. However, its main drawback is that it does not accurately represent the refractive index when there is a contribution arising from free carriers in narrow-bandgap or doped semiconductors.

1.3.1 Cauchy dispersion relation

In the Cauchy relationship, the dispersion relationship between the refractive index (n) and wavelength of light (λ) is commonly stated in the following form:

$$n = A + \frac{B}{\lambda^2} + \frac{C}{\lambda^4} \qquad (1.11)$$

where A, B, and C are material-dependent specific constants. Equation (1.11) is known as Cauchy's formula and it is typically used in the visible spectrum region for various optical glasses and is applies to *normal dispersion*, when n decreases with increasing λ [1.26, 1.27]. The third term is sometimes dropped for a simpler representation of n versus λ behaviour. The original expression was a series in terms of the wavelength, λ, or frequency, ω, or photon energy $\hbar\omega$ of light as:

$$n = a_0 + a_2\lambda^{-2} + a_4\lambda^{-4} + a_6\lambda^{-6} + \ldots \lambda > \lambda_{th} \qquad (1.12a)$$

or

$$n = n_0 + n_2(\hbar\omega)^2 + n_4(\hbar\omega)^4 + n_6(\hbar\omega)^6 + \ldots \hbar\omega < \hbar\omega_{tt} \qquad (1.12b)$$

where $\hbar\omega$ is the photon energy, $\hbar\omega_{th} = hc/\lambda_{th}$ is the optical excitation threshold (e.g., bandgap energy), $a_0, a_2, \ldots$ and $n_0, n_2, \ldots$ are constants. It has been found that a Cauchy relation in the following form [1.28]:

$$n = n_{-2}(\hbar\omega)^{-2} + n_0 + n_2(\hbar\omega)^2 + n_4(\hbar\omega)^4 \qquad (1.13)$$

can be used satisfactorily over a wide range of photon energies. The dispersion parameters of Equation (1.13) are listed in Table 1 for a few selected materials over specific photon energy ranges.

Cauchy's dispersion relations given in Equations (1.11–1.13) were originally called the elastic-ether theory of the refractive index. It has been widely used for many materials although, in recent years, many researchers have preferred to use the Sellmeier equation described below.

1.3.2 Sellmeier dispersion equation

The Sellmeier equation is an empirical relation between the refractive index n of a substance and wavelength λ of light in the form of a series of single-dipole oscillator terms each of which has the usual $\lambda^2/(\lambda^2 - \lambda_i^2)$ dependence as in

$$n^2 = 1 + \frac{A_1\lambda^2}{\lambda^2 - \lambda_1^2} + \frac{A_2\lambda^2}{\lambda^2 - \lambda_2^2} + \frac{A_3\lambda^2}{\lambda^2 - \lambda_3^2} + \cdots \qquad (1.14)$$

where λ_i is a constant, and A_1, A_2, A_3, λ_1, λ_2 and λ_3 are called *Sellmeier coefficients*, which are determined by fitting this expression to the experimental data. The actual Sellmeier

Table 1.1   Approximate Cauchy dispersion parameters of Equation (1.13) for a few materials from various sources. n_0 and n_e denote, respectively, the ordinary and extraordinary refractive indices of KDP

Material	$\hbar\omega$(eV) Min	$\hbar\omega$(eV) Max	n_{-2}(eV^2)	n_0	n_2(eV^{-2})	n_4(eV^{-4})
Diamond	0.0500	5.4700	-1.0700×10^{-5}	2.3780	0.00801	0.0001
Si	0.0020	1.08	-2.0400×10^{-8}	3.4189	0.0815	0.0125
Ge	0.0020	0.75	-1.0000×10^{-8}	4.0030	0.2200	0.1400
AlSb	0.0620	1.24	-6.1490×10^{-4}	3.1340	0.5225	0.2186
GaP	0.0571	3.60	-1.7817×10^{-3}	3.0010	0.0784	0.0058
GaAs	0.0496	2.90	-5.9737×10^{-4}	3.3270	0.0779	0.0151
InAs	0.0496	2.40	-6.1490×10^{-4}	3.4899	0.0224	0.0284
InP	0.08	3.18	-3.0745×10^{-4}	3.0704	0.1788	-0.0075
ZnSe	0.113	1.24	-4.6117×10^{-4}	2.4365	0.0316	0.0026
KDP, (n_o)	0.62	6.2	-1.7364×10^{-2}	1.5045	0.00181	0.000033
KDP, (n_e)	0.62	6.2	-4.3557×10^{-3}	1.4609	0.001859	0.000021

Table 1.2   Sellmeier coefficients of a few materials, where $\lambda_1, \lambda_2, \lambda_3$ are in μm. (From various sources and approximate values)

Material	A_1	A_2	A_3	λ_1	λ_2	λ_3
SiO$_2$ (fused silica)	0.696749	0.408218	0.890815	0.0690660	0.115662	9.900559
86.5%SiO$_2$–13.5%GeO$_2$	0.711040	0.451885	0.704048	0.0642700	0.129408	9.425478
GeO$_2$	0.80686642	0.71815848	0.85416831	0.068972606	0.15396605	11.841931
BaF$_2$	0.63356	0.506762	3.8261	0.057789	0.109681	46.38642
Sapphire	1.023798	1.058264	5.280792	0.0614482	0.110700	17.92656
Diamond	0.3306	4.3356		0.175	0.106	
Quartz, n_o	1.35400	0.010	0.9994	0.092612	10.700	9.8500
Quartz, n_e	1.38100	0.0100	0.9992	0.093505	11.310	9.5280
KDP, n_o	1.2540	0.0100	0.0992	0.09646	6.9777	5.9848
KDP, n_e	1.13000	0.0001	0.9999	0.09351	7.6710	12.170

formula is more complicated. It has more terms of similar form, e.g., $A_i \lambda^2/(\lambda^2 - \lambda_i^2)$, where $i = 4, 5, \ldots$, but these can generally be neglected in representing n versus λ behaviour over typical wavelengths of interest and ensuring that three terms included in Equation (1.14) correspond to the most important or relevant terms in the summation. Sellmeier coefficients for some materials as examples, including pure silica (SiO$_2$) and 86.5 mol.% SiO$_2$–13.5 mol.% GeO$_2$, are given in Table 2.

There are two methods for determining the refractive index of silica–germania glass $(SiO_2)_{1-x}(GeO_2)_x$: First is a simple, but approximate, linear interpolation of the refractive index between known compositions, e.g., $n(x) - n(0.135) = (x - 0.135)[n(0.135) - n(0)]/0.135$, where $n(x)$ is for $(SiO_2)_{1-x}(GeO_2)_x$; $n(0.135)$ is for 86.5 mol.% SiO$_2$–13.5 mol.% GeO$_2$; $n(0)$ is for SiO$_2$. Second is an interpolation for coefficients A_i and λ_i between SiO$_2$ and GeO$_2$ as reported in [1.29]:

$$n^2 - 1 = \frac{\{A_1(S) + X[A_1(G) - A_1(S)]\}\,\lambda^2}{\lambda^2 - \{\lambda_1(S) + X[\lambda_1(G) - \lambda_1(S)]\}^2} + \ldots \tag{1.15}$$

where X is the mole fraction of germania, S and G in parentheses refer to silica and germania, respectively. The theoretical basis of the Sellmeier equation lies in representing the solid as a sum of N lossless (frictionless) Lorentz oscillators such that each has the usual form of $\lambda^2/(\lambda^2 - \lambda_i^2)$ with different λ_i and each has different strengths, or weighting factors; A_i, $i = 1$ to N [1.30, 1.31]. Such dispersion relationships are essential in designing photonic devices such as waveguides. (Note that although A_i weight different Lorentz contributions, they do not sum to 1 since they include other parameters besides the oscillator strength f_i.) The refractive indices of most optical glasses have been extensively modelled by the Sellmeier equation. Various optical glass manufacturers such as Schott Glass normally provide the Sellmeier coefficients for their glasses [1.32]. Optical dispersion relations for glasses have been discussed by a number of authors [1.6, 1.18, 1.33]. The Sellmeier coefficients normally depend on the temperature and pressure; their dependences for optical glasses have been described by Ghosh [1.34–1.36].

There are other Sellmeier–Cauchy-like dispersion relationships that inherently take account of various contributions to the optical properties, such as the electronic and ionic polarization and interaction of photons with free electrons. For example, for many semiconductors and ionic crystals, two useful dispersion relations are:

$$n^2 = A + \frac{B\lambda^2}{\lambda^2 - C} + \frac{D\lambda^2}{\lambda^2 - E} \tag{1.16}$$

and

$$n^2 = A + \frac{B}{\lambda^2 - \lambda_o^2} + \frac{C}{(\lambda^2 - \lambda_o^2)^2} + D\lambda^2 + E\lambda^4 \tag{1.17}$$

where A, B, C, D, E, and λ_0 are constants particular to a given material. Table 3 provides a few examples. Both the Cauchy and the Sellmeier equations are strictly applicable in wavelength regions where the material is transparent, that is, the extinction coefficient is relatively small. There are many application-based articles in the literature that provide empirical dispersion relations for a variety of materials; a recent example on far-infrared substrates (Ge, Si, ZnSe, ZnS, ZnTe) is given in reference [1.37].

Table 1.3 Parameters of Equations (1.16) and (1.17) for some selected materials {Si data from D.F. Edwards and E. Ochoa, *Appl. Optics*, **19**, 4130 (1980); others from ref. [1.1]}

Material	λ_o (µm)	A	B (µm)^2	C (µm)$^{-4}$	D (µm)$^{-2}$	E (µm)$^{-4}$
Silicon	0.167	3.41983	0.159906	−0.123109	1.269×10^{-6}	-1.951×10^{-9}
MgO	0.11951	2.95636	0.021958	0	-1.0624×10^{-2}	-2.05×10^{-5}
LiF	0.16733	1.38761	0.001796	-4.1×10^{-3}	-2.3045×10^{-3}	-5.57×10^{-6}
AgCl	0.21413	4.00804	0.079009	0	-8.5111×10^{-4}	-1.976×10^{-7}

1.3.3 Refractive index of semiconductors

A. Refractive index of crystalline semiconductors

A particular interest in the case of semiconductors is in n and K for photons energies greater than the bandgap E_g for optoelectronics applications. Owing to various features and singularities in the E–**k** diagrams of crystalline semiconductors (where **k** is the electron's wave vector), the optical constants n and K for $\hbar\omega > E_g$ are not readily expressible in simple terms. Various authors, for example, Forouhi and Bloomer [1.38, 1.39], Chen et al. [1.40] have nonetheless provided useful and tractable expressions for modelling n and K in this regime. In particular, Forouhi–Bloomer (FB) equations express n and K in terms of the photon energy $\hbar\omega$ in a consistent way that obey the Kramers–Kronig relations [1.39], i.e.

$$K = \sum_{i=1}^{q} \frac{A_i(\hbar\omega - E_g)^2}{(\hbar\omega)^2 - B_i(\hbar\omega) + C_i} \quad \text{and} \quad n = n(\infty) + \sum_{i=1}^{q} \frac{B_{oi}(\hbar\omega) + C_{oi}}{(\hbar\omega)^2 - B_i(\hbar\omega) + C_i} \quad (1.18)$$

where q is an integer that represents the number of terms needed to suitably model experimental values of n and K, E_g is the bandgap, and A_i, B_i, C_i, B_{oi}, C_{oi} are constants, B_{oi} and C_{oi} depending on A_i, B_i, C_i, and E_g; only the latter four are independent parameters, $B_{oi} = (A_i/Q_i)[-(1/2)B_i^2 + E_g B_i - E_g^2 + C_i]$; $C_{oi} = (A_i/Q_i)[(1/2)(E_g^2 + C_i)B_i^2 - 2E_g C_i]$; $Q_i = (1/2)(4C_i - B_i^2)^{1/2}$. Forouhi and Bloomer provide a table of FB coefficients, A_i, B_i, C_i, and E_g, for four terms in the summation in Equation (1.18) [1.39] for a number of semiconductors; an example that shows an excellent agreement between the FB dispersion relation and the experimental data is shown in Figure 1.3. Table 4 provides the FB coefficients for a few selected semiconductors. The reader is referred to Adachi's recent book and his papers for further discussions and other models on the refractive index of crystalline and amorphous semiconductors [1.10, 1.11, 1.41–1.44]; the optical properties of amorphous semiconductors are treated in a later chapter of this book.

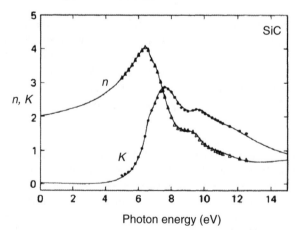

Photon energy (eV)

Figure 1.3 n and K versus photon energy for crystalline SiC. The solid line is obtained from the FB equation with four terms with appropriate parameters, and points represent the experimental data. See original reference [1.39] for details [Reprinted with permission Fig. 2c, A.R. Forouhi and I. Bloomer, *Phys. Rev. B*, **38**, 1865. Copyright (1988) by the American Physical Society]

Table 1.4 FB coefficients for selected semiconductors [1.39] for four terms $i = 1$ to 4. First entry in the box is for $i = 1$, and the fourth is for $i = 4$

	A_i	B_i (eV)	C_i (eV^2)	$n(\infty)$	E_g(eV)
Si	0.00405	6.885	11.864	1.950	1.06
	0.01427	7.401	13.754		
	0.06830	8.634	18.812		
	0.17488	10.652	29.841		
Ge	0.08556	4.589	5.382	2.046	0.60
	0.21882	6.505	11.486		
	0.02563	8.712	19.126		
	0.07754	10.982	31.620		
GaP	0.00652	7.469	13.958	2.070	2.17
	0.14427	7.684	15.041		
	0.13969	10.237	26.567		
	0.00548	13.775	47.612		
GaAs	0.00041	5.871	8.619	2.156	1.35
	0.20049	6.154	9.784		
	0.09688	9.679	23.803		
	0.01008	13.232	44.119		
GaSb	0.00268	4.127	4.267	1.914	0.65
	0.34046	4.664	5.930		
	0.08611	8.162	17.031		
	0.02692	11.146	31.691		
InP	0.20242	6.311	10.357	1.766	1.27
	0.02339	9.662	23.472		
	0.03073	10.726	29.360		
	0.04404	13.604	47.602		
InAs	0.18463	5.277	7.504	1.691	0.30
	0.00941	9.130	20.934		
	0.05242	9.865	25.172		
	0.03467	13.956	50.062		
InSb	0.00296	3.741	3.510	1.803	0.12
	0.22174	4.429	5.447		
	0.06076	7.881	15.887		
	0.04537	10.765	30.119		

B. Bandgap and temperature dependence

The refractive index of a semiconductor (typically for $\hbar\omega < E_g$) typically decreases with increasing energy bandgap E_g. There are various empirical and semi-empirical rules and expressions that relate n to E_g. Based on an atomic model, Moss has suggested that n and E_g are related by $n^4 E_g = K = $ constant (K is about ~100 eV). In the Hervé–Vandamme relationship [1.45],

$$n^2 = 1 + \left(\frac{A}{E_g + B} \right)^2 \qquad (1.19)$$

where A and B are constants as $A \approx 13.6\,\text{eV}$ and $B \approx 3.4\,\text{eV}$. The temperature dependence of n arises from the variation of E_g with the temperature T and typically it increases with increasing temperature. The temperature coefficient of refractive index (TCRI) of semiconductors can be found from the Hervé–Vandamme relationship as:

$$\text{TCRI} = \frac{1}{n} \cdot \frac{dn}{dT} = \frac{(n^2 - 1)^{3/2}}{13.6n^2} \left[\frac{dE_g}{dT} + \frac{dB}{dT} \right] \tag{1.20}$$

where $dB/dT \approx 2.5 \times 10^{-5}\,\text{eV}\,\text{K}^{-1}$. TCRI is typically found to be in the range of 10^{-6} to $10^{-4}\,\text{K}^{-1}$.

1.3.4 Gladstone–Dale formula and oxide glasses

The Gladstone–Dale formula is an empirical equation that allows the average refractive index n of an oxide glass to be calculated from its density ρ and its constituents as:

$$\frac{n-1}{\rho} = p_1 k_1 + p_1 k_1 + \ldots = \sum_{i=1}^{N} p_i k_i = C_{\text{GD}} \tag{1.21}$$

where the summation is for various oxide components (each a simple oxide), p_i is the weight fraction of the i-th oxide in the compound, and k_i is the refraction coefficient that represents the polarizability of the i-th oxide. The right-hand side of Equation (1.21) is called the *Gladstone–Dale* coefficient C_{GD}. In more general terms, as a mixture rule for the overall refractive index, the Gladstone–Dale formula is frequently written as:

$$\frac{n-1}{\rho} = \frac{n_1 - 1}{\rho_1} w_1 + \frac{n_2 - 1}{\rho_2} w_2 + \cdots \tag{1.22}$$

where n and ρ are the effective refractive index and effective density, respectively, of the whole mixture, n_1, n_2, ... are the refractive indices of the constituents, and ρ_1, ρ_2, ... represent the densities of each constituent. Gladstone–Dale equations for the polymorphs of SiO_2 and TiO_2 give the average n, respectively, as [1.46, 1.47]:

$$n(SiO_2) = 1 + 0.21\rho \quad \text{and} \quad n(TiO_2) = 1 + 0.40\rho \tag{1.23}$$

1.3.5 Wemple–DiDomenico dispersion relation

Based on the single-oscillator model, the Wemple–DiDomenico is a semi-empirical dispersion relation for determining the refractive index at photon energies below the interband absorption edge in a variety of materials. It is given by:

$$n^2 = 1 + \frac{E_o E_d}{E_o^2 - (h\nu)^2} \tag{1.24}$$

where v is the frequency, h is the Planck constant, E_o is the single-oscillator energy, E_d is the dispersion energy which is a measure of the average strength of interband optical transitions; $E_d = \beta N_c Z_a N_e$ (eV), where N_c is the effective coordination number of the cation nearest-neighbour to the anion (e.g., $N_c = 6$ in NaCl, $N_c = 4$ in Ge), Z_a is the formal chemical valency of the anion ($Z_a = 1$ in NaCl; 2 in Te; and 3 in GaP), N_e is the effective number of valence electrons per anion excluding the cores ($N_e = 8$ in NaCl, Ge; 10 in TlCl; 12 in Te; $9\frac{1}{3}$ in As_2Se_3), and β is a constant that depends on whether the interatomic bond is ionic (β_i) or covalent (β_c): $\beta_i = 0.26 \pm 0.04\,\mathrm{eV}$ for (e.g., halides NaCl, ThBr, etc. and most oxides, Al_2O_3, etc.), $\beta_c = 0.37 \pm 0.05\,\mathrm{eV}$ for (e.g., tetrahedrally bonded $A^N B^{8-N}$ zinc blende- and diamond-type structures, GaP, ZnS, etc., and wurtzite crystals have a β-value that is intermediate between β_i and β_c). Further, empirically, $E_o = CE_g(D)$, where $E_g(D)$ is the *lowest* direct bandgap and C is a constant; typically $C \approx 1.5$. E_o has been associated with the main peak in the $\varepsilon_r''(hv)$ versus hv spectrum. The parameters required for calculating n from Equation (1.24) are listed in Table 5 [1.48]. While it is apparent that the Wemple–DiDomenico relation can only be approximate, it has nonetheless found wide acceptance among experimentalists due to its straightforward simplicity.

Table 1.5 Examples of parameters for Wemple–DiDomenico dispersion relationship [Equation (1.24)] in various materials [1.48]

Material	N_c	Z_a	N_e	E_o (eV)	E_d (eV)	β (eV)	β	Comment
NaCl	6	1	8	10.3	13.6	0.28	β_i	Halides, LiF, NaF, etc
CsCl	8	1	8	10.6	17.1	0.27	β_i	CsBr, CsI, etc
TlCl	8	1	10	5.8	20.6	0.26	β_i	TlBr
CaF$_2$	8	1	8	15.7	15.9	0.25	β_i	BaF$_2$, etc
CaO	6	2	8	9.9	22.6	0.24	β_i	Oxides, MgO, TeO$_2$, etc
Al$_2$O$_3$	6	2	8	13.4	27.5	0.29	β_i	
LiNbO$_3$	6	2	8	6.65	25.9	0.27	β_i	
TiO$_2$	6	2	8	5.24	25.7	0.27	β_i	
ZnO	4	2	8	6.4	17.1	0.27	β_i	
ZnSe	4	2	8	5.54	27.0	0.42	β_c	II–VI, Zinc blende, ZnS, ZnTe,CdTe
GaAs	4	3	8	3.55	33.5	0.35	β_c	III–V, Zinc blende, GaP, etc
Si (Crystal)	4	4	8	4.0	44.4	0.35	β_c	Diamond, covalent bonding; C (diamond), Ge, β-SiC etc
SiO$_2$ (Crystal)	4	2	8	13.33	18.10	0.28	β_i	Average crystalline form
SiO$_2$ (Amorphous)	4	2	8	13.38	14.71	0.23	β_i	Fused silica
CdSe	4	2	8	4.0	20.6	0.32	β_i–β_c	Wurtzite

1.3.6 Group index

Group index is a factor by which the group velocity of a group of waves in a dielectric medium is reduced with respect to propagation in free space. It is denoted by N_g and defined by $N_g = v_g/c$, where v_g is the group velocity, defined by $v_g = d\omega/dk$ where k is the wave vector or the propagation constant. The group index can be determined from the ordinary refractive index n through:

$$N_g = n - \lambda \frac{dn}{d\lambda} \tag{1.25}$$

where λ is the wavelength of light. Figure 1.4 illustrates the relation between N_g and n in SiO_2. The group index N_g is the quantity that is normally used in calculating dispersion in optical fibers since it is N_g that determines the group velocity of a propagating light pulse in a glass or transparent medium. It should be remarked that although n versus λ can decrease monotonically with λ over a range of wavelengths, N_g can exhibit a minimum in the same range where the dispersion, $dN_g/d\lambda$, becomes zero. The point $dN_g/d\lambda = 0$ is called the zero-material *dispersion wavelength*, which is around 1300 nm for silica as is apparent in Figure 1.4.

1.4 THE SWANEPEL TECHNIQUE: MEASUREMENT OF n AND α

1.4.1 Uniform-thickness films

In many instances, the optical constants are conveniently measured by examining the transmission through a thin film of the material deposited on a transparent glass or other (e.g.,

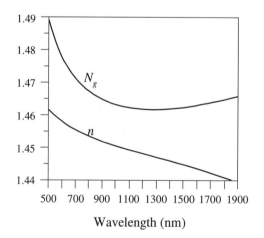

Figure 1.4   Refractive index n and the group index N_g of pure SiO_2 (silica) glass as a function of wavelength [Reproduced with permission from S.O. Kasap, *Principles of Electronic Materials and Devices*, 3rd Edition, McGraw-Hill, Boston, 2005]

sapphire) substrate. The classic reference on the optical properties of thin films has been the book by Heavens [1.49]; the book is still useful in clearly describing what experiments can be carried out, and has a number of useful derivations such as the reflectance and transmittance through thin films in the presence of multiple reflections. Since then numerous research articles and reviews have been published. Poelmen and Smet [1.50] have critically reviewed how a single transmission spectrum measurement can be used to extract the optical constants of a thin film. In general, the amount of light that gets transmitted through a thin film material depends on the amount of reflection and absorption that takes place along the light path. If the material is a thin film with a moderate absorption coefficient α then there will be multiple interferences at the transmitted side of the sample, as illustrated in Figure 1.5.

In this case, some interference fringes will be evident in the transmission spectrum obtained from a spectrophotometer, as shown in Figure 1.6. One very useful method that makes use of these interference fringes to determine the optical properties of the material is called the *Swanepoel method* [1.51].

Swanepoel has shown that the optical properties of a uniform thin film of thickness d, refractive index n, and absorption coefficient α, deposited on a substrate with a refractive index s, as shown in Figure 1.5, can be obtained from the transmittance T given by:

$$T = \frac{Ax}{B - Cx\cos\varphi + Dx^2} \tag{1.26}$$

where $A = 16n^2s$, $B = (n + 1)^3(n + s^2)$, $C = 2(n^2 - 1)(n^2 - s^2)$, $D = (n^2 - 1)^3(n - s^2)$, $\varphi = 4\pi nd/\lambda$, $x = \exp(-\alpha d)$ is the absorbance, and n, s, and α are all function of wavelength, λ. What is very striking and useful is that all the important optical properties can be determined from the application of this equation; these will be introduced in the subsequent paragraphs. Before optical properties of any thin films can be extracted, the refractive index of

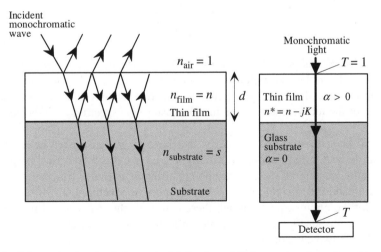

Figure 1.5 Schematic sketch of the typical behavior of light passing through a thin film on a substrate. On the left, oblique incidence is shown to demonstrate the multiple reflections. In most measurements, the incident beam is nearly normal to the film as shown on the right

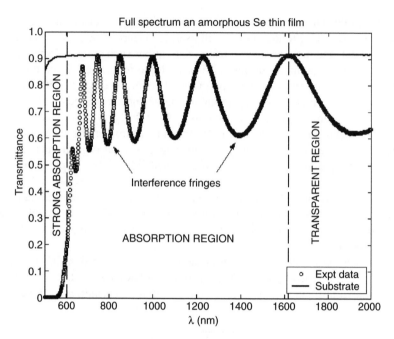

Figure 1.6 An example of a typical transmission spectrum of a 0.969 μm thick amorphous Se thin film that has been vacuum coated onto a glass substrate held at a substrate temperature of 50 °C during the deposition

their substrate must first be calculated. For a glass substrate with very negligible absorption, $K \leq 0.1$ and $\alpha \leq 10^{-2}\,\mathrm{cm}^{-1}$, in the range of the operating wavelengths, the refractive index s is:

$$s = \frac{1}{T_s} + \sqrt{\left(\frac{1}{T_s^2} - 1\right)} \tag{1.27}$$

where T_s is the transmittance value measured from a spectrophotometer. This expression can be derived from the transmittance equation for a bulk sample with little attenuation. With this refractive index s, the next step is to construct two envelopes around the maxima and minima of the interference fringes in the transmission spectrum as indicated in Figure 1.7.

There will altogether be two envelopes that have to be constructed before any of the expressions derived from Equation (1.26) can be used to extract the optical properties. This can be done by locating all the extreme points of the interference fringes in the transmission spectrum and then making sure that the respective envelopes, $T_M(\lambda)$ for the maxima and $T_m(\lambda)$ for the minima, pass through these extremes, the maxima and minima, of $T(\lambda)$ tangentially. From Equation (1.26), it is not difficult to see that at $\varphi = \pm 1$, the expressions that describe the two envelopes are:

$$T_M = \frac{Ax}{B - Cx + Dx^2} \tag{1.28a}$$

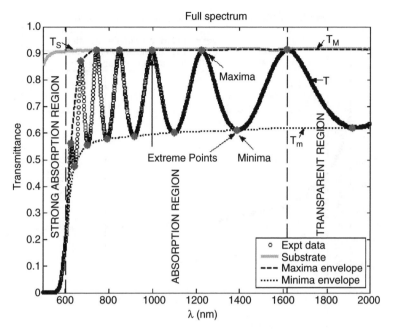

Figure 1.7 The construction of envelopes in the transmission spectrum of the thin amorphous Se film in Figure 1.5

$$T_m = \frac{Ax}{B + Cx + Dx^2} \tag{1.28b}$$

Figure 1.7 shows two envelopes constructed for a transmission spectrum of an amorphous Se thin film. It can also be seen that the transmission spectrum has been divided into three special regions according to their transmittance values: (i) the *transparent region*, where $T(\lambda) \geq 99.99\%$ of the substrate's transmittance value of $T_s(\lambda)$, (ii) the *strong-absorption region*, where $T(\lambda)$ is typical smaller than 20% of $T_s(\lambda)$, and (iii) the *absorption region*, in between the two latter regions as shown in Figure 1.7:

The refractive index of the thin film can be calculated from the two envelopes, $T_M(\lambda)$ and $T_m(\lambda)$, and the refractive index of the substrate s through

$$n = \left[N + (N^2 - s^2)^{1/2} \right]^{1/2}; \quad N = 2s \left[\frac{T_M - T_m}{T_M T_m} \right] + \frac{s^2 + 1}{2} \tag{1.29}$$

where N is defined by the second equation above. Since the equation is not valid in the strong-absorption region, where there are no maxima and minima, the calculated refractive index has to be fitted to a well established dispersion model for extrapolation to shorter wavelengths before it can be used to obtain other optical constants. Usually either the Sellmeier or the Cauchy dispersion equation is used to fit n versus λ experimental data in this range. Figure 1.8 shows the refractive indices extracted from the envelopes and a fitted Sellmeier dispersion model with two terms.

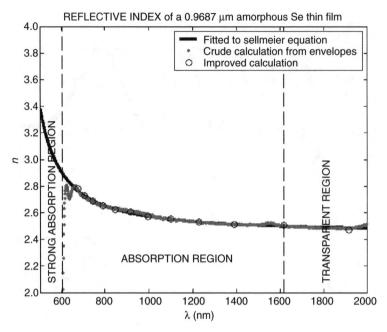

Figure 1.8 Determination of the refractive index from the transmission spectrum maxima and minima shown in Figure 1.6. The fitted Sellmeier n vs λ curve follows $n^2 = 3.096 + 2.943\lambda^2/[\lambda^2 - (4.02310^{-7})^2]$

With the refractive index of the thin film corresponding to two adjacent maxima (or minima) at points 1 and 2 given as n_1 at λ_1 and n_2 at λ_1, the thickness can be easily calculated from the basic interference equation of waves as follows:

$$d_{\text{crude}} = \frac{\lambda_1\lambda_2}{2(\lambda_1 n_2 - \lambda_2 n_1)} \quad (1.30)$$

where d_{crude} refers to the thickness obtained from the maxima (minima) at points 1, 2. As other adjacent pairs of maxima or minima points are used, more thickness values can be deduced, and hence an average value calculated. It is assumed the film has an ideal uniform thickness.

The absorption coefficient α can be obtained once the absorbance x is extracted from the transmission spectrum. This can be done as follows:

$$\alpha = -\frac{\ln(x)}{d_{\text{ave}}} \quad (1.31)$$

where $\quad x = \dfrac{E_M - \sqrt{E_M^2 - (n^2-1)^3(n^2-s^4)}}{(n-1)^3(n-s)^2}; E_M = \dfrac{8n^2 s}{T_M} + (n^2-1)(n^2-s^2) \quad$ and d_{ave} is the average thickness of d_{crude}.

The accuracy of the thickness, the refractive index, and the absorption coefficient can all be further improved in the following manner. The first step is to determine a new set of interference orders number, m', for the interference fringes from the basic interference equation of waves, that is:

$$m' = \frac{2n_e d_{\text{ave}}}{\lambda_e} \qquad (1.32a)$$

where n_e and λ_e are values taken at any extreme points, and m' is an integer if the extremes taken are maxima, or a half-integer if the extremes taken are minima.

The second step is to get a new corresponding set of thickness, d', from this new set of order numbers m', by rearranging Equation (1.32a) as:

$$d' = \frac{m'\lambda_e}{2n_e} \qquad (1.32b)$$

From this new set of thicknesses, d', a new average thickness, d_{new}, must be calculated before it can applied to improve the refractive index. With this new average thickness, a more accurate refractive index can be obtained from the same equation:

$$n'_e = \frac{m'\lambda_e}{2d_{\text{new}}} \qquad (1.32c)$$

This new refractive index can then be fitted to the previous dispersion model again so that an improved absorption coefficient α can be calculated from Equation (1.31). All these parameters can then be used in Equation (1.26) to regenerate a transmission spectrum $T_{\text{cal}}(\lambda)$ so that the root mean square error (RMSE) can be determined from the experimental spectrum T_{exp}. The RMSE is calculated as follows:

$$\text{RMSE} = \sqrt{\frac{\sum\limits_{i=1}^{q}(T_{\text{exp}} - T_{\text{cal}})^2}{q}} \qquad (1.33)$$

where T_{exp} is the transmittance of the experimental or measured spectrum, T_{cal} is the transmittance of the regenerated spectrum using the Swanepoel calculation method, and q is the range of the measurement. Figure 1.9 shows the regenerated transmission spectrum of the amorphous Se thin film that appeared in Figure 1.6 using the optical constants calculated from the envelopes.

1.4.2 Thin films with nonuniform thickness

For a film with a wedge-like cross-section as shown in Figure 1.10, Equation (1.26) must be integrated over the thickness of the film in order for it to more accurately describe the transmission spectrum [1.52]. The transmittance then becomes

$$T_{\Delta d} = \frac{1}{\varphi_2 - \varphi_1} \int_{\varphi_1}^{\varphi_2} \frac{Ax}{B - Cx\cos\varphi + Dx^2} \, dx \qquad (1.34)$$

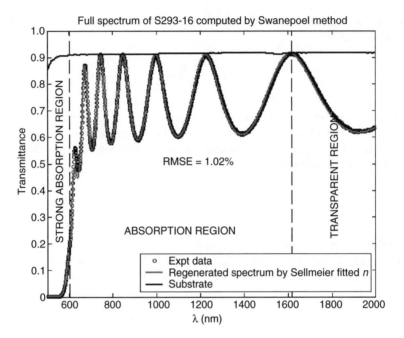

Figure 1.9   Regenerated transmission spectrum of the sample in Figure 1.6

with $\varphi_1 = \dfrac{4\pi n(\bar{d} - \Delta d)}{\lambda}$, and $\varphi_2 = \dfrac{4\pi n(\bar{d} + \Delta d)}{\lambda}$, where $A = 16n^2 s$, $B = (n + 1)^3(n + s^2)$,

$C = 2(n^2 - 1)(n^2 - s^2)$, $D = (n - 1)^3(n - s^2)$, $x = \exp(-\alpha d)$ is the absorbance, n and s are the refractive index of the film and substrate respectively, α the absorption coefficient, $\bar{d}$ is the average thickness of the film, and Δd is the *thickness variation* throughout the illumination area, which has been called the *roughness* of the film. (This nomenclature is actually confusing since the film may not be truly 'rough' but may just have a continuously increasing thickness as in a wedge from one end to the other.)

The first parameter to be extracted before the rest of the optical properties is Δd. Since the integration in Equation (1.34) cannot be carried out from one branch of the tangent to another, it cannot be used directly in this form. The equation was thus modified by considering the maxima and minima, which are both continuous function of λ, in a case-by-case basis. In this way, we have

Maxima:
$$T_{Md} = \frac{\lambda}{2\pi n \Delta d} \frac{a}{\sqrt{1 - b^2}} \tan^{-1}\left[\frac{1 + b}{\sqrt{1 - b^2}} \tan\left(\frac{2\pi n \Delta d}{\lambda}\right)\right] \tag{1.35a}$$

Minima:
$$T_{md} = \frac{\lambda}{2\pi n \Delta d} \frac{a}{\sqrt{1 - b^2}} \tan^{-1}\left[\frac{1 - b}{\sqrt{1 - b^2}} \tan\left(\frac{2\pi n \Delta d}{\lambda}\right)\right] \tag{1.35b}$$

where $a = \dfrac{A}{B + D}$, and $b = \dfrac{C}{B + D}$. As long as $0 < \Delta d < \lambda/4n$, the refractive index, n, and Δd can both be obtained simultaneously by solving Equation (1.35a) and (1.35b) numerically.

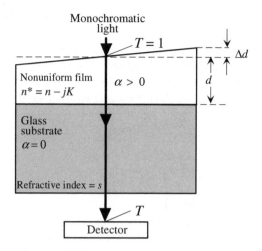

Figure 1.10   System of an absorbing thin film with a variation in thickness on a thick finite transparent substrate

Since Equations (1.35a) and (1.35b) are only valid in the region of zero absorption, the refractive index outside the transparent region must be obtained in another way. Theoretically, a direct integration of Equation (1.34) over both Δd and x can be performed, though this would be analytically too difficult; nevertheless, an approximation to the integration is also possible as follows:

Maxima:
$$T_{Mx} = \frac{\lambda}{2\pi n\Delta d} \frac{a_x}{\sqrt{1-b_x^2}} \tan^{-1}\left[\frac{1+b_x}{\sqrt{1-b_x^2}}\tan\left(\frac{2\pi n\Delta d}{\lambda}\right)\right] \tag{1.36a}$$

Minima:
$$T_{mx} = \frac{\lambda}{2\pi n\Delta d} \frac{a_x}{\sqrt{1-b_x^2}} \tan^{-1}\left[\frac{1-b_x}{\sqrt{1-b_x^2}}\tan\left(\frac{2\pi n\Delta d}{\lambda}\right)\right] \tag{1.36b}$$

where $a = \dfrac{Ax}{B+Dx^2}$, and $b = \dfrac{Cx}{B+Dx^2}$. As long as $0 < x \le 1$, numerically, there will only be one unique solution. Therefore the two desired optical properties, refractive index, n, and the absorbance, x, can both be obtained when Equations (1.36a) and (1.36b) are solved simultaneously using the calculated average Δd.

As before, the calculated refractive index can be fitted to a well established dispersion model, such as the Cauchy or Sellmeier equation, for extrapolation to shorter wavelengths and the thickness is calculated from any two adjacent maxima (or minima) using Equation (1.30). Given that the absorbance, from Equations (1.36a) and (1.36b) is not valid in the strong-absorption region, the absorption coefficient outside this region is calculated differently from those in the strong region as:

$$\alpha_{\text{out}} = -\frac{\ln(x_{\text{out}})}{d_{\text{ave}}} \tag{1.37}$$

where x_{out} is the absorbance obtained from Equations (1.36a) and (1.36b) and d_{ave} is the average thickness.

According to Swanepoel, in the region of strong absorption, the interference fringes are smaller and the spectrum approaches the interference-free transmission sooner. Since the transmission spectra in this region are the same for any film with the same average thickness, regardless of its uniformity, the absorption coefficient in the strong region will thus be

$$\alpha_{strong} = -\frac{\ln(x_{strong})}{d_{ave}} \qquad (1.38)$$

where $\quad x_{strong} = \dfrac{A - \sqrt{(A^2 - 4T_i^2 BD)}}{2T_i D}, \quad T_i = \dfrac{2T_M T_m}{T_M + T_m}, \quad$ and T_M and T_m are the envelopes

constructed from the measured spectrum.

The accuracy of the thickness and refractive index can be further improved in exactly the same way as for a uniform thickness film and used for the computation of the new refractive index and absorption coefficient using Equations (1.36a) and (1.36b). Figure 1.11 shows the regenerated transmission spectrum of a simulated sample with nonuniform thickness using the optical constants calculated from the envelopes. Marquez et al. [1.53] have discussed the application of the Swanepoel technique to wedge-shaped As_2S_3 thin films and made use of the fact that a nonuniform wedge-shaped thin film has a compressed transmission spectrum.

Various computer algorithms that can be used to obtain n and K based on the Swanepoel technique are available in the literature [1.54]. Further discussions and enhancements are

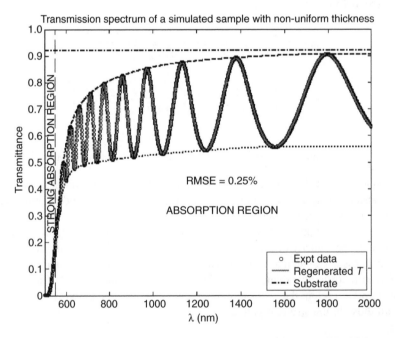

Figure 1.11 A regenerated transmission spectrum of a sample with an average thickness of $1\,\mu m$, average Δd of $30\,nm$, and a refractive index fitted to a Cauchy equation in Figure 1.12

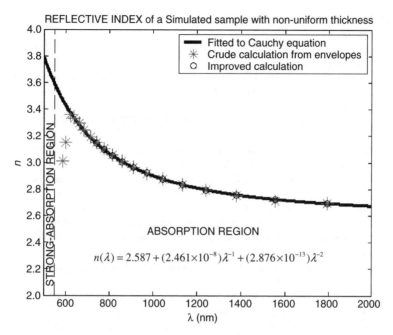

Figure 1.12   The refractive index of a sample with $\bar{d} = 1\,\mu m$ and $\Delta d = 30\,nm$, and n fitted to a Cauchy equation

also available in the literature [1.50, 1.55, 1.56]. There are numerous useful applications of the Swanepoel technique for extracting the optical constants of thin films; some selected recent examples are given in [1.57–1.67].

1.5 CONCLUSIONS

This chapter has provided a semiquantitative explanation and discussion of the complex refractive index $n^* = n - jK$, the relationship between the real n and imaginary part K through the Kramers–Kronig relationships, various common dispersion (n versus λ) relationships such as the Cauchy, Sellmeier, Wemple–DiDomenico dispersion relations, and the determination of the optical constants of a material in thin film form using the popular Swanepoel technique. Examples are given to highlight the concepts and provide applications. Optical constants of various selected materials have also been provided in tables to illustrate typical values and enable comparisons to be made.

REFERENCES

[1.1] W.L. Wolfe, in *The Handbook of Optics*, edited by W.G. Driscoll and W. Vaughan (McGraw-Hill, New York, 1978).

[1.2] P. Klocek, *Handbook of Infrared Optical Materials* (Marcel Dekker, New York, 1991).

[1.3] E.D. Palik, *Handbook of Optical Constants of Solids* (Academic Press, San Diego, 1985) (now Elsevier).

[1.4] E.D. Palik, *Handbook of Optical Constants of Solids II* (Academic Press, San Diego, 1991) (now Elsevier).

[1.5] L. Ward, *The Optical Constants of Bulk Materials and Films* (Institute of Physics Publishing, Bristol, 1994) (reprint 1998).

[1.6] A.M. Efimov, *Optical Constants of Inorganic Glasses* (CRC Press, Boca Raton, 1995).

[1.7] E.D. Palik and G.K. Ghosh, *Handbook of Optical Constants of Solids*, Vols 1–5 (Academic Press, San Diego, 1997) (now Elsevier).

[1.8] D. Nikogosyan, *Properties of Optical and Laser-Related Materials: A Handbook* (John Wiley & Sons, Inc., New York, 1997).

[1.9] J.H. Weaver and H.P.R. Frederikse, in *CRC Handbook of Chemistry and Physics*, Editor in Chief D.R. Lide (CRC Press, Boca Raton, 1999), Ch. 12.

[1.10] S. Adachi, *Physical Properties of III-V Semiconductor Compounds* (John Wiley & Sons, Inc., New York, 1992).

[1.11] S. Adachi, *Optical Constants in Crystalline and Amorphous Semiconductors: Numerical Data and Graphical Information* (Kluwer Academic Publishers, Boston, 1999).

[1.12] S. Adachi, *Properties of Group-IV, III-V and II-VI Semiconductors* (John Wiley & Sons, Ltd, Chichester, 2005).

[1.13] O. Madelung, *Semiconductors: Data Handbook*, 3rd Edition (Springer-Verlag, New York, 2004).

[1.14] H.S. Nalwa, *Handbook of Advanced Electronic and Photonic Materials and Devices*, Vols 1–10 (Academic Press, San Diego, 2001) (now Elsevier).

[1.15] M.J. Weber, *Handbook of Optical Materials* (CRC Press, Boca Raton, 2003).

[1.16] W. Martienssen and H. Walimont, *Springer Handbook of Condensed Matter and Materials Data* (Springer, Heidelberg, 2005), Ch. 3.4, 4.1, 4.4.

[1.17] S.O. Kasap and P. Capper, *Springer Handbook of Electronic and Photonic Materials* (Springer, Heidelberg, 2006), Ch. 3.

[1.18] J.H. Simmons and K.S. Potter, *Optical Materials* (Academic Press, San Diego, 2000) (now Elsevier).

[1.19] R.W. Collins, 'Ellipsometry' in *The Optics Encyclopedia*, Vol. 1, edited by T.G. Brown, K. Creath, H. Kogelnik, M.A. Kriss, J. Schmit, and M.J. Weber (Wiley-VCH, Weinheim, 2004), p. 609.

[1.20] S.O. Kasap, *Principles of Electronic Materials and Devices*, 3rd Edition (McGraw-Hill, Boston, 2005), Ch. 7 and Ch. 9.

[1.21] M. Fox, *Optical Properties of Solids* (Oxford University Press, Oxford, 2001).

[1.22] Y. Toyozawa, *Optical Processes in Solids* (Cambridge University Press, Cambridge, 2003).

[1.23] R. Kronig, *J. Opt. Soc. Am.*, **12**, 547 (1926).

[1.24] H.A. Kramers, *Estratto Dagli Atti del Congresso Internazionale de Fisici*, **2**, 545 (1927).

[1.25] D.Y. Smith and E. Shiles, *Phys. Rev. B*, **17**, 4689 (1978).

[1.26] A.L. Cauchy, *Bull. Sci. Math.*, **14**, 6 (1830).

[1.27] A.L. Cauchy, *M'emoire sur la Dispersion de la Lumiere* (Calve, Prague, 1836).

[1.28] D.Y. Smith, M. Inokuti, and W. Karstens, *J. Phys.: Cond. Matt.*, **13**, 3883 (2001).

[1.29] J.W. Fleming, *Appl. Optics*, **23**, 4486 (1984).

[1.30] K.L. Wolf and K.F. Herzfeld, *Handbooch der Physik*, edited by H. Geiger and K. Scheel (Springer Verlag, Berlin, 1928), Vol. 20, Ch. 10.

[1.31] M. Herzberger, *Opt. Acta*, **6**, 197 (1959).

[1.32] H. Bachs and N. Neuroth, *Schott Series on Glass and Glass Ceramics* (Springer, Heidelberg, 1995).

[1.33] N.J. Kreidl and D.R. Uhlmann, *Optical Properties of Glass* (The American Ceramic Society, 1991).

[1.34] G. Ghosh, M. Endo, and T. Iwasaki, *J. Light Wave Technol.*, **12**, 1338 (1994).

[1.35] G. Ghosh, *Appl. Optics*, **36**, 1540 (1997).

[1.36] G. Ghosh, *Phys. Rev. B*, **14**, 8178 (1998).

[1.37] G. Hawkins and R. Hunneman, *Infrared Phys. Technol.*, **45**, 69 (2004).

[1.38] A.R. Forouhi and I. Bloomer, *Phys. Rev. B*, **34**, 7018 (1986).

[1.39] A.R. Forouhi and I. Bloomer, *Phys. Rev. B*, **38**, 1865 (1988).

[1.40] Y.F. Chen, C.M. Kwei, and C.J. Tung, *Phys. Rev. B*, **48**, 4373 (1993).

[1.41] S. Adachi, *Phys. Rev. B*, **35**, 123161 (1987).

[1.42] S. Adachi, *Phys. Rev. B*, **38**, 12966 (1988).

[1.43] S. Adachi, H. Mori, and S. Ozaki, *Phys. Rev. B*, **66**, 153201 (2002).

[1.44] S. Adachi, *Phys. Rev. B*, **43**, 123161 (1991).

[1.45] P.J.L. Hervé and L.K.J. Vandamme, *J. Appl. Phys.*, **77**, 5476 (1996).

[1.46] D. Dale and F. Gladstone, *Phil. Trans.*, **148**, 887 (1858).

[1.47] D. Dale and F. Gladstone, *Phils. Trans.*, **153**, 317 (1863).

[1.48] S.H. Wemple and M. DiDomenico, *Phys. Rev. B*, **3**, 1338 (1971).

[1.49] O.S. Heavens, *Optical Properties of Thin Solid Films* (Dover Publications, New York, 1965 and 1991).

[1.50] D. Poelmen and P.F. Smet, *J. Phys. D: Appl. Phys.*, **36** 1850 (2003).

[1.51] R. Swanepoel, *J. Phys. E: Sci. Instrum.*, **16**, 1214 (1983).

[1.52] R. Swanepoel, *J. Phys. E: Sci. Instrum.*, **17**, 896 (1984).

[1.53] E. Marquez, J.B. Ramirez-Malo, P. Villares, R. Jimenez-Garay, and R. Swanepoel, *Thin Solid Films*, **254**, 83 (1995).

[1.54] A.P. Caricato, A. Fazzi, and G. Leggieri, *Appl. Surf. Sci.*, **248**, 440 (2005) and references therein.

[1.55] I. Chambouleyron, S.D. Ventura, E.G. Birgin, and J.M. Martínez, *J. Appl. Phys.*, **92**, 3093 (2002) and references therein.

[1.56] K. Ayadi and N. Haddaoui, *J. Mater Sci.: Mater. Electron.*, **11**, 163 (2000).

[1.57] E. Marquez, J. Ramirez-Malo, P. Villares, R. Jimenez-Garay, P.J.S. Ewen, and A.E. Owen, *J. Phys. D*, **25**, 535 (1992).

[1.58] E. Márquez, J.M. González-Leal, R. Prieto-Alcón, M. Vlcek, A. Stronski, T. Wagner, and D. Minkov, *Appl. Phys. A: Mater. Sci. Process.*, **67**, 371 (1998).

[1.59] J.M. Gonzalez-Leal, A. Ledesma, A.M. Bernal-Oliva, R. Prieto-Alcon, E. Marquez, J.A. Angel, and J. Carabe, *Mater. Lett.*, **39**, 232 (1999).

[1.60] E. Marquez, A.M. Bernal-Oliva, J.M. González-Leal, R. Prieto-Alcon, A. Ledesma, R. Jimenez-Garay, and I. Martil, *Mater. Chem. Phys.*, **60**, 231 (1999).

[1.61] A.H. Moharram, A.A. Othman, and M.A. Osman, *Appl. Surf. Sci.*, **200**, 143 (2002).

[1.62] N.A. Bakr, H. El-Hadidy, M. Hammam, and M.D. Migahed, *Thin Solid Films*, **424**, 296 (2003).

[1.63] J.M. González-Leal, R. Prieto-Alcon, J.A. Angel, and E. Marquez, *J. Non-Crystalline Solids*, **315**, 134 (2003).

[1.64] S.M. El-Sayed and G.A.M. Amin, *ND&E International*, **38**, 113 (2005).

[1.65] N. Tigau, V. Ciupina, and G. Prodan, *J. Cryst. Growth*, **277**, 529 (2005).

[1.66] S.A. Fayek and S.M. El-Sayed, *ND&E International*, **39**, 39 (2006).

[1.67] J. Sanchez-Gonzalez, A. Diaz-Parralejo, A.L. Ortiz, and F. Guiberteau, *Appl. Surf. Sci.*, in press (2006).

2 Fundamental Optical Properties of Materials II

K. Koughia[1], J. Singh[2], S.O. Kasap[1], and H.E. Ruda[3]

[1]*University of Saskatchewan, Saskatoon, Canada*
[2]*Faculty of Technology, B-41 Charles Darwin University, Darwin, NT 0909, Australia*
[3]*Centre for Nanotechnology, University of Toronto, 170 College Street, Toronto, Ontario M5S 3E4, Canada*

2.1 INTRODUCTION

Optical properties of semiconductors typically consist of their refractive index n and extinction coefficient K or absorption coefficient α (or equivalently the real and imaginary parts of the relative permittivity) and their dispersion relations, that is their dependence on the wavelength, λ, of the electromagnetic radiation or photon energy $h\nu$, and the changes in the dispersion relations with temperature, pressure, alloying, impurities, etc. A typical relationship between the absorption coefficient and photon energy observed in a crystalline semiconductor is shown in Figure 2.1, where various possible absorption processes are

Optical Properties of Condensed Matter and Applications   Edited by J. Singh

illustrated. The important features in the α vs $h\nu$ behaviour as the photon energy increases can be classified in the following types of absorptions: (a) Reststrahlen or lattice absorption in which the radiation is absorbed by vibrations of the crystal ions, (b) free-carrier absorption due to the presence of free electrons and holes, an effect that decreases with increasing photon energy, (c) an impurity absorption band (usually narrow) due the various dopants, (d) exciton absorption peaks that are usually observed at low temperatures and are close to the fundamental absorption edge, and (e) band-to-band or fundamental absorption of photons, which excites an electron from the valence to the conduction band. The type (e) absorption has a large absorption coefficient and occurs when the photon energy reaches the bandgap energy E_g. It is probably the most important absorption effect; its characteristics for $h\nu > E_g$ can be predicted using the results of Section 2.4. The values of E_g, and its temperature shift, dE_g/dT, are therefore important factors in semiconductor-based optoelectronic devices. In nearly all semiconductors E_g decreases with temperature, hence shifting the fundamental absorption to longer wavelengths. The refractive index n also changes with temperature. dn/dT depends on the wavelength, but for many semiconductors $(dn/dT)/n \approx 5 \times 10^{-5}\,\mathrm{K}^{-1}$; e.g., for GaAs, $(dn/dT)/n = 4 \times 10^{-4}\,\mathrm{K}^{-1}$ at $\lambda = 2\,\mu\mathrm{m}$. There is a good correlation between the refractive index and the bandgap of semiconductors in which, typically, n decreases as E_g increases; wider-bandgap semiconductors have lower refractive indices. The refractive index n and the extinction coefficient K (or α) are related by virtue of the Kramers–Kronig relations, described in Chapter 1. Thus, large increases in the absorption coefficient for $h\nu$ near and above the bandgap energy E_g also result in increases in the refractive index n vs $h\nu$ in this region. Optical and some structural properties of various semiconductors are listed in Table 2.1. The characteristics of some of these absorptions are described in the following sections. While these topics have been covered in extensive detail

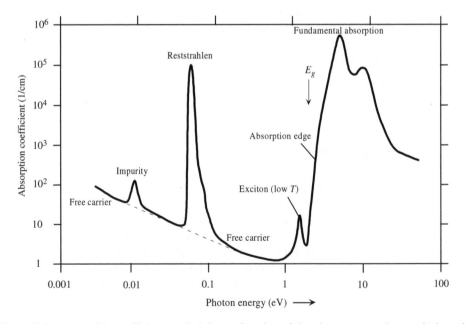

Figure 2.1   Absorption coefficient is plotted as a function of the photon energy in a typical semiconductor to illustrate various possible absorption processes

Table 2.1 Crystal structure, lattice parameter a, bandgap energy E_g at 300 K, type of bandgap (D = Direct and I = Indirect), change in E_g per unit temperature change (dE_g/dT) at 300 K, bandgap wavelength λ_g, and refractive index n close to λ_g. (A = Amorphous, D = Diamond, W = Wurtzite, ZB = Zinc blende)

Semiconductors	Crystal	a nm	E_g eV	Type	dE_g/dT meV K^{-1}	λ_g (μm)	n (λ_g)	dn/dT $\times 10^{-5}$ K^{-1}
Group IV								
Diamond	D	0.3567	5.48	I	-0.05	0.226	2.74	1.1
Ge	D	0.5658	0.66	I	-0.37	1.87	4	27.6 / 42.4 (4 μm)
Si	D	0.5431	1.12	I	-0.25	1.11	3.45	13.8 / 16 (5 μm)
a-Si:H	A		1.7–1.8			0.73		
III–V Compounds								
AlAs	ZB	0.5661	2.16	I	-0.50	0.57	3.2	15
AlP	ZB	0.5451	2.45	I	-0.35	0.52	3	11
AlSb	ZB	0.6135	1.58	I	-0.3	0.75	3.7	
GaAs	ZB	0.5653	1.42	D	-0.45	0.87	3.6	15
GaAs$_{0.88}$Sb$_{0.12}$	ZB		1.15	D		1.08		
GaN	W	0.3190 a / 0.5190 c	3.44	D	-0.45	0.36	2.6	6.8
GaP	ZB	0.5451	2.26	I	-0.54	0.40	3.4	
GaSb	ZB	0.6096	0.73	D	-0.35	1.7	4	33
In$_{0.53}$Ga$_{0.47}$As on InP	ZB	0.5869	0.75	D		1.65		
In$_{0.58}$Ga$_{0.42}$As$_{0.9}$P$_{0.1}$ on InP	ZB	0.5870	0.80	D		1.55		
In$_{0.72}$Ga$_{0.28}$As$_{0.62}$P$_{0.38}$ on InP	ZB	0.5870	0.95	D		1.3		
InP	ZB	0.5869	1.35	D	-0.36	0.91	3.4–3.5	9.5
InAs	ZB	0.6058	0.36	D	-0.28	3.5	3.8	2.7
InSb	ZB	0.6479	0.18	D	-0.3	7	4.2	29
II–VI Compounds								
ZnSe	ZB	0.5668	2.7	D	-0.50	0.46	2.3	6.3
ZnTe	ZB	0.6101	2.3	D	-0.45	0.55	2.7	

in various graduate-level textbooks in the past [2.1, 2.2], the approach here is to provide concise descriptions with more insight from an experimentalist's point of view. In addition, we have included the electro-optic effects since these have become particularly important in the last two decades with the advent of optical communications [2.3].

2.2 LATTICE OR RESTSTRAHLEN ABSORPTION AND INFRARED REFLECTION

In the infrared wavelength region, ionic crystals reflect and absorb light strongly due to the resonance interaction of the electromagnetic (EM) wave field with the transverse optical (TO) phonons. The simplest dipole oscillator model based on ions driven by an EM wave gives the complex relative permittivity as:

$$\varepsilon_r = \varepsilon_r' - j\varepsilon_r'' = \varepsilon_{r\infty} + \frac{\varepsilon_{r\infty} - \varepsilon_{ro}}{\left(\dfrac{\omega}{\omega_T}\right)^2 - 1 - j\dfrac{\gamma}{\omega_T}\left(\dfrac{\omega}{\omega_T}\right)} \tag{2.1}$$

where ε_{ro} and $\varepsilon_{r\infty}$ are the relative permittivity at $\omega = 0$ (very low frequencies or dc) and $\omega = \infty$ (very high frequencies) respectively, γ is the loss coefficient per unit reduced mass representing the rate of energy transfer from the EM wave to optical phonons, and ω_T is a transverse optical phonon frequency, that is related to the nature of the bonding between the ions in the crystal, i.e., $\omega_T^2 = \omega_o^2(\varepsilon_{r\infty} + 2)/(\varepsilon_{ro} + 2)$ in which $\omega_o^2 = \beta/M_r$, β is the force constant in 'restoring force $= -\beta \times$ displacement', and M_r is the reduced mass of the negative and positive ions in the crystal. The loss, ε_r'', and the absorption are maxima when $\omega = \omega_T$, and the wave is attenuated by the transfer of energy to the transverse optical phonons, thus the EM wave couples to the transverse optical phonons. At $\omega = \omega_L$, the wave couples to the longitudinal optical phonons. The refractive index n from ionic polarization vanishes, and the reflectance is minimum. Figure 2.2 shows the optical properties of AlSb in terms of n, K, and R vs wavelength. The extinction coefficient K and reflectance R peaks occur over about the same wavelength region, corresponding to the coupling of the EM wave to the transverse optical phonons. At wavelengths close to $\lambda_T = 2\pi/\omega_T$, n and K peak, and there is a strong absorption of light which corresponds to the EM wave resonating with the TO lattice vibrations, then R rises sharply. ω_T and ω_L are related through

$$\omega_T^2 = \omega_L^2\,(\varepsilon_{r\infty}/\varepsilon_{ro}) \tag{2.2}$$

which is called the *Lyddane–Sachs–Teller relation*. It relates the high- and low-frequency refractive indices $n_\infty = \sqrt{\varepsilon_{r\infty}}$ and $n_o = \sqrt{\varepsilon_{ro}}$ to ω_T and ω_L. The complex refractive index, $n^{*-} = n - jK$, then becomes:

$$n^{*2} = (n - jK)^2 = \varepsilon_r' - j\varepsilon_r'' = \varepsilon_{r\infty}\left[1 + \frac{\omega_L^2 - \omega_T^2}{\omega_T^2 - \omega^2 + j\gamma\omega}\right]. \tag{2.2a}$$

(Note that $n^2 - K^2 = \varepsilon_r'$ and $2nK = \varepsilon_r''$).

Taking CdTe as an example, and substituting the values for $\varepsilon_{r\infty}$, ω_T, ω_L, and γ from Table

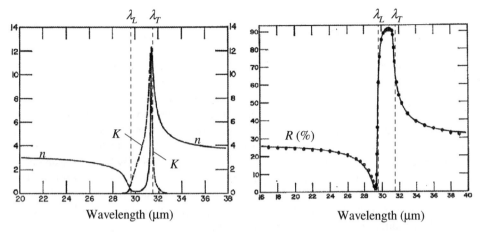

Figure 2.2 Infrared refractive index n, extinction coefficient K (left), and reflectance R (right) of AlSb. Note: The wavelength axes are not identical, and wavelengths λ_T and λ_L corresponding to ω_T and ω_L, respectively, are shown as dashed vertical lines [Data extracted from Turner and Reese [2.4]]

Table 2.2 Values of the quantities required for calculating n and K from Equation (2.2a) for some selected crystals; values adapted from reference [2.5]. ω_T, ω_L and γ are in rad s^{-1}

Sample	$\varepsilon_{r\infty}$	$\omega_T \times 10^{12}$	$\omega_L \times 10^{12}$	$\gamma \times 10^{12}$
CdTe	7.1	26.58	31.86	1.24
GaAs	11.0	50.65	55.06	0.45
InAs	11.7	41.09	45.24	0.75
InP	9.61	57.25	65.03	0.66
SiC	6.7	149.48	182.65	0.90

2.2 into the above expression, at $\lambda = 70\,\mu$m or $\omega = 2.6909 \times 10^{12}\,$rad s^{-1}, one gets $n = 3.20$ and $K = 0.00235$. Although the above expression is usually sufficient to predict n^* in the infrared for many compound semiconductors and ionic crystals, for low-bandgap semiconductors one should also include the contribution from the free carriers.

2.3 FREE CARRIER ABSORPTION (FCA)

An electromagnetic wave with sufficiently low-frequency oscillations can interact with free carriers in a material and thereby drift the carriers. This interaction results in an energy loss from the EM wave to the lattice vibrations through the carrier scattering processes. Based on the Drude model, the relative permittivity $\varepsilon_r(\omega)$ due to N free electrons per unit volume is given by:

$$\varepsilon_r = \varepsilon_r' - j\varepsilon_r'' = 1 - \frac{\omega_p^2}{\omega^2 - j\omega/\tau}; \quad \omega_p^2 = \frac{Ne^2}{\varepsilon_o m_e} \tag{2.3}$$

where ω_p is a plasma frequency which depends on the electron concentration and τ is the relaxation time of the scattering process, i.e., the mean scattering time. For metals where the electron concentration is very large, ω_p is of the order of $\sim10^{16}$ rad s^{-1}, in the range of UV frequencies, and for $\omega > \omega_p$ $\varepsilon_r \approx 1$, the reflectance becomes very small. Metals lose their free-electron reflectance in the UV range, thus becoming UV transparent. The reflectance does not fall to zero because there are other absorption processes such as those due to interband electron excitations or excitations from core levels to energy bands. Plasma-edge transparency where the reflectance diminishes can also be observed in doped semiconductors. For example, the reflectance of doped InSb has a plasma-edge wavelength that decreases with increasing free-carrier concentration [2.6]. Equation (2.3) can be written in terms of the conductivity σ_o at low frequencies (dc) as:

$$\varepsilon_r = \varepsilon_r' - j\varepsilon_r'' = 1 - \frac{\tau\sigma_o}{\varepsilon_o\left[(\omega\tau)^2 + 1\right]} - j\frac{\sigma_o}{\varepsilon_o\omega\left[(\omega\tau)^2 + 1\right]}. \tag{2.4}$$

In metals, σ_o is high. At frequencies where $\omega < 1/\tau$, the imaginary part $\varepsilon_r'' = \sigma_o/\varepsilon_o\omega$, is normally much more than 1, and $n = K \approx \sqrt{(\varepsilon_r''/2)}$, so that the free-carrier attenuation coefficient α is then given by:

$$\alpha = 2k_oK \approx \frac{2\omega}{c}\left(\frac{\varepsilon_r''}{2}\right)^{1/2} \approx (2\sigma_o\mu_o)^{1/2}\,\omega^{1/2} \tag{2.5}$$

Furthermore, the reflectance can be calculated also using $n = K \approx \sqrt{(\varepsilon_r''/2)}$, which leads to the well-known *Hagen–Rubens relationship* [2.7]:

$$R \approx 1 - 2\left(\frac{2\omega\varepsilon_o}{\sigma_o}\right)^{1/2} \tag{2.6}$$

In semiconductors one encounters typically $\sigma_o/\varepsilon_o\omega < 1$, since the free-electron concentration is small, and we can treat n as constant due to various other polarization mechanisms, e.g., electronic polarization. Since $2nK = \varepsilon_r''$, the absorption coefficient becomes [2.8]:

$$\alpha = 2k_oK \approx \frac{2\omega}{c}\left(\frac{\varepsilon_r''}{2n}\right) = \frac{\sigma_o}{nc\varepsilon_o\left[(\omega\tau)^2 + 1\right]} \tag{2.7}$$

At low frequencies where $\omega < 1/\tau$, we have $\alpha(\lambda) \sim \sigma_o/n(\lambda)$ so that α should be controlled by the dc conductivity, and hence the amount of doping. Furthermore, α will exhibit the frequency dependence of the refractive index n, i.e., $\alpha(\lambda) \sim 1/n(\lambda)$, thus n would typically be determined by the electronic polarization of the crystal.

At high frequencies where $\omega > 1/\tau$, α becomes proportional to N, the free-carrier concentration, and λ^2 as:

$$\alpha \sim \sigma_o/\omega^2 \sim N\lambda^2 \tag{2.8}$$

Experimental observations on FCA in doped semiconductors are, in general, in agreement with these predictions. For example, α increases with N, whether N is increased by doping or by carrier injection [2.9]. However, not all semiconductors show the simple $\alpha \propto \lambda^2$ behaviour. A proper account of the field-driven electron motion and scattering must consider the fact that τ will depend on the electron's energy. The correct approach is to use the Boltzmann transport equation [2.10] with the appropriate scattering mechanism. FCA can be calculated by using a quantum mechanical approach based on second-order time-dependent perturbation theory with Fermi–Dirac statistics [2.11].

Absorption due to free carriers is commonly written as $\alpha \propto \lambda^p$, where the index p depends primarily on the scattering mechanism, though it is also influenced by the compensation doping ratio, if the semiconductor has been doped by compensation, and the free-carrier concentration. In the case of lattice scattering one has to consider scattering from acoustic and optical phonons. For acoustic phonon scattering, $p \approx 1.5$; for optical phonon scattering, $p \approx 2.5$; and for impurity scattering, $p \approx 3.5$. Accordingly, the observed free-carrier absorption coefficient will then have all three contributions, as

$$\alpha = A_{\text{acoustic}}\lambda^{1.5} + A_{\text{optical}}\lambda^{2.5} + A_{\text{impurity}}\lambda^{3.5} \tag{2.9}$$

Inasmuch as α for FCA depends on the free-carrier concentration N, it is possible to evaluate the latter from the experimentally measured α, given its wavelength dependence and p as discussed by Ruda [2.12]. Free-carrier absorption coefficient $\alpha(\text{mm}^{-1})$ for GaP, n-type PbTe, and n-type ZnO are shown in Figure 2.3.

FCA coefficient, 1/mm

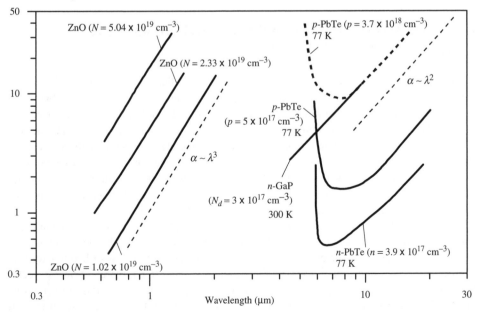

Figure 2.3 Free-carrier absorption in *n*-GaP at 300 K (data extracted from [2.13]), *p*- and *n*-type PbTe (data extracted from [2.14]) at 77 K, and In-doped *n*-type ZnO at room temperature [Data extracted from [2.15]]

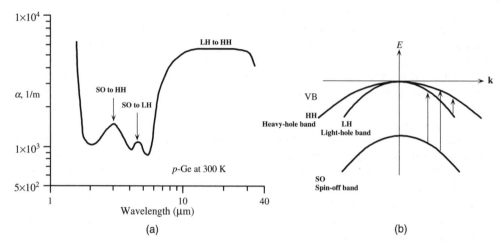

Figure 2.4   (a) Free-carrier absorption due to holes in p-Ge (data extracted from [2.16]). (b) The valence band of Ge has three bands; heavy-hole, light-hole, and spin-off bands

Free-carrier absorption in p-type Ge demonstrates how the FCA coefficient α can be dramatically different from what is expected from Equation (2.9). Figure 2.4(a) shows the wavelength dependence of the absorption coefficient for p-Ge over the wavelength range from about 2 to 30 µm [2.16]. The observed absorption is due to excitations of electrons from the spin-off band to the heavy-hole band, and from spin-off band to the light-hole band, and from the light-hole band to the heavy-hole band as marked in the Figure 2.4(b).

2.4 BAND-TO-BAND OR FUNDAMENTAL ABSORPTION (CRYSTALLINE SOLIDS)

Band-to-band absorption or fundamental absorption of radiation occurs due to the photo-excitation of an electron from the valence band to the conduction band. Thus, absorption of a photon creates an electron in the conduction band and a hole in the valence band and requires the energy and momentum conservation of the excited electron, hole, and photon. In crystalline solids, as the band structures depend on the electron wavevector $\mathbf{k}$, there are two types of band-to-band absorptions corresponding to direct and indirect transitions. In contrast, in amorphous solids, where no long-range order exists only direct transitions are meaningful. The band-to-band absorption in crystalline solids is described below and that in amorphous or disordered solids will be presented in Chapter 3.

First the direct and then indirect transitions will be described here. A direct transition is a photoexcitation process in which no phonons are involved. As the photon momentum is negligible compared with the electron momentum when the photon is absorbed to excite an electron from the valence band (VB) to the conduction band (CB), the electron's $\mathbf{k}$-vector does not change. A direct transition on the E–$\mathbf{k}$ diagram is a vertical transition from an initial energy E and wavevector $\mathbf{k}$ in the VB to a final energy E' and wavevector $\mathbf{k}'$ in the CB where $k' = k$ as shown in Figure 2.5(a), where ($k = |\mathbf{k}|$). The energy ($E' - E_c$) is the kinetic energy $(\hbar k)^2/(2m_e^*)$ of the electron with an effective mass m_e^*, and ($E_v - E$) is the kinetic

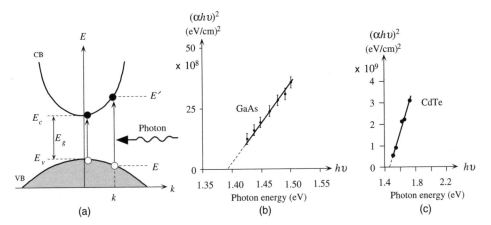

Figure 2.5 (a) A direct transition from the valence band (VB) to the conduction band (CB) by the absorption of a photon. Absorption behavior represented as $(\alpha h\upsilon)^2$ vs photon energy $h\upsilon$ near the band edge for single crystals of (b) p-type GaAs [Data extracted from Kudman and Seidel [2.17] and (c) CdTe, data extracted from Rakhshani [2.18]]

energy $(\hbar k)^2/(2m_h^*)$ of the hole left behind in the VB. The ratio of the kinetic energies of the photogenerated electron and hole depends inversely on the ratio of their effective masses.

The absorption coefficient α is derived from the quantum mechanical transition probability from E to E', the occupied density of states at E in the VB from which electrons are excited, and the unoccupied density of states in the CB at $E + h\upsilon$. Thus α depends on the joint density of states at E and $E + h\upsilon$, and we have to suitably integrate this joint density of states. Near the band edges, the density of states can be approximated by a parabolic band, and the absorption coefficient α is obtained as a function of the photon energy as:

$$\alpha h\upsilon = A(h\upsilon - E_g)^{1/2} \qquad (2.10)$$

where the constant $A \approx [(e^2/(nch^2m_e^*)](2\mu^*)^{3/2}$ in which μ^* is a reduced electron and hole effective mass, n is the refractive index, and E_g is the direct bandgap, minimum $E_c - E_v$ at the same $\mathbf{k}$ value. Experiments indeed show this type of behavior for photon energies above E_g and close to E_g at room temperature as shown Figure 2.5(b) for a GaAs crystal [2.17] and in (c) for a CdTe crystal [2.18]. The extrapolation to zero photon energy gives the direct bandgap E_g, which is about 1.40 eV for GaAs and 1.46–1.49 eV for CdTe. For photon energies very close to the bandgap energy, the absorption is usually due to exciton absorption, especially at low temperatures, and it will be discussed later in this chapter.

In indirect bandgap semiconductors such as crystalline Si and Ge, the photon absorption for photon energies near E_g requires the absorption or emission of phonons during the absorption process as illustrated in Figure 2.6(a). The absorption onset corresponds to a photon energy of $(E_g - h\vartheta)$, which represents the absorption of a phonon with energy $h\vartheta$. In this case, α is proportional to $[h\upsilon - (E_g - h\vartheta)]^2$. Once the photon energy reaches $(E_g + h\vartheta)$, then the photon-absorption process can also occur by phonon emission for which the absorption coefficient is larger than that for phonon absorption. The absorption coefficients for the phonon absorption and emission processes are given by [2.19]:

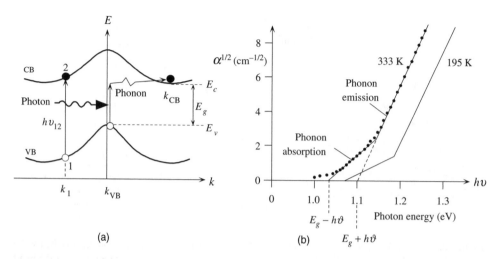

Figure 2.6 (a) Indirect transitions across the bandgap involve phonons. Direct transitions in which dE/dk in the CB is parallel to dE/dk in the VB lead to peaks in the absorption coefficient. (b) Fundamental absorption in Si at two temperatures. The overall behavior is well described by Equations (2.11) and (2.12)

$$\alpha_{\text{absorption}} = A[f_{\text{BE}}(h\vartheta)][h\nu - (E_g - h\vartheta)]^2 \; ; h\nu > (E_g - h\vartheta) \qquad (2.11)$$

and

$$\alpha_{\text{emission}} = A[1 - f_{\text{BE}}(h\vartheta)][h\nu - (E_g + h\vartheta)]^2 \; ; h\nu > (E_g + h\vartheta) \qquad (2.12)$$

where A is a constant, and $f_{\text{BE}}(h\vartheta)$ is the Bose–Einstein distribution function at the phonon energy $h\vartheta$, i.e., $f_{\text{BE}}(h\vartheta) = [(\exp(h\vartheta/k_BT) - 1]^{-1}$, where k_B is the Boltzmann constant and T is the temperature.

As we increase the photon energy in the range $(E_g - h\vartheta) < h\nu < (E_g + h\vartheta)$, the absorption is controlled by $\alpha_{\text{absorption}}$ and the plot of $\alpha^{1/2}$ vs $h\nu$ has an intercept of $(E_g - h\vartheta)$.

For photon energies $h\nu > (E_g + h\vartheta)$, the overall absorption coefficient is $\alpha_{\text{absorption}} + \alpha_{\text{emission}}$, but at slightly higher photon energies than $(E_g + h\vartheta)$, α_{emission} quickly dominates over $\alpha_{\text{absorption}}$ since $[f_{\text{BE}}(h\vartheta)] >> [1 - f_{\text{BE}}(h\vartheta)]$. Figure 2.6(b) shows the behaviour of $\alpha^{1/2}$ vs photon energy for Si at two temperatures for $h\nu$ near band-edge absorption. At low temperatures, $f_{\text{BE}}(h\vartheta)$ is small and $\alpha_{\text{absorption}}$ decreases with decreasing temperature as apparent in Figure 2.6(b). Equations (2.11) and (2.12) intersect the photon energy axis at $(E_g - h\vartheta)$ and $(E_g + h\vartheta)$, which can be used to determine E_g.

Examination of the extinction coefficient K or ε_r'' vs photon energy for Si in Figure 1.2 (Ch. 1) shows that the absorption peaks at certain photon energies, $h\nu \approx 3.5$ and $4.3\,\text{eV}$. These peaks are due to the fact that the joint density of states function peaks at these energies. The absorption coefficient peaks whenever there is a direct transition in which the E vs k curve in the VB is parallel to the E vs k curve in the CB as schematically illustrated in Figure 2.6(a) where a photon of energy $h\nu_{12}$ excites an electron from state 1 in the VB to state 2 in the CB in a direct transition $\mathbf{k}_1 = \mathbf{k}_2$. Such transitions where E vs $\mathbf{k}$ curves are parallel at a photon energy $h\nu_{12}$ result in a peak in the absorption vs photon energy behaviour and can be represented by the condition that

$$(\nabla_{\mathbf{k}}E)_{CB} - (\nabla_{\mathbf{k}}E)_{VB} = 0 \qquad (2.13)$$

The above condition is normally interpreted as the joint density of states reaching a peak value at certain points in the Brillouin zone called van Hove singularities. Identification of peaks from the plat of the extinction coefficient K vs photon energy $h\nu$ involves the examination of all E vs $\mathbf{k}$ curves of a given crystal that can participate in a direct transition. In silicon, ε_r'' peaks at $h\nu \approx 3.5\,\text{eV}$ and $4.3\,\text{eV}$ correspond to Equation (2.13) being satisfied at points L, along $\langle 111 \rangle$ in $\mathbf{k}$-space, and X along $\langle 100 \rangle$ in $\mathbf{k}$-space, at the edges of the Brillouin zone.

In degenerate semiconductors, the Fermi level E_F lies in a band; for example, it lies in the CB for a degenerate n-type semiconductor. In these semiconductors, electrons in the VB can only be excited to states above E_F in the CB rather than to the bottom of the CB. The absorption coefficient then depends on the free carrier concentration since the latter determines E_F. Fundamental absorption is then said to depend on band filling, and there is an apparent shift in the absorption edge, called the *Burstein–Moss shift*. Furthermore, in degenerate indirect semiconductors, the indirect transition may involve a non-phonon scattering process, such as impurity or electron–electron scattering, which can change the electron's wavevector $\mathbf{k}$. Thus, in degenerate indirect bandgap semiconductors, absorption can occur without phonon assistance and the absorption coefficient becomes:

$$\alpha \sim [h\nu - (E_g + \Delta E_F)]^2 \qquad (2.14)$$

where ΔE_F is the energy depth of E_F into the band measured from the band edge.

Heavy doping of degenerate semiconductors normally leads to a phenomenon called bandgap narrowing and bandtailing. A bandtailing means that the band edges at E_v and E_c are no longer well defined cut-off energies and there are electronic states above E_v and below E_c whose density of states falls sharply with energy away from the band edges. Consider a degenerate direct bandgap p-type semiconductor. One can excite electrons from states below E_F in the VB where the band is nearly parabolic to tail states below E_c where the density of states decreases exponentially with energy into the bandgap, away from E_c. Such excitations lead to α depending exponentially on $h\nu$, a dependence that is usually called the Urbach rule [2.20, 2.21], given by:

$$\alpha = \alpha_0 \exp[(h\nu - E_o)/\Delta E] \qquad (2.15)$$

where α_0 and E_o are material-dependent constants, and ΔE, called the Urbach width, is also a material-dependent constant. The Urbach rule was originally reported for alkali halides. It has been observed for many ionic crystals, degenerately doped crystalline semiconductors, and almost all amorphous semiconductors. While exponential bandtailing can explain the observed Urbach tail of the absorption coefficient vs photon energy, it is also possible to attribute the absorption tail behaviour to strong internal fields arising, for example, from ionized dopants or defects. Temperature-induced disorder in the crystal is yet another important mechanism that leads to an Urbach exponential absorption tail.

2.5 IMPURITY ABSORPTION

Impurity absorption can be registered as the peaks of absorption coefficient lying below the fundamental (band-to-band) and excitonic absorption. Mostly they can be related to the presence of ionized impurities or, simply, ions. The origin of these peaks lies in the electronic transitions either between electronic states of an ion and conduction/valence band or intra-ionic transitions (e.g., within d or f shells, or between s and d shells, etc.). In the first case the appearing features are intense and broad lines while in the latter case their appearance strongly depends on whether these transitions are allowed or not by the parity selection rules. For allowed transitions the appearing absorption peaks are quite intense and broad while the forbidden transitions produce weak and narrow peaks. General reviews of this topic may be found in Blasse and Grabmaier [2.22], Henderson and Imbusch [2.23], and DiBartolo [2.24]. In the following sections we first introduce the absorption cross-section concept, and then we concentrate primarily on the properties of rare earth ions, which are of prime importance for modern optoelectronics.

2.5.1 Optical absorption of trivalent rare earth ions: Judd–Ofelt analysis

Rare earth (RE) is the common name for the elements from Lanthanum (La) to Lutetium (Lu). They have atomic numbers from 57 to 71 and form a separate group in the Periodic Table. The specific feature of these elements is the incompletely filled 4f shell. The electronic configurations of RE elements are listed in Table 2.3. The RE may be embedded in different host materials in the form of divalent or trivalent ions. As divalent ions, RE elements exhibit broad absorption-emission lines related to allowed 4f→5d transitions. In the trivalent form, RE elements lose two 6s electrons and one 4f or 5d electron. As a result of Coulombic interaction between 4f electron(s) with the positively charged core the 4f level becomes split into a complicated set of manifolds whose position, as a first approximation,

Table 2.3   Occupation of outer electronic shells for rare earth elements

57	**La**	$4s^2$	$4p^2$	$4d^{10}$		$5s^2$	$5p^6$	$5d^1$	$6s^2$
58	**Ce**	$4s^2$	$4p^2$	$4d^{10}$	$4f^1$	$5s^2$	$5p^6$	$5d^1$	$6s^2$
59	**Pr**	$4s^2$	$4p^2$	$4d^{10}$	$4f^3$	$5s^2$	$5p^6$		$6s^2$
60	**Nd**	$4s^2$	$4p^2$	$4d^{10}$	$4f^4$	$5s^2$	$5p^6$		$6s^2$
...									
68	**Er**	$4s^2$	$4p^2$	$4d^{10}$	$4f^{12}$	$5s^2$	$5p^6$		$6s^2$
...									
70	**Yb**	$4s^2$	$4p^2$	$4d^{10}$	$4f^{14}$	$5s^2$	$5p^6$		$6s^2$
71	**Lu**	$4s^2$	$4p^2$	$4d^{10}$	$4f^{14}$	$5s^2$	$5p^6$	$5d^1$	$6s^2$

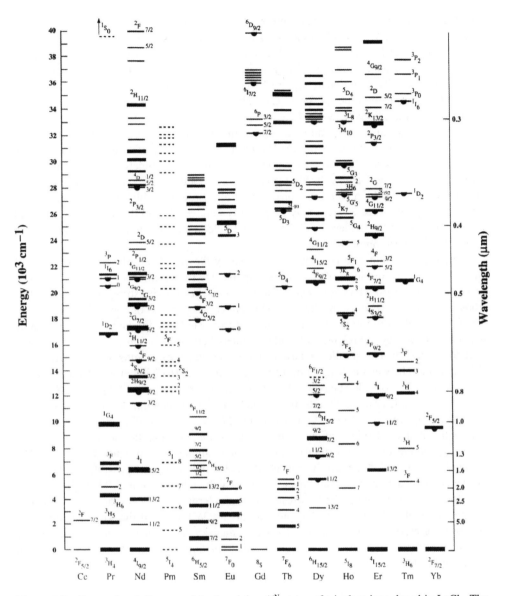

Figure 2.7   Energy level diagram of the low-lying $4f^N$ states of trivalent ions doped in $LaCl_3$. The pendant semicircles indicate fluorescent levels [From Becker et al. [2.25], Henderson and Imbusch [2.23], and Hüfner [2.26]]

is virtually independent of the host matrix because the 4f level is well screened by 5s and 5p shells from outer influences [2.27]. The optical absorption due to electronic transitions between manifolds is very often described using an *absorption cross-section* whose definition is discussed in the next sub-section. Figure 2.7 shows the energy level diagram of the low-lying $4f^N$ states of trivalent ions embedded in $LaCl_3$. As a second approximation, the exact construction and precise energy position of manifolds depend on the host material via the crystal field and via the covalent interaction with *ligands* surrounding the RE ion. The ligand is an atom (or molecule, radical, or ion) with one or more unshared pairs of electrons that can attach to a central metallic ion (or atom) to form a coordination complex. Examples of ligands are ions (F^-, Cl^-, Br^-, I^-, S^{2-}, CN^-, NCS^-, OH^-, NH_2^-) or molecules (NH_3, H_2O, NO, CO) that donate a pair of electrons to a metal atom or ion. Some ligands that share electrons with metals form very stable complexes.

The optical transitions between 4f manifold levels are forbidden by the parity selection rule which states that for a permitted atomic (ionic) transition wave functions of initial and final states must have different parity. The parity is a property of a quantum mechanical state that describes the function after mirror reflection. *Even* functions (states) are symmetric; that is, identical after reflection (like a cosine function), while *odd* functions (states) are antisymmetric (like a sine function). For an ion (or atom) which is embedded in host material the parity selection rule may be partially removed due to the action of the crystal field giving rise to 'forbidden lines'. The crystal field is the electric field created by a host material at the position of the ion.

The parity selection rule is somewhat relaxed by the admixture of 5d state with the 4f state and by the disturbance of RE ions symmetry due to the influence of the host, which increases with the growth of *covalency*. A higher covalency implies stronger sharing of electrons between the RE ions and ligands. This effect is known as the nephelauxetic effect. The resulting absorption-emission lines are characteristic for individual RE ions and quite narrow because they are related to forbidden inner-shell 4f transitions.

Judd–Ofelt (JO) analysis allows the calculation of oscillator strengths of an electric dipole (ED) transition between two states of a trivalent rare-earth (RE) ion embedded in different host lattices. The possible states of RE ions are often referred to as $^{2S+1}L_J$, where $L = 0, 1, 2, 3, 4, 5, 6 \ldots$ determines the electron's total angular momentum and is conventionally represented by letters S, P, D, F, G, I. The term $(2S + 1)$ is called spin multiplicity and represents the number of spin configurations, while J is the total angular momentum, which is the vector sum of the overall (total) angular momentum and overall spin ($\mathbf{J} = \mathbf{L} + \mathbf{S}$). The value $(2J + 1)$ is called the term's multiplicity and is the number of possible combinations of overall angular momentum and overall spin, which yield the same J. Thus, the notation $^4I_{15/2}$ for the ground state of Er^{3+} corresponds to the term $(J, L, S) = (15/2, 6, 3/2)$, which has a multiplicity $2J + 1 = 16$ and a spin multiplicity $2S + 1 = 4$. If the wave functions $|\psi_i\rangle$ and $|\psi_f\rangle$ correspond to the initial ($^{2S+1}L_J$) and final ($^{2S+1}L_{J'}$) states of an electric dipole transition of the RE ion, the line strength of this transition, according to the JO theory, can be calculated using:

$$S_{ED} = |\langle \psi_f | H_{ED} | \psi_i \rangle|^2 = \sum_{k=2,4,6} \Omega_k |\langle f_\gamma^N S'L'J' | U^{(k)} | f_\gamma^N SLJ \rangle|^2 \qquad (2.16)$$

where H_{ED} is ED interaction Hamiltonian, Ω_k are the coefficients reflecting the influence of host material, and $U^{(k)}$ are reduced tensor operator components, which are virtually independent of the host material and their values are calculated using the so-called intermediate coupling approximation, e.g., Weber [2.28]. The theoretical values of S_{ED} thus calculated are compared with the values derived from experimental data using:

$$S_{exp} = \frac{3hcn}{8\pi^3 \langle \lambda \rangle} \frac{2J+1}{\chi_{ED}} \int_{Band} \frac{\alpha(\lambda)}{\rho} d\lambda \qquad (2.17)$$

where $\langle \lambda \rangle$ is the mean wavelength of the transition, h is the Plank constant, c is the speed of light, e is the elementary electronic charge, $\alpha(\lambda)$ is absorption coefficient, ρ is RE ion density, n is the refractive index, and the factor $\chi_{ED} = (n^2 + 2)^2/9$ is a so-called *local field correction*. The key idea of JO analysis is to minimize the discrepancy between experimental and calculated values of line strengths, S_{ED} and S_{exp}, by the appropriate choice of coefficients Ω_k, which are used to characterize the optical transition and compare different materials. The complete analysis should also include the magnetic dipole transition. The latter tend to be typically neglible compared with the electric dipole transitions except in certain, but important, cases such as the transitions between $^4I_{15/2}$ and $^4I_{13/2}$ in the Er^{3+} ion. The Ω_2 value is of prime importance because it is the most sensitive to the local structure and material composition and is correlated with the degree of *covalence*. The values of Ω_k are used to calculate the probabilities of radiative transitions and appropriate radiative lifetimes of excited states, which are very useful for numerous optical applications. More detailed analysis may be found in, e.g., Desurvire [2.29]. The values of Ω_k for different ions and host materials may be found in, e.g., Gschneidner, Jr. and Eyring [2.30].

2.5.2 Optical absorption cross-section

Consider the absorption of radiation due to dopants or impurities in a material system, for example Er^{3+} ions in a glass host such as a silica–germania–alumina glass. Let N (m^{-3}) be the number of dopants per unit volume, α be the absorption coefficient for dopant excitations from a manifold centered at energy E_1 to a manifold centered at E_2. The integrated or *total absorption cross-section* σ_a is defined as integrated absorption within one absorption band per ion, i.e.

$$\sigma_a = \frac{1}{N} \int_{AB} \alpha(\lambda)\,d\lambda \qquad (2.18)$$

where AB stands for the 'absorption band'. The σ_a may vary substantially from one absorption band to another. Thus, for Er^{3+} ions in various silica–germania–alumina glasses, typically, σ_a is ~$6 \times 10^{-21}\,cm^2$ at 1.53 μm, and ~$2.5 \times 10^{-21}\,cm^2$ at 980 nm.

2.6 EFFECT OF EXTERNAL FIELDS

2.6.1 Electro-optic effects

The application of an external field (**E** or **B**) can change the optical properties in a number of ways that depends not only on the material but also on its crystal structure. *Electro-optic effects* refer to changes in the refractive index of a material induced by the application of an external electric field, which therefore 'modulates' the optical properties; the applied field is not the electric field of any light wave, but a separate external field. The application of such a field distorts electronic motions in atoms or molecules of a substance and/or the crystal structure resulting in changes in the optical properties. For example, an applied external field can cause an optically isotropic crystal such as GaAs to become *birefringent*. Typically, changes in the refractive index are small due to the applied electric field. The frequency of the applied field has to be such that it appears to be static over the time a medium takes to change its properties, as well as the time the light requires to pass through the substance. The electro-optic effects are classified according to first- and second-order effects. If we consider the refractive index as a function of the applied electric field E, that is $n = n(E)$, we can, of course, expand it in a Taylor series. Denoting the electric-field-dependent refractive index by n' one can write:

$$n' = n + a_1 E + a_2 E^2 + \ldots \tag{2.19}$$

where n represents the electric-field-independent refractive index and the coefficients a_1 and a_2 are called the *linear* and *second*-order electro-optic effect coefficients, respectively. Although one may consider including even higher terms in the expansion of Equation (2.19), these are generally very small and have negligible effects within highest practical fields. The change in n due to the linear E-dependent term is called the *Pockels effect*, and that due to the E^2 term is called the *Kerr effect*. The coefficient a_2 is generally written as λK, where K is called the *Kerr coefficient*. Thus, the contributions to n' from the two effects can be written as:

$$\Delta n_1 = a_1 E \tag{2.20a}$$

$$\Delta n_2 = a_2 E^2 = (\lambda K) E^2 \tag{2.20b}$$

All materials exhibit the Kerr effect but not the Pockels effect, because only a few crystals have non zero a_1. For all noncrystalline materials (such as glasses and liquids) and those crystals that have a center of symmetry (centrosymmetric crystals such as NaCl) $a_1 = 0$. Only crystals that are noncentrosymmetric exhibit the Pockels effect. For example, an NaCl crystal (centrosymmetric) exhibits no Pockels effect but a GaAs crystal (noncentrosymmetric) does.

The Pockels effect involves examining the effect of the field on the indicatrix, that is, the index ellipsoid, and requires the electro-optic tensor. For example, the change in the principal refractive index n_1 along the principal axis x (where **D** and **E** are parallel) of the indicatrix is written as:

$$\Delta n_1 = -\frac{1}{2} n_1^3 r_{11} E_1 - \frac{1}{2} n_1^3 r_{12} E_2 - \frac{1}{2} n_1^3 r_{13} E_3 \tag{2.21}$$

where E_j is the field along j, $j = 1–3$ corresponding to x, y, and z, and r_{ij} are the elements of the electro-optic tensor **r**, a 6×3 matrix. The Pockels coefficients for various crystals can be found in references [2.31, 2.32, 2.33]. The Pockels effect has found important applications in optical communications in Pockels cell modulators that typically use lithium niobate ($LiNbO_3$) [2.3].

2.6.2 Electro-absorption and Franz–Keldysh effect

Electro-absorption is a change of the absorption spectrum caused by an applied electric field. There are fundamentally three types of electro-absorption processes: The Franz–Keldysh effect, field-injected free-carrier absorption, and the confined Stark effect. In the Franz–Keldysh effect (FKE) [2.34, 2.35], a strong constant applied electric field induces changes in the band structure which change the photon-assisted probability of an electron tunneling from the maxima of the valence band to the minima of the conduction band. The FKE can be detected as a variation of absorption and/or reflection of light with photon energies slightly less than the corresponding bandgaps. This effect was initially observed in CdS, where the absorption edge was red-shifted with the applied electric field, causing the increase of absorption which could be readily detected [2.36]. Later, the FKE was observed and investigated in Ge [2.37], Si [2.38], and other semiconducting materials. In the dynamic Franz–Keldysh effect (DFKE), ultrafast band structure changes are induced in a semiconductor in the presence of a strong laser electric field. These changes include absorption below the band edge and oscillatory behavior in the absorption above. The steady-state effect is normally quite small but the dynamic effect can be up to 40% as found in thin films of GaAs [2.39]. In the presence of an electric field, the absorption coefficient α can be written as [2.1]:

$$\alpha(\omega) \simeq \alpha_0(\omega)\exp\left[-\frac{4}{3}\sqrt{\frac{2m^*}{\hbar^2}}\frac{(E_g - \hbar\omega)^{3/2}}{eF}\right] \qquad (2.22)$$

where $\alpha_0(\omega)$ is the absorption coefficient in the absence of electric field, m^* is the effective mass, E_g is the optical gap, e is the electron charge, and F is the applied electric field. It should be mentioned that the electric-field-induced change in the absorption coefficient implies a change in the refractive index as well [2.40].

In free-carrier absorption, the concentration of free carriers N in a given band is changed (modulated) by an applied voltage, to change the extent of photon absorption. In this case the absorption coefficient is proportional to N and to the light wavelength λ raised to some power, typically 2–3 as shown in Equation 2.8.

In the confined Stark effect, the applied electric field modifies the energy levels in a quantum well (QW). A QW is a thin crystalline semiconductor between two crystalline semiconductors as barriers, which can confine electrons or holes in the dimension perpendicular to the layer surface, while the movement in the other dimensions is not restricted. It is often made with a thin crystalline layer of a semiconductor medium, embedded between other crystalline semiconductor layers of wider bandgap, e.g., GaAs embedded in AlGaAs. The thickness of such a quantum well is typically about 5–20 nanometers. Such thin crystalline layers can be fabricated by molecular beam epitaxy (MBE) or metal-organic chemical vapor deposition. The energy levels are reduced by an amount proportional to the square of the applied field. Without any applied bias, light with photon energy just less than the QW exciton excitation energy will not be significantly absorbed. When a field is applied, the energy levels are lowered and the incident photon energy is now sufficient to excite an electron-and-hole pair in the QWs. Therefore the relative transmission decreases with the reverse bias as shown in Figure 2.8.

In practice the effect of electroabsorption is used to construct an *electroabsorption modulator* (EAM) which is a semiconductor device to control the intensity of a laser beam via an applied electric voltage. Most EAMs are made in the form of a waveguide with electrodes for applying an electric field in a direction perpendicular to the modulated light beam. For achieving a high extinction ratio, one usually exploits the quantum confined Stark effect in a quantum well structure. They can be operated at very high speed; a modulation bandwidth of tens of gigahertz can be achieved, which makes these devices potentially useful for optical fiber communications.

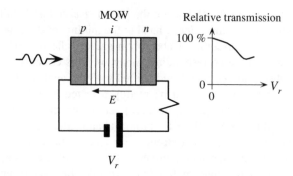

Figure 2.8   A schematic illustration of an electroabsorption modulator using the quantum confined Stark effect in multiple quantum wells (MQWs). The i-region has MQWs. The transmitted light intensity can be modulated by the applied reverse bias to the *pin* device, because the electric field modifies the exciton energy in the QWs

2.6.3 Faraday effect

The Faraday effect, originally observed by Michael Faraday in 1845, is the rotation of the plane of polarization of a light wave as it propagates through a medium subjected to a magnetic field parallel to the direction of propagation of light. When an optically inactive material such as glass is placed in a strong magnetic field and then plane-polarized light is propogated along the direction of the magnetic field, it is found that the emerging light's plane of polarization has been rotated. The magnetic field can be applied, for example, by inserting the material into the core of a magnetic coil – a solenoid. The induced specific rotatory power, given by θ/L, where L is the length of the medium and θ is the angle by which the rotation occurs, is found to be proportional to the magnitude of the applied magnetic field, B, which gives the amount of rotation as:

$$\theta = \vartheta BL \tag{2.23}$$

where ϑ is the proportionality constant, the so-called Verdet constant, and it depends on the material and wavelength of light. The Faraday effect is typically small. For example, a magnetic field of ~0.1 T causes a rotation of about 1° through a glass rod of length 20 mm. It seems to appear that an 'optical activity' has been induced by the application of a strong magnetic field to an otherwise optically inactive material. There is, however, an important distinction between the natural optical activity and the Faraday effect. The sense of rotation θ in the Faraday effect, for a given material (Verdet constant), depends only on the direction of the magnetic field B. If ϑ is positive, for light propagating parallel to **B**, the optical field **E** rotates in the same sense as an advancing right-handed screw pointing in the direction of **B**. The direction of light propagation does not change the absolute sense of rotation of θ. If we reflect the wave to pass through the medium again, the rotation increases to 2θ. The Verdet constant depends not only on the wavelength λ but also on the charge to mass ratio of the electron and the refractive index, $n(\lambda)$, of the medium through:

Table 2.4   Verdet constants of some materials

Material wavelength	Quartz 589.3 nm	Flint glass 632 nm	Tb–Ga Garnet 632 nm	Tb–Ga Garnet 1064 nm	ZnS 589.3 nm	Crown glass 589.3 nm	NaCl 589.3 mm
ϑ $(\mathrm{rad\,m^{-1}\,T^{-1}})$	4.0	4.0	134	40	82	6.4	9.6

$$\vartheta = -\frac{(e/m_e)}{2c} \lambda \frac{dn}{d\lambda} \tag{2.24}$$

The values of Verdet constants for some materials are listed in Table 2.4. The Faraday effect has found useful application in photonics, for example in optical isolators [2.3].

2.7 CONCLUSIONS

Selected optical properties of solids are briefly and semiquantitatively reviewed in this chapter. The emphasis has been on physical insight rather than mathematical rigor. Appropriate examples have been given with typical values for various constants. A classical approach is used for describing lattice or Reststrahlen absorption, infrared reflection and free carrier absorption, whereas a quantum approach is used for band-to-band absorption of photons. Direct and indirect absorption and the corresponding absorption coefficients are discussed with typical examples, but without invoking the quantum mechanical transition matrices. Impurity absorption that occurs in rare-earth-doped glasses is also covered with a brief description of the Judd–Ofelt analysis. Owing to their increasing importance in photonics, the Kerr, Pockels, Franz–Keldysh, and Faraday effects are also described, but without delving into their difficult mathematical formalisms.

REFERENCES

[2.1] K.W. Boer, *Survey of Semiconductor Physics* (Van Nostrand Reinhold, New York, 1990).

[2.2] M. Balkanski and R.F. Wallis, *Semiconductor Physics and Applications* (Oxford University Press, Oxford, 2000).

[2.3] S.O. Kasap, 'Optoelectronics' in *The Optics Encyclopedia*, edited by T.G. Brown, K. Creath, H. Kogelnik, M.A. Kriss, J. Schmit, and M.J. Weber (Wiley-VCH, Weinheim, 2004), Vol. 4, p. 2237.

[2.4] W.J. Turner and W.E. Reese, *Phys. Rev.*, **127**, 126 (1962).

[2.5] Handbook of Optical Constants of Solids, edited by E.D. Palik, (Academic Press, New York, 1985).

[2.6] W.G. Spitzer and H.Y. Fan, *Phys. Rev.*, **106**, 882 (1957).

[2.7] E. Hagen and H. Rubens, *Ann. Phys.*, **14**, 986 (1904).

[2.8] R.J. Elliott and A.F. Gibson, *An Introduction to Solid State Physics and Its Applications* (MacMillann Press Ltd., London, 1974).

[2.9] H.B. Briggs and R.C. Fletcher, *Phys. Rev.*, **91**, 1342 (1953).

[2.10] C.R. Pidgeon, in *Handbook on Semiconductors*, Vol. 2, edited by M. Balkanski (North Holland Publishing, Amsterdam, 1980), Vol. 2, Ch. 5, pp. 223–328.

[2.11] H.E. Ruda, *J. Appl. Phys.*, **72**, 1648 (1992).

[2.12] H.E. Ruda, *J. Appl. Phys.*, **61**, 3035 (1987).

[2.13] J.D. Wiley and M. DiDomenico, *Phys. Rev. B*, **1**, 1655 (1970).

[2.14] H.R. Riedl, *Phys. Rev.*, **127**, 162 (1962).

[2.15] R.L. Weihler, *Phys. Rev.*, **152**, 735 (1966).

[2.16] W. Kaiser, R.J. Collins, and H.Y. Fan, *Phys. Rev.*, **91**, 1380 (1953).

[2.17] I. Kudman and T. Seidel, *J. Appl. Phys.*, **33**, 771 (1962).

[2.18] A.E. Rakhshani, *J. Appl. Phys.*, **81**, 7988 (1997).

[2.19] R.H. Bube, *Electronic Properties of Crystalline Solids* (Academic Press, San Diego, 1974), Ch. 11.

[2.20] F. Urbach, *Phys. Rev.*, **92**, 1324 (1953).

[2.21] J. Pankove, *Phys. Rev.*, **140**, A2059 (1965).

[2.22] G. Blasse and B.C. Grabmaier, *Luminescent Materials* (Springer-Verlag, Berlin, 1994).

[2.23] B. Henderson and G.F. Imbusch, *Optical Spectroscopy of Inorganic Solids* (Clarendon Press, Oxford, 1989).

[2.24] B. DiBartolo, *Optical Interactions in Solids* (John Wiley & Sons, Inc., New York, 1968).

[2.25] P.C. Becker, N.A. Olsson, and J.R. Simpson, *Erbium Doped Fiber Amplifiers. Fundamentals and Technology* (Academic Press, San Diego, London, Boston, New York, Sydney, Tokyo, Toronto, 1999).

[2.26] S. Hüfner, *Optical Spectra of Rare Earth Compounds* (Academic Press, New York, 1978).

[2.27] W.T. Carnall, G.L. Goodman, K. Rajnak, and R.S. Rana, *J. Chem. Phys.*, **90**, 3443 (1989).

[2.28] M.J. Weber, *Phys. Rev. B*, **157**, 262 (1967).

[2.29] E. Desurvire, Erbium-doped fibre amplifiers (John Wiley & Sons, Inc., New York, Chichester, Brisbane, Toronto, Singapore, 1994).

[2.30] K.A. Gschneidner, Jr., and L. Eyring, (Eds.), *Handbook on the Physics and Chemistry of Rare Earths* (Elsevier, Amsterdam, Lausanne, New York, Oxford, Shannon, Singapore, Tokyo, 1998), Vol. 25.

[2.31] A. Yariv, *Optical Electronics*, 3rd Edition (Holt Rinehart and Winston Inc., New York, 1985), Ch. 9.

[2.32] A. Yariv, *Optical Electronics in Modern Communications*, 5th Edition (Oxford University Press, Oxford, 1997).

[2.33] C.C. Davis, *Laser and Electro-Optics* (Cambridge University Press, Cambridge, 1996), Chs. 18, 19.

[2.34] W. Franz, *Z. Naturforsch., Teil A*, **13**, 484 (1958).

[2.35] L.V. Keldysh, *Zh. Eksperim. Teor. Fiz.*, **34**, 1138 (1958); [English transl.: *Soviet Phys. – JETP*, **7**, 788 (1958)].

[2.36] K.W. Böer, *IBM Journal of R&D*, Vol. 13, No. 5, p. 573 (1969).

[2.37] B.O. Seraphin and R.B. Hess, *Phys. Rev. Lett.*, **14**, 138 (1965).

[2.38] A. Frova, P. Handler, F.A. Germano, and D.E. Aspnes, *Phys. Rev.*, **145**, 575 (1966).

[2.39] A. Srivastava, R. Srivastava, J. Wang, and J. Kono, *Phys. Rev. Lett.*, **93**, 157401 (2004).

[2.40] B.O. Seraphin and N. Bottka, *Phys. Rev.*, **139**, A565 (1965).

3 Optical Properties of Disordered Condensed Matter

K. Shimakawa[1], J. Singh[2], and S.K. O'Leary[3]

[1]*Department of Electrical and Electronic Engineering, Gifu University, Gifu 501-1193, Japan*
[2]*Faculty of Technology, B-41, Charles Darwin University, Darwin, NT 0909, Australia*
[3]*Faculty of Engineering, University of Regina, Regina, Canada*

3.1 INTRODUCTION

In a defect-free crystalline semiconductor, there exists a well defined energy gap between the valence and conduction bands. In contrast, in an amorphous semiconductor, the distributions of conduction- and valence-band electronic states do not terminate abruptly at the band edges. Instead, some of the electronic states, referred to as tail states, encroach into the otherwise empty gap region [3.1]. In addition to tail states, there are other localized states deep within the gap region [3.2]. These localized tail states in amorphous semiconductors arise as a consequence of defects. The defects in amorphous semiconductors are considered to be all cases of departure from the normal nearest-neighbor coordination (or normal valence requirement). Examples of defects are broken and dangling bonds (typical for amorphous silicon), over- and under-coordinated atoms (such as 'valence alternation pairs' in chalcogenide glasses), voids, pores, cracks, and other macroscopic defects. As these tail and deep defect states are localized, and there exist mobility edges, which separate these localized states from their extended counterparts [3.3–3.5]. These localized tail and deep defect states are responsible for many of the unique properties exhibited by amorphous semiconductors.

Optical Properties of Condensed Matter and Applications   Edited by J. Singh
© 2006 John Wiley & Sons, Ltd

Despite years of intensive investigation, the exact form of the distribution of electronic states associated with amorphous semiconductors remains a matter for debate. While there are still some unresolved issues, there is general consensus that the tail states arise as a consequence of the disorder (weak and dangling bonds) present within the amorphous network, and that the breadth of these tails reflects the amount of disorder present [3.6]. Experimental results from Tiedje et al. [3.7] and Winer and Ley [3.8], suggest exponential distributions for the valence- and conduction-band tail states in hydrogenated amorphous silicon a-Si:H, although other possible functional forms [3.9] cannot be ruled out. Singh and Shimakawa [3.5, 3.10] have derived different effective masses for the charge carriers in their extended and tail states. This implies that the density-of-states (DOS) of the extended and tail states can be represented in two different parabolic forms. A representation of the current understanding of the distribution of electronic states, for the case of a-Si:H, is schematically presented in Figure 3.1 [3.5, 3.11–13].

The existence of tail states in amorphous solids has a profound impact upon the band-to-band optical absorption. Unlike the case of a crystalline solid, the absorption of photons in an intrinsic amorphous solid can also occur for photon energies below the optical gap, i.e., $\hbar\omega \le E_0$, due to the presence of the tail states in the forbidden gap, E_0 denoting the optical gap, which is usually close to the mobility gap, i.e., the energy difference between the conduction- and valence-band mobility edges.

In a crystalline semiconductor, the energy and crystal momentum of an electron involved in an optical transition must be conserved. In an amorphous semiconductor, however, only the energy needs to be conserved [3.4, 3.5]. As a result, for optical transitions caused by photons of energy $\hbar\omega \ge E_0$ in amorphous semiconductors, the joint density of states approach is not applicable [3.5, 3.14]. One has to consider the product of densities of both the conduction and the valence bands in calculating the corresponding absorption coefficient [3.5, 3.15]. Two approaches are presented here, which are used for calculating the absorption coefficient in amorphous semiconductors. In the first approach, one assumes that the transition matrix element is independent of the photon energy that is absorbed. In the second approach, contrary to the first one, using the dipole approximation, the transition matrix element is found to be photon-energy-dependent. Applying the first approach, one obtains

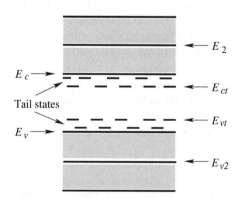

Figure 3.1   Schematic illustration of the electronic energy states, E_2, E_c, E_{ct}, E_{vt}, E_v, and E_{v2}, in amorphous semiconductors. The shaded region represents the extended states. Energies E_2 and E_{v2} correspond to the center of the conduction band and valence band extended states, and E_{ct} and E_{vt} represent the end of conduction band and valence band tail states, respectively

the well known Tauc's relation for the absorption coefficient of amorphous semiconductors, i.e., $(\alpha\hbar\omega)^{1/2} \propto (\hbar\omega - E_0)$, where α is the absorption coefficient. However, applying the second approach, one obtains $[\alpha/\hbar\omega]^{1/2} \propto (\hbar\omega - E_0)$, i.e., a slightly different functional dependence. The first approach has been used widely and successfully to interpret the experimental measurements of examples, which support Tauc's relation as well as which deviate from it. For the latter, as described below, the concept of fractals and effective medium have been used to account for the deviation from the traditional Tauc's relation.

In this chapter, first we will review the application and results of the first approach and then consider those of the second. It will be shown that through the first approach both the fractal and effective medium theories are useful for explaining the optical properties of disordered condensed matter.

3.2 FUNDAMENTAL OPTICAL ABSORPTION (EXPERIMENTAL)

Here we will consider the first approach, where it is assumed that the transition matrix element is independent of the photon energy. As the application of this approach to amorphous semiconductors, like the hydrogenated amorphous silicon (a-Si:H) is well known and can be found in many books [3.4, 3.5], the examples of amorphous chalcogenides (a-Chs) and hydrogenated nanocrystalline silicon (a-ncSi:H) will be presented here.

3.2.1 Amorphous chalcogenides

Analysis of the optical absorption spectra is one of the most useful tools for understanding the electronic structure of solids in any form, crystalline or amorphous. As found in a free electron gas, the DOS for both the conduction and the valence bands is expected to be proportional to the square root of the energy in 3D materials [3.5]. Applying this to amorphous structures with the first approach leads to the well known Tauc plot for the optical absorption coefficient α as a function of photon energy $\hbar\omega$, giving $(\alpha\hbar\omega)^{1/2} \propto (\hbar\omega - E_0)$ [3.16]. However, this quadratic energy dependence of the absorption coefficient on the photon energy is not always observed; e.g., a linear energy dependence $[(\alpha\hbar\omega) \propto (\hbar\omega - E_0)]$ for amorphous Se (a-Se) and a cubic energy dependence $[(\alpha\hbar\omega)^{1/3} \propto (\hbar\omega - E_0)]$ for multicomponent chalcogenide glasses have been observed [3.4]. The deviations from the simple Tauc relation may be regarded as arising from the deviations of the DOS functions from a simple power law.

The DOS in disordered matter, in general, may be described by taking into account the *fractals* that are known to dominate many physical properties in amorphous semiconductors [3.17, 3.18]. We revisit the classical problem for interpreting the optical properties of amorphous semiconductors on the basis of the form of DOS applicable to amorphous chalcogenides. We find that the fundamental optical absorption in amorphous chalcogenides can be written as $(\alpha\hbar\omega)^n \propto (\hbar\omega - E_0)$, where the value of n deviates from 1/2 [3.19]. Typical examples are shown in Figures 3.2 and 3.3 for obliquely deposited amorphous As_2S_3 and Se (hereafter a-As_2S_3 and a-Se), respectively, which are given in the plot of $(\alpha\hbar\nu)^n$ vs $\hbar\nu$. The fitting to the experimental data for a-As_2S_3 [Figure 3.2 (a) and (b)] produces $n = 0.70$ before annealing (as deposited) and $n = 0.59$ after annealing near the glass transition temperature for film. Note that similar values of n (0.73 for as deposited and 0.58 after annealing) are also obtained for a-As$_2$Se$_3$. The fitting to the data for a-Se (Figure 3.3) produces $n = 1$ before annealing (as deposited) and it remains unchanged after annealing at 30°C for 2h.

Here, a brief derivation of the fundamental optical absorption coefficient is given in order to facilitate the interpretation of the experimental results given above. For interband electronic transitions, the optical absorption coefficient through the first approach can be written as [3.4]:

(a)

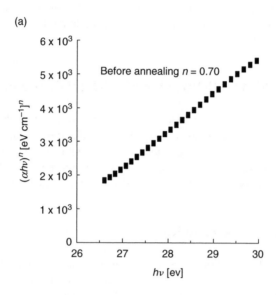

(b)

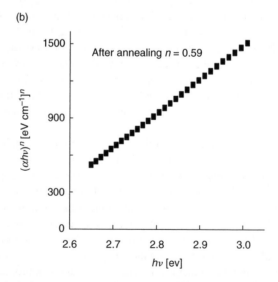

Figure 3.2   The optical absorption spectra of obliquely deposited a-As$_2$S$_3$, where $(\alpha h\nu)^n$ vs $h\nu$ is plotted: (a) $n = 0.70$ before annealing and (b) $n = 0.59$ after annealing at 170°C for 2h

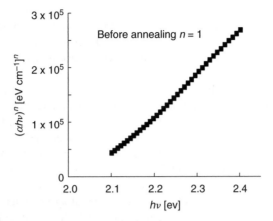

Figure 3.3   The optical absorption spectra of obliquely deposited a-Se, where $(\alpha h v)^n$ vs $h v$ is plotted with $n = 1$ and it remains unchanged before and after annealing

$$\alpha(\omega) = B \int \frac{N_v (E - \hbar\omega) N_c (E) \mathrm{d}E}{\hbar\omega} \qquad (3.1)$$

where B is a constant which includes the square of the transition matrix element as a factor and the integration is over all pairs of states in the valence $N_v(E)$ and conduction states $N_c(E)$. If the density of states for the conduction and valence states is assumed to be $N_c(E) = \text{const}$ $(E - E_C)^s$ and $N_v(E) = \text{const}$ $(E_V - E)^p$, respectively, then Equation 3.1 produces [3.19]:

$$\alpha(\omega)\hbar\omega = B' (\hbar\omega - E_0)^{p+s+1} \qquad (3.2)$$

where B' is another corresponding constant. This gives

$$[\alpha(\omega)\hbar\omega]^n = B'^{\frac{1}{n}} (\hbar\omega - E_0) \qquad (3.3)$$

where $1/n = p + s + 1$. If the form of both $N_c(E)$ and $N_v(E)$ is parabolic, i.e., $p = s = 1/2$ for 3D, then the photon-energy dependence of the absorption coefficient obtained from Equation (3.3) becomes:

$$[\alpha(\omega)\hbar\omega]^{1/2} = B'^{1/2} (\hbar\omega - E_0) \qquad (3.4)$$

which is the well known Tauc's relation obtained for the absorption coefficient.

Let us consider Equation (3.3) and first discuss the simple case of a-Se, where $n = 1$ is obtained. In this case, the sum of $(p + s)$ should be zero for $n = 1$. This is only possible if the product of the DOS functions is independent of the energy. The origin of such DOS functions was argued a long time ago but was unclear. A chain-like structure is basically expected in a-Se. The top of the valence states is known to be formed by p-lone pair (LP) orbitals (lone-pair interaction) of Se atoms. The interaction between lone-pair electrons should be 3D in nature, and therefore the parabolic DOS near the valence band edge can be expected, i.e., $p = 1/2$. The bottom of the conduction-band states, on the other hand, is formed by the anti-bonding states of Se atoms. If the interaction between chains is ignored, the DOS near the conduction-band states may be 1D in nature, i.e., $s = -1/2$. Hence, we obtain $n = 1/(p + s + 1) = 1$, producing a linear dependence of energy, i.e., $(\alpha \hbar \omega) \propto (\hbar \omega - E_0)$ [3.19].

Next, we discuss $As_2S(Se)_3$ binary systems. These systems are suggested to have layered structures. The top of valence-band states are formed from the LP band, and hence, the parabolic DOS near the top of the valence band can also be expected in these systems, since LP–LP interactions occur in 3D space, as was already mentioned. Unlike a-Se, however, the bottom of the conduction band arises from a 2D structure in nature, if the layer–layer interactions can be ignored for the anti-bonding states. This means that the corresponding DOS is independent of energy ($s = 0$). The value of n, in this case, should be given by 2/3, since $p = 1/2$ and $s = 0$ are predicted from the argument of space dimensions, and it is close to those observed for as-deposited oblique films of a-$As_2S(Se)_3$. It should be noted, however, that the layer–layer interactions are not ignored in the DOS of the conduction band [3.20].

The deviations from $n = 2/3$ or $n = 1/2$ may be attributed to the fractional nature of the DOS functions, i.e., p or s cannot be given only values such as 1/2 (3D), 0 (2D), and $-1/2$ (1D). In obliquely deposited As_2Se_3, for example, $n = 0.70$ (before annealing) produces $p + s = 0.43$. In order to interpret this result we may need to discuss the DOS for fractal structures. The DOS for the extended states with energy E on d-space dimension in usual Euclid space is given as:

$$N(E)dE \propto \rho^{d-1}d\rho \tag{3.5}$$

where ρ is defined as $(2m_e^* E)^{1/2}/\hbar$, instead of the wave vector κ, which is not a good quantum number in disordered matter, m_e^* is the electronic effective mass. In a fractal space, on the other hand, a fractal dimension, D, is introduced, instead of d. The DOS for the extended states in the fractal space D can be given by:

$$N(E)dE \propto \rho^{D-1}d\rho \propto E^{\frac{D-2}{2}}dE \tag{3.6}$$

Note that D is introduced as $M(r) \propto r^D$, where M is the 'mass' of an object of radius r in a space, and hence D can take any fractional value (even larger than 3). A similar argument of fractional dimensionality on interband optical transitions has also been presented in anisotropic crystals applicable to low-dimensional structures [3.21].

As we have discussed already, the energy dependence of the DOS for the conduction band is expected to be different from that for the valence band in amorphous chalcogenides, because usually the space dimensionality for the valence band is larger than that for the conduction band. Therefore, we introduce D_v and D_c for the dimensionality of the valence band and the conduction band, respectively. Then $p + s + 1$ in Equation (3.2) is replaced by

$$p+s+1=\frac{D_v+D_c-2}{2} \tag{3.7}$$

for fractional-dimension systems. From the value of n, i.e., $2/(D_v + D_c - 2)$, $D_v + D_c$ can be deduced. In As_2S_3, for example, $D_v + D_c$ is 4.86. After thermal annealing, the value of n tends to 0.5 for both the oblique and flat samples, i.e., $D_v + D_c \approx 6$, indicating that each D_v and D_c approach 3-dimensions. This is due to the fact that thermal annealing produces a more ordered and dense structural network. A similar argument has also been given for flatly deposited materials.

A cubic energy dependence, $n = 1/3$, is often observed for multicomponents (e.g., Ge–As–Te–Si) [3.4], which gives $D_v + D_c = 8$ according to the above analysis. This higher fractal dimension may be related to 'branching' or 'cross-linking' between Te chains by introducing As, Ge, and Si atoms. The 'branching' may be equivalent to a 'Bethe lattice' (or 'Cayley tree'), resulting in an increase in the spatial dimensions [3.18].

In summary, the fundamental optical absorption, empirically presented by the relation $(\alpha \hbar \omega)^n \propto (\hbar \omega - E_0)$, where n deviates from 1/2 in amorphous chalcogenides, is interpreted by introducing the DOS of fractals. The energy-dependent DOS form is not the same for the conduction band and the valence band. The presence of disorder can greatly influence the nature of the electronic DOS even for the extended states. Accordingly, the concept of DOS on fractal structures has successfully been applied to interpret the fundamental optical absorption spectra.

Finally, we should briefly discuss the validity of Equation (3.1) itself in which the transition matrix element is assumed to be independent of energy, i.e., B is independent of energy. Dersch et al. [3.22], on the other hand, suggested that the transition matrix element is energy-dependent and has a peaked nature near the bandgap energy. The absorption coefficient with and without the energy-dependent matrix element will be briefly discussed in section 3.3.

3.2.2 Hydrogenated nanocrystalline silicon (nc-Si:H)

In this section, we discuss the optical properties of nanocrystalline Si prepared by the plasma-enhanced chemical vapor deposition (PECVD) method. Since this material consists of both amorphous and crystalline phases, its structure is very complex and nonuniform. Therefore, it may be expected that the optical absorption cannot be described by Tauc's relation [Equation (3.4)] [3.16]. One of the interesting properties these materials have is the *excess of optical absorption* in the fundamental absorption region. An optical absorption coefficient for nc-Si:H larger than for crystalline silicon (c-Si) has been reported in the infrared to blue region. This is an advantage for using nc-Si:H to fabricate solar cells, because many more photons can be absorbed in the films. An example of this difference is shown in Figure 3.4. Three solid lines represent the experimental data for a-Si:H, nc-Si:H, and c-Si. According to Figure 3.4, a-Si:H has the highest optical absorption coefficient and c-Si has the lowest in the energy range $E > 1.7\,\text{eV}$. Therefore a sample of nc-Si:H (a mixture of both) should be expected to have an absorption in between the two. However, how to calculate the absorption in hydrogenated microcrystalline silicon (μc-Si:H) is still a matter of debate. Although the scattering of light is suggested to be an origin for the enhanced optical absorption, Shimakawa [3.23] took the effective medium approximation (EMA), in an alternative way, to explain the excess of absorption.

Here, first we briefly introduce the EMA before proceeding with the discussion on enhanced absorption within nc-Si:H. The EMA predicts that the total network conductance σ_m for composite materials in D dimensions follows the following condition:

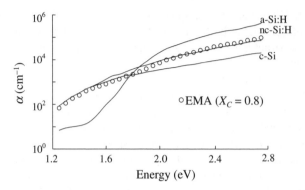

Figure 3.4   Comparison of the optical absorption coefficient of a-Si:H, nc-Si:H, and c-Si. The EMA results are given by solid circles

$$\left\langle \frac{\sigma - \sigma_m}{\sigma + (D-1)\sigma_m} \right\rangle = 0 \tag{3.8}$$

where σ is a random variable of conductivity and $\langle \ldots \rangle$ denotes spatial averaging. Assuming that a random mixture of particles of *two* different conductivities, e.g., one volume fraction, C, has conductivity of σ_0, and the remainder has conductivity of σ_1, σ_1 being substantially less than σ_0, simple analytical expressions for the dc conductivity and Hall mobility as a function of C have been derived (see the pioneering works by Kirkpatrik [3.24] and Cohen and Jortner [3.25]). EMA has also been extended to calculate the ac conductivity in which case σ in Equation (3.1) becomes a complex admittance ($\sigma^* = \sigma_1 + i\sigma_2$) [3.26]. As the dielectric constant, $\varepsilon^* = \varepsilon_1 - i\varepsilon_2 = \sigma_2/\omega - i\sigma_1/\omega$, is closely related to σ^*, optical absorption coefficient $\alpha(\omega)$ can be calculated using $\alpha(\omega) = 4\pi\sigma(\omega)/cn$, where c is the speed of light and n the refractive index [3.23] of the material.

The results obtained from the EMA for the crystalline volume fraction $X_c = 0.8$ and $D = 3$, are shown by open circles in Figure 3.4. The frequency (energy)-independent refractive index, $n_0 = 3.9$ for c-Si and $n_1 = 3.2$ for a-Si:H, which gives the square root of the corresponding real part of their optical dielectric constants, are used in the calculation. The calculated results agree very well with the experimental data, except at an energy of around 1.7 eV (see Figure 3.4). This suggests that a mean field constructed through a mixture of amorphous and crystalline states dominates the optical absorption in nc-Si:H. It may be noted that the multiple light scattering suggested above seems to be not so important in this energy range.

3.3 ABSORPTION COEFFICIENT (THEORY)

In the previous section we discussed the absorption coefficient in relation to the band-to-band absorption under the assumption that the transition matrix element was independent of the absorbed photon energy. The constant coefficient B' in Equations (3.3) and (3.4) was not derived and it could only be determined from fitting of the experimental data. In this section, the theoretical derivation of the absorption coefficient from both approaches, photon energy independent and dependent transition matrix element, will be given. An effort is made to clearly identify the differences between the two approaches and their applications. The absorption coefficient in Equation (3.1), derived from the rate of absorption, can be written as in reference [3.5]:

$$\alpha = \frac{1}{ncV\omega}\left(\frac{2\pi e}{m_e^*}\right)^2 \int\limits_{E_c}^{E_v+\hbar\omega} |p_{cv}|^2 \, N_c(E)\,N_v(E-\hbar\omega)\,\mathrm{d}E \qquad (3.9)$$

where n is the real part of the refractive index and V is the illuminated volume of the material; p_{cv} is the transition matrix element of the electron–photon interaction between the conduction and valence bands. Thus, the integrand consists of a product of three factors, all three of which depend on the photon energy and integration variable of energy E. Therefore, it becomes very difficult to evaluate the integral analytically. For crystalline materials, p_{cv} is evaluated under the dipole approximation, but for amorphous solids the first approximation introduced was to consider p_{cv} as being independent of the photon energy ($\hbar\omega$) [3.4] as described in section 3.2. Then Equation (3.9) reduces to the form of Equation (3.1) and the whole analysis of the absorption properties presented in section 2 is based on this approach. However, it is shown here that the resulting constant coefficient, B', which is an unknown in Equations (3.3) and (3.4), can now be determined. For the constant p_{cv}, a popular form used is that shown in reference [3.4]:

$$p_{cv} = -i\hbar\pi (L/V)^{1/2} \qquad (3.10)$$

where L is the average interatomic spacing in the sample. Assuming that both the valence band and conduction band DOS functions have square-root dependencies on energy (as in a free-electron gas), one gets Tauc's relation given in Equation (3.4) with B' as given in reference [3.5]:

$$B' = \left[\frac{1}{nc\varepsilon_0}\left(\frac{e}{m_e^*}\right)^2\left(\frac{L(m_e^* m_h^*)^{3/2}}{2^2\hbar^3}\right)\right] \qquad (3.11)$$

where m_h^* is the effective mass of a hole in the valence band. The advantage of Equation (3.11) is that if one knows the effective masses, then the so called Tauc's coefficient, B', can be determined theoretically without having to fit Equation (3.4) to any experimental data. As stated in section 3.2, the absorption coefficient of many amorphous semiconductors, including chalcogenides, fit to Equation (3.4) very well, but not all. Deviations from Tauc's relation have been observed, and if one assumes a constant transition matrix element, p_{cv} in Equation (3.9), then the only way these deviations can be explained is from the deviations in the squared-root form of the density of states, N_c and N_v, as discussed in section 3.2 for various examples of chalcogenides. For instance, some experimental data for a-Si:H fit much better to a cubic root relation given by a formula in reference [3.4]:

$$(\alpha\hbar\omega)^{1/3} = C(\hbar\omega - E_0) \qquad (3.12)$$

and therefore the cubic root has been used to determine the optical gap E_0. Here C is another constant. If one considers that the optical transition matrix element is photon-energy-independent [3.5] one finds that the cubic root dependence on photon energy can be obtained only when the valence band and conduction band DOS depend linearly on energy. Using such DOS functions, the cubic root dependence has been explained by Mott and Davis [3.4]. Another approach to arrive at the cubic root dependence has been suggested by Sokolov et al. [3.27]. Using Equation (3.12), they have modeled the cubic root dependence on photon energy by considering the fluctuations in the optical bandgap due to structural disorders. For arriving at the cubic root dependence, they finally assume that the fluctuations are constant over the range of integration and then the integration of Equation (3.12) over the optical gap energy produces a cubic root dependence on the photon energy. Although their approach shows a method of achieving the cubic root dependence, as the integration over the optical gap is carried out by assuming constant fluctuations Sokolov et al.'s model is little different from the linear DOS model suggested by Mott and Davis [3.4].

Let us now consider the second approach, where p_{cv} is not assumed to be a constant. As stated above, for studying the atomic and crystalline absorptions one uses the dipole approximation to evaluate the transition matrix element. This yields (see reference [3.5]):

$$p_{cv} = im_e^* \omega r_{eh}$$ (3.13)

where r_{eh} is the average separation between the excited electron-and-hole pair. It may also be noted that in the case where an electron-and-hole pair is excited by the absorption of a photon, m_e^* should be replaced by their reduced mass μ, where $\mu^{-1} = m_e^{*(-1)} + m_h^{*(-1)}$.

Let us use Equation (3.13) for amorphous solids as well. Inserting Equation (3.13) into Equation (3.9), Cody [3.28] has derived the absorption coefficient in amorphous semiconductors. Accordingly, the absorption coefficient is obtained as shown in reference [3.5]:

$$[\alpha \hbar \omega] = B'' (\hbar \omega)^2 (\hbar \omega - E_0)^2$$ (3.14)

where

$$B'' = \frac{e^2}{nc\varepsilon_0} \left[\frac{(m_e^* m_h^*)^{3/2}}{2\pi^2 \hbar^7 v \rho_A} \right] r_{eh}^2$$ (3.15)

Here v denotes the number of valence electrons per atom and ρ_A represents the atomic density per unit volume.

Equation (3.14) suggests that $[\alpha \hbar \omega]$ depends on the photon energy in the form of a polynomial of order 4. Then, depending on which term of the polynomial may be more significant in which material, one can get square, cubic, fourth, or any other root of the dependence of $[\alpha \hbar \omega]$ on the photon energy. In this case, Equation (3.14) may be expressed as:

$$[\alpha \hbar \omega]^x \propto (\hbar \omega - E_0)$$ (3.16)

where $x \leq 1/2$. Thus, in a way, any deviation from the square root or Tauc's plot may be attributed to the energy-dependent matrix element [3.5, 3.10]. However, this is rather a difficult issue to resolve unless one can determine the form of the DOS functions associated with the conduction and valence bands unambiguously. As a result, as the first approach of the constant transition matrix element has been successful for many samples over the last few decades by many experimental groups, there appears to be a kind of prejudice in the literature in its favor.

Let us now discuss how to determine the constants B' in Equation (3.11) and B'' in Equation (3.15), which involve the effective masses of electrons and holes, m_e^* and m_h^*, respectively. Recently, a simple approach [3.5, 3.29] has been developed to calculate the effective mass of charge carriers in amorphous solids. Accordingly, different effective masses of the charge carriers are obtained in the extended and tail states. The approach applies the concepts of tunneling and effective medium and one obtains the effective mass of an electron in the conduction extended states, denoted by m_{ex}^*, and in the tail states, denoted by m_{et}^*, as in references [3.5, 3.29]:

$$m_{ex}^* \approx \frac{E_L}{2(E_2 - E_c)a^{1/3}} m_e \qquad (3.17)$$

and

$$m_{et}^* \approx \frac{E_L}{(E_c - E_{ct})b^{1/3}} m_e \qquad (3.18)$$

where:

$$E_L = \frac{\hbar^2}{m_e L^2} \qquad (3.19)$$

Here $a = \dfrac{N_1}{N} < 1$, N1 is the number of atoms contributing to the extended states,

$b = \dfrac{N_2}{N} < 1$, N2 is the number of atoms contributing to the tail states, such that a + b =

1 (N = N1 + N2), and m_e is the free-electron mass. The energy E_2 in Equation (3.17) corresponds to the energy of the middle of the conduction extended states at which the imaginary part of the dielectric constant becomes a maximum and E_{ct} is the energy corresponding to the end of the conduction tail states (see Figure 3.1).

Likewise, the hole effective masses m_{hx}^* and m_{ht}^* in the valence extended and tail states are obtained, respectively, as:

$$m_{hx}^* \approx \frac{E_L}{2(E_v - E_{v2})a^{1/3}} m_e \qquad (3.20)$$

and

$$m_{ht}^* \approx \frac{E_L}{2(E_{vt} - E_v)b^{1/3}} m_e \qquad (3.21)$$

where E_{v2} and E_{vt} are energies corresponding to the half width of valence extended states and the end of the valence tail states, respectively (see Figure 3.1).

Using Equations (3.17) and (3.18) and the values of parameters involved, different effective masses of an electron are obtained in the extended and tail states. Considering, for example, the density of weak bonds contributing to the tail states as 1 at.%, i.e., $b = 0.01$ and $a = 0.99$, the effective mass and energy E_L thereby calculated for hydrogenated amorphous silicon (a-Si:H) and germanium (a-Ge:H) are given in Table 3.1. For sp^3 hybrid amorphous semiconductors such as a-Si:H and a-Ge:H, the energies E_{ct} and E_{vt} can be approximated as: $E_{ct} = E_{vt} = E_c/2$.

According to Equations (3.17), (3.18), (3.20), and (3.21), for sp^3 hybrid amorphous semiconductors such as a-Si:H and a-Ge:H, the electron and hole effective masses are expected to be the same. In these semiconductors, as the conduction and valence bands are two equal halves of the same electronic band, their widths are the same and that gives equal effective masses for electron and hole [3.5, 3.29]. This is one of the reasons for using $E_{ct} = E_{vt} = E_c/2$, which gives equal effective masses for electrons and holes in the tail states as well. This is different from crystalline solids where m_e^* and m_h^* are usually not the same. This difference between amorphous and crystalline solids is similar to, for example, having direct and indirect crystalline semiconductors but only direct amorphous semiconductors.

Using the effective masses from Table 3.1 and Equation (3.15), B' can be calculated for a-Si:H and a-Ge:H. The values thus obtained with the refractive index $n = 4$ for a-Si:H and a-Ge:H are $B' = 6.0 \times 10^6 \, \mathrm{cm^{-1} eV^{-1}}$ for a-Si:H and $B' = 4.1 \times 10^6 \, \mathrm{cm^{-1} eV^{-1}}$ for a-Ge:H, which are an order of magnitude higher than those estimated by fitting the experimental data [3.4]. However, considering the quantities involved in B' [Equation (3.11)] this can be regarded as a good agreement.

In a recent paper, Malik and O'Leary [3.33] have studied the distributions of conduction- and valence-bands electronic states associated with a-Si:H. They have noted that the effective masses associated with a-Si:H are material parameters which had yet to be experimentally determined. In order to remedy this deficiency, they have fitted square-root DOS functions to experimental DOS data and found $m_h^* = 2.34 \, m_e$ and $m_e^* = 2.78 \, m_e$.

The value of the constant B'' in Equation (3.15) can also be calculated theoretically, provided r_{eh} is known. Using the atomic density of crystalline silicon and four valence electrons per atom, Cody [3.28] has estimated $r_{eh}^2 = 0.9 \, \text{Å}^2$, which gives $r_{eh} \approx 0.095 \, \mathrm{nm}$, less than half of the interatomic separation of $0.235 \, \mathrm{nm}$ in a-Si:H, but of the same order of magnitude. Using $v = 4$, $\rho_A = 5 \times 10^{28} \, \mathrm{m^{-3}}$, $r_{eh}^2 = 0.9 \, \text{Å}^2$, and extended-state effective masses, we get $B'' = 4.6 \times 10^3 \, \mathrm{cm^{-1} eV^{-3}}$ for a-Si:H and $1.3 \times 10^3 \, \mathrm{cm^{-1} eV^{-3}}$ for a-Ge:H. Cody has estimated an optical gap, $E_0 = 1.64 \, \mathrm{eV}$, for a-Si:H, from which, using Equation (3.14), we get $\alpha = 1.2 \times 10^3 \, \mathrm{cm^{-1}}$ at a photon energy of $\hbar\omega = 2 \, \mathrm{eV}$. This agrees reasonably well with the value $\alpha = 6.0 \times 10^2 \, \mathrm{cm^{-1}}$ used by Cody. If we use the interatomic spacing, L, in place of r_{eh}

Table 3.1 Effective mass of electrons in the extended and tail states of a-Si:H and a-Ge:H calculated using Equations (3.17) and (3.18) for $a = 0.99$, $b = 0.01$ and $E_{ct} = E_{vt} = E_c/2$. E_L is calculated from Equation (3.19). All energies are given in eV. Note that as the optical absorption coefficient is measured in $\mathrm{cm^{-1}}$, the value for the speed of light is used in cm/s. Sources: [a]Morigaki [3.30], [b]Ley [3.31], [c]Street [3.1], and [d]Aoki et al. [3.32]

	L(nm)	E_2	E_c	E_L	E_c–E_{ct}	m_{ex}^*	m_{et}^*
a-Si:H	0.235[a]	3.6[b]	1.80[c]	1.23	0.9	$0.34 m_e$	$6.3 m_e$
a-Ge:H	0.245[a]	3.6	1.05[d]	1.14	0.53	$0.22 m_e$	$10.0 m_e$

in Equation (3.14), we get $B'' = 2.8 \times 10^4 \mathrm{cm}^{-1}\mathrm{eV}^{-3}$, and then the corresponding absorption coefficient becomes $3.3 \times 10^3 \mathrm{cm}^{-1}$. This suggests that for an estimate one may use the inter-atomic spacing in place of r_{eh}, if the latter is unknown. Thus, both the constants B' and B'' can be determined theoretically, a task not possible earlier due to our lack of knowledge of the effective masses in amorphous semiconductors.

The absorption of photons of energy less than the band gap energy, $\hbar\omega < E_0$, in amorphous solids involves the localized tail states and hence follows neither Equation (3.4) nor Equation (3.12). Instead, the absorption coefficient depends on photon energy exponentially as shown in Chapter 2, giving rise to Urbach's tail. Abe and Toyozawa [3.34] have calcu-lated the interband absorption spectra in crystalline solids, introducing the Gaussian site diagonal disorder and applying the coherent potential approximation. They have shown that Urbach's tail occurs due to static disorder (structural disorder). However, the current stage of understanding is that Urbach's tail in amorphous solids occurs due to both thermal and structural disorder [3.28]. More recent issues in this area have been addressed by Orapunt and O'Leary [3.35].

Keeping the above discussion in mind, the optical absorption in amorphous semicon-ductors near the absorption edge is usually characterized by three types of optical transi-tions corresponding to transitions between tail and tail states, tail and extended states, and extended and extended states. The first two types correspond to $\hbar\omega \leq E_0$ while the third corresponds to $\hbar\omega \geq E_0$. Thus, the absorption coefficient vs photon energy (α vs. $\hbar\omega$) depen-dence has the corresponding three different regions A, B, and C that correspond to these three characteristic optical transitions shown in Figure 3.5.

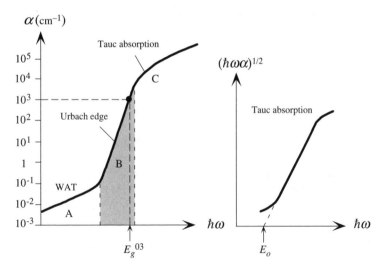

Figure 3.5 Typical spectral dependence of the optical absorption coefficient in amorphous semi-conductors. In the A and B regions, the optical absorption is controlled by the optical transitions between tail and tail, and tail and extended states, respectively, and in the C region it is dominated by transitions from extended to extended states. In domain B, the optical absorption coefficient follows Urbach rule [Equation (2.15) Chapter 2]. In region C, the optical absorption coefficient follows Tauc's relation [Equation (3.4)] in a-Si:H as shown on the right-hand-side figure [Reproduced by permission of Professor S. Kasap]

In the small optical absorption coefficient range A (also called the weak absorption tail (WAT)), where $\alpha < 10^{-1}\,\text{cm}^{-1}$, the optical absorption is controlled by optical transitions from tail-to-tail states. As stated above, the localized tail states in amorphous semiconductors arise from defects. To some extent, the absolute value of the absorption in region A may be used to estimate the density of defects in the material. In region B, where *typically* $10^{-1} < \alpha < 10^{4}\,\text{cm}^{-1}$, the optical absorption is related to transitions from the localized tail states above the valence band edge to extended states in the conduction band and/or from extended states in the valence band to localized tail states below the conduction band. Usually, the spectral dependence of α follows the so-called *Urbach rule*, given in Equation (2.15) in Chapter 2. For many amorphous semiconductors, ΔE has been related to the breadth of the valence- or conduction-band tail states, and may be used to compare the 'breadth' of such localized tail states in different materials; ΔE typically ranges from 50 to 100 meV for the case of a-Si:H. In region C, the optical absorption is controlled by optical transitions from extended to extended states. For many amorphous semiconductors, the α vs $\hbar\omega$ behavior follows Tauc's relation given in Equation (3.4). The optical bandgap, E_0, determined for a given material from the α vs $\hbar\omega$ relations obtained in Equations (2.15) (Chapter 2), (3.4) and (3.12) can vary as shown in Table 3.2 for a-Si:H alloys.

3.4 COMPOSITIONAL VARIATION OF THE OPTICAL BANDGAP IN AMORPHOUS CHALCOGENIDES

In sections 3.2 and 3.3 the behavior of the optical absorption coefficient vs photon energy has been discussed. In this section we discuss the effect of the compositional variation on the optical gap E_0 of amorphous chalcogenide alloys. The bandgap varies with the composition and often exhibits extrema at stoichiometric compositions, e.g., minima in $As_2S(Se)_3$ and a maximum in $GeSe_2$ (see, for example, ref. [3.36]). Applying the virtual crystal approach proposed by Phillips [3.37], Shimakawa has accounted for such compositional variations in the following form [3.36]:

$$E_\text{o} = xE_\text{o}(A) + (1-x)E_\text{o}(B) - \gamma x(1-x) \qquad (3.22)$$

where A and B are composite elements in an A_xB_{1-x} alloy and γ is referred to as the bowing parameter.

Through the study of the optical gap in composite chalcogenides, amorphous chalcogenides can be classified into three types: (1) random bond network (RBN) type, (2) chemically ordered bond network (CON) type, and (3) nondefinite network type. For the RON

Table 3.2 The optical bandgap of a-Si$_{1-x}$C$_x$:H films obtained from Tauc's [Equation (3.4)], Sokolov et al.'s [Equation (3.12)] and Cody's [Equation(3.14)] relations [3.28]

	E_g at $\alpha = 10^3\,\text{cm}^{-1}$	E_g at $\alpha = 10^4\,\text{cm}^{-1}$	$E_g = E_o$ (Tauc)	$E_g = E_o$ (Cody)	$E_g = E_o$ (Sokolov)	ΔE meV
a-Si:H	1.76	1.96	1.73	1.68	1.60	46
a-Si$_{0.88}$C$_{0.18}$:H	2.02	2.27	2.07	2.03	1.86	89

type, A and B are taken as composite elements, e.g., A = Sb and B = Se for Sb_xSe_{1-x}. For the CON type, A and B are taken as the stoichiometric composition and the element in excess, respectively, e.g., A = As_2Se_3 and B = Se in a As_xSe_{1-x} system.

It is known that the optical gap of a-Ge_xSi_{1-x}:H can also be represented by Equation (3.22) with $\gamma = 0$ [3.38]. Note also that the validity of Equation (3.22), proposed by Shimakawa [3.36], has been confirmed in many a-Chs [3.39]. The main conclusions here are as follows: (1) the optical bandgap for amorphous semiconducting alloys is determined by the volume fraction and the optical gap of each element of the alloy, leading to the conjecture that a *modified* virtual crystal approach for mixed crystals is acceptable for an amorphous system. (2) The classification into the three types presented above is supported by the effective medium approach (EMA) which has been applied to electronic transport in amorphous chalcogenides [3.40].

3.5 CONCLUSIONS

The current understanding of the fundamental optical properties in some disordered semiconductors are briefly reviewed. Fundamental optical absorption in amorphous semiconductors, including the chalcogenides, cannot always be expressed by the well-known Tauc's relation. The deviation from Tauc's relation has been discussed through two approaches, first with the energy-independent transition matrix element and second with the energy-dependent matrix element. In the first approach, deviations from Tauc's relation are overcome by introducing the density-of-states in fractal structures of amorphous chalcogenides. This fractal nature may be attributed to the clustered layer structures in amorphous chalcogenides. The applicability of an effective medium approach (EMA) to the fundamental optical absorption spectra of nanocrystalline silicons is confirmed and the effect of the compositional variation on the optical gap in amorphous semiconductors can also be connected to the validity of EMA.

REFERENCES

[3.1] R.A. Street, *Hydrogenated Amorphous Silicon* (Cambridge University Press, Cambridge, 1991).

[3.2] D.A. Papaconstantopoulos and E.N. Economou, *Phys. Rev. B*, **24**, 7233 (1981).

[3.3] M.H. Cohen, H. Fritzsche, and S.R. Ovshinsky, *Phys. Rev. Lett.*, **22**, 1065 (1969).

[3.4] N.F. Mott and E.A. Davis, *Electronic Processes in Non-crystalline Materials* (Clarendon Press, Oxford, 1979).

[3.5] J. Singh and K. Shimakawa, *Advances in Amorphous Semiconductors* (Taylor & Francis, London and New York, 2003).

[3.6] S. Sherman, S. Wagner, and R.A. Gottscho, *Appl. Phys. Lett.*, **69**, 3242 (1996).

[3.7] T. Tiedje, J.M. Cebulla, D.L. Morel, and B. Abeles, *Phys. Rev. Lett.*, **46**, 1425 (1981).

[3.8] K. Winer and L. Ley, *Phys. Rev. B*, **36**, 6072 (1987).

[3.9] D.P. Webb, X.C. Zou, Y.C. Chan, Y.W. Lam, S.H. Lin, X.Y. Lin, K.X. Lin, and S.K. O'Leary, *Solid State Commun.*, **105**, 239 (1998).

[3.10] J. Singh, *J. Mater. Sci.: Mater. Electron.*, **14**, 171 (2003).

[3.11] W.B. Jackson, S.M. Kelso, C.C. Tsai, J.W. Allen, and S.-H. Oh, *Phys. Rev. B*, **31**, 5187 (1985).

[3.12] S.K. O'Leary, S.R. Johnson, and P.K. Lim, *J. Appl. Phys.*, **82**, 3334 (1997).

[3.13] S.M. Malik and S.K. O'Leary, *J. Non-Cryst. Solids*, **336**, 64 (2004).

[3.14] S.R. Elliott, *The Physics and Chemistry of Solids* (John Wiley & Sons, Ltd, Chichester, 1998).

[3.15] J. Singh, *Nonlinear Optics, Principles, Materials, Phenomena, Devices*, **29**, 119 (2002).

[3.16] J. Tauc, *The Optical Properties of Solids*, edited by F. Abeles (North-Holland, Amsterdam, 1979), p. 277.

[3.17] B.B. Mandelbrot, *The Fractal Geometry of Nature* (Freeman, New York, 1982).

[3.18] R. Zallen, *The Physics of Amorphous Solids* (John Wiley & Sons, Inc., New York, 1983), p. 135.

[3.19] M. Nessa, K. Shimakawa, A. Ganjoo, and J. Singh, *J. Optoelectron. Adv. Mater.*, **2**, 133 (2000).

[3.20] Y. Watanabe, H. Kawazoe, and M. Yamane, *Phys. Rev. B*, **38**, 5677 (1988).

[3.21] X.-F. He, *Phys. Rev. B*, **42**, 11751 (1990).

[3.22] U. Dersch, M. Grunnewald, H. Overhof, and P. Thomas, *J. Phys. C: Solid State Phys.*, **20**, 121 (1987).

[3.23] K. Shimakawa, *J. Non-Cryst. Solids*, **266–269**, 223 (2000); K, Shimakawa, *Encyclopedia of Nanoscience and Nanotechnology* (American Scientific Publishes, Valencia, 2004), Vol. 4, p. 35; K. Shimakawa, *J. Mater. Sci.: Materials in Electronics*, **15**, 63 (2004).

[3.24] S. Kirkpatrick, *Rev. Mod. Phys.*, **45**, 574 (1973).

[3.25] M.H. Cohen and J. Jortner, *Phys. Rev. Lett.*, **30**, 699 (1973).

[3.26] B.E. Springett, *Phys. Rev. Lett.*, **31**, 1463 (1973).

[3.27] A.P. Sokolov, A.P. Shebanin, O.A. Golikova, and M.M. Mezdrogina, *J. Phys.: Condens. Matter*, **3**, 9887 (1991).

[3.28] G.D. Cody, *Semiconductors and Semimetals B*, **21**, 11 (1984).

[3.29] J. Singh, T. Aoki, and K. Shimakawa, *Philos. Mag. B*, **82**, 855 (2002).

[3.30] K. Morigaki, *Physics of Amorphous Semiconductors* (World Scientific, London, 1999).

[3.31] L. Ley, 1984, *The Physics of Hydrogenated Amorphous Silicon II*, edited by J.D. Joannopoulos and G. Lukovsky (Springer, Berlin, 1984), p. 61.

[3.32] T. Aoki, H. Shimada, N. Hirao, N. Yoshida, K. Shimakawa, and S.R. Elliott, *Phys. Rev. B*, **59**, 1579 (1999).

[3.33] S.M. Malik and S.K. O'Leary, *J. Mater. Sci.: Mater. Electron.* **16**, 177 (2005); S.K. O'Leary, *J. Mater. Sci.: Mater. Electron.*, **15**, 401 (2004).

[3.34] S. Abe and Y. Toyozawa, *J. Phys. Soc. Jpn.*, **50**, 2185 (1981).

[3.35] F. Orapunt and S.K. O'Leary, *Appl. Phys. Lett.*, **84**, 523 (2004).

[3.36] K. Shimakawa, *J. Non-Cryst. Solids*, **43**, 229 (1981).

[3.37] J.C. Phillips, *Bond and Band in Semiconductors* (Academic Press, New York, 1973).

[3.38] G.H. Bauer, *Solid-State Phenomena*, **44–46**, 365 (1995).

[3.39] L. Tichy, A. Triska, C. Barta, H. Ticha, and M. Frumar, *Philos. Mag. B*, **46**, 365 (1982).

[3.40] K. Shimakawa and S. Nitta, *Phys. Rev. B*, **17**, 3950 (1978).

4 Concept of Excitons

J. Singh[1] and H.E. Ruda[2]

[1]*Faculty of Technology, Charles Darwin University, Darwin, NT 0909, Australia*
[2]*Centre for Nanotechnology, and Electronic and Photonic Materials Group, Department of Materials Science, University of Toronto, Toronto, Ontario Canada*

4.1 INTRODUCTION

An optical absorption in semiconductors and insulators can create an exciton, which is an electron – hole pair excited by a photon and bound together through their attractive Coulomb interaction. The absorbed optical energy remains held within the solid for the lifetime of an exciton. Because of the binding energy between the excited electron and hole, excitonic states lie within the band gap near the edge of the conduction band. There are two types of excitons that can be formed in nonmetallic solids: Wannier or Wannier–Mott excitons and Frenkel excitons. The concept of Wannier–Mott excitons is valid for inorganic semiconductors like Si, Ge, and GaAs, because in these materials the large overlap of interatomic electronic wave functions enables electrons and holes to be far apart but bound in an excitonic state. For this reason these excitons are also called large-radii orbital excitons. Excitons formed in organic crystals are called Frenkel excitons. In organic semiconductors/insulators or molecular crystals, the intermolecular separation is large and hence the overlap of intermolecular electronic wave functions is very small and electrons remain tightly bound to individual molecules. Therefore, the electronic energy bands are very narrow and closely related to individual molecular electronic energy levels. In such solids, the absorption of photons occurs close to the individual molecular electronic states and excitons are also formed within the molecular energy levels (see, e.g., reference [4.1]). Such excitons are therefore also called molecular excitons. For details of the theory of Wannier

Optical Properties of Condensed Matter and Applications Edited by J. Singh
© 2006 John Wiley & Sons, Ltd

and Frenkel excitons, readers may like to refer to the book by Singh [4.1]. The excitonic concept from the theory point of view was initially developed only for crystalline solids and it used to be believed that excitons cannot be formed in amorphous semiconductors. However, several experiments on photoluminescence measurements have revealed the existence of excitons in amorphous semiconductors as well, and the theory of excitons in such solids has subsequently been developed. Here the concept of excitons in crystalline solids is reviewed briefly first as it is well established, and then amorphous semiconductors will be considered.

4.2 EXCITONS IN CRYSTALLINE SOLIDS

In Wannier–Mott excitons, the Coulombic interaction between the hole and electron can be viewed as an effective hydrogen atom with, for example, the hole establishing the coordinate reference frame about which the reduced-mass electron moves. If the effective masses of the isolated electron and hole are m_e^* and m_h^*, respectively, their reduced mass, μ_x, is given by:

$$\mu_x^{-1} = (m_e^*)^{-1} + (m_h^*)^{-1} \tag{4.1}$$

Note that in the case of so-called hydrogenic impurities in semiconductors (i.e., both shallow donor and acceptor impurities), the reduced mass of the nucleus takes the place of one of the terms in Equation (4.1) and hence the reduced mass is given to a good approximation by the effective mass of the appropriate carrier. In the case of an exciton as the carrier effective masses are comparable and hence the reduced mass is markedly lower – accordingly, the exciton binding energy is markedly lower than that for hydrogenic impurities. The energy of a Wannier–Mott exciton is given by (e.g., see reference [4.1]):

$$E_x = E_g + \frac{\hbar^2 K^2}{2M} - E_n \tag{4.2}$$

where E_g is the bandgap energy, $\hbar K$ the linear momentum, and M ($= m_e^* + m_h^*$) the effective mass associated with the center of mass of an exciton, and E_n is the exciton binding energy given by:

$$E_n = \frac{\mu_x e^4 \kappa^2}{2\hbar^2 \varepsilon^2} \frac{1}{n^2} = \frac{R_y^x}{n^2} \tag{4.3}$$

where e is the electronic charge, $\kappa = \dfrac{1}{4\pi\varepsilon_0}$, ε is the static dielectric constant of the solid, and n is the principal quantum number associated with the internal excitonic states $n = 1$ (s), 2 (p), etc., as shown in Figure 4.1. R_y^x is the so-called effective Rydberg constant of an exciton given by $R_y(\mu_x/m_e)/\varepsilon^2$, where $R_y = 13.6\,\mathrm{eV}$. According to Equation (4.3), as stated above, the excitonic states are formed within the bandgap near the conduction band edge. However, as the exciton binding energy is very small, e.g., a few meV in bulk Si and Ge crystals, exciton absorption peaks can be observed only at very low temperatures. For bulk

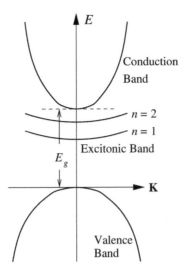

Figure 4.1   Schematic illustration of excitonic bands for $n = 1$ and 2 in semiconductors. E_g represents the energy gap

GaAs, the exciton binding energy ($n = 1$) corresponds to about 5 meV. Following from the hydrogen-atom model, the extension of the excitonic wave function can be found from an effective exciton Bohr radius a_x given in terms of the Bohr radius as $a_0 = h^2/\pi e^2$: that is $a_x = a_0(\varepsilon/\mu_x)$. For GaAs this corresponds to about 12 nm or about 21 lattice constants – that is, the spherical volume of the exciton radius contains $\sim(a_x/a)^3$ or ~9,000 unit cells, where a is the lattice constant of GaAs. As $R_y^* \ll E_g$ and $a_x \gg a$, excitons in GaAs are large-radii orbital excitons, as stated above. It should be noted that the binding energy of excitons in semiconductors tends to be a strong function of the bandgap. In Figures 4.2(a) and (b) are shown the dependencies of R_y^* (exciton binding energy) and a_x/a on the bandgap of semiconductors, respectively. In the case of excitons having large binding energy and correspondingly small radius (i.e., approaching the size of about one lattice parameter), the excitons become localized on a lattice site as observed in most organic semiconductors. As stated above, such excitons are commonly referred to as Frenkel excitons or molecular excitons. Unlike Wannier–Mott excitons, which typically are dissociated at room temperature, these excitations are stable at room temperature. For the binding energy of Frenkel excitons one may refer to, e.g., Singh [4.1].

Excitons may recombine radiatively, emitting a series of hydrogen-like spectral lines as described by Equation (4.3). In bulk crystalline (3D) semiconductors such as Si, Ge, and GaAs, exciton lines can be observed only at low temperatures. As the binding energy is small, typically ≤ 10 meV, excitons in bulk are easily dissociated by thermal fluctuations.

The above discussion refers to so-called free excitons formed between conduction-band electrons and valence-band holes of a crystalline semiconductor or insulator. According to Equation (4.2), such an exciton is able to move throughout a material with a center of mass kinetic energy (second term on the right-hand side). It should be noted, however, that free electrons and holes move with a velocity $\hbar(dE/dk)$ where the derivative is taken for the appropriate band edge. To move through a crystal, both electron and hole must have identical

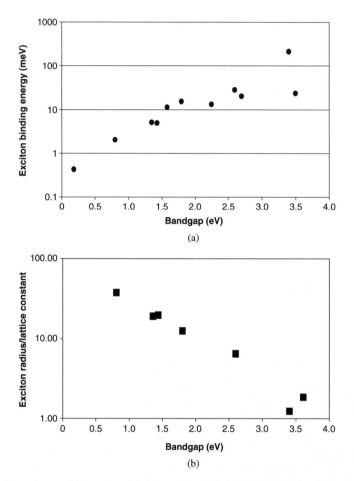

Figure 4.2 Dependence of (a) exciton binding energy (R_y^*) [Equation (4.3)] and (b) size in terms of the ratio of the excitonic Bohr radius to lattice constant (a_B^*/a_0) as a function of the semiconductor bandgap. Exciton binding energy increases concurrently with marked diminishment of exciton spreading as the bandgap increases. Above a bandgap of about 2 eV the Wannier-based description is not appropriate

translational velocity, thereby restricting the regions in k-space where these excitations can occur to those with $(dE/dk)_{electron} = (dE/dk)_{hole}$, commonly referred to as critical points.

In quantum wells and other structures of reduced dimensionality, the spatial confinement of both the electron and hole wave functions in the same layer ensures strong excitonic transitions of a few meV below the bandgap, even at room temperature. The binding energy of excitons, $E_{bx}(\alpha,n)$, in confined systems of dimension α is given by [4.2, 4.3]:

$$E_{bx}(\alpha,n) = \frac{R_y^x}{[n+(\alpha-3)/2]^2} \qquad (4.4)$$

Thus, the binding energy of excitons in quantum wells ($\alpha \approx 2$) and $n = 1$ increases to four times the value in 3D ($\alpha = 3$). For quantum wires, $\alpha = 1$, the binding energy becomes infinitely large and for quantum dots ($\alpha = 0$) it becomes the same as in 2D. It is this enhancement in the binding energy due to confinement that allows excitonic absorption and photoluminescence to be observed even at room temperature in quantum wells. Furthermore, the observation of biexcitons [4.3, 4.4] and excitonic molecules also becomes possible due to the large binding energy. The ratio of the binding energy of biexcitons, E_b^{xx}, to that of excitons, E_b^x for $n = 1$ is usually constant in quantum wells, and for GaAs quantum wells one gets $E_b^{xx} = 0.228\, E_b^x$ [4.2, 4.3].

4.2.1 Excitonic absorption in crystalline solids

As exciton states lie below the conduction-band edges in crystalline solids, absorption to excitonic states is observed below the conduction-band edge. According to Equation (4.2), the difference of energy between the bandgap and excitonic absorption gives the binding energy. As the exciton–photon interaction operator and pair of excited electron and hole and photon interaction operator depend only on their relative motion momentum, the form of these interactions is the same for band-to-band and excitonic absorption. Therefore, for calculating the excitonic absorption coefficient one can use the same form of interaction as that for band-to-band absorption in Chapter 2 but one has to use the joint density of states for crystalline solids. Thus, the absorption coefficient associated with the excitonic states in crystalline semiconductors is obtained as (e.g., [4.5]):

$$\alpha \hbar \omega = \frac{4e^2}{\sqrt{\varepsilon} c \hbar^2} \left(\frac{2}{\mu_x} \right)^{1/2} |p_{xv}|^2 \left(\hbar \omega + E_n - E_g \right)^{1/2}, \quad n = 1, 2, \ldots \tag{4.5}$$

where $\hbar \omega$ is the energy of the absorbed photon, p_{xv} is the transition matrix element between the excitonic state and the valence band, and E_n is the exciton binding energy corresponding to a state with $n = 1$, 2, etc. [see Equation (4.3)]. Equation (4.5) is similar to the case of direct band-to-band transitions discussed in Chapter 2 and it is valid only for the photon energies, $\hbar \omega \geq E_x$. There is no absorption below the excitonic ground state in pure crystalline solids. The absorption of photons to the excitonic energy levels are possible either by exciting electrons to higher energy levels in the conduction band and then by their non-radiative relaxation to the excitonic energy level or by exciting an electron directly to the exciton energy level. Excitonic absorption occurs in both direct as well as indirect semiconductors. If the valence hole is a *heavy* hole, the exciton is called a heavy-hole exciton; conversely, if the valence hole is light, the exciton is a light-hole exciton.

For large exciton binding energy, say corresponding to $n = 1$, the excitonic state will be well separated from the edge, E_g, of the conduction band and then one can observe a sharp excitonic peak at the photon energy $\hbar \omega = E_g - E_1$. The absorption to higher excitonic states, corresponding to $n > 1$, may not be observed in materials with small binding energies as these states will be located within the conduction band. The excitonic absorption and photoluminescence in GaAs quantum wells are shown in Figure 4.3(a) and (b), respectively [4.3]. In Figure 4.3(b) are also shown the biexcitonic photoluminescence peaks (with superscript xx) observed in GaAs quantum wells. In the absorption spectra [Figure 4.3(a)], both light-hole (LHh^x) and heavy-hole (HH^x) excitons are observed in quantum wells of different well width indicated by the corresponding subscript. For example, HH$_{100}^x$ and HH$_{100}^{xx}$ mean heavy-hole exciton and biexciton peaks, respectively, in a quantum well of width 100 Å.

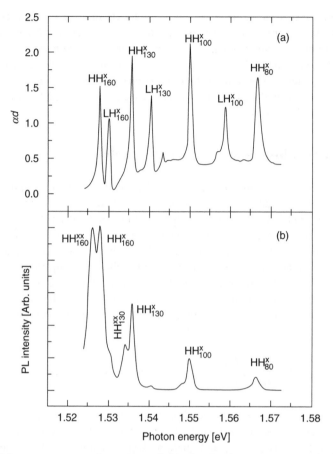

Figure 4.3 (a) Low-temperature absorption and (b) photoluminescence spectra in GaAs quantum wells of different well widths. (HH^x_{100} and HH^{xx}_{100} denote heavy-hole exciton and biexciton, respectively, in a quantum well of width 100 Å.) The photoluminescence data are obtained using HeNe laser excitation [4.3]

It may be noted that the transition matrix element, p_{xv}, in Equation (4.5) is assumed to be photon-energy independent. It is the average of the linear relative momentum between electron and hole in an exciton. However, if one applies the dipolar approximation, the transition matrix element thus obtained depends on the photon energy [4.5, 4.6], which gives a different photon-energy dependence in the absorption coefficient. This aspect of the absorption coefficient is described in detail in Chapter 2.

Excitonic absorption is spectrally well located and very sensitive to optical saturation. For this reason, it plays an important role in nonlinear semiconductor devices (nonlinear Fabry–Perot resonator, nonlinear mirror, saturable absorber, and so on). For practical purposes, the excitonic contribution to the overall susceptibility around the resonance frequency ν_{ex} can be written as:

$$\chi_{exc} = -A_0 \frac{(\nu - \nu_{ex}) + j\Gamma_{ex}}{(\nu - \nu_{ex})^2 + \Gamma_{ex}^2(1+S)} \tag{4.6}$$

with Γ_{ex} being the line width and $S = I/I_S$ the *saturation parameter* of the transition. For instance, in GaAs multiple quantum wells (MQW), the saturation intensity I_S is as low as $1\,kW/cm^2$, and Γ_{ex} ($\approx 3.55\,meV$ at room temperature) varies with the temperature according to:

$$\Gamma_{ex} = \Gamma_0 + \frac{\Gamma_1}{\exp(\hbar\omega_{LO}/kT) - 1} \tag{4.7}$$

where $\hbar\Gamma_0$ is the inhomogeneous broadening ($\approx 2\,meV$), $\hbar\Gamma_1$ the homogeneous broadening ($\approx 5\,meV$), and $\hbar\omega_{LO}$ the longitudinal optical-phonon energy ($\approx 36\,meV$). At high carrier concentrations (provided either by electrical pumping or by optical injection), an efficient mechanism of saturating the excitonic line is the *screening* of the Coulombic attractive potential by free electrons and holes.

A number of more complex pairings of carriers can also occur, which may also include fixed charges or ions. For example, for the case of three charged entities with one being an ionized donor impurity (D+) the following possibilities can occur: (D+)(+)(−), (D+)(−)(−), and (+)(+)(−) as excitonic ions, and (+)(+)(−)(−) and (D+)(+)(−)(−) as biexcitons or even bigger excitonic molecules (see, e.g., references [4.2, 4.3, and 4.7]). Complexity abounds in these systems as each electronic level possesses a fine structure corresponding to allowed rotational and vibrational levels. Moreover, the effective mass is often anisotropic. Note that when the exciton or exciton complex is bound to a fixed charge such as an ionized donor or acceptor center in the material, the exciton or exciton complex is referred to as a bound exciton. Indeed, bound excitons may also involve neutral fixed impurities. It is usual to relate the exciton in these cases to the center binding them – thus if an exciton is bound to a donor impurity, it is usually termed a donor-bound exciton.

4.3 EXCITONS IN AMORPHOUS SEMICONDUCTORS

As stated above, the concept of excitons is traditionally valid only for crystalline solids. However, several observations in the photoluminescence spectra of amorphous semiconductors have revealed the occurrence of photoluminescence associated with the singlet and triplet excitons (see, e.g., reference [4.5]). Applying the effective-mass approach, a theory for the Wannier–Mott excitons in amorphous semiconductors has recently been developed in real coordinate space [4.8–4.11]. The energy of an exciton thus derived is obtained as:

$$W_x = E_0 + \frac{P^2}{2M} - E_n(S) \tag{4.8}$$

where E_0 is the optical gap, P is the linear momentum associated with the exciton's center of mass motion, and $E_n(S)$ is the binding energy of excitons given by:

$$E_x(S) = \frac{\mu_x e^4 \kappa^2}{2\hbar^2 \varepsilon'(S)^2 n^2} \tag{4.9}$$

where

$$\varepsilon'(S) = \varepsilon \left[1 - \frac{(1-S)}{A} \right]^{-1} \tag{4.10}$$

with S being the spin of an exciton ($S = 0$ for singlet and $= 1$ for triplet) and A is a material-dependent constant representing the ratio of the magnitude of the Coulomb and exchange interactions between the electron and hole of an exciton. Equation (4.9) is analogous to Equation (4.3) obtained for excitons in crystalline solids for $S = 1$. This is because Equation (4.3) is derived within the large-radii orbital approximation, which neglects the exchange interaction and hence it is valid only for triplet excitons [4.1, 4.12]. As amorphous solids lack long-range order, the exciton binding energy is found to be larger in amorphous solids than in their crystalline counterparts; for example, in hydrogenated amorphous silicon (a-Si:H) the binding energy is higher than in crystalline silicon (c-Si). This is the reason that it is possible to observe the photoluminescence of both singlet and triplet excitons in a-Si:H [4.13] but not in c-Si.

According to Equation (4.9), the singlet exciton binding energy corresponding to $n = 1$ becomes:

$$E_1(S = 0) = \frac{(A-1)^2 \mu_x e^4 \kappa^2}{2A^2 \hbar^2 \varepsilon^2} \tag{4.11}$$

and the triplet exciton binding energy becomes:

$$E_1(S = 1) = \frac{\mu_x e^4 \kappa^2}{2\hbar^2 \varepsilon^2} \tag{4.12}$$

From Equations (4.11) and (4.12) we get the relation between the singlet and triplet exciton binding energies as:

$$E_1(S = 0) = \frac{(A-1)^2}{A^2} E_1(S = 1) \tag{4.13}$$

For hydrogenated amorphous silicon (a-Si:H), A is estimated to be about 10 [4.5], which gives from Equation (4.13) the result that the singlet exciton binding energy is about 90% of the triplet exciton binding energy. As μ_x is different in the extended and tail states, there are four possibilities through which an exciton can be formed in amorphous solids: (I) both the excited electron and hole are in their extended states, (II) the excited electron is in the extended states and the hole is in the tail states, (III) the excited electron is in the tail states and the hole is in the extended states, and (IV) both the excited electron and hole are in their tail states.

Using the expressions of effective mass of charge carriers derived in the extended and tail states in Chapter 2, we get the effective mass of the electron and hole in the extended states in a-Si:H as $m_{ex}^* = 0.34 m_e$ and in the tail states as $m_{et}^* = 7.1 m_e$. This gives $\mu_x = 0.17 m_e$ for possibility (I) when both the excited carriers are in their extended states, $\mu_x = 0.32 m_e$ for possibilities (II) and (III) when one of the excited charge carriers is in the tail and the other in the extended states, and $\mu_x = 3.55 m_e$ for possibility (IV) when both electron and hole are in their tail states in a-Si:H. Using then $\varepsilon = 12$, one obtains the singlet

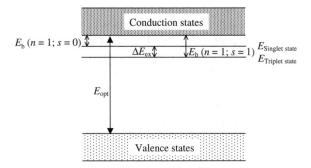

Figure 4.4   Schematic illustration of singlet and triplet excitonic states in amorphous solids. ΔE_{ex} represents the energy difference between singlet and triplet excitonic states

exciton binding energy in a-Si:H as $E_1(S = 0) = 16$ meV for possibility (I), 47 meV for possibilities (II) and (III), and 0.33 eV for possibility (IV). The corresponding binding energies for a triplet exciton will be 18 meV for possibility (I), 52 meV for possibilities (II) and (III), and 0.37 eV for possibility (IV). These energies are measured from the conduction-band edge or the conduction-band mobility edge, E_c, and are schematically shown in Figure 4.4. Accordingly, the triplet exciton states lie below the singlet exciton states.

The excitonic Bohr radius is also found to be different for singlet and triplet excitons in amorphous semiconductors. Writing the exciton energy as:

$$E_1(S) = -\frac{\kappa e^2}{2a_x(S)} \tag{4.14}$$

we get from Equations (4.11)–(4.13):

$$a_x(S = 0) = \frac{A^2}{(A-1)^2} a_x(S = 0) \tag{4.15}$$

which gives for a-Si:H with $A = 10$, $a_x(S = 0) \approx \frac{5}{4}a_x(S = 0)$. Accordingly, for a singlet exciton, we get for possibility (I) $a_x(S = 0) \approx 4.67$ nm, for possibilities (II) and (III) $a_x(S = 0) \approx 2.5$ nm, and for possibility (IV) $a_x(S = 0) \approx 0.233$ nm. The excitonic Bohr radius plays a very significant role in the radiative recombination of excitons because this is the average separation between the electron and hole in an exciton prior to their recombination. Therefore the rates of spontaneous emission depends on the excitonic Bohr radius (see Chapter 5 on photoluminescence).

4.3.1 Excitonic absorption in amorphous solids

In amorphous semiconductors, the excitonic absorption and photoluminescence can be quite complicated. According to Equation (4.8), the excitonic energy level lies below the optical bandgap by an energy equal to the binding energy given in Equation (4.9). However, as stated in the previous section, there are four possibilities of transitions for absorption in amorphous semiconductors: (I) valence-extended to conduction-extended states, (II) valence-tail to conduction-extended states, (III) valence-extended to conduction-tail states, and (IV) valence-tail to conduction-tail states. These possibilities will have different optical gap energies, E_0, and different binding energies. Possibility (I) will give rise to absorption as in the free exciton states, possibilities (II) and (III) will give absorption in the bound exciton states because one of the charge carriers is localized in the tail states, and the absorption through possibility (IV) will create localized excitonic geminate pairs. This can be visualized as follows: If an electron – hole pair is excited by a high-energy photon through possibility (I) and forms an exciton, initially its excitonic energy level and the corresponding Bohr radius will have a reduced mass corresponding to both charge carriers being in the extended states. As such an exciton relaxes downward nonradiatively, the binding energy and excitonic Bohr radius will change because the effective mass changes in the tail states. When both charge carriers reach the tail states, possibility (IV), although the pair is localized its excitonic Bohr radius will be maintained. In this situation, the excitonic nature breaks down as both charge carriers are localized and the excitonic wave function cannot be used to calculate any physical property of these localized carriers. One has to use the individual localized wave functions of the electron and hole.

For calculating the excitonic absorption coefficient in amorphous semiconductors one can use the same approach as presented in Chapter 3, and similar expressions, such as Equations (3.4) and (3.14), will be obtained. This is because (i) the transition matrix element remains the same for excitonic absorption and for band-to-band free-carrier absorption and (ii) the concept of the joint density of states applied to excitonic absorptions in crystalline solids is not applicable to excitonic absorptions in amorphous solids. Therefore, by replacing the effective masses of charge carriers by the excitonic reduced mass and the distance between the excited electron and hole by the excitonic Bohr radius, one can use Equations (3.4) and (3.14) for calculating the excitonic absorption coefficients for transitions corresponding to the above four possibilities in amorphous semiconductors. Thus, one obtains two types of excitonic absorption coefficients for amorphous solids. The first type is obtained by assuming that the transition matrix element is independent of the photon energy but depends on the excitonic Bohr radius as:

$$[\alpha \hbar \omega]^{1/2} = B_x^{1/2}[\hbar \omega - E_0] \tag{4.16}$$

where

$$B_x = \frac{\mu_x e^2}{4nc\varepsilon_0} \frac{a_x}{\hbar^3} \tag{4.17}$$

The coefficient B_x in Equation (4.17) is obtained by replacing m_e^* and m_h^* in Equation (2.17) by μ_x, and L by a_x. The absorption coefficient derived in Equation (4.16) has the same photon-energy dependence as Tauc's relation obtained for the band-to-band free carrier absorption (see Chapter 3). The second expression is obtained by applying the dipole approximation as:

$$[\alpha \hbar \omega] = B_x'(\hbar \omega)^2 (\hbar \omega - E_0)^2 \tag{4.18}$$

where

$$B'_x = \frac{e^2}{nc\varepsilon_0} \frac{\mu_x^3 a_x^2}{2\pi^2 \hbar^7 v \rho_A} \tag{4.19}$$

Here again r_{eh} in Equation (3.14) has been replaced by a_x. Thus, the excitonic absorption coefficient depends on the photon energy in the same way as does the band-to-band absorption coefficient described in Chapter 2. This is probably the reason that distinct excitonic absorption peaks in amorphous semiconductors have not yet been observed to the best of our knowledge, but excitonic photoluminescence peaks have been observed [4.12].

4.4 CONCLUSIONS

Concepts of excitons in both crystalline and amorphous solids are presented. Excitonic absorption in crystalline solids is reviewed. It is shown that, in amorphous solids, the excitonic absorption spectrum is similar to the band-to-band absorption spectrum. For example, the excitonic absorption in amorphous semiconductors also satisfies Tauc's relation, as does band-to-band absorption. This is because: (i) the transition matrix element remains the same for excitonic absorption and for band-to-band free-carrier absorption and (ii) the concept of the joint density of states applied to excitonic absorptions in crystalline solids is not applicable to excitonic absorptions in amorphous solids.

REFERENCES

[4.1] J. Singh, *Excitation Energy Transfer Processes in Condensed Matter* (Plenum, New York, 1994).
[4.2] J. Singh, D. Birkedal, V.G. Lyssenko, and J.M. Hvam, *Phys. Rev. B*, **53**, 15909 (1996), and references therein.
[4.3] D. Birkedal, J. Singh, V.G. Lyssenko, J. Erland, and J.M. Hvam, *Phys. Rev. Lett.*, **76**, 672 (1996).
[4.4] R. Miller, D. Kleinman, A. Gossard, and O. Monteanu, *Phys. Rev. B*, **25**, 6545 (1982).
[4.5] J. Singh and K. Shimakawa, *Advances in Amorphous Semiconductors* (Taylor & Francis, London and New York, 2003).
[4.6] J. Singh and I.-K. Oh, *J. Appl. Phys.*, **95**, 063516 (2005).
[4.7] J. Singh, *Nonlinear Optics*, **18**, 171 (1997).
[4.8] J. Singh, T. Aoki, and K. Shimakawa, *Philos. Mag. B*, **82**, 855 (2002).
[4.9] J. Singh, *J. Non-Cryst. Solids*, **299–302**, 444 (2002).
[4.10] J. Singh, *Nonlinear Optics*, **29**, 119 (2002).
[4.11] J. Singh, *J. Mater. Sci.*, **14**, 171 (2003).
[4.12] R.J. Elliot, in *Polarons and Excitons*, edited by K.G. Kuper and G.D. Whitfield (Oliver & Boyd, Edinburgh and London, 1962), p. 269.
[4.13] T. Aoki, S. Koomedoori, S. Kobayashi, T. Shimizu, A. Ganjoo, and K. Shimakawa, *Nonlinear Optics*, **29**, 273 (2002).

5 Photoluminescence

T. Aoki

Department of Electronics and Computer Engineering & Joint Research Center of High-technology, Tokyo Polytechnic University, Atsugi 243-0297, Japan
e-mail: *aoki@ee.t-kougei.ac.jp*

5.1 INTRODUCTION

A radiative emission process in condensed matter is called *luminescence*. Luminescence can occur through a variety of electronic processes, among which *photoluminescence* (PL) and

Optical Properties of Condensed Matter and Applications   Edited by J. Singh
© 2006 John Wiley & Sons, Ltd

electroluminescence (EL) are most popularly used. The radiative process requires a non-equilibrium carrier concentration in the electronic band of solids or in the electronic state of an impurity or defect. If the nonequilibrium state is created by photoexcitation, the resulting luminescence is called PL, and if it is obtained by carrier injection through an electric field, it is called EL [5.1, 5.2].

The study of luminescence from condensed matter is not only of scientific but also of technological interest because it forms the basis of solid-state lasers and it is important for display panels in electronic equipment, lighting, and paints. Moreover, PL frequently provides a nondestructive technique for material characterization or research in materials science as well. PL spectroscopy is a sensitive tool for investigating both intrinsic electronic transitions between energy bands and extrinsic electronic transitions at impurities and defects of organic molecules, semiconductors, and insulators.

As a comprehensive survey of PL spectroscopy used on all condensed-matter systems is beyond the scope of this chapter, only a limited number of topics on PL recombination in disordered materials, in particular, amorphous semiconductors, are addressed, and readers interested in other topics or further details should refer to the relevant references herein. Section 5.2 provides a brief outline of the fundamental aspects of PL in condensed matter. Section 5.3 is concerned with the experimental aspects of PL, where quadrature frequency-resolved spectroscopy (QFRS) for the measurement of broad PL lifetime distributions of amorphous semiconductors is described in detail, and finally Section 5.4 provides examples of recent progress in PL lifetime spectroscopy of amorphous semiconductors, in particular, hydrogenated amorphous Si (a-Si:H) and hydrogenated amorphous Ge (a-Ge:H).

5.2 FUNDAMENTAL ASPECTS OF PHOTOLUMINESCENCE (PL) IN CONDENSED MATTER

Conceptually PL is the inverse of the absorption process of photons; photons absorbed into a material transfer their energy to electrons in the ground state and excite them to excited states. If there are multiple numbers of excited states, then electrons excited to higher excited states rapidly relax nonradiatively to the *lowest* excited sate, S_1, and subsequently recombine radiatively to the ground state, S_0, by emitting photons as shown in Figure 5.1. Therefore, the absorbed photons are usually of higher energy than the emitted photons by the energy of thermalization or emitted phonons. The difference in energy of the absorbed and emitted photons is called Stokes' shift.

PL includes *fluorescence* and *phosphorescence*, which are preferably used in the field of organic chemistry as well as biochemistry. Fluorescence and phosphorescence depend on whether a radiative transition from the excited state to the ground state is spin-allowed, which means both the ground and excited sates have the same spin multiplicity (singlet–singlet S_1–S_0 or sometimes triplet–triplet), or whether it is spin-forbidden, where the ground and excited states have different spin multiplicities (triple–singlet T_1–S_0). A spin-forbidden transition can only occur through spin–orbit coupling, which is usually weak, and hence a radiative recombination of such states has a longer lifetime than does a spin-allowed transition. As shown in Figure 5.1 for organic molecules, fluorescence and phosphorescence arise from radiative recombination of singlet and triplet Frenkel excitons, respectively [5.3–5.5]. The fluorescence lifetime is usually much shorter ($0.1 \sim 10\,\mathrm{ns}$) than the phosphorescence lifetime ($1\,\mathrm{ms} \sim 10\,\mathrm{s}$) due to the spin selection rule. However, delayed fluo-

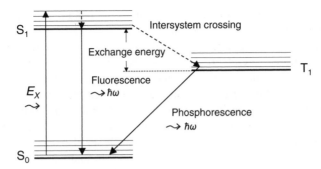

Figure 5.1 Illustration of radiative recombination processes in organic materials. S_0 and S_1 denote ground and first excited singlet states, respectively. T_1 denotes the first excited triplet state. E_X is photoexcitation energy, and $\hbar\omega$ is emitted photon energy. Here, competing nonradiative recombination processes are omitted

rescence possesses an exceptionally longer lifetime, similar to the phosphorescence lifetime. It should be noted that the distinction between fluorescence and phosphorescence is not always clear [5.6]. Sometimes, therefore, delayed fluorescence is also called phosphorescence.

Metal–ligand complexes (MLCs), which contain a transition metal and one or more organic ligands, possess rather large intersystem crossing coefficient due to a large spin–orbit coupling with the heavy metals and therefore they display strong phosphorescence due to mixed singlet–triplet sates (Figure 5.1). Accordingly these MLCs display intermediate lifetimes of 10 ns ~ 10 μs [5.6], and, in particular, *fac*-tris(2-phenylpyridine)iridium [Ir(ppy)$_3$] demonstrates a very high internal quantum efficiency (QE). In this case, the emitting state is assigned to be a metal-to-ligand charge-transfer (MLCT) triplet state, which gives the high emission QE due to the three-fold degeneracy of a triplet exciton [5.7, 5.8]. These organometallic triplet emitters exhibit very high EL efficiencies and are likely candidates of guest materials for organic light emitting divces (OLEDs) [5.9].

In inorganic semiconductors, PL is classified into two categories, intrinsic PL and extrinsic PL [5.1, 5.10–5.12]. Intrinsic PL occurs mainly due to band-to-band radiative transition in a highly pure semiconductor even at a relatively high temperature, where by the absorption of a photon an electron is excited into the conduction band and a hole is formed in the valence band and then they radiatively recombine to give rise to intrinsic PL. The band-to-band transition, which occurs in indirect-gap semiconductors like Si and Ge, is called an indirect transition, and that in direct-gap semiconductors such as GaAs is called direct transition. The latter transition has a much larger radiative recombination probability, i.e., high QE, and is extensively applied to light-emitting diodes (LEDs) and semiconductor lasers.

At low temperatures, an excited electron – hole (e–h) pair can form a free Wannier exciton through their Coulomb attractive force and then the PL occurs through the excitonic recombination in place of the band-to-band transition in the pure semiconductors. Besides the PL from free Wannier excitons, various other excitonic recombinations also occur. These are due to radiative recombination of bound excitons, excitonic polarons, self-trapped excitons, and excitonic molecules, which usually takes place at low temperatures. Excitons play even more important roles in semiconductor quantum wells, where the excitonic PL can be observed even at much higher temperatures because the quantum confinement increases exciton-binding energy [5.13–5.15] (see Chapter 2).

Extrinsic PL is caused by impurities or defects and classified into two types: delocalized and localized [5.1]. Impurities intentionally incorporated into luminescent materials such as ionic crystals and semiconductors, mostly metallic impurities or defects, are called *activators* and materials made luminescent in this way are called *phosphors*. In the delocalized type, PL occurs between free carriers and impurity states, but in the localized type the photoexcitation and photoemission processes are confined to a localized luminescence center.

The most important impurities are donors (D), which are also sometimes called *activators*, and acceptors (A), sometimes called *coactivators*, in ionic crystals and semiconductors. In compensated semiconductors both donors and acceptors are present and normally exist in ionized or charged form: donors as D^+ and acceptors as A^- with their respective ionization energies as ED and EA. The two cases distinguished in donor–acceptor pair (DAP) recombination are: (i) *distant* DAP recombination when the separation r between a D^+ ion and an A^- ion is much greater than the internal dimensions (the effective Bohr radius a) of either the neutral donor (D) or the neutral acceptor (A), and (ii) *associated* DAP recombination when r is comparable or smaller than the internal dimension of either D or A [5.11]. Both cases (i) and (ii) of DAP recombination are schematically represented in Figure 5.2.

In the case (i) of the distant DAP recombination, an excited free electron in the conduction band is bound on a D^+ ion and a free excited hole in the valence band is bound to an A^- ion after band-to-band excitations. Thus, both the donor and acceptor become neutral (D and A). Then recombination between the bound e–h pair occurs radiatively, leaving charged D^+ and A^- [see case (i) in Figure 5.2]. In this case, the DAP recombination is rather a delocalized type of impurity luminescence because the excited e–h pair is free before being bound to the ionized impurities [5.1].

In the type (ii) DAP recombination, however, the electron on an A^- ion is directly excited to a D^+ ion by a photon and then they subsequently recombine with photoemission, which is rather close to the localized type. In both types of DAP recombination the donors and acceptors return to ionized states after the radiative recombination as $D + A \rightarrow D^+ + A^- +$ photon, and a Coulomb potential energy $-e^2/\kappa r$ remains, where κ is a dielectric constant of host material for case (i) and a differently deduced constant for case (ii) [5.11]. Therefore

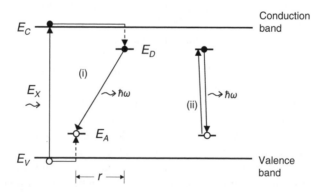

Figure 5.2   Schematic optical transition models of donor–acceptor pair (DAP) recombination in semiconductors; (i) distant DAP recombination and (ii) associated DAP recombination. r: donor (D)–acceptor (A) separation. E_X: photoexcitation energy. $\hbar\omega$: emitted photon energy

the emitted photon energy $\hbar\omega$, i.e., energy difference between initial and final states of the radiative transition, is higher by the Coulomb energy for each DAP and is given by:

$$\hbar\omega = E_G - E_D - E_A + \frac{e^2}{\kappa r} \qquad (5.1)$$

Since the D–A distance, r, can be multiples of the crystallographic lattice constant, the last term in Equation (5.1) gives rise to a long series of sharp lines at the higher photon energy end of the PL spectrum. This was experimentally observed in the indirect gap semiconductor GaP and was first analysed by Hopfield et al. [5.16].

The transition probability of a DAP recombination is proportional to the square of the overlap of hydrogen-like donor and acceptor wave functions. Thus, the PL lifetime, τ, of DAP recombination depends on the DAP separation r as:

$$\tau = \tau_0 \exp\left(\frac{2r}{a}\right) \qquad (5.2)$$

where τ_0 is an electric-dipole transition time usually shorter than 10 ns, and a is the an effective Bohr radius.

Distant DAP recombination shows the following interesting features. First, as the recombination rate (being inversely proportional to τ) increases with a decrease of the D–A separation r in Equation (5.2), PL intensity should increase as r decreases. However, the number of DAPs also decreases as r decreases and thus the PL intensity should exhibit a maximum at a certain r. Since the PL photon energy is related to r by Equation (5.1), the D–A separation r is reflected in the PL spectrum. It has been observed in GaP crystals at 1.6 K that the PL spectrum is broad at $r > 4$ nm with a maximum at $r \approx 5$ nm and discrete at $1 < r < 4$ nm with a number of peaks corresponding to the D–A separations [5.17]. Since the recombination rate is slower for a DAP of larger r, PL emission intensity at the lower-energy end of the PL spectrum is easily saturated and decreases in comparison with the higher-energy part by increasing PL excitation intensity.

Secondly, if we observe the time dependence of the PL spectrum, a DAP with smaller r will decay faster than that with a larger r, so that the PL peak energy shifts to lower energy as time elapses. This suggests that a detailed study on the DAP recombination needs to use not only static PL spectroscopy but also PL lifetime spectroscopy (described below).

Defects and metallic impurities intentionally incorporated in ionic crystals and semiconductors often act as efficient luminescence centers [5.1, 5.2]. Color- or F-centers are optically active vacancies in ionic crystals such as alkali halides. Paramagnetic metal ions incorporated into host materials act as localized activators; actually these are transition metal ions such as from the iron group, Mn^{+2}, Cr^{+2}, etc., and rare earth ions of Nd^{+3}, Eu^{+3}, Tb^{+3}, Er^{+3}, etc. For the iron-group metal ions, transitions occur between the unfilled d–d states, and for the rare earth ions they occur between the unfilled f–f states. Both d–d and f–f transitions are forbidden according to the selection rules for electric dipole transitions but their forbidden character is altered by the crystalline electric field, and both transitions become more or less allowed.

Since 3d orbitals have relatively large radius, they are not so shielded by outer filled shells. Therefore the electronic states are rather sensitive to the crystalline environment; the energy or color of the PL emission is somewhat dependent on the host materials. On the other hand, rare earth ions are formed when the outermost 6s electrons are removed, leaving the optically active 4f orbitals inside the filled 5s and 5p shells. This makes the unfilled 4f orbitals smaller in radius and less sensitive to the crystal field, so that the optical spectra of doped rare earth ions are generally similar to those of free ions with narrow emission lines. A good example of this type of luminescence is the YAG laser, in which Nd^{+3} rare earth ion is incorporated as an activator for $1.064\,\mu m$ emission into the host of yttrium aluminum garnet ($Y_3Al_5O_{12}$ or YAG).

5.3 EXPERIMENTAL ASPECTS

5.3.1 Static PL spectroscopy

Static PL spectroscopy [5.18, 5.19] apparatus is usually assembled as shown in Figure 5.3. The PL excitation source can be any laser with emission energy larger than the PL emission energy, which is usually close to the bandgap energy of the sample. The sample is often installed into a cryostat when temperature dependence of PL is measured. For a thin film sample on a substrate, the substrate surface should be roughened and the thickness of the film should be as thick as possible, because the internal PL reflection causes interference effects in the PL spectrum. A substrate which can absorb PL should be avoided, since it will significantly affect the PL spectrum and reduce the intensity [5.20]. In some cases, the self-absorption of PL in the film also modifies the spectrum, which therefore needs some corrections [5.10].

A grating monochromator is used to measure the PL spectrum, i.e., intensity vs wavelength λ, but a light of wavelength of higher-order frequency harmonics, such as $\lambda/2$, $\lambda/3$, etc., must be blocked by a long-pass filter (LPF). However, care must be taken in ultravio-

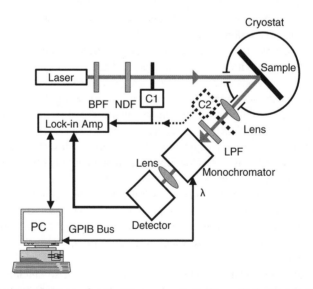

Figure 5.3 Experimental set-up of static PL measurement. C1: optical chopper at normal position. C2: optical chopper for residual PL decay measurement. BPF: band-pass filter. NDF: neutral density filter. LPF: long-pass filter. PC: personal computer

let (UV) photoexcitation in that the blocking filter as well as other filters, e.g., a neutral density filter (NDF) used in adjusting the intensity, often fluoresce when they are irradiated by any stray or reflected UV light. This is also the case even with UV-laser notch filters. The situation becomes more serious with a glass filter, which fluoresces even at under visible-light excitation. Sometimes such an effect can be avoided or reduced by putting the filters at the output side of the monochromator.

The throughput of the monochromator as well as the sensitivity of the detector are important in increasing the PL sensitivity; a monochromator with low *f*-number is efficient at the expense of spectral resolution. Usually, by focusing the light source at an input slit with the same *f*-number, we get the maximum throughput, i.e., optical matching. Since PL is not always a point source of light and it is collected by a lens of a small *f*-number, e.g., *f*/1.0 compared with that of the monochromator, e.g., *f*/4.0, the image of the light source focused at the slit under optical matching conditions is magnified (by 4 in this example) compared with the slit area and some intensity is lost, which is called a *vignetting loss*. We also have to trade between the throughput and the spectral resolution by adjusting the slit width. Therefore insertion of a grating monochromator between sample and detector often decreases the throughput to about 1/1000th of the original value even if it has a low *f*-number.

A set of band-pass filters (BPFs) is an inexpensive and more efficient alternative for the grating monochromator, but sacrifices the arbitrary choice of λ. Another alternative in the near-infrared (NIR) and far-infrared (FIR) ranges is Fourier transform infrared (FTIR) spectrometry based on a Michelson interferometer, which is commercially available from Bruker Optics or Oriel Instruments. The method is called Fourier transform photoluminescence (FTPL) [5.19, 5.21]. Unlike the dispersive spectrometer, the FTIR spectrometer collects all wavelengths simultaneously (*Felgett Advantage*) and does not use the entrance and exit slits. Hence the PL sensitivity is increased and measuring time is reduced by the FTPL. Bignazzi et al. [5.22] have, however, given some warnings about limitations of the FTPL.

PL can be detected by various detectors such as a photomultiplier tube (PMT), photodiode (PD), and photoconductor. A PMT is the most sensitive detector with high-speed response and hence is usable for time-resolved spectroscopy (TRS) as introduced later. However, its spectral response is usually limited below a wavelength of $\approx$900 nm. Though there exist numerous types of near-infrared (NIR) detectors, the NIR PMT, e.g., R5509-72 and its series of InGaAsP/InP photocathodes having sensitivity up to $\lambda \approx 1700$ nm, are now produced by Hamatsu Photonics. This type of PMT is superior to other NIR detectors such as the formerly used Ge diode, both in terms of the sensitivity and the time response.

Occasionally the output of the PMT is directly measured using an electronic dc current meter (ammeter). Usually, however, the excitation light is modulated at a rather low frequency, typically a few Hz to tens kHz, by setting a optical chopper C1 in front of the sample (Figure 5.3) and thereby the PL is lock-in detected, resulting in a significant increase in both sensitivity and signal-to-noise ratio (S/N). However care must be taken in choosing the chopping frequency for the reason described in section 5.3.5, when PL lifetime is comparable with a chopping period. Incidentally, by setting the chopper C2 between the sample and the detector we can avoid the above problem, and also can observe the residual decay of PL with high sensitivity and high S/N after turning off the PL excitation light (laser) as shown in Figure 5.3 and described in section 5.4.4.

At very low PL intensity, one can operate the PMT in the digital mode (called the *photon-counting* method) instead of the analogue mode [5.6]. When the light intensity is very low, incident photons on the photocathode of the PMT are individually separated and thereby

PMT output is composed of separated electric pulses corresponding to the respective incident photons; this condition is called a single photoelectron state. Since the number of output pulses is directly proportional to the amount of incident light, we can measure the light intensity by electronically counting the output pulses. Thus photon-counting is the digital mode, in contrast to the analogue mode where the PMT output pulses overlap each other and eventually can be regarded as an electric current of superposed shot noises. Even in the dark the PMT has dark current or noise made up of contributions from various sources. The amplitude of a noise pulse is generally lower than that of the incident photons. The photon-counting system equipped with electronic discriminator for the PMT output pulse significantly enhances S/N.

Usually PL is anisotropic emission that is dependent on the configuration of the sample, and sometime it is polarized; in particular, PL emission from a film of large refractive index is significantly anisotropic. An integrating sphere is used to determine the photon yield or quantum efficiency (QE) defined as the number of PL photons divided by that of absorbed photons [5.18]. However, laser beams transmitted through and reflected from a film made of organic luminescent materials can again excite PL in the integrating sphere, which causes some errors in QE measurements. This can be eliminated by using two detectors for both the beams; the system for precisely measuring the QE of organic electroluminescent films is now commercially available from Sumitomo Heavy Industries Advanced Machinery.

5.3.2 Photoluminescence excitation (PLE) spectroscopy and photoluminescence absorption spectroscopy (PLAS)

If we replace the laser of fixed wavelength with a dye laser or lamplight dispersed by another monochromator in Figure 5.3, the apparatus serves as an alternative technique of photoluminescence excitation (PLE) spectroscopy carried out by varying the exciting photon energy E_X and setting the monochromator at a fixed wavelength [5.18]. PLE is known to be successful in observing a resonant excitation such as an exciton [5.23].

Since the PL signal is generally proportional to the number of photons absorbed at the PL excitation wavelength, the absorption coefficient can be deduced by dividing the PL emission intensity by the photoexcitation intensity. This might appear to be a complicated method but it is very useful in measuring the absorption coefficient or spectrum of a thin film with weak absorption where the intensity change in the transmitted light can be hardly measured. Also this technique is applicable to measurement of the absorption coefficient of a thin layer on an opaque substrate [5.19]. When we remove the monochromator in front of the detector in Figure 5.3 and measure the total or spectrally integrated PL intensity I as a function of excitation photon energy E_X with the excitation intensity I_0, for a film thickness d, I is given by

$$I = I_0 \left[1 - \exp(-\alpha d)\right] \eta \tag{5.3}$$

where α and η are the total absorption coefficient and QE at E_X, respectively [5.24]. The QE η can be expressed as

$$\eta = \alpha_L / \alpha \tag{5.4}$$

where α_L represents the contribution to α from all processes that excite the PL. When $\alpha d \ll 1$, we obtain from Equations (5.3) and (5.4):

$$I/I_0 \approx \alpha_L d \tag{5.5}$$

Thus the PLE reflects the absorption coefficient, which does not include nonradiative but does include radiative recombination.

Photoluminescence absorption spectroscopy (PLAS) is an alternative technique, which uses PL as a built-in light source in a sample of thin film to determine a weak absorption coefficient [5.25]. In PLAS the thin film must have a large refractive index compared with the surroundings, e.g., a substrate. In this case PL propagates and attenuates along the sample length in an optical guided-wave mode and hence we can use the length instead of the film's thickness to calculate the absorption coefficient at well below the bandgap energy. The method needs analysis of the guided-wave mode affected by the sample thickness, refractive index, and adjacent materials.

5.3.3 Time-resolved spectroscopy (TRS)

When luminescent condensed matter is photoexcited with a generation rate G_0, the rate equation for photocarrier density n is given by

$$\frac{dn}{dt} = G_0 - \frac{n}{\tau} \tag{5.6}$$

where τ is the photocarrier lifetime. If we can ignore the nonradiative recombination, e.g., at low temperatures, PL intensity $I(t)$ is expressed by $n(t)/\tau$. Therefore measurement of the photocarrier lifetime τ, which is related to $I(t)$, is very important for studying PL or radiative recombination mechanisms.

TRS is a rather straightforward method for measuring the lifetime, τ, in materials by applying light pulses of a short duration $T \ll \tau$. Then the constant term G_0 in Equation (5.6) is replaced by a rectangular pulse of height G_0 and duration T, and the solution after a time $t = T$ gives a simple exponential decay $\propto \exp(-t/\tau)$.

For convenience in analysing TRS as well as frequency-resolved spectroscopy (FRS) described later, consider an RC low-pass filter (LPF) simulating a system experiencing a simple exponential decay with a lifetime $\tau = RC$ as shown in Figure 5.4(a) [5.26]. When a pulse of voltage of duration T and amplitude V_i is applied, a charge accumulation $q(t)$ in the capacitor C can be formulated [Equation (5.7)] by substituting $q(t)$ and V_i/R for $n(t)$ and G_0, respectively. An ordinary circuit analysis gives an output $V_o/R = q(t)/\tau$ corresponding to $I(t) = n(t)/\tau$ after $t = T$ as:

$$\frac{q(t)}{\tau} = \frac{V_o}{R} \approx \frac{V_i T}{R\tau} \exp\left(-\frac{t}{\tau}\right) = \frac{G_0 T}{\tau} \exp\left(-\frac{t}{\tau}\right) \tag{5.7}$$

where $T \ll \tau$. For an impulse of light, letting $T \to 0$ and $G_0 \to \infty$ with a finite product $G_0 T (G_0 T \to 1)$, we can express the impulse PL response, $I(t)$, as:

$$I(t) \propto \frac{1}{\tau} \exp\left(-\frac{t}{\tau}\right) \tag{5.8}$$

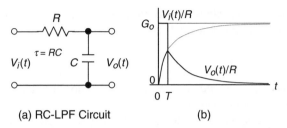

(a) RC-LPF Circuit   (b)

Figure 5.4 (a) An RC low-pass filter (LPF) with time constant $\tau = RC$ to represent a system experiencing a simple exponential decay. (b) Output response $V_o(t)/R$ against a rectangular input pulse $V_i(t)/R$ with an amplitude G_o and a short duration $T \ll \tau$ [modified from ref 5.26]

$I(t)$ in Equation (5.8) is measured after the PL excitation of a delta function, and the lifetime τ is obtained from the slope of a plot of $\log_{10} I(t)$ vs t for a simple PL mechanism with a single lifetime. Luminescent materials with multiple decay lifetimes do not show a single straight line in the $\log_{10} I(t)$ vs t plot. In such cases, $I(t)$ is represented by multi-exponential decays as:

$$I(t) = \sum_i \frac{\alpha_i}{\tau_i} \exp\left(-\frac{t}{\tau_i}\right) \tag{5.9}$$

where lifetimes τ_i and amplitudes α_i are usually obtained by curve fitting.

A product of the expression in Equation (5.8) and t gives a peak at $t = \tau$ and likewise we get peak lifetimes τ_i from $t \cdot I(t)$ from the expression in Equation (5.9), provided the different τ_i values are widely different from one another. Disordered materials generally show broader lifetime distributions, which are sometimes associated with multiple lifetimes. In what follows, we introduce the probability density of lifetime distribution, $P(\tau)$ and substitute it for α_i in Equation (5.9) and then replacing the summation Σ by an integral $\int$ we get:

$$t \cdot I(t) \propto \int_0^\infty P(\tau) \frac{t}{\tau} \exp\left(-\frac{t}{\tau}\right) d\tau \tag{5.10}$$

Substituting x for $\log_{10} t$ (or $t = 10^x$) and u for $\log_{10} \tau$ in Equation (5.10), we get:

$$t \cdot I(t) = 10^x I(10^x) = \int_{-\infty}^\infty 10^u P(10^u) \ln 10 \cdot 10^{x-u} \exp(-10^{x-u}) du$$

$$= \int_{-\infty}^\infty 10^u P(10^u) \lambda(x-u) du \tag{5.11}$$

The impulse response function $\lambda(x) = \ln 10 \cdot 10^x \exp(-10^x)$ peaks at $x = 0$ with a full width at half maximum (FWHM) ≈ 1.06 on the logarithmic scale of $x = \log_{10} \tau$ and the integral of the variable x from $-\infty$ to ∞ is normalized to unity as shown in Figure 5.5 [5.26, 5.27]. If the lifetime distribution $P(\tau)$ is sufficiently broad compared with $\lambda(x)$, we consider $\lambda(x)$ as a delta function and then the lifetime distribution $10^x P(10^x)$ will approximately be equal to $10^x I(10^x)$ on the logarithmic scale, $x = \log_{10} t$. If it is not so broad, we put the acquired data into the left-hand side of Equation (5.11) as $t \cdot I(t) = 10^x I(10^x)$ with $x = \log_{10} t$ and carry out

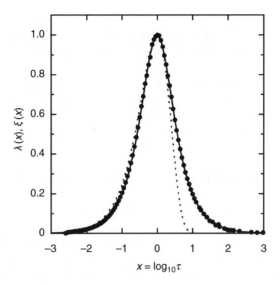

Figure 5.5 Calculated system response functions $\lambda(x)$ for TRS (dotted curve) and $\xi(x)$ for QFRS (solid curve) where the variable $x = \log_{10}\tau$ is used. Dots (•) are electronically measured by our experimental set-up for the RC low-pass filter having time constant $\tau = RC = 0.1\,\text{ms}$, where the variable x is modified as $x = \log_{10}(\tau/RC)$ [modified from ref 5.26]

the deconvolution for recovery of the true distribution $10^x P(10^x)$ to reconcile the right-hand side with the data of the left-hand side.

TRS is conventionally carried out by two methods [5.6]: the first is the stroboscopic method, i.e., sampling the intensity decay, $I(t)$, repetitively by following the pulsed excitation. The PL signal can be detected during the duration of the high-voltage pulse applied to the PMT in pulsed operation. The sampling gate is also achieved by a boxcar integration, in which case the photon-counting method can also be applied with a high sensitivity and high S/N. The second method is the measurement of the decay form of $I(t)$ by a high-frequency digital oscilloscope. The former stroboscopic method can sample the decay with a resolution of a few ns and the latter with that of sub-nanoseconds. In TRS, one needs a short and intense impulse light for PL excitation. When the PL excitation is not short enough to satisfy the condition of $T \ll \tau$ or the instrumental response function is not a delta function, one should deconvolute the measured $I(t)$ with the excitation pulse shape or the instrumental response to recover the exact PL decay [5.6, 5.28].

In the case of a multi-exponential PL decay, the long-lived component appears around the same time as its own PL lifetime, when the PL would have already decayed significantly. Therefore, the excitation pulse must be intense enough to measure the long-lifetime component with a sufficient S/N and the repetition rate must be so slow as to avoid overlapping of PL response signals. Thus, the slow repetition prolongs the time of measurements.

However, intense photoexcitation sometimes causes nonlinear effects in PL recombination and damages the sample. An example of the nonlinear effect in PL recombination is the nongeminate e–h recombination, where an electron created by a photon recombines with a hole created by another photon. This is contrary to the geminate recombination, where an e–h pair created by the same photon recombines radiatively. Whether the recombination kinetics is geminate or nongeminate has been a controversial issue for a long time in field of PL of amorphous semiconductors in particular for hydrogenated amorphous silicon (a-Si:H), as described later.

In order to avoid the occurrence of nonlinear effects or damage, one must carry out TRS measurements at a sufficiently low generation rate G. This can be realized by the time-correlated single-photon counting (TCSPC) method, described below, although it is limited only to rather short decay times at the present state of the art.

5.3.4 Time-correlated single-photon counting (TCSPC)

TCSPC [5.6, 5.28–5.30] is suitable for measurements of short PL decay from sub-nanosecond to hundreds of ns at a very low excitation intensity. The method involves statistical measurements of the probability distribution of a single-photon emission as a function of the time after an excitation pulse impinged on a sample.

Figure 5.6 illustrates a block diagram of TCSPC, which is composed of start and stop channels. The light source, which is usually a mode-locked laser, emits a light pulse. The trigger (start) pulse of the start channel is synchronously generated by the electronic driver pulsing the light source, or by PMT or high-speed photodiode (PD), which monitors a split of the light pulse by a beam splitter. The start pulse is fed through a constant-fraction discriminator (CFD), which minimizes the contribution of pulse-height distribution, or a leading-edge discriminator (LED), which is suitable for a stable laser system exciting light pulses that are uniform in height. Then the start pulse triggers a time amplitude converter (TAC), which initiates a voltage ramp.

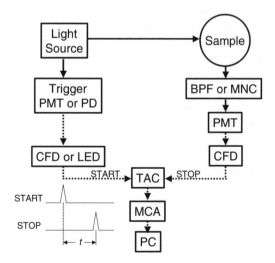

Figure 5.6 Block diagram of a conventional time-correlated single-photon counting (TCSPC) system. PMT: photomultiplier tube. PD: photodiode. CFD: constant-fraction timing discriminator. LED: leading-edge timing discriminator. BPF: band-pass filter. MNC: monochromator. TAC: time-to-amplitude converter. MCA: multi-channel analyzer. PC: personal computer. Solid line: optical pass. Dotted line: electrical pass

Meanwhile the photoexcitation pulse on the stop channel or PL channel excites the sample, which subsequently emits photons, i.e., PL. The PL is transmitted to the PMT through an optical band-pass filter (BPF) or monochromator (MNC) and its intensity should be decreased to be low enough for *at most one photon* to be detected by the PMT for each photoexcitation event in the TCSPC (Figure 5.6). An output pulse from the PMT on the stop channel is discriminated by the CFD, in favor of the pulse generated by a single-photon event among unwanted PMT pulses such as dark noise. Then the discriminated pulse stops the TAC, which gives a voltage output proportional to the time difference t between the start and stop pulses to multi-channel analyser (MCA). A channel number of the MCA is assigned for each voltage corresponding to t. By measuring t between excitation and emission of a huge number of single photons, a histogram of number of photoemission events against each channel, i.e., the time t for each event, can be constructed. This is expected to reflect the PL decay curve of the sample. Usually it takes several μs for the TAC to be reset by discharging the capacitor, which leads to a dead time for the TCSPC system. Thus the TAC behaves as a rate-limiting component in high-repetition operations such as photoexcitation by a mode-locked laser.

A variant of the above method is a reverse-mode interchanging of start and stop pulses, i.e., the PMT output pulse corresponding to a single-photon emission from the sample is routed to the start input of the TAC, which is stopped by a trigger pulse synchronized with photoexcitation [5.31]. The advantage of the reverse mode is a reduction in the measuring time since the TAC does not work in the absence of a single-photon emission. On the other hand, in a normal mode, the TAC is initiated by the photoexcitation pulse in every repetition, but the frequent occurrence of dead times during which the PMT detects no single photon prolongs the measuring time.

The shortest instrumental response function is of a sub-nanosecond order for the TCSPC method using laser pulses of ≈5 ps width at a high repetition rate of ≈80 MHz and very high speed microchannel plate (MCP) PMT or avalanche PD as a detector. On the other hand, a streak camera, which is a completely different method in spite of using photoelectron emission, can provide simultaneous measurements of both PL spectral- and time-dependencies with a resolution of several ps and some even with sub-picosecond resolution. The streak camera, however, has disadvantages such as a low dynamic range for measurable intensity of nearly 3 orders and also low S/N [5.6, 5.28].

5.3.5 Frequency-resolved spectroscopy (FRS)

In FRS measurements [5.6, 5.28] the excitation source is sinusoidally modulated with a frequency ω comparable to the inverse of lifetime τ and hence the generation rate of photoexcited carriers is written as $G_0 (1 + ae^{i\omega t})$ with a modulation depth a. Setting the complex exponential generation rate $G_0 ae^{i\omega t}$ in correspondence with the input V_i/R and the modulated PL intensity $I(t)$ to output V_o/R, then by a simple circuit analysis for the LPF of Figure 5.4(a), we obtain:

$$I(t) \propto \frac{1}{1+i\omega\tau} G_0 ae^{i\omega t} = \frac{G_0 a}{\sqrt{1+(\omega\tau)^2}} e^{i[\omega t - \theta(\omega)]} \tag{5.12}$$

According to Equation (5.12), the amplitude of $I(t)$ is proportional to $R(\omega) = 1/\sqrt{1+(\omega\tau)^2}$ and the phase delay is $\theta(\omega) = \tan^{-1}(\omega\tau)$ for PL of a single lifetime τ. In this case, τ is obtained from the phase delay $\theta(\omega)$ and/or the amplitude $R(\omega)$ at a fixed frequency ω. Usually only $\theta(\omega)$ is used and hence the method is sometimes called *phase-modulated fluorometry* (PMF).

It should be note that since the PL amplitude $R(\omega)$ is a decreasing function of the modulation frequency ω, the PL intensity $I(t)$ decreases with increasing ω, in particular when $\omega\tau > 1$. Therefore, in static PL spectroscopy with lock-in detection of the chopper frequency ω, one should be careful that any component with a long lifetime is significantly reduced in the PL spectrum when $\omega\tau \gg 1$. If the optical chopper is put in the position of C2, i.e., between the sample and the detector (Figure 5.3), however, such a problem can be avoided.

The lifetime τ can be more precisely deduced by fitting $R(\omega)$ and $\theta(\omega)$ to experimentally measured values of amplitude and phase delay as functions of frequency ω. Materials having multi-exponential decay as given by Equation (5.9) are usually analyzed by applying the method of nonlinear least squares [5.32] to the data of $R(\omega)$ and $\theta(\omega)$ but it is not so easy to recover the lifetimes τ_i and the amplitudes α_i exactly. Such FRS with varying frequency ω is called *variable-frequency fluorometry*, and it is used for analyzing luminescent materials such as organic and biological molecules and a variety of condensed materials [5.6].

Another method, a variant of FRS, is photon-counting PMF (PCPMF), which is the FRS version of TCSPC and thus can be used at very low generation rates G [5.33]. In place of the pulsed light source, the PCPMF uses a light source sinusoidally modulated by a sine-wave generator with outputs of synchronous trigger pulses to start the TAC. The intensity of sinusoidally modulated light or the generation rate of e–h pairs G is kept very low to realize the situation of single-photon events. A pulse corresponding to the photoelectron detected by PMT stops the TAC in a similar way as in TCSPC. Thus, a histogram statistically obtained by numerously repeating such photoexcitations and detections reflects the sinusoidally modulated PL intensity $I(t)$ as given in Equation (5.12). The lifetime is then deduced from the amplitude $R(\omega)$ and phase delay $\theta(\omega)$. However, the method is limited to rather high frequencies in the order of MHz or more and hence is not applicable to longer-lifetime events.

If a luminescent material has a broad lifetime distribution like amorphous semiconductors, it will be more difficult to obtain the lifetime distribution by the above-mentioned FRS. In order to overcome this, as well as the disadvantages of TRS, Depinna and Dunstan [5.34] have modified FRS by devising the Quadrature frequency-resolved spectroscopy (QFRS), described below. The theoretical principle of QFRS was fully established by Stachowitz et al. [5.27].

5.3.6 Quadrature frequency-resolved spectroscopy (QFRS)

QFRS [5.27, 5.34] is more suitable for studying amorphous or disordered luminescent materials having broad PL lifetime distributions. Here we again employ the technique based on the circuit analysis of Figure 5.4(a) to explain the QFRS method. The transfer function of the LPF given in Equation (5.12) is separated into real and imaginary parts as follows:

$$\frac{1}{1+i\omega\tau} = \frac{1}{1+(\omega\tau)^2} - i\frac{\omega\tau}{1+(\omega\tau)^2} \tag{5.13}$$

where the real part corresponds to the in-phase output, and the imaginary part to the quadrature output for the input signal $G_0 a e^{i\omega t}$.

The quadrature part, $\omega\tau/[1 + (\omega\tau)^2]$, becomes a maximum at $\omega = \tau^{-1}$. Therefore, by varying the frequency ω, we can find τ by searching for that value of ω, that gives the maximum quadrature output $I_Q(\omega)$ in Equation (5.12). Following Equation (5.10), here again we introduce the probability density $P(\tau)$ to write $I_Q(\omega)$ of the quadrature part as:

$$I_Q(\omega) \propto aG_0 \int_0^{\infty} P(\tau)\frac{\omega\tau}{1+(\omega\tau)^2}d\tau \tag{5.14}$$

The function $\omega\tau/[1 + (\omega\tau)^2]$ is single-peaked but it cannot be approximated as a delta function, because its integral from $\tau = 0$ to ∞ diverges and it is not a sharply peaked function. In order to solve this problem we substitute x for $\log_{10}\omega^{-1}$, and u for $\log_{10}\tau$, to get:

$$I_Q(\omega) \propto \frac{\pi}{2}aG_0 \int_{-\infty}^{\infty} 10^u P(10^u)\frac{\ln 10 \cdot \mathrm{sec}\,h[\ln 10 \cdot (x-u)]}{\pi}du$$

$$= \frac{\pi}{2}aG_0 \int_{-\infty}^{\infty} 10^u P(10^u)\xi(x-u)\,du \tag{5.15}$$

where the frequency response function of a single exponential decay $\xi(x) = \dfrac{\ln 10 \cdot \mathrm{sec}\,h(\ln 10 \cdot x)}{\pi}$ can be normalized for its integration from $x = -\infty$ to ∞ to be unity and sharply peaked at $x = 0$ with an FWHM of ≈ 1.14 on the logarithmic scale as shown in Figure 5.5. Thus, if the distribution function $10^x P(10^x)$ is broad enough for $\xi(x)$ to be regarded as a delta function, we get the approximate lifetime distribution in the logarithm scale of reciprocal ω as:

$$I_Q(\omega) = I_Q(10^{-x}) \propto aG_0 10^x P(10^x) \tag{5.16}$$

with $x = \log_{10}\omega^{-1}$. Therefore, $I_Q(\omega)$ gives a proportion of the distribution at the lifetime ω^{-1} and we obtain the lifetime distribution by plotting $I_Q(\omega)$ against the logarithm of lifetime $x = \log_{10}\omega^{-1}$ [5.27].

The function $10^x P(10^x)$ of Equation (5.16) shows the distribution for the logarithm of lifetime x and the multiplication of 10^x mathematically comes from changing the uniform scale of ω^{-1} to the logarithmic scale of $x = \log_{10}\omega^{-1}$. On the logarithmic scale, the range for $\omega^{-1} < 1$ ($x < 0$) is expanded, which a decrease of 10^x compensates for, and the reverse holds for $\omega^{-1} > 1$ ($x > 0$). Hence an area beneath the probability density $P(\omega^{-1})$ in the uniform scale of a certain range of ω^{-1} is equal to that beneath $10^x P(10^x)$ in the logarithmic scale of the corresponding range of $x = \log_{10}\omega^{-1}$.

When $10^x P(10^x)$ is not broad enough for $\xi(x)$ to be regarded as a delta function, we must recover the true lifetime distribution $10^x P(10^x)$ by deconvoluting Equation (5.15) with $\xi(x)$. However the deconvolution is not always easy because the experimental noise is often exaggerated after the deconvolution, leading to an unusable deconvoluted signal. Therefore, two approximate methods of deconvolution are given below.

Since the integration of $\mathrm{sech}\,x$ is $\tan^{-1}(\sinh x)$, we can analytically convolute a simple rectangular function of x with the response function $\xi(x)$. Assuming the lifetime distribution $10^x P(10^x)$ to be a superposition of rectangular functions like piled up building blocks, we get an analytical expression of its convolution. Then fitting it to the data of the QFRS spectrum by non commercial curve-fitting software, we can recover approximate $10^x P(10^x)$ with a shape resembling piled up building blocks [5.35]. The other method is completely numerical; at first we assume a linear combination of Gaussians as the true lifetime distribution and then numerically convolute it with $\xi(x)$ to be reconciled with the data for determining its parameters.

As mentioned above, it is rather difficult to determine the absolute QE for the PL measurements but relative QE for a component in the QFRS spectrum is deduced by dividing the area of the deconvoluted component by the total area of all the components. TRS and QFRS methods using the system response functions $\lambda(x)$ and $\xi(x)$, respectively, are mathematically equivalent. However, QFRS has a decisive advantage in a wide and broad lifetime distribution ranging from $\approx 1\,\mu s$ to $\approx 0.1\,s$, i.e., over a range of 5 decades (orders of magnitude), corresponding to frequencies from $1\,Hz$ to $100\,kHz$ in a conventional lock-in amplifier. Moreover the lock-in detection technique makes it possible to measure the lifetime distribution at a very low generation rate G, where the first-order recombination kinetics prevails.

5.4 PHOTOLUMINESCENCE LIFETIME SPECTROSCOPY OF AMORPHOUS SEMICONDUCTORS BY QFRS TECHNIQUE

5.4.1 Overview

Using the QFRS techinique, Depinna and Dunstan [5.34] first demonstrated the PL lifetime distribution of hydrogenated amorphous silicon a-Si:H more precisely than could be obtained from TRS measurements, and also observed emissions from singlet- and triplet-bound excitons in GaP crystals, well separated in lifetime as well as in emission energy. Although static PL spectroscopy of a-Si:H was nearly established in the 1970s and early 1980s [5.20, 5.36], considerable debate over its exact mechanisms still continues even today [5.37, 5.38]. It has been generally agreed that, in an intrinsic a-Si:H at a low temperature T, photoexcited electrons and holes immediately thermalize in extended and band-tail states by the hopping process, and that intrinsic PL arises from the radiative tunneling (RT) transition between electrons and holes localized in their respective tail states [5.20, 5.36]. The PL lifetime is governed by the RT lifetime identically with the DAP recombination as:

$$\tau = \tau_0 exp\left(\frac{2R}{a}\right) \tag{5.17}$$

where R is the electron–hole (e–h) separation. The pre-exponential factor τ_0 is the electrical-dipole transition time usually expected to be $\approx 10^{-8}\,s$ and a is the extent of the larger of the electron and hole wave functions, usually the electron Bohr radius of $\approx 1\,nm$.

However, as stated above, whether the recombination is *geminate*, i.e., recombination of the e–h pair created by a single photon, or *nongeminate*, i.e., recombination of an e–h pair created by different photons, is still a controversial issue. This may be partly due to the rather featureless distribution of PL lifetime as well as to the luminescence spectrum induced by the disorder in amorphous semiconductors. In fact, the optical absorption due to an exciton being a typical *geminate* e–h pair has not been observed in a-Si:H even at low temperatures, unlike in crystalline semiconductors. Thus the precise measurement of lifetime is important for studying PL mechanisms. QFRS is a powerful technique for analysing the lifetime distribution of photoexcited carriers at sufficiently low generation rates G [5.27, 5.34].

Bort et al. [5.39] have used the QFRS technique to observe the G-independent lifetime distribution in a-Si:H at low T below a generation rate $G \approx 10^{19}\,\mathrm{cm^{-3}\,s^{-1}}$. They have proposed a geminate recombination model where the average lifetime τ is constant under the condition $G \leq 10^{19}\,\mathrm{cm^{-3}\,s^{-1}}$. By contrast, the light-induced electron-spin resonance (LESR) intensity, known to be proportional to n, depends on G sublinearly [5.39–5.42]. The distant-pair (DP) or nongeminate model based on Equation (5.17) predicts the sublinear G-dependence of the LESR intensity, while it fails to explain the G-independence of τ at $G \leq 10^{19}\,\mathrm{cm^{-3}\,s^{-1}}$ [5.43]. The DP recombination occurs between an electron localized in a conduction-band tail state and a hole localized in a valence-band tail state; this is similar to the DAP recombination in crystals but the tail states before capturing the e–h pair are considered to be *neutral* in a-Si:H unlike the DAP recombination.

Meanwhile, under the geminate condition, Boulitrop and Dunstan [5.44] were the first to identify a double-peaked lifetime distribution of PL consisting of short-lived (~µs) and long-lived (~ms) components in a-Si:H by QFRS. QFRS of PL in a-Si:H was later studied in further detail by Ambros et al. [5.45], expanding the high-frequency limit to 2 MHz corresponding to a lifetime of $\tau \approx 0.1\,\mathrm{\mu s}$. It is difficult to identify the two lifetime components on the basis of the RT model [5.44, 5.45]. Stachowitz et al. [5.46] have proposed the exciton's involvement in the double-peak phenomenon, attributing the short- and long-lived components to singlet- and triplet-excitons, respectively. We observed double-peak lifetime behavior in the similar tetrahedral amorphous semiconductor a-Ge:H, also as shown in Figure 5.7, supporting the exciton model [5.38, 5.47, 5.48]. However, identifying the short-lived component of ~µs to singlet excitonic recombination raised concerns because the true

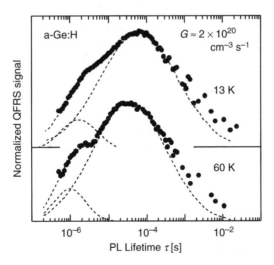

Figure 5.7   QFRS PL spectra of a-Ge:H at $T = 13\,\mathrm{K}$ and $60\,\mathrm{K}$ with the generation rate $G \approx 2 \times 10^{20}\,\mathrm{cm^{-3}\,s^{-1}}$ [5.47]. Dashed curves are separated two gaussians and their summation [Reprinted from S. Ishii, M. Kurihara, T. Aoki, K. Shimakawa, and J. Singh, *J. Non-Cryst. Solids*, **266–269**, 721 (2000) by permission of Elsevier]

lifetime of a singlet exciton should be less than 10 ns as obtained from the usual fluorescence measurements, which is far below the shorter lifetime limit feasible with the conventional QFRS technique.

Indeed, one of the disadvantages of QFRS is the minimum lifetime limited to µs due to the upper limit frequency of the conventional lock-in amplifiers. Since only the rf digital lock-in amplifier (SR844, Stanford Research System) is commercially available for use from 25 kHz up to 200 MHz, we have developed a nanosecond QFRS system named *dual-phase double lock-in* (DPDL) QFRS by using the rf lock-in amplifier, described below.

5.4.2 Dual-phase double lock-in (DPDL) QFRS technique

As illustrated in Figure 5.8, we have developed the DPDL technique [5.35, 5.49, 5.50] to measure PL lifetime distribution from 2 ns to 5 µs, employing the SR844 rf digital lock-in amplifier and an electro-optic modulator (EOM) available from Conoptics to modulate the laser beam from 0.5 Hz to 80 MHz instead of an acousto-optic modulator (AOM), which has an acoustic delay much larger than the PL delay. Recently, however, the AOM also has been found to be usable for the present purpose if we adjust the length of the delay line, i.e., rf coaxial cable between the function generator (FG) and the rf lock-in amplifier shown in Figure 5.8, to compensate for the acoustic delay. Since electromagnetic disturbance on the input signal into the rf lock-in amplifier from the EOM driver (D) in Figure 5.8 becomes serious at rf frequencies (ω) above 10 MHz, the laser beam was also chopped at a low frequency of $\omega_m \ll \omega$ to discriminate the PL signal from the interference by the *double lock-*

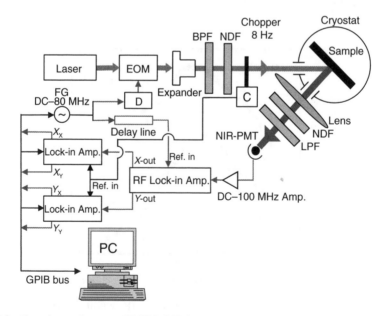

Figure 5.8 Experimental set-up of DPDL QFRS system. C: optical chopper. D: electronic driver for EOM (electro-optic modulator). FG: function generator. BPF: band-pass filter. LPF: long-pass filter. NDF: neutral density filter. NIR-PMT: near-infrared photomultiplier tube. PC: personal computer

in detection method [5.35, 5.49]. The signal flow in DPDL is schematically represented in Figure 5.9. Using a sinusoidal instead of a complex exponential $e^{i[\omega t - \theta(\omega)]}$ in Equation (5.12), we express the doubly modulated PL signal as:

$$S(t) = R(\omega)\sin[\omega t - \theta(\omega)] \cdot \sin(\omega_m t) \tag{5.18}$$

where $R(\omega)$ and $\theta(\omega)$ are the PL amplitude and phase at ω, respectively. When $\omega_m \ll \omega$, the quadrature part of Equation (5.18) is given by $R(\omega)\sin\theta(\omega)$, which has been to be proportional to $I_Q(\omega)$ in Equation (5.12). However, as $R(\omega)$ and $\theta(\omega)$ include components of instrumental responses due to the EOM, the PL detecting system, and the optical and electrical lengths, the quadrature signal should be calibrated as mentioned below.

The time constant of LPF of the rf lock-in amplifier is set much greater than ω^{-1} and much less than ω_m^{-1}. Thus the rf lock-in amplifier outputs an in-phase signal $X(t)$ on the X-channel and a quadrature signal $Y(t)$ on the Y-channel, which are sinusoidal at the frequency of ω_m as:

$$\left. \begin{aligned} X(t) &= \frac{1}{2}R(\omega)\cos\theta(\omega)\sin(\omega_m t) \\ Y(t) &= \frac{1}{2}R(\omega)\sin\theta(\omega)\sin(\omega_m t) \end{aligned} \right\} \tag{5.19}$$

These signals are again synchronously detected at ω_m using the two digital lock-in amplifiers (SR830) with LPF time constants much greater than ω_m^{-1}. An instrumental phase shift ψ is inserted between the chopped light and the synchronous signal of the chopper driver (see Figure 5.9). Hence the X- and Y-channel outputs of the two lock-in amplifiers (X_X, X_Y, Y_X and Y_Y) are given by:

$$\left. \begin{aligned} X_X &= \frac{1}{4}R(\omega)\cos\theta(\omega)\cos\psi \\ X_Y &= \frac{1}{4}R(\omega)\cos\theta(\omega)\sin\psi \\ Y_X &= \frac{1}{4}R(\omega)\sin\theta(\omega)\cos\psi \\ Y_Y &= \frac{1}{4}R(\omega)\sin\theta(\omega)\sin\psi \end{aligned} \right\} \tag{5.20}$$

From Equation (5.20), by eliminating ψ, we obtain $R(\omega)$ and $\theta(\omega)$ as follows:

$$\left. \begin{aligned} R(\omega) &= 4\sqrt{X_X^2 + X_Y^2 + Y_X^2 + Y_Y^2} \\ \theta(\omega) &= \frac{1}{2}\left(\tan^{-1}\frac{Y_X + X_Y}{X_X - Y_Y} + \tan^{-1}\frac{Y_X - X_Y}{X_X + Y_Y} \right) \end{aligned} \right\} \tag{5.21}$$

Denoting the instrumental values of PL amplitude and phase of the modulated laser light reflected by a roughened Al plate instead of the sample by $R'(\omega)$ and $\theta'(\omega)$, respectively, and by measuring them, the intrinsic QFRS signal of PL can be obtained as [5.50]:

$$R(\omega)\sin[\theta(\omega) - \theta'(\omega)]/R'(\omega) \tag{5.22}$$

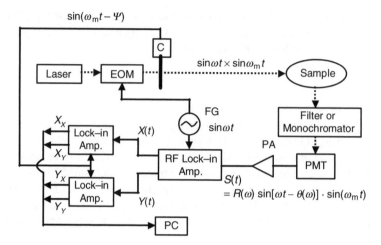

Figure 5.9 Block diagram of DPDL QFRS system. C: optical chopper with chopping frequency of ω_m. FG: function generator. PA: preamplifier. PMT: photomultiplier tube. PC: personal computer [Reproduced from T. Aoki, T. Shimizu, D. Saito, and K. Ikeda, *J. Optoelectron. Adv. Mater.*, **7**, 137 (2005) with permission from INOE]

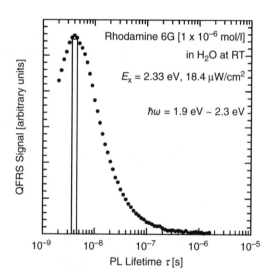

Figure 5.10 QFRS Spectrum of Rhodamine 6G having a fluorescence lifetime of 4 ns at RT. The rectangle represents an actual range of lifetimes recovered by the building-block type of deconvolution [Modified from T. Aoki et al., *Nonlinear Optics*, **29**, 273, 2002. Copyright Taylor & Francis]

The DPDL QFRS technique has been confirmed to resolve PL lifetimes of nanosecond order through observations of the fluorescence of Rhodamine 6G in 1×10^{-6} mol/l aqueous solution, which has a PL lifetime of ≈ 4 ns at room temperature as shown in Figure 5.10, in good agreement with that obtained by using sampled single-photon detection [5.49, 5.51].

5.4.3 Exploring broad PL lifetime distribution in a-Si:H and a-Ge:H by wideband QFRS

As PL lifetime of an amorphous semiconductor is widely distributed due its disorder, the range measurable by QFRS should also be expanded to longer lifetimes as well. This can be accomplished by coupling an AOM with the Stanford SR830 digital lock-in amplifier having the frequency span from 1 mHz to 102 kHz, corresponding to a lifetime distribution 1.6 μs ~ 160 s. The lock-in amplifier is operated in *internal reference* mode together with synchronous filtering in order to avoid the phase noise [5.52]. A measurement of lifetime ≥30 s in this mode needs a time constant of 3 ks for the lock-in amplifier and thus it takes 6 h to obtain one data point. The wideband QFRS system combining DPDL QFRS with the internal reference mode allows analysis of lifetimes over almost 11 decades from 2 ns to 160 s. Using this system it was revealed that the lifetime distribution is triple-peaked, with the third lifetime peak in a lifetime range of 0.1 ~ 160 s in PL of a-Si:H and a-Ge:H under the geminate conditions. The third component exhibited the characteristic features of DP recombination [5.52]. Thus, we claimed that geminate and nongeminate (DP) recombination coexist in PL of a-Si:H and a-Ge:H at low T and low G. Moreover, similar phenomena were observed in chalcogenide amorphous semiconductors, indicating that the triple-peak QFRS spectrum, i.e., the coexistence of geminate and nongeminate recombination, is universal among amorphous semiconductors [5.53]. In addition, PL recombination of a-Si:H for lifetimes greater than 160 s was investigated by observing residual PL decay after cessation of PL excitation at low G [5.54].

In the following, previously published results of our group are introduced for demonstrating the exciton involvement as well as DP recombination in the triple-peak lifetime structure recently observed in the intrinsic a-Si:H and a-Ge:H. Also the residual PL decay persisting for more than 10^4 s is presented, showing the kinetics of DP recombination and steady-state photocarrier concentration, consistent with the LESR results [5.54].

Films of intrinsic a-Si:H were deposited on roughened Al substrates, with a thickness of 1 ~ 9 μm and defect density ≤2.0 × 10^{16} cm^{-3}, and intrinsic a-Ge:H with a thickness of ≈1 μm and defect density ≈10^{16} cm^{-3}. Spectrally-integrated PL signals were detected by a *Hamamatsu* R5509-42 NIR PMT at photon energies ranging from 0.9 to 1.7 eV for a-Si:H and an R5509-72 NIR PMT from 0.7 to 1.5 eV for the narrower band-gap material a-Ge:H. An optical system of f/1.0 ~ 2.0 optics was carefully designed to collect PL emission into the PMTs. The QFRS spectra of dispersed PL were measured by placing a 10 cm and f/3.0 monochromator having a resolution of ≈30 meV in the optical path between the sample and the PMT as shown in Figure 5.9.

Figure 5.11 shows QFRS spectra of a-Si:H from 2 ns to 160 s excited at the PL excitation energy E_X of 2.33 eV for various generation rates, G, from 2.5 × 10^{15} to 5.0 × 10^{22} cm^{-3} s^{-1} [5.55]. We can see that the long- and short-lived components are fixed at τ_T ≈ 4 ms and at τ_S ≈ 3 μs, respectively, even though G changes from 2.5 × 10^{15} to 4.1 × 10^{19} cm^{-3} s^{-1}; this is the well-known double-peak lifetime distribution observed under the so-called geminate condition $G \leq 10^{19}$ cm^{-3} s^{-1} [5.44, 5.45].

By extending the longer lifetime limit of the QFRS technique, however, a third peak higher than the other two peaks is observed at τ_D ≈ 20 s for G ≈ 2.5 × 10^{15} cm^{-3} s^{-1} (see Figure 5.11). This peak might have been overlooked earlier due to the lack of very low frequency QFRS (the internal reference mode). As G increases, τ_D continuously shifts to shorter lifetimes and the peak merges with the τ_T-component at G ≈ 1.3 × 10^{18} cm^{-3} s^{-1}. The

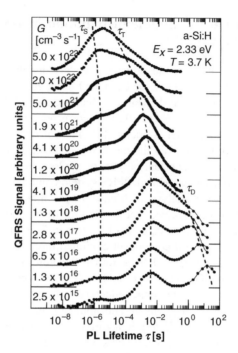

Figure 5.11 QFRS spectra from 2 ns to 160 s for a-Si:H at 3.7 K and $E_X = 2.33$ eV with various G from 2.5×10^{15} to 5.0×10^{22} cm^{-3} s^{-1}. The two data at G of ~10^{22} cm^{-3} s^{-1} were taken by laser light condensed through a lens [Reproduced from T. Aoki, *J. Non-Cryst. Solids*, **352**, 1188 (2006) by permission of Elsevier]

continuous shortening of τ_D with increasing G is a distinctive feature of DP recombination based on the RT model, and a plot of τ_D vs G corresponds well to the curve calculated from the balance equation [5.50, 5.56, 5.57]. Thus, the lifetime distribution of an intrinsic a-Si:H is triple-peaked at a low temperature of 3.7 K under the geminate condition $G \leq 10^{19}$ cm^{-3} s^{-1} and at photoexcitation energy $E_X = 2.33$ eV. The third component has a peak lifetime $\tau_D \approx 0.1 \sim 160$ s.

At sufficiently low G, the three peaks have well separated lifetimes, suggesting that the recombination events at τ_S, τ_T, and τ_D occur via three independent channels, i.e., noncompeting recombination [5.45]. However, when the τ_D-component begins to merge with the τ_T-component as G approaches $\approx 10^{18}$ cm^{-3} s^{-1} (Figure 5.11), the two recombination events at τ_T and τ_D no longer occur independently. Indeed, a further increase in G to $\approx 1.2 \times 10^{20}$ cm^{-3} s^{-1} shifts the combined component of τ_T and τ_D to shorter lifetimes and merges it with the τ_S-component at around $G \approx 10^{22}$ cm^{-3} s^{-1}, leading to a single-peak structure. Similar triple-peaked lifetime structures are also observed in the G-dependent QFRS spectra of a-Ge:H excited by $E_X = 1.81$ eV under the geminate condition $G \ll 10^{19}$ cm^{-3} s^{-1} at $T = 3.7$ K as shown in Figure 5.12 [5.52].

QFRS spectra of a-Si:H at various monochromatized PL energies, $\hbar\omega$, are shown in Figure 5.13, at 3.7 K, $E_X = 2.33$ eV and $G \approx 2.3 \times 10^{17}$ cm^{-3} s^{-1}. In the QFRS spectra of dispersed PL, the two lifetime peaks τ_S and τ_T tend to get shorter with increasing $\hbar\omega$, whereas τ_D remains constant. Plots of recombination rates τ_S^{-1} and τ_T^{-1} against $\hbar\omega$ are almost proportional to $(\hbar\omega)^3$ [5.58]. These results suggest excitonic involvement in the τ_S- and τ_T-

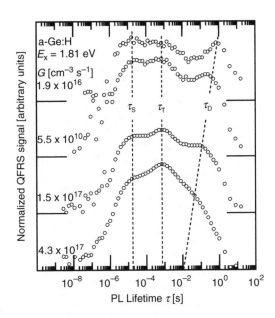

Figure 5.12 QFRS spectra ranging from 2 ns to 20 s for a-Ge:H excited at 3.7 K and $E_X = 1.81$ eV with various generation rates $G < 10^{19}$ cm^{-3} s^{-1} under the geminate conditions [Reproduced from T. Aoki, T. Shimizu, S. Komedoori, S. Kobayashi, and K. Shimakawa, *J. Non-Cryst. Solids*, **338–340**, 456 (2004) by permission of Elsevier]

recombination events, since the recombination rates of localized excitons or self-trapped excitons (STEs) obey the $(\hbar\omega)^3$ law [5.59–5.61]. The reason for observing a rather slow lifetime τ_S of μs for the singlet recombination is explained by the difference in e–h orbital sizes of the STEs. In this case, the hole orbit is much smaller than the electron orbit due to the self-trapped hole in a-Si:H [5.58]. By contrast, DP recombination is responsible for the τ_D-component because Equation (5.17), underlying the DP recombination, is independent of $\hbar\omega$.

By fixing the frequency at 39 kHz, 58 Hz, and 1.1 Hz, corresponding to $\tau_S \approx 4.1$ μs, $\tau_T \approx$ 2.7 ms, and $\tau_D \approx 0.14$ s, respectively, PL spectra of QFRS signals of the three components were taken at $G \approx 2.8 \times 10^{17}$ cm^{-3} s^{-1} and $T = 3.7$ K for a thicker sample of thickness ≈9 μm to avoid the interference effects (see Figure 5.14). Although the lifetime peaks for τ_S- and τ_T-components do not remain constant and become shorter as $\hbar\omega$ increases as seen in Figure 5.13, the PL spectrum of the QFRS signal of the τ_S-component is similar to that of τ_T and shifts to higher $\hbar\omega$ by ≈40 meV despite its smaller magnitude. This was confirmed by deconvoluting QFRS spectra of Figure 5.13 and plotting the areas of the τ_S- and τ_T-components as functions of $\hbar\omega$ [5.58]. On the other hand, the peak of the τ_D-component is still lower in energy, suggesting that the emission of this component comes from deeper states (Figure 5.14).

As PL excitation energy, E_X, increases at a constant G of ≈2.0 × 10^{15} cm^{-3} s^{-1}, the QFRS spectra for the τ_S- and τ_T-components do not change but the τ_D-component increases [see Figure 5.15(a)]. The results of Figures 5.14 and 5.15(a) suggest that both the τ_S- and τ_T-component originate from similar recombination processes but that they are different in PL emission energy, which can presumably be attributed to the excitonic recombination

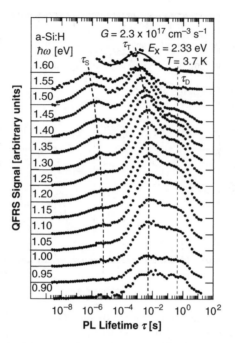

Figure 5.13   QFRS spectra of the monochromatized PL for $\hbar\omega$ from 0.90 to 1.60 eV under the geminate conditions $G = 2.3 \times 10^{17}\,cm^{-3}\,s^{-1}$ for a-Si:H at 3.7 K and $E_x = 2.33$ eV [Reproduced from T. Aoki, *J. Non-Cryst. Solids*, **352**, 1138 (2006) by permission of Elsevier]

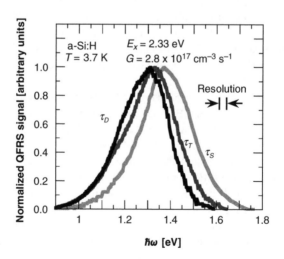

Figure 5.14   QFRS PL spectra of a-Si:H with thickness $\approx 9.3\,\mu m$ fixed at lifetime peaks $\tau_S \approx 4.1\,\mu s$, $\tau_T \approx 2.7\,ms$, and $\tau_D \approx 0.14\,s$ at 3.7 K and $E_x = 2.33$ eV with $G = 2.8 \times 10^{17}\,cm^{-3}\,s^{-1}$. All the peaks of the spectra are normalized to unity; actually the peaks for τ_S and τ_D are ~30% and ~80% of that for τ_T, respectively [Reproduced from T. Aoki, *J. Non-Cryst. Solids*, **352**, 1138 (2006) by permission of Elsevier]

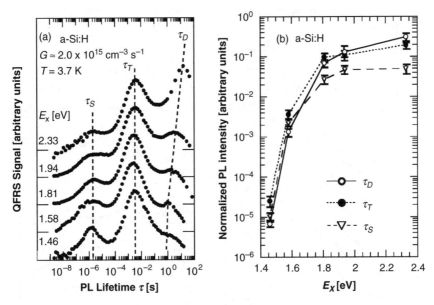

Figure 5.15 (a) QFRS spectra of a-Si:H excited at various PL excitation energy E_x with $G \approx 2.0 \times 10^{15}\,\text{cm}^{-3}\,\text{s}^{-1}$ at 3.7 K. (b) PL excitation spectra (PLE) for three QFRS components τ_S (∇), τ_T ($\bullet$) and τ_D ($\bigcirc$); each component deconvoluted and normalized by PL excitation intensity [Reproduced from T. Aoki, *J. Non-Cryst. Solids*, **352**, 1138 (2006) by permission of Elsevier]

processes with an exchange energy of $\approx 40\,\text{meV}$ between singlet- and triplet-excitons [5.15, 5.34, 5.48, 5.61]. The increase in the τ_D-component with increasing E_x is due to the thermalization or diffusion of DPs excited at the higher E_X in extended or band-tail states [5.52].

PL excitation (PLE) spectra for the three components are shown in Figure 5.15(b), by dividing the QE of each component deconvoluted from the spectrum of Figure 5.15(a) by the PL excitation intensity. The findings that the τ_D-component declines faster into the bandgap than do the τ_S- and τ_T-component in PLE, and that the PL peak energy of the τ_D-component is the lowest in Figure 5.14, indicate that the PL of the τ_D-component has the largest Stokes shift, resulting from the recombination of DPs deeply trapped in tail states [5.55]. The theory of such a large Stokes shift is given in Chapter 6.

Figure 5.16(a) shows QFRS spectra of a-Si:H excited at 2.33 eV with $G \approx 1.0 \times 10^{17}\,\text{cm}^{-3}\,\text{s}^{-1}$ at various temperatures T [5.50, 5.52]. As T is raised from 3.7 K, the τ_T-component shifts to shorter lifetimes and disappears at $T \approx 85\,\text{K}$. The third peak, at τ_D, on the other hand, persists up to ~133 K with a continuous shift of τ_D to shorter lifetimes. The τ_S-component has faded out already at $T \approx 30\,\text{K}$ and, concomitantly, another shoulder emerges at $\tau_G \approx 0.1\,\text{ms}$ between τ_S and τ_T, growing into a hump at higher T. Thus, a new double-peak structure with maxima at τ_D and τ_G is established in QFRS spectra for a-Si:H at $\approx 100\,\text{K}$ [5.50, 5.52].

The disappearance of the τ_S-component at the lower T of $\approx 30\,\text{K}$ and that of the τ_T-component at $T \approx 85\,\text{K}$ are explained by attributing the τ_S- and τ_T-component to singlet and triplet excitons, respectively, since the binding energy of a singlet exciton is smaller than that of a triplet exciton by the exchange energy of $\approx 40\,\text{meV}$. Moreover, the temperature at

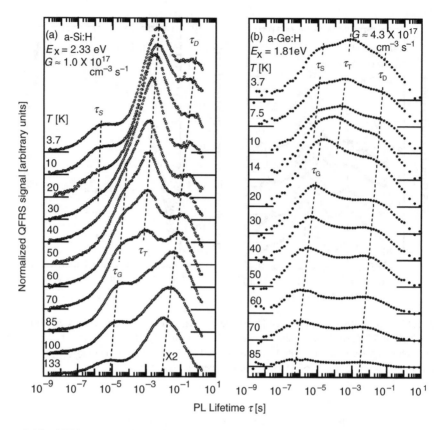

Figure 5.16   QFRS spectra of (a) a-Si:H photoexcited at $E_X = 2.33\,\text{eV}$ with $G \approx 1.0 \times 10^{17}\,\text{cm}^{-3}\text{s}^{-1}$ and (b) a-Ge:H photoexcited at $E_X = 1.81\,\text{eV}$ with $G \approx 4.3 \times 10^{17}\,\text{cm}^{-3}\text{s}^{-1}$ for various temperatures T [Reproduced from T. Aoki, T. Shimizu, S. Komedoori, S. Kobayashi, and K. Shimakawa, *J. Non-Cryst. Solids*, **338–340**, 456 (2004) by permission of Elsevier]

which the τ_T-component disappears (~85 K) corresponds to that for the onset of instability of self-trapped holes, to which electrons are bound individually, forming STEs [5.62]. By contrast, the τ_D-component is selectively enhanced by elevating T, accompanied by the shortening of τ_D due to nonradiative competing recombination.

Figure 16(b) demonstrates the T-dependence of QFRS spectra of a-Ge:H excited at 1.81 eV with $G \approx 4.3 \times 10^{17}\,\text{cm}^{-3}\text{s}^{-1}$ [5.52]. The τ_S- and τ_T-component become featureless at ~14 K, whereas the τ_D-component survives up to ~85 K. Moreover, another peak becomes noticeable at $\tau_G \approx 20\,\mu\text{s}$ between τ_S and τ_T when $T \approx 20$ K. Thus the QFRS spectrum for a-Ge:H represents a new double-peak structure with maxima at τ_G and τ_D in the range of ≈ 20–85 K. The small exciton binding energy of a-Ge:H compared with that of a-Si:H [5.48] is responsible for the low threshold of T for the disappearance of the τ_S- and τ_T-component in a-Ge:H.

At elevated T in Figure 5.16(a), the thermal energy prevents photogenerated e–h pairs from forming excitons. However, another shoulder followed by a peak appears at $\tau_G \approx 10^{-5}$–10^{-4} s between τ_S and τ_T. Moreover, τ_G remains constant as G increases up to

$\approx 10^{20}\,\mathrm{cm^{-3}\,s^{-1}}$, whereas τ_D continues to decrease [5.50, 5.52]. These observations suggest that at 100 K, geminate pairs are not in the form of excitons, but rather, nonexcitonic geminate pairs continue to be generated. Using the classical Onsager model, we showed the occurrence of geminate recombination even above 100 K at above-gap excitation [5.50]. Presumably, the spin effect on geminate pairs fades out at ~100 K, but the Coulombic effect persists even above 100 K. Figure 5.17 shows the steady-state carrier concentration $n_D = \eta_D G \tau_D$ with the QE η_D of the τ_D-component derived by deconvoluting the G-dependent QFRS spectra of Figure 5.11 and the LESR spin density as functions of G for a-Si:H [5.39, 5.41, 5.42]; the QFRS results agree with the sublinear G-dependence of $n_D \propto G^{0.2}$ without any fitting parameters [5.52]. The steady-state carrier concentrations $n_S = \eta_S G \tau_S$ and $n_T = \eta_T G \tau_T$ with the QEs η_S and η_T of the respective component obtained by deconvoluting the QFRS spectra of Figure 5.11 are also plotted as functions of G. The steady-state carrier concentration n_T of the τ_T-component increases almost in proportion to the generation rate G under the geminate condition $G \leq 10^{19}\,\mathrm{cm^{-3}\,s^{-1}}$, since τ_T and η_T remain nearly constant with G.

As G exceeds $10^{19}\,\mathrm{cm^{-3}\,s^{-1}}$, the n_T is absorbed into the sublinear curve of $n_D \propto G^{0.2}$ with $n_T \approx n_D \approx 10^{17}\,\mathrm{cm^{-3}}$. At around $G \approx 10^{19}\,\mathrm{cm^{-3}\,s^{-1}}$, the two components are observed to merge (Figure 5.11). This implies that the recombination of triplet exciton starts to compete with that of DPs at around the steady-state carrier concentration $n_T \approx 10^{17}\,\mathrm{cm^{-3}}$. Similarly, an extrapolation of the plot of n_S vs G, which deviates from a straight line at $G \approx 10^{22}\,\mathrm{cm^{-3}\,s^{-1}}$ presumably due to the irradiation condensed by a lens, intersects the sublinear curve at around $n_S \approx n_D \approx 10^{18}\,\mathrm{cm^{-3}}$ $(G \approx 10^{23}\text{–}10^{24}\,\mathrm{cm^{-3}\,s^{-1}})$; such a coalescence of all the components is also seen in Figure 5.11 at $G \approx 5.0 \times 10^{22}\,\mathrm{cm^{-3}\,s^{-1}}$. At each coalescence, the average

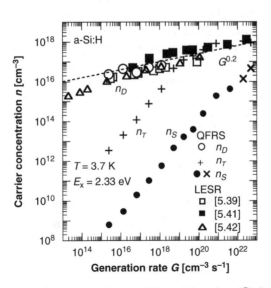

Figure 5.17   Steady-state carrier concentrations n_S (●), n_T (+), and n_D (○) for three QFRS components τ_S, τ_T, and τ_D, respectively, vs generation rate G at 3.7 K with $E_x = 2.33\,\mathrm{eV}$ [5.52, 5.55]; two data n_S (×) were taken by condensed light [5.55]. LESR spin densities vs G for a-Si:H; the plots (□) from ref. [5.39], (■) from ref. [5.41] and (△) from ref. [5.42]. Dashed line indicates the sublinear G-dependence of $n_D \propto G^{0.2}$ [Reproduced from T. Aoki, *J. Non-Cryst. Solids*, **352**, 1138 (2006) by permission of Elsevier]

inter-pair distance between triplet excitons $\approx 0.5\, n_T^{-1/3}$ works out to be $\approx 10\,\text{nm}$ when $n_T \approx n_D \approx 10^{17}\,\text{cm}^{-3}$ and that between the singlet excitons $\approx 5\,\text{nm}$ obtained by $n_S \approx n_T \approx 10^{18}\,\text{cm}^{-3}$.

The spatial wavefunction of a singlet exciton is symmetric due to the anti-parallel spins, while that of a triplet exciton having parallel spins is antisymmetric. The exciton radius is, on average, smaller in the case of the symmetric spatial function than in the case of the anti-symmetric function [5.4, 5.55]. Thus, the triplet exciton, which possesses a larger radius, is absorbed into DPs at a lower concentration n_T compared with the singlet exciton of the concentration n_S. If a singlet exciton radius a_{ex} is close to the inter-pair distance $0.5\, n_S^{-1/3} \approx 5\,\text{nm}$, the exciton binding energy $e^2/2\kappa a_{ex}$ works out to be $\approx 5\,\text{meV}$ with a dielectric constant $\kappa \approx 12$, which explains the disappearance of the τ_S component at the low temperature $T \approx 30\,\text{K}$.

Since STEs are localized, in particular, holes are self-trapped to weak Si–Si bonds in a-Si:H [5.37, 5.62], their wave functions are contracted compared with those of free excitons. Therefore the small orbitals of the STEs give the large exchange energy of $\approx 40\,\text{meV}$ [5.4, 5.15]. The three recombination processes of the τ_S-, τ_T- and τ_D-components are schematically drawn in Figure 5.18.

A magnetic field of up to 0.9 T had little influence on the τ_S-component, but enhanced the τ_T-component and reduced the τ_D-component. These magnetic field effects also support the assignment of three components to the excitonic and DP recombination processes [5.50].

5.4.4 Residual PL decay of a-Si:H

The spectrally integrated PL intensity, $I(t)$, was measured by turning off the 2.33 eV laser light by an electromechanical shutter with a speed of ~30 ms after irradiation of the sample for a sufficiently long time to allow $I(t)$ to reach its steady-state value. The subsequent or residual PL decay of $I(t)$ was lock-in detected by chopping PL emitted light in the position of C2 (Figure 5.3) with a chopping frequency of ~190 Hz using a programmable current amplifier; in that position of the chopper C2 the amplitude $I(t)$ is independent of the chopping frequency regardless of PL lifetime. All experimental procedures were performed electronically, including shuttering the laser beam, changing the time constant from 10 ms to 30 s interlocked with the sensitivity setting of the lock-in amplifier over 7 orders of dynamic

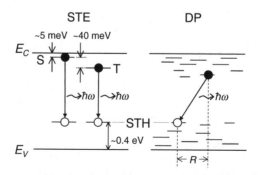

Figure 5.18 Schematic models of three types of radiative recombination in a-SiH at low temperatures. STE: self-trapped exciton. S: singlet. T: triplet. STH: self-trapped hole. R: intra-distance of a distant-pair (DP)

range, and acquiring data on a computer [5.54]. Consequently, we could analyze the PL recombination processes of a-Si:H in conjunction with the wideband QFRS technique over 13 decades with ultrahigh sensitivity.

Figure 5.19 demonstrates that the residual PL intensity $I(t)$ persists for a very long time after cessation of the PL excitation of 2.33 eV at $G \approx 2.0 \times 10^{16} \, \mathrm{cm}^{-3} \, \mathrm{s}^{-1}$ under the geminate conditions at two different temperatures, $T \approx 3.7 \, \mathrm{K}$ and $100 \, \mathrm{K}$. The PL decay at $100 \, \mathrm{K}$ ceases at around $1.5 \times 10^{3} \, \mathrm{s}$, while that for $3.7 \, \mathrm{K}$ persists over $2.0 \times 10^{4} \, \mathrm{s}$. Unlike the QFRS, the residual PL decay $I(t)$ obeys recombination kinetics over the full range from an initial carrier concentration to zero. We have measured $I(t)$ for various G values and integrated from the cessation of illumination to infinity to determine the steady-state photocarrier concentration. Although this technique requires one fitting parameter, the integral of $I(t)$ with respect to G agrees well with n_D as well as LESR densities obtained by others [5.54]. Furthermore, the integrated $I(t)$ plot against T matches the LESR data as well as n_D obtained from the wideband QFRS (Figure 5.20) [5.55, 5.63, 5.64].

Decays of PL at ~3.7 K and ~100 K last more than 2×10^{4} and $1.5 \times 10^{3} \, \mathrm{s}$, respectively, whereas the lifetimes τ_D obtained by QFRS at the same T and G correspond to ~8 and ~0.2 s, respectively. This is because the QFRS measures an *effective* lifetime at a certain density of quasi-steady-state or metastable photocarriers n_D, while the residual PL decay reflects the recombination kinetics of the metastable carrier concentration changing from n_D to 0. In fact, log–log plots of $I(t)$ fit the derivatives of a stretched exponential function well, for various T and G, indicating that the recombination kinetics of the DP is monomolecular, in particular at low T and low G [5.54]. The monomolecular kinetics arises from the immobility of trapped carriers localized in the tail states, into which photocarriers are thermalized immediately after their photoexcitation.

5.5 CONCLUSIONS

The wide-band QFRS technique has revealed the triple-peak lifetime distribution in PL of a-Si:H as well as a-Ge:H at low temperatures, T, and low generation rates, G. The

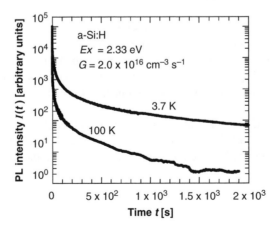

Figure 5.19 PL decay for a-Si:H with $G = 2.0 \times 10^{16} \, \mathrm{cm}^{-3} \, \mathrm{s}^{-1}$ and $E_X = 2.33 \, \mathrm{eV}$ at 3.7 and 100 K [Reproduced from T. Aoki, *J. Non-Cryst. Solids*, **352**, 1138 (2006) by permission of Elsevier]

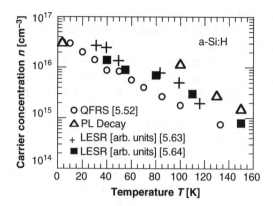

Figure 5.20 Plots of steady-state carrier concentrations n; the plots ($\triangle$) estimated from PL decay, ($\bigcirc$) from QFRS [5.52], and the LESR intensities in arbitrary units as functions of T for a-Si:H; (+) from ref. [5.63], ($\blacksquare$) from [5.64] [Reproduced from T. Aoki, *J. Non-Cryst. Solids*, **352**, 1138 (2006) by permission of Elsevier]

dependence of the three components on G, T, PL emission energy $\hbar\omega$, PL excitation energy E_X, and magnetic field are explained by attributing the well known double-peak components to excitonic recombination, and the newly discovered third component to DP recombination. At high T and low G, nonexcitonic geminate recombination coexists with nongeminate (DP) recombination. Steady-state carrier concentrations obtained from the DP component agree with the G- and T-dependencies of LESR density. The residual PL decay of a-Si:H extending to $\sim 10^4$ s is identified with the third component of QFRS spectra originating in DP recombination and shows that recombination kinetics of DPs is monomolecular at low T and low G.

ACKNOWLEDGEMENTS

The author wishes to acknowledge that the present study was conducted in cooperation with K. Ikeda, D. Saito, T. Shimizu, S. Komedoori, S. Ishii, and Dr S. Kobayashi of Tokyo Polytechnic University, and Dr A. Ganjoo and Prof. K. Shimakawa of Gifu University. The author is also grateful to Prof. J. Singh of Charles Darwin University for valuable comments and discussions on excitonic recombination as well as a critical reading of the manuscript.

The author thanks the Japan Private School Promotion Foundation for financial aid. This work has also been supported by a Grant-in-Aid for Scientific Research (C) (2000–2002) from the Ministry of Education, Culture, Sport, Science and Technology of Japan.

REFERENCES

[5.1] S. Shionoya, *Luminescence of Solids*, edited by D.R. Vij (Plenum Press, New York, 1998), p. 95.

[5.2] M. Fox, *Optical Properties of Solids* (Oxford University Press, 2001).

[5.3] R.S. Becker, *Theory and Interpretation of Fluorescence and Phosphorescence* (John Wiley & Sons, Inc., New York, 1969).

[5.4] S.P. Mcglynn, T. Azumi, and M. Kinoshita, *Molecular Spectroscopy of The Triplet State* (Prentice-Hall, Englewood Cliffs, 1969).

[5.5] M. Pope and C.E. Swenberg, *Electronic Processes in Organic Crystals and Polymers 2nd Edition* (Oxford University Press, NY, 1999).

[5.6] J.R. Lakowicz, *Principles of Fluorescence Spectroscopy*, 2nd Edition (Kluwer Academic/Plenum Publishers, New York, 1999).

[5.7] M.A. Baldo, S. Lamansky, P.E. Burrows, M.E. Thompson, and S.R. Forrest, *Appl. Phys. Lett.*, **75**, 4 (1999).

[5.8] W.J. Finkenzeller and H. Yersin, *Chem. Phys. Lett.*, **377**, 299 (2003).

[5.9] *Physics of Organic Semiconductors*, edited by W. Brütting (Wiley-VCH, Weinheim, 2005).

[5.10] J.I. Pankove, *Optical Processes in Semiconductors* (Dover Publishers, New York, 1971).

[5.11] H.B. Bebb and E.W. Williams, *Smiconductors and Semimetals*, Vol. 8, edited by R.K. Willardson and A.C. Beer (Academic Press, New York, 1972) Vol. 8, p. 181.

[5.12] K.W. Böer, *Survey of Semiconductor Physics*, 2nd Edition (John Wiley & Sons, Inc., New York, 2002), Vol. 1.

[5.13] J. Singh, *Excitation Energy Transfer Processes in Condensed Matter* (Plenum Press, New York, 1994).

[5.14] P.K. Basu, *Theory of Optical Processes in Semiconductors* (Clarendon Press, Oxford, 1997).

[5.15] Y. Toyozawa, *Optical Processes in Solids* (Cambridge University Press, 2003).

[5.16] J.J. Hopfield, D.G. Thomas, and M. Gershenzon, *Phys. Rev. Lett.*, **10**, 162 (1963).

[5.17] D.G. Thomas, M. Gershenzon, and F.A. Trumbore, *Phys. Rev. A*, **133**, 269 (1964).

[5.18] *Optical Radiation Measurement*, edited by K.D. Mielenz (Academic Press, New York, 1982), Vol. 3: Measurement of Photoluminescence.

[5.19] E.C. Lightowlers, *Growth and Characterization of Semiconductors*, edited by R.A. Stradling and P.C. Klipstein (Adam Hilger, Bristol and New York, 1990), p. 135.

[5.20] R.A. Street, *Adv. Phys.*, **30**, 593 (1981).

[5.21] S. Perkowitz, *Optical Characterization of Semiconductors* (Academic Press, London, 1993).

[5.22] A. Bignazzi, E. Grilli, M. Radice, M. Guzzi, and E. Castiglioni, *Rev. Sci. Instrum.*, **67**, 666 (1996).

[5.23] R.C. Miller, A.C. Gossard, D.A. Kleinman, and O. Munteanu, *Phys. Rev. B*, **29**, 3740 (1984).

[5.24] R.A. Street, *Philos. Mag. B*, **37**, 35 (1978).

[5.25] N. Gal, R. Ranganathan, and P.C. Taylor, Jr., *J. Non-Cryst. Solids*, **77–78**, 543 (1985).

[5.26] T. Aoki, *Materials for Information Technology in the New Millennium*, edited by J.M. Marshall et al. (Bookcraft, UK, 2001), p. 58.

[5.27] R. Stachowitz, M. Schubert, and W. Fuhs, *Philos. Mag. B*, **70**, 1219 (1994).

[5.28] *Topics in Fluorescence Spectroscopy*, edited by J.R. Lakowicz (Plenum Press, New York, 1991), Vol. 1: Techniques.

[5.29] D.V. O'Connor and D. Philips, *Time-correlated Single Photon Counting* (Academic Press, London, 1984).

[5.30] W. Becker, *Advanced Time-Correlated Single Photon Counting Techniques* (Springer Series, Chemical Physics, Vol. 81) (Springer, Berlin, 2005).

[5.31] G.R. Haugen, B.W. Wallin, and F.F. Lytle, *Rev. Sci. Instrum.*, **50**, 64 (1979).

[5.32] *Topics in Fluorescence Spectroscopy*, edited by J.R. Lakowicz (Plenum Press, New York, 1992), Vol. 2: Principles.

[5.33] T. Iwata, A. Hori, and T. Kamada, *Opt. Rev.*, **8**, 326 (2001).

[5.34] S.P. Depinna and D.J. Dunstan, *Philos. Mag. B*, **50**, 579 (1984).

[5.35] T. Aoki, S. Komedoori, S. Kobayashi, C. Fujihashi, A. Ganjoo, and K. Shimakawa, *J. Non-Cryst. Solids*, **299–302**, 642 (2002).

[5.36] R.A. Street, *Hydrogenated Amorphous Silicon* (Cambridge University Press, 1991), p. 276.

[5.37] K. Morigaki, *Physics of Amorphous Semiconductors* (World Scientific, Imperial College Press, London, 1999).

[5.38] J. Singh and K. Shimakawa, *Advances in Amorphous Semiconductors* (CRC Press, Boca Raton, 2003).

[5.39] M. Bort, W. Fuhs, S. Liedtke, R. Stachowitz, and R. Carius, *Philos. Mag. Lett.*, **64**, 227 (1991).

[5.40] R.A. Street and D.K. Biegelsen, *Solid State Commun.*, **44**, 501 (1982).

[5.41] F. Boulitrop and D.J. Dunstan, *Solid State Commun.*, **44**, 841 (1982).

[5.42] B. Yan, N.A. Shultz, A.L. Efros, and P.C. Taylor, *Phys. Rev. Lett.*, **84**, 4180 (2000).

[5.43] D.J. Dunstan, *Philos. Mag. B*, **52**, 111 (1985).

[5.44] F. Boulitrop and D.J. Dunstan, *J. Non-Cryst. Solids*, **77–78**, 663 (1985).

[5.45] S. Ambros, R. Carius, and H. Wagner, *J. Non-Cryst. Solids*, **137–138**, 555 (1991).

[5.46] R. Stachowitz, M. Schubert, and W. Fuhs, *J. Non-Cryst. Solids*, **227–230**, 190 (1998).

[5.47] S. Ishii, M. Kurihara, T. Aoki, K. Shimakawa, and J. Singh, *J. Non-Cryst. Solids*, **266–269**, 721 (2000).

[5.48] J. Singh, T. Aoki, and K. Shimakawa, *Philos. Mag. B*, **82**, 855 (2002).

[5.49] T. Aoki, S. Komedoori, S. Kobayashi, T. Shimizu, A. Ganjoo, and K. Shimakawa, *Nonlinear Optics*, **29**, 273 (2002).

[5.50] T. Aoki, T. Shimizu, D. Saito, and K. Ikeda, *J. Optoelectron. Adv. Mater.*, **7**, 137 (2005).

[5.51] J.M. Harris and F.E. Lytle, *Rev. Sci. Instrum.*, **11**, 1469 (1977).

[5.52] T. Aoki, T. Shimizu, S. Komedoori, S. Kobayashi, and K. Shimakawa, *J. Non-Cryst. Solids*, **338–340**, 456 (2004).

[5.53] T. Aoki, D. Saito, K. Ikeda, S. Kobayashi, and K. Shimakawa, *J. Optoelectron. Adv. Mater.*, **7**, 1749 (2005).

[5.54] T. Aoki, K. Ikeda, S. Kobayashi, and K. Shimakawa, *Philos. Mag. Lett.*, **86**, 137 (2006).

[5.55] T. Aoki, *J. Non-Cryst. Solids*, **352**, 1138 (2006).

[5.56] B.I. Shklovskii, H. Fritzsche, and S.D. Baranovskii, *Phys. Rev. Lett.*, **62**, 2989 (1989).

[5.57] T.M. Searle, *Philos. Mag. Lett.*, **61**, 251 (1990).

[5.58] T. Aoki, *J. Mater. Sci. – Materi. Electron.*, **14**, 697 (2003).

[5.59] B.A. Wilson and T.P. Kerwin, *Phys. Rev. B*, **25**, 5276 (1982).

[5.60] K.S. Song and R.T. Williams, *Self-Trapped Excitons*, 2nd Edition (Springer, Berlin, 1996).

[5.61] J. Singh and I.-K. Oh, *J. Appl. Phys.*, **97**, (2005) 063516.

[5.62] K. Morigaki, M. Yamaguchi, and I. Hirabayashi, *J. Non-Cryst. Solids*, **164–166**, 571 (1993).

[5.63] R.A. Street and D.K. Biegelsen, *Solid State Commun.*, **33**, 1159 (1980).

[5.64] S. Yamasaki, H. Okushi, A. Matsuda, K. Tanaka, and J. Isoya, *Phys. Rev. Lett.*, **65**, 756 (1990).

6 Photoluminescence and Photoinduced Changes in Noncrystalline Condensed Matter

J. Singh

School of Engineering and Logistics, Faculty of Technology, B-41,
Charles Darwin University, Darwin, NT 0909, Australia

6.1 INTRODUCTION

Photoluminescence (PL) provides direct information about the electronic states and carrier dynamics in a material. In noncrystalline or disordered condensed matter, where the theoretical methods are less advanced in comparison with the experimental developments, many optical properties, including PL, have not been fully understood. One of the most studied amorphous materials in the last 2–3 decades is hydrogenated amorphous silicon (a-Si:H)

Optical Properties of Condensed Matter and Applications   Edited by J. Singh
© 2006 John Wiley & Sons, Ltd

[6.1–6.5]. Both the evolution of PL peak structure with the decay time and time-resolved behavior have been studied [6.6–6.12]. In the PL of a-Si:H, two problems have puzzled researchers for decades: 1) the occurrence of a double- and triple-peak structures in the PL of a-Si:H (see Chapter 5 by Aoki), and 2) the controversy over the radiative lifetime measured by time-resolved spectroscopy (TRS) and quadrature frequency-resolved spectroscopy (QFRS). For 1), the occurrence of the double-peak structure of PL has been resolved and the fact that the double-peak structure occurs from the singlet (short time) and triplet (long time) excitons has been established [6.10, 6.12–6.19]. Using the QFRS technique, as shown in Chapter 5, Aoki has observed additional peaks in the PL of a-Si:H, which have also been studied theoretically and are attributed to arise from geminate and nongeminate radiative recombinations [6.20].

The controversy mentioned in 2) has appeared as follows. Using TRS, Wilson et al. [6.11] have observed PL peaks with their radiative lifetime in the nanosecond (ns), microsecond (μs), and millisecond (ms) time ranges in a-Si:H at a temperature of 15 K. In contrast to this, using QFRS, other groups [6.13–6.16] have observed only a double-peak structure PL in a-Si:H at the liquid helium temperature. One peak appears at a short time in the μs range and the other in the ms range [6.7, 6.12]. Results of recent measurements [6.17–6.19] of the normalized QFRS PL signals vs PL lifetime in a-Si:H at different PL energies (See Figure 5.13) clearly illustrate the appearance of the two peaks. These measurements do not exhibit any PL peak in the nanosecond time range in a-Si:H. It has therefore become puzzling why no PL peaks are found in the nanosecond time range by QFRS although they have been observed by TRS.

In addition to the above well known puzzling problems in PL of a-Si:H, photoluminescence in amorphous chalcogenides (a-Chs), e.g., As_2S_3, occurs in a band peaking at an energy close to that of half the optical gap [6.21]. It has not yet been resolved unambiguously why the PL peak in a-Chs occurs at an energy equal to half of the bandgap.

Finally, it is known that most amorphous semiconductors show photoinduced changes. There are three phenomena that are observed when a sample of a-Chs is illuminated by bandgap light: creation of light-induced defects, photodarkening, and volume expansion. These are not observed in crystalline chalcogenides. The creation of light-induced defects was first observed in a-Si:H, and is usually known as the Staebler–Wronski Effect (SWE) [6.22]. The covalent bonds in the amorphous structure are broken due to bandgap illumination or light-soaking and hence the number of dangling (broken) bonds increases after the illumination. This is also called the generation of light-induced metastable defects (LIMD) [6.3–6.5]. SWE causes a reduction in the initial efficiency of a photovoltaic device made of a-Si:H, e.g., a-Si:H solar cells, after a prolonged exposure to radiation, because the amorphous sample degrades due to an enhanced number of broken bonds. Several models [6.23] have been put forward for the mechanism of SWE in a-Si:H. The most recent one, the hydrogen-collision model proposed by Branz [6.24], emphasizes the role of hydrogen-hopping motion to induce metastable dangling bonds of Si, and is regarded to be very successful in explaining the creation of LIMD in a-Si:H. However, according to the model, only Si-H bonds are broken in a-Si:H and therefore the model can only be applied to hydrogenated amorphous solids, not to chalcogenides, which contain no hydrogen.

In a-Chs, along with the creation of LIMD, one also observes photodarkening (PD) – a reduction in the optical bandgap occurs due to bandgap illumination [6.5, 6.25]. In a-As_2S_3, PD has also been observed by below-bandgap illumination [6.26]. Also, a volume expansion (VE) in the samples of a-Chs occurs by bandgap illumination [6.5] (see Chapter 7). Several models have been proposed for the occurrence of PD in a-Chs in the last two decades [6.5, 6.27]; however, none has been successful in resolving all aspects observed in materials exhibiting PD. PD was first observed to be metastable, i.e., it disappeared by thermal annealing but remained even if the illumination was stopped. Now transient PD [6.28] has also been observed in a-As$_2$S$_3$, a-As$_2$Se$_3$, and a-Se. It disappears as soon as the illumination is stopped, which means that the material reverts back to its original state once the illumination has stopped. It has not been established what causes the two types of PD, one that is metastable and is reversed only by thermal annealing and the other that reverses after stopping of the illumination. A molecular dynamics simulation for the mechanism of VE is presented in Chapter 7. In this chapter, theories of PL, PD, and VE are presented and results are discussed in the light of experimental observations.

In section 2, a theory of calculating the rates of spontaneous emission is first derived within the two-level approximation, and then under nonthermal equilibrium and thermal equilibrium. The radiative lifetime is calculated from the inverse of the maximum rates obtained at the thermal equilibrium and is compared with experimental ones in a-Si:H. Finally in section 3, a theory of PD and photo-volume expansion is presented.

6.2 PHOTOLUMINESCENCE

A theory is presented here for calculating the rate of spontaneous emission and radiative lifetime of excitons and excited pairs of electrons and holes in amorphous semiconductors (a-semiconductors). It is assumed that an exciton or excited pair of an electron (e) and a hole (h) is created by a photon of energy equal to or higher than the optical gap energy so that initially both the charge carriers are excited in their extended states. In the case of an exciton so created, as it relaxes downward to tail states, it remains an exciton with its identified excitonic Bohr radius and binding energy until the charge carriers recombine radiatively by emitting a photon. Thus, the excitonic relaxation is restricted by the excitonic internal energy quantum states and hence it is not as fast as the thermal relaxation of free carriers (not bound in excitons). This results in a peaked excitonic photoluminescence (see, e.g., Figure 5.13) as has been observed in several experiments [6.8–6.9, 6.11–6.19]. In the case of a free electron and hole pair, there is no such restriction on the energy and interparticle separation and such pairs relax down to their tail states very fast before recombining radiatively as geminate or nongeminate pairs.

As an exciton initially excited in the extended states relaxes down to the tail states, although it maintains its excitonic Bohr radius it loses its excitonic character, because both e and h are localized in their tail states and are not moving around each other in an hydrogenic state. Thus, it becomes a geminate pair. A geminate pair is defined to be a pair of an electron and a hole excited by the absorption of the same single photon but not bound in an excitonic state. Accordingly, there are two types of geminate pairs possible in noncrystalline semiconductors: type I, a pair of free charge carriers excited by a single photon and relaxed to the tail states, and type II, an exciton relaxed to the tail states. Type I geminate pairs have been studied extensively [6.6, 6.29–6.31], the type II has been identified only recently [6.20, 6.32]. Type II geminate pairs can be in singlet or triplet states depending on their original excitonic spin states, and the average separation between the electron and hole is governed by the corresponding exciton Bohr radius. However, type I pairs are the conventional geminate pairs and have no spin correlations. In addition, photoluminescence from the tail states can also occur from nongeminate pairs, in which the electron–hole separation is usually much larger [6.30–6.32]. A nongeminate pair is not excited by the same single photon. Thus, when both the charge carriers have relaxed to the tail states, three kinds of PL are possible to originate from the tail states: Type I and II geminate pairs, and nongeminate pairs.

It is therefore very important to know where (in energy and inter-particle separation) the excited charge carriers are in noncrystalline materials prior to their radiative recombination. This is different from the case of crystalline solids where most excited carriers relax down to their band edges and then recombine radiatively. In noncrystalline materials there are four possibilities: (i) both of the excited particles, electron and hole, are in their extended states, (ii) electron is in the extended state and hole in the tail state, (iii) electron is in the tail and hole in the extended state, and (iv) both in their tail states. In view of the discussion presented above, only an excitonic PL can occur from possibilities (ii)–(iii), because for other charge carriers nonradiative relaxation is much faster than radiative recombination, and hence the PL from type I and II geminate pairs and nongeminate pairs can only originate from possibility (iv).

Here, we have first derived the rates of spontaneous emission within two-level approximation, and then for the above four possibilities we have derived them under both non-thermal equilibrium and thermal equilibrium conditions. It is found that the rates derived within the two-level approximation and under nonequilibrium are not applicable for studying the PL radiative lifetime in amorphous semiconductors. Rates derived under equilibrium are used to calculate the PL radiative lifetime in a-Si:H associated with all the four possibilities. It is found that the rates depend on the initial energy state of recombination, emitted photon energy, and temperature.

6.2.1 Radiative recombination operator and transition matrix element

Let us consider an excited pair of an electron and a hole such that the electron (e) is in the conduction states and hole (h) in the valence states, and then they recombine radiatively by emitting a photon due to their interaction with photons. The interaction operator between a pair of excited e and h and a photon can be written as:

$$\hat{H}_{xp} = -\left(\frac{e}{m_e^*} \mathbf{p}_e - \frac{e}{m_h^*} \mathbf{p}_h \right) \cdot \mathbf{A} \tag{6.1}$$

where m_e^* and $\mathbf{p}_e$ and m_h^* and $\mathbf{p}_h$ are the effective masses and linear momenta of the excited electron and hole, respectively, and $\mathbf{A}$ is the vector potential given by:

$$\mathbf{A} = \sum_{\lambda} \left(\frac{\hbar}{2\varepsilon_0 n^2 V \omega_\lambda} \right)^{1/2} [c_\lambda^+ \hat{\varepsilon}_\lambda + c.c.] \tag{6.2}$$

where n is the refractive index, V is the volume of the material, ω_λ is the frequency, c_λ^+ is the creation operator of a photon in a mode λ, $\hat{\varepsilon}_\lambda$ is the unit polarization vector of photons and c.c. is the complex conjugate of the first term. The second term of $\mathbf{A}$ corresponds to the absorption and will not be considered here onward.

Using the centre of mass, $\mathbf{R}_x = \dfrac{m_e^* \mathbf{r}_e + m_h^* \mathbf{r}_h}{M}$ and relative $\mathbf{r} = \mathbf{r}_e - \mathbf{r}_h$ coordinate transformations, the interaction operator $\hat{H}_{xp}$ [Equation (6.1)] can be transformed into:

$$\hat{H}_{xp} = -\frac{e}{\mu_x} \mathbf{A} \cdot \mathbf{p} \qquad (6.3)$$

where $\mathbf{p} = -i\hbar \nabla_r$ is the linear momentum associated with the relative motion between e and h, and μ_x is their reduced mass ($\mu_x^{-1} = m_e^{*-1} + m_h^{*-1}$). The operator in Equation (6.3) does not depend on the center of mass motion of e and h. Therefore, the operator in Equation (6.1) is the same for the exciton–photon interaction and for an interaction between a pair of e and h and a photon.

For amorphous solids, it is important to distinguish whether the excited charge carriers are created in the extended states or tail states [6.10, 6.20, 6.32–6.35]. Therefore, one should identify which one of the above four possibilities one is dealing with. This is because charge carriers have different wave functions, effective masses, and hence different excitonic Bohr radii in their extended and tail states [6.10, 6.20, 6.32–6.35].

It may be emphasized here that the thermal relaxation of the excited free charge carriers from extended to tail states occurs in the picosecond (ps) time scale. That means the majority of excited charge carriers can relax down to the tail states before recombining radiatively and giving rise to PL, which occurs in the ns time scale or longer. In this case, it is only meaningful to consider the possibility (iv) for studying PL in a-semiconductors. This is true, however, only for the radiative recombination of the free excited charge carriers, which do not form excitons. In an exciton, the excited electron and hole are bound through their Coulomb interaction in hydrogen-like energy states of a certain excitonic Bohr radius and energy, and hence cannot thermalize like free excited charge carriers. They can only relax down nonradiatively by maintaining their Bohr radius and internal energy and recombining radiatively through quantum transitions. Therefore for the excitonic radiative recombination, other possibilities, i.e., (i)–(iii), are also relevant and should be considered. This is also supported by the observed Stokes shifts in a-Si:H [6.8, 6.9, 6.17, 6.18], which will be discussed later.

The field operator $\hat{\psi}_c(\mathbf{r}_e)$ of an electron in the conduction states can be written as:

$$\hat{\psi}_c(\mathbf{r}_e) = N^{-1/2} \sum_{l,\sigma_e} \exp(i\mathbf{t}_e \cdot \mathbf{R}_l^e) \phi_l(\mathbf{r}_e, \sigma_e) a_{cl}(\sigma_e) \qquad (6.4)$$

where N is the number of atoms in the sample, $\mathbf{R}_l^e$ is the position vector of an atomic site at which the electron is present, $\phi_l(\mathbf{r}_e)$ is the wave function of an electron at the excited site l, $\mathbf{r}_e$ is the position coordinate of the electron with respect to site l, and $\mathbf{t}_e$ is given by:

$$|\mathbf{t}_e| = t_e = \sqrt{2m_e^*(E_e - E_c)}\big/\hbar \qquad (6.5)$$

where E_e is the energy of the electron and E_c is that of its mobility edge. $a_{cl}(\sigma)$ is the annihilation operator of an electron with energy E_e and spin σ at a site l in the conduction c states. According to Equation (6.5), if the electron energy E_e is above the mobility edge then it moves as a free particle in the conduction extended states, but if $E_e < E_c$ then it is localized because t_e becomes imaginary and the envelope function becomes exponentially decreasing. Likewise the field operator, $\hat{\psi}_v(\mathbf{r}_h)$, of a hole excited with an energy E_h in the valence states can be written as:

$$\hat{\psi}_v(\mathbf{r}_h) = N^{-1/2} \sum_{l,\sigma_h} \exp(-i t_h \cdot \mathbf{R}_l^e) \phi_l(\mathbf{r}_h, \sigma_h) d_{vl}(\sigma_h), \quad d_{vl}(\sigma_h) = a_{vl}^+(-\sigma_h) \tag{6.6}$$

where

$$|\mathbf{t}_h| = t_h = \sqrt{2 m_h^* (E_v - E_h)} \big/ \hbar \tag{6.7}$$

where E_v is the energy of the hole mobility edge, and $d_{vl}(\sigma)$ is the annihilation operator of a hole in the valence states, v, with energy E_h and spin σ. Here again the hole behaves like a free particle for $E_v > E_h$ but gets localized in the tail states for $E_v < E_h$.

Using Equations (6.2), (6.4), and (6.6), the interaction operator $\hat{H}_{xp}$ [Equation (6.3)] can be written in the second quantized form as:

$$\hat{H}_{xp} = -\frac{e}{\mu_x} \sum_\lambda \left(\frac{\hbar}{2 \varepsilon_0 n^2 V \omega_\lambda} \right)^{1/2} Q_{cv} c_\lambda^+ \tag{6.8}$$

where

$$Q_{cv} = N^{-1} \sum_l \sum_m \exp[-i t_e \cdot \mathbf{R}_l^e] \exp[-i t_h \cdot \mathbf{R}_m^h] Z_{lm\lambda} B_{cvlm}(S) \tag{6.9}$$

where $B_{cvlm}(S) = \sum_{\sigma_e, \sigma_h} a_{cl}(\sigma_e) d_{vm}(\sigma_h)$ is the annihilation operator of an exciton of spin $S = 0$ (singlet) or 1 (triplet) by annihilating an electron in the conduction c states at site l and a hole in the valence v states at site m, and the summations over σ_e and σ_h represent combinations of spins to form singlet and triplet excitons [6.5]. $Z_{lm\lambda}$ is given by:

$$Z_{lm\lambda}(S) = Z_{lm\lambda} \delta_{\sigma_e, -\sigma_h}, \quad Z_{lm\lambda} = \int \phi_l^*(r_e) \hat{\varepsilon}_\lambda \cdot \mathbf{p} \phi_m(r_h) d\mathbf{r}_e d\mathbf{r}_h \tag{6.10}$$

It may be noted that it is not necessary to use the exciton operator $B_{cvlm}(S)$ in Equation (6.9), one can also use the product of fermion operators $a_{cl}(\sigma_e) d_{vm}(\sigma_h)$ instead.

We now consider a transition from an initial state with one singlet exciton created at a site, say, l (both e and h created on the same site) and where any other site is assumed to have zero excitons. We also assume that there are no photons in the initial state. The transition matrix element is then obtained as [6.20]:

$$\langle f | \hat{H}_{xp} | i \rangle = -\frac{e}{\mu_x} \sum_\lambda \left(\frac{\hbar}{2 \varepsilon_0 n^2 V \omega_\lambda} \right)^{1/2} p_{cv} \tag{6.11}$$

where

$$p_{cv} = N^{-1} \sum_l \sum_m \exp[-it_e \cdot \mathbf{R}_l^e] \exp[-it_h \cdot \mathbf{R}_m^h] Z_{lm\lambda} \delta_{l,m} \qquad (6.12)$$

It is to be noted that the transition matrix element in Equation (6.11) vanishes for a triplet exciton due to the condition on electron and hole spins in Equation (6.10). Also the same transition matrix element is obtained even if one considers the electron and hole field operators in Equations (6.4) and (6.6), respectively, to be spin independent. That means the transition matrix element of the radiative recombination of a singlet exciton or an excited free electron–hole pair remains the same as given in Equation (6.11) but that it is zero for a triplet spin configuration. Therefore, the formalism developed hereonward for calculating the rates of spontaneous emission is applicable to singlet excitons, type I and singlet type II geminate pairs, and nongeminate pairs.

Let us first derive $Z_{lm\lambda}$. There are two approaches, which are used in amorphous solids, to evaluate this integral. The integral actually determines the average value of the relative momentum between the excited electron–hole pair. In the first approach it is assumed to be a constant and independent of the photon energy as [6.1, 6.5]:

$$Z_{lm\lambda} = \int \phi_l^*(r_e) \hat{\varepsilon}_\lambda \cdot \mathbf{p} \phi_m(r_h) \, d\mathbf{r}_e d\mathbf{r}_h = Z_1 = \pi h \left(\frac{L}{V}\right)^{1/2} \qquad (6.13)$$

where L is the average bond length in a sample and Z_1 denotes the matrix element obtained from the first approach. This form was first introduced by Mott and Davis [6.1] and has been used widely since then.

In the second approach the integral is evaluated using the dipole approximation as [6.5, 6.34–6.36]:

$$Z_{lm\lambda} = \int \phi_l^*(r_e) \hat{\varepsilon}_{ex} \cdot \mathbf{p} \phi_m(r_h) \, d\mathbf{r}_e d\mathbf{r}_h = Z_2 = i\omega\mu_x |r_e - h| \qquad (6.14)$$

where Z_2 denotes the transition matrix element obtained from approach 2, $\hbar\omega = E_c' - E_v'$ is the emitted photon energy, and $|r_{e-h}| = \langle l|\hat{\varepsilon}_\lambda \cdot \mathbf{r}|m\rangle$ is the average separation between the excited electron–hole pair, which also can be assumed to be site independent. Thus, in both approaches the integral becomes site independent and can be taken out of the summation in Equation (6.12).

In the case of singlet excitons and type II geminate pairs, it can be easily assumed that $|r_{e-h}| = a_{ex}$, the excitonic Bohr radius, but for the case of type I geminate and nongeminate pairs, $|r_{e-h}|$ is the average separation between e and h. It may be noted here that the value of the integral obtained from the first approach, Z_1, leads to the well known Tauc's relation for the absorption coefficient observed in amorphous semiconductors, and that obtained from the second approach is used to explain the deviations from Tauc's relation in the absorption coefficient observed in some amorphous semiconductors [6.1, 6.34, 6.35] (Also see Chapter 3 of this volume).

Using Equations (6.13) and (6.14), p_{cv} in Equation (6.6) can be written as:

$$p_{cv} = Z_i N^{-1} \sum_l \sum_m \exp[-it_e \cdot \mathbf{R}_l^e] \exp[-it_h \cdot \mathbf{R}_m^h] \delta_{l,m}, \quad i = 1,2 \qquad (6.15)$$

Now the derivation of p_{cv} in Equation (6.15) depends on the four possibilities, which will be considered below separately.

(i) Extended-to-extended states transitions

Rearranging the exponents in Equation (6.15) as:

$$\exp[-it_e \cdot \mathbf{R}_l^e - it_h \cdot \mathbf{R}_m^h] = \exp[-it_e \cdot (\mathbf{R}_l^e - \mathbf{R}_m^h) - i(t_e + t_h) \cdot \mathbf{R}_m^h] \qquad (6.16)$$

and identifying the fact that $\mathbf{R}_l^e - \mathbf{R}_l^h = \mathbf{a}_{ex}$, the excitonic Bohr radius (the separation between e and h in an exciton prior to their recombination, which is assumed to be site independent), the first exponential becomes site independent. It can be taken out of the summation signs and then the matrix element in Equation (6.15) becomes:

$$p_{cv} = Z_i \exp[-it_e \cdot \mathbf{a}_{ex}] \delta_{t_e, -t_h}, \quad i = 1, 2 \qquad (6.17)$$

where $\delta_{t_e, -t_h}$ represents the momentum conservation in the transition. The square of the transition matrix element then becomes:

$$|p_{cv}|^2 = |Z_i^* Z_i| = |Z_i|^2, \quad i = 1, 2 \qquad (6.18)$$

(ii) Transitions from extended to tail states

Here we consider the electron e excited in the extended state and the hole h in the tail states. For this case Equation (6.15) can be written as:

$$p_{cv} = Z_i N^{-1} \sum_l \sum_m \exp[-it_e \cdot \mathbf{R}_l^e] \exp[-t_h' \cdot \mathbf{R}_m^h] \delta_{l,m} \qquad (6.19)$$

where

$$|t_h'| = t_h' = \sqrt{2m_h^*(E_h - E_v)}/\hbar \qquad (6.20)$$

Rewriting Equation (6.19) as:

$$p_{cv} = Z_i N^{-1} \sum_l \sum_m \exp[-it_e \cdot (\mathbf{R}_l^e - \mathbf{R}_m^h)] \exp[-i(t_e - it_h') \cdot \mathbf{R}_m^h] \delta_{l,m} \qquad (6.21)$$

and simplifying it in the same way as Equation (6.16), we get:

$$p_{cv} = Z_i \exp[-it_e \cdot \mathbf{a}_{ex}] \delta_{t_e, it_h'}, \quad i = 1, 2 \qquad (6.22)$$

Equation (6.16) gives the same expression for $|p_{cv}|^2$ as in Equation (6.18) for the possibility (i), and in this case also the momenta remain conserved.

(iii) Transitions from tail to extended states

We consider an electron e excited in the tail and the hole h in the extended states. For this case the transition matrix element can be derived in a way analogous to that for the extended-to-tail states. p_{cv} is obtained as:

$$p_{cv} = Z_i \exp[-i t_h \cdot \mathbf{a}_{ex}] \delta_{it'_e, t_h}, \quad i = 1, 2 \tag{6.23}$$

where

$$|\mathbf{t}'_e| = t'_e = \sqrt{2 m^*_e (E_c - E_e)} \big/ \hbar \tag{6.24}$$

Here again $|p_{cv}|^2$ remains the same as in Equation (6.18).

(iv) Transitions from tail to tail states

Here we consider that both electron e and hole h are excited in the tail states. In this case, an exciton loses its usual excitonic character and behaves like a type II geminate pair, as explained above. This is because the localized form of wave functions of e and h does not give rise to an exciton. For this case, using the localized form of the electron and hole wave functions, p_{cv} is obtained as:

$$p_{cv} = Z_i N^{-1} \sum_l \sum_m \exp[-\mathbf{t}'_e \cdot \mathbf{R}^e_l] \exp[-\mathbf{t}'_h \cdot \mathbf{R}^h_m] \delta_{l,m} \tag{6.25}$$

Here again, one can rearrange the exponential as:

$$p_{cv} = Z_i N^{-1} \sum_l \sum_m \exp[-\mathbf{t}'_e \cdot (\mathbf{R}^e_l - \mathbf{R}^h_m)] \exp[-(\mathbf{t}'_h + \mathbf{t}'_e) \cdot \mathbf{R}^h_m] \tag{6.26}$$

which becomes:

$$p_{cv} = Z_i \exp[-\mathbf{t}'_e \cdot \mathbf{a}_{ex}] \delta_{t'_e, -t'_h}, \quad i = 1, 2 \tag{6.27}$$

Assuming that the relative momentum of the electron, $\hbar \mathbf{t}'_e$, will be along the direction of $\mathbf{a}_{ex}$ at the time of recombination, $|p_{cv}|^2$ from Equation (6.27) becomes:

$$|p_{cv}|^2 = |Z_i|^2 \exp[-2 t'_e a_{ex}] \delta_{t'_e, -t'_h}, \quad i = 1, 2 \tag{6.28}$$

6.2.2 Rates of spontaneous emission

Using Equation (6.11) and applying Fermi's golden rule, the rate R_{sp} (s^{-1}) of spontaneous emission can be written as [6.20, 6.37–6.41]:

$$R_{sp} = \frac{2\pi e^2}{\mu_x^2} \left(\frac{1}{2\varepsilon_0 n^2 V \omega_\lambda} \right) \sum_{E_c', E_v'} |p_{cv}|^2 f_c f_v \delta(E_c' - E_v' - \hbar\omega_\lambda) \qquad (6.29)$$

where f_c and f_v are the probabilities of occupation of an electron in the conduction state and a hole in the valence state, respectively. The rate of spontaneous emission in Equation (6.29) can be evaluated under several conditions, which will be described below:

(i) Rate of spontaneous emission under two-level approximation

Here only two energy levels are considered as in atomic systems. An electron takes a downward transition from an excited state to the ground state; no energy bands are involved. In this case $f_c = f_v = 1$, and then denoting the corresponding rate of spontaneous emission by R_{sp12}, we get:

$$R_{sp12} = \frac{\pi e^2}{\varepsilon_0 n^2 \omega \mu_x^2} |p_{12}|^2 \delta(\hbar\omega - \hbar\omega_\lambda) \qquad (6.30)$$

where $\hbar\omega = E_2 - E_1$, which is the energy difference between the excited (E_2) and ground (E_1) states. In this case p_{12} is derived using the dipole approximation [Equation (6.14)] as:

$$p_{12} = i\omega\mu_x \langle \hat{\varepsilon}_\lambda \cdot \mathbf{r} \rangle \qquad (6.31)$$

where $\mathbf{r}$ is the dipole length, and $\langle \ldots \rangle$ denotes integration over all photon modes λ. Substituting Equation (6.31) into Equation (6.30), and then integrating over the photon wave vector $\mathbf{k}$ using $\omega = kc/n$, we get [6.8, 6.38]:

$$R_{sp12} = \frac{4\kappa e^2 \sqrt{\varepsilon} \omega^3 |r_{e-h}|^2}{3\hbar c^3} \qquad (6.32)$$

where $\varepsilon = n^2$ is the static dielectric constant, $|r_{e-h}|$ is the mean separation between the electron and hole, and $\kappa = 1/(4\pi\varepsilon_0)$. It may be noted that for deriving the rate of spontaneous emission within the two-level approximation, p_{cv} with $Z_{em\lambda}$ in the form of Equation (6.13) has not been used to the author's knowledge.

The expression of the rate of spontaneous emission obtained in Equation (6.32) is well known [6.8, 6.11, 6.37] and it is independent of the electron and hole masses and temperature. As only two discrete energy levels are considered, the density of states is not used in the derivation. Therefore, although by using $|r_{e-h}| = a_{ex}$, the excitonic Bohr radius, the rate R_{sp12} has been applied to a-Si:H [6.8, 6.11] it should only be used for calculating the radiative recombination in isolated atoms, not in condensed matter.

(ii) Rates of spontaneous emission in amorphous solids

In applying Equation (6.29) for any condensed-matter system, it is necessary to determine f_c and f_v and the density of states. On the one hand, it may be argued that short-time photoluminescence can occur before the system reaches thermal equilibrium and therefore no equilibrium distribution functions can be used for the excited charge carriers. In this case one should use $f_c = f_v = 1$ in Equation (6.29) for condensed matter as well. On the other hand, as the carriers are excited by the same energy photons, even in a short time delay, they may be expected to reach thermal equilibrium among themselves but not necessarily with the lattice. Therefore, they will relax according to an equilibrium distribution. As the electronic states of amorphous solids include the localized tail states, it is more appropriate to use the Maxwell–Boltzmann distribution for this situation. We will consider here radiative recombination under both nonequilibrium and equilibrium conditions.

(A) At nonthermal equilibrium

Considering $f_c = f_v = 1$ and substituting $|p_{cv}|^2$ [Equation (6.12)] derived above from the two approaches in Equation (6.29), we get the rates, R_{spni}, for the possibilities (i)–(iii) as:

$$R_{spni} = \frac{2\pi e^2}{\mu_x^2} \left(\frac{1}{2\varepsilon_0 n^2 V \omega_\lambda} \right) |Z_i|^2 \sum_{E_c', E_v'} \delta(E_c' - E_v' - \hbar\omega_\lambda), \quad i = 1, 2 \tag{6.33}$$

where the subscript *spn* in R_{spni} denotes rates of spontaneous emission derived at nonequilibrium. For evaluating the summation over E_c' and E_v' in Equation (6.33), the usual approach is to convert it into an integral by using the excitonic density of states, which can be obtained as follows. Using the effective-mass approximation, the electron energy in the conduction band and hole energy in the valence band can be written as:

$$E_c' = E_c + \frac{p_e^2}{2m_e^*} \tag{6.34}$$

and

$$E_v' = E_v - \frac{p_h^2}{2m_h^*} \tag{6.35}$$

Subtracting Equation (6.35) from Equation (6.34) and then applying the centre of mass and relative coordinate transformations [see above, Equation (6.3)], we get the excitonic energy E_x in the parabolic form as:

$$E_x = E_0 + \frac{P^2}{2M} + \frac{P^2}{2\mu_x} \tag{6.36}$$

where $E_0 = E_c - E_v$ is the optical gap, P and $M = m_e^* + m_h^*$ are linear momentum and mass of an exciton associated with its center of mass motion, and the last term is the kinetic energy of the relative motion between e and h, which contributes to the exciton binding energy through the attractive Coulomb interaction potential between them [6.43]. The exciton density of states then comes from the parabolic form of the second term associated with the center of mass motion as:

$$g_x = \frac{V}{2\pi^2} \left(\frac{2M}{\hbar^2} \right)^{3/2} (E_x - E_0)^{1/2} \tag{6.37}$$

Using this joint density of states in Equation (6.37), the summation in Equation (6.33) can be converted into an integral as:

$$\sum_{E_x'} \delta(E_x' - \hbar\omega) = I_j = \int_{E_x'} g_x \delta(E_x' - \hbar\omega) dE_x' \tag{6.38}$$

which gives:

$$I_j = \frac{V}{2\pi^2} \left(\frac{2M}{\hbar^2}\right)^{3/2} (\hbar\omega - E_o)^{1/2} \tag{6.39}$$

Substituting Equation (6.39) into Equation (6.33) we get:

$$R_{spni}^{j} = \frac{2\pi e^2}{\mu_x^2} \left(\frac{1}{2\varepsilon_0 n^2 V\omega_\lambda}\right) |Z_i|^2 I_j \Theta(\hbar\omega - E_0), \quad i = 1,2 \tag{6.40}$$

where $\Theta(\hbar\omega - E_0)$ is a step function used to indicate that there is no radiative recombination for $\hbar\omega < E_0$, and j denotes that these rates are derived through the joint density of states. Having derived the rates of spontaneous emission using the excitonic density of states, it is important to remember that the use of such a joint density of states for amorphous semiconductors does not give the well known Tauc's relation [6.1, 6.5] in the absorption coefficient [6.35]. Therefore, by using it in calculating the rate of spontaneous emission, one would violate the Van Roosbroeck and Shockley relation [6.43] between the absorption and emission. For this reason, the product of individual electron and hole densities of states is used in evaluating the summation in Equation (6.27) for amorphous semiconductors. This approach has proven to be very useful in amorphous solids as it gives the correct Tauc's relation [6.1, 6.5, 6.34, 6.44].

Using the product of individual densities of states, the summations over E_c' and E_v' in Equation (6.33) can be evaluated by converting these into integrals as:

$$I = \int_{E_c}^{E_v + \hbar\omega} \int_{E_c' - E_v}^{E_v} g_c(E_c') g_v(E_v') \delta(E_c' - E_v' - \hbar\omega) dE_c' dE_v' \tag{6.41}$$

where $g_c(E_c')$ and $g_v(E_v')$ are the densities of states of the conduction and valence states, respectively, and within the effective-mass approximation these can be written as:

$$g_q(E') = \frac{V}{2\pi^2} \left(\frac{2m^*}{\hbar^2}\right)^{3/2} E_q^{1/2}, \quad q = c, v \tag{6.42}$$

where m^* is the effective mass of the corresponding charge carrier and $q = c$ (conduction) and $q = v$ (valence) states. Substituting Equation (6.42) into Equation (6.41), the integral can be evaluated analytically to give [6.5]:

$$I = \frac{V\left(m_e^* m_h^*\right)^{3/2}}{4\pi^3 \hbar^6} (\hbar\omega - E_0)^2 \tag{6.43}$$

Using Equation (6.43) in Equation (6.33) and substituting the corresponding Z_i, we get the two rates in nonthermal equilibrium for possibilities (i)–(iii) as:

$$R_{spn1} = \frac{e^2 L \left(m_e^* m_h^*\right)^{3/2}}{4\varepsilon_0 \hbar^3 n^2 \mu_x^2 (\hbar\omega)} (\hbar\omega - E_0)^2 \, \Theta(\hbar\omega - E_0) \tag{6.44}$$

and

$$R_{spn2} = \frac{\left(m_e^* m_h^*\right)^{3/2} e^2 |r_{e-h}|^2}{2\pi^2 \varepsilon_0 n^2 \hbar^7} \hbar\omega (\hbar\omega - E_0)^2 \, \Theta(\hbar\omega - E_0) \tag{6.45}$$

For excitonic transitions it is more appropriate to replace m_e^* and m_h^* in Equations (6.44) and (6.45) by the excitonic reduced mass μ_x and use $|r_{e-h}| = a_{ex}$. It may also be noted that the volume V is appearing in the formula for R_{spn2} [Equation (6.45)]. This is inevitable through the second approach and it has been tackled earlier by Cody [6.36] by defining $2N_0/V = \nu\rho_A$, where N_0 is the number of single spin states in the valence band and thus $2N_0$ becomes the total number of valence electrons occupying N_0 states, ν is the number of coordinating valence electrons per atom, and ρ_A is the atomic density per unit volume. Thus replacing V by V/N_0 in Equation (6.45) one can get around this problem. We thus obtain:

$$R_{spn1} = \frac{e^2 L \mu_x}{4\varepsilon_0 \hbar^3 n^2 (\hbar\omega)} (\hbar\omega - E_0)^2 \, \Theta(\hbar\omega - E_0) \tag{6.46}$$

and

$$R_{spn2} = \frac{\mu_x^3 e^2 a_{ex}^2}{2\pi^2 \varepsilon_0 n^2 \hbar^7 \nu\rho_A} \hbar\omega (\hbar\omega - E)^2 \, \Theta(\hbar\omega - E_0) \tag{6.47}$$

It may be emphasized here that E_0, defined as the energy of the optical gap, is not always the same in amorphous solids. It depends on the lowest-energy state within the conduction band from where the radiative recombination occurs. As rates derived in Equations (6.46) and (6.47) do not have any peak value, it is not possible to determine E_0 from these rates.

Likewise, using $|p_{cv}|^2$ [Equation (6.28)] in Equation (6.33) for possibility (iv), we get the rates of spontaneous emission from the two approaches for the tail-to-tail states transitions as:

$$R_{spnti} = R_{spni} \exp(-2t_e' |r_{e-h}|), \quad i = 1, 2 \tag{6.48}$$

where the subscript $spnt$ of R_{spnt} stands for the spontaneous emission at nonequilibrium from tail-to-tail states. r_{e-h} is the average separation between e and h and, for excitons, $r_{e-h} = a_{ex}$, where a_{ex} is the excitonic Bohr radius in the tail states, and for a singlet exciton it is given by [6.6, 6.10]:

$$a_{ex} = \frac{5\mu\varepsilon}{4\mu_x} a_0 \tag{6.49}$$

where μ is the reduced mass of an electron in the hydrogen atom and $a_0 = 0.529\,\text{Å}$ is the Bohr radius. Results for the rates of spontaneous emission obtained above are valid in the nonthermal equilibrium condition as stated above. However, unless one knows the relevant effective masses of charge carriers and the value of E_0, these rates cannot be applied to determine the lifetime of PL. The effective mass of charge carriers in their extended and tail states can be determined [6.5], as also described in Chapter 3 of this volume, but not E_0. For this reason, it is useful first to derive these rates under thermal equilibrium as well, as shown below:

(B) At thermal equilibrium

Assuming that the excited charge carriers are in thermal equilibrium among themselves, the distribution functions, f_c and f_v, can be given by the Maxwell–Boltzmann distribution function as given in reference [6.45]:

$$f_c = \exp[-(E_e - E_{Fn})/\kappa_B T] \tag{6.50}$$

and

$$f_v = \exp[-(E_{Fp} - E_h)/\kappa_B T] \tag{6.51}$$

where E_e and E_h are the energies of an electron in the conduction and a hole in the valence state, respectively, and E_{Fn} and E_{Fp} are the corresponding Fermi energies. κ_B is the Boltzmann constant and T temperature of the excited charge carriers. The product $f_c f_v$ is then obtained as:

$$f_c f_v \approx \exp[-(\hbar\omega - E_0)/\kappa_B T] \tag{6.52}$$

where $E_e - E_h = \hbar\omega$ and $E_{Fn} - E_{Fp} = E_0$ are used. It may be noted that Equation (6.52) is also an approximate form of the Fermi–Dirac distribution obtained for $E_e > E_{Fn}$ and $E_h < E_{Fp}$ and it has been used widely [6.37, 6.39–6.41] for calculating the rate of spontaneous emission in semiconductors. Substituting Equation (6.52) in the rate in Equation (6.29), the integral in Equation (6.41) is obtained as:

$$I = \frac{V\left(m_e^* m_h^*\right)^{3/2}}{4\pi^3 \hbar^6}(\hbar\omega - E_0)^2 \exp[-(\hbar\omega - E_0)/\kappa_B T] \tag{6.53}$$

Using Equation (6.53), the rates of spontaneous emission at thermal equilibrium for possibilities (i)–(iii) are obtained as:

$$R_{spi} = R_{spni} \exp[-(\hbar\omega - E_0)/\kappa_B T], \quad i = 1,2 \tag{6.54}$$

and for possibility (iv) we get:

$$R_{spti} = R_{spnti} \exp[-(\hbar\omega - E_0)/\kappa_B T], \quad i = 1,2 \tag{6.55}$$

Rates in Equations (6.54) and (6.55) have a maximum value, which can be used to determine E_0 as described below.

(iii) Determining E_0

For determining E_0, we assume that the peak of the observed PL intensity occurs at the same energy as that of the rate of spontaneous emission obtained in Equations (6.54) and (6.55). The PL intensity as a function of $\hbar\omega$ has been measured in a-Si:H. From these measurements the photon energy corresponding to the PL peak maximum can be determined. By comparing the experimental energy thus obtained with the energy corresponding to the maximum of the rate of spontaneous emission, we can determine E_0. For this purpose, we need to determine the energy at which the rates in Equations (6.54) and (6.55) become maximum.

Defining $x = \hbar\omega - E_0$, $(x > 0)$ and $\beta = 1/\kappa_B T$ and then setting $dR_{spi}/dx = 0$ $(i = 1, 2)$, we get x_{01} and x_{02} from Equation (6.54) or Equation (6.55) at which the rates are maximum, respectively, as [6.20]:

$$x_{01} = \frac{1 - \beta E_0}{2\beta} \pm \sqrt{\frac{(1 - \beta E_0)^2}{4\beta^2} + \frac{2E_0}{\phi}}$$
(6.56)

and

$$x_{02} = \frac{3 - \beta E_0}{2\beta} \pm \sqrt{\frac{(1 - \beta E_0)^2}{4\beta^2} + \frac{2E_0}{\phi}}$$
(6.57)

where only the + sign produces $x > 0$. It may be noted that both Equations (6.54) and (6.55) give the same expression for x_{01} and x_{02} as derived in Equations (6.56) and (6.57). Using $x_0 = E_{mx} - E_0$, where $E_{mx} = \hbar\omega_{max}$ is the emission energy at which the PL peak intensity is observed, the corresponding E_0 value is obtained from Equations (6.56) and (6.57) as:

$$E_{01} = \frac{E_{mx}(-1 + \beta E_{mx})}{1 + \beta E_{mx}}$$
(6.58)

$$E_{02} = \frac{E_{mx}(-3 + \beta E_{mx})}{1 + \beta E_{mx}}$$
(6.59)

Using this in Equations (6.54) and (6.55) one can calculate the rates of spontaneous emission from the two approaches for all four possibilities. The time of radiative recombination or radiative lifetime, τ_{ri}, is then obtained from the inverse of the maximum rate ($\tau_{ri} = 1/R_{spi}$, $i = 1, 2$) obtained from Equations (6.54) and (6.55) calculated at $\hbar\omega = E_{mx}$.

6.2.3 Results of spontaneous emission and radiative lifetime

Applying the above theory, rates of spontaneous emission can be calculated (a) within two-level approximation, (b) under nonequilibrium with the joint and product density of states, and (c) under equilibrium with the product density of states. Although the rate derived within the two-level approximation is strictly valid only for isolated atomic systems, for applying it to a condensed-matter system one requires ω and $|r_{e-h}|$. For applying it to the excitonic radiative recombination, one may be able to assume $|r_{e-h}| = a_{ex}$, but it is not clear at what value of ω one should calculate the rate to determine the radiative lifetime. For this reason, the rate R_{sp12} in Equation (6.32) derived within the two-level approximation will be calculated later at the end of this section.

Rates in Equations (6.46)–(6.48), derived under nonthermal equilibrium with the product density of states, cannot be used to calculate the radiative lifetime unless one knows the value of E_0, which cannot be determined from these expressions. Therefore, here we can only present the results for the rates of spontaneous emission derived under the thermal equilibrium in Equations (6.54) and (6.55). We will use Equation (6.54) to calculate the rates, R_{spi}, $i = 1$, 2, for possibilities (i)–(iii), and Equation (6.55) to calculate R_{spt1} and R_{spt2} for possibility (iv) in a-Si:H. As the non-radiative relaxation is very fast (in ps), only excitonic recombinations can be expected to contribute to PL from possibilities (i)–(iii). All free excited carriers (type I and II geminate pairs and nongeminate pairs) are expected to have relaxed to their tail states before recombining radiatively and hence they can contribute to PL only through possibility (iv).

For calculating the effective masses of charge carriers, we consider a sample of a-Si:H with 1 at % weak bonds contributing to the tail states (i.e., $a = 0.99$ and $b = 0.01$), using $L = 0.235$ nm [6.4], $E_2 = 3.6$ eV [6.46, 6.47], $E_c = 1.80$ eV [6.2], and $E_c - E_{ct} = 0.8$ eV [6.48]. Using these in Equations (3.17)–(3.18) we arrive at $m_{ex}^* = 0.34\ m_e$ and $m_{et}^* = 7.1\ m_e$, respectively, for a-Si:H.

For determining E_0 from Equations (6.58) and (6.59), we need to know the values of E_{mx} and the carrier (exciton) temperature T. Wilson et al. [6.8] have measured PL intensity as a function of the emission energy for three different samples of a-Si:H at 15 K and at two different decay times of 500 ps and 2.5 ns. Stearns [6.9] and Aoki [6.12, 6.50, 6.51] have also measured it at 20 K and 3.7 K, respectively. E_{mx} estimated from these spectra is given in Table 6.1.

The next problem is to find the temperature of the excited carriers before they recombine radiatively. This can also be done on the basis of the three experimental results on PL [6.9, 6.11, 6.49–6.51]. The measured energy, E_{mx}, of maximum PL intensity is below the mobility edge E_c by from 0.4 eV [6.8, 6.11] to 0.44 eV [6.51]. That means that most excited charge carriers have relaxed down below their mobility edges and are not hot carriers any more before

Table 6.1 Values of E_{mx} estimated from the maximum of the observed PL intensity from three different experiments (Figure 6.2) and the corresponding values of E_0 calculated from Equations (6.58) and (6.59). (Note that both expressions produce the same value)

Experiment	T (K)	E_{mx} (eV)		E_0 (eV)	
		500 ps	2.5 ns	500 ps	2.5 ns
Sample 1 [6.11]	15	1.428	1.401	1.425	1.398
Sample 2 [6.11]	15	1.444	1.400	1.441	1.397
Sample 3 [6.11]	15	1.448	1.405	1.445	1.402
Sample 4 [6.9]	20	1.450		1.447	
Sample 5 [6.51]	3.7	1.360		1.359	

(a)

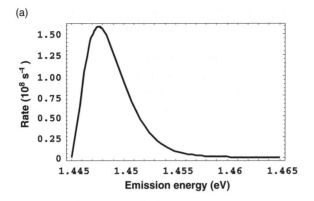

(b)

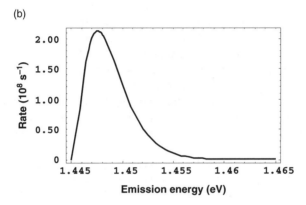

Figure 6.1   Rates of spontaneous emission plotted as a function of the emission energy for $E_0 =$ 1.445 eV at a temperature of 15 K and calculated from (a) method 1 and (b) method 2 for calculating the matrix elements

the radiative recombination. It is therefore only logical to assume that the excited charge carriers in these experiments are in thermal equilibrium with the lattice and E_0 should be calculated at the lattice temperature. The assumption is very consistent with the established fact that the carrier–lattice interaction is much stronger in a-Si:H [6.5] than in crystalline Si.

Using the experimental values of E_{mx} and the corresponding lattice temperature, the values of E_{01} calculated from Equation (6.58) and those of E_{02} calculated from Equation (6.59) are also listed in Table 6.1. Both the approaches, Equation (6.58) and (6.59), produce identical results for E_0. It should be noted, however, that E_0 increases slightly with the lattice temperature. Having determined the effective mass and E_0, we can now calculate the rates of spontaneous emission for transitions through possibilities (i)–(iv).

(i) Possibility (i) – Both e and h excited in extended states

As explained above, possibilities (i)–(iii) are only applicable to the singlet excitonic recombinations. In possibility (i), the excitonic reduced mass is obtained as $\mu_x = 0.17\ m_e$ and the excitonic Bohr radius as 4.67 nm. Using $n = 4$, the calculated R_{sp1} is plotted as a function of the emission energy and at a temperature $T = 15$ K for one of the samples of a-Si:H of

Wilson et al. [6.11] in Figure 6.1(a) and R_{sp2} for the same sample in Figure 6.1(b). Similar curves are obtained for all the samples at other temperatures as well but the magnitudes of the rates vary.

The radiative lifetime can be obtained from the inverse of the maximum value of the rate at any temperature and E_0. The maximum value of the rates is obtained at the emission energy, E_{mx}, at a given temperature. The rates and the corresponding radiative lifetimes thus calculated are listed in Table 6.2. As can be seen from Table 2, the radiative lifetime is found to be in the ns time range at 15 K and 20 K and in the μs range at 3.7 K.

(ii) Possibilities (ii)–(iii) – Extended-to-tail states recombinations

For possibilities (ii) and (iii), where one of the charge carriers of a singlet exciton is in its extended state and the other in its tail state, we get $\mu_x = 0.32\ m_e$ and $a_{ex} = 2.5$ nm. Using these and the other quantities as in possibility (i) above, we get from Equation (6.54) ($i = 1$), $R_{sp1} = 2.94 \times 10^8\ s^{-1}$, which gives $\tau_{r1} = 3.41$ ns at 15 K, $5.22 \times 10^8\ s^{-1}$ giving $\tau_{r1} = 1.92$ ns at 20 K, and at 3.7 K we get $R_{sp1} = 1.47 \times 10^7\ s^{-1}$, which gives $\tau_{r1} = 0.1\ \mu$s. Likewise from Equation (6.54) ($i = 2$), we get $R_{sp2} = 3.45 \times 10^8\ s^{-1}$ and $\tau_{r2} = 2.9$ ns at 15 K, and $R_{sp2} = 1.08 \times 10^9\ s^{-1}$ giving $\tau_{r2} = 0.92$ ns at 20 K. At 3.7 K, we obtain $R_{sp2} = 1.62 \times 10^7\ s^{-1}$ and $\tau_{r2} = 0.1\ \mu$s.

Thus, the radiative lifetimes for all the three possibilities (i)–(iii) are in the ns time range at 15 K and 20 K, but in the μs time range at 3.7 K. It is not possible from these results to determine where exactly the PL is originating from in these experiments, i.e., is it from the extended

Table 6.2 The maximum rates, R_{sp1} and R_{sp2}, of spontaneous emission and the corresponding radiative lifetime calculated using the values of E_{mx} and E_0 for the 5 samples in Table 6.1. Results are given for transitions involving extended-extended states [possibility (i)], extended-tail states [possibilities (ii) and (iii)] and tail-tail states [possibility (iv)]

Sample	T (K)	R_{sp1} (s^{-1})	τ_{r1} (s)	R_{sp2} (s^{-1})	τ_{r2} (s)
1[a]	15	1.58×10^8	6.34×10^{-9}	2.05×10^8	4.88×10^{-9}
2[a]	15	1.56×10^8	6.42×10^{-9}	2.07×10^8	4.82×10^{-9}
3[a]	15	1.55×10^8	6.43×10^{-9}	2.00×10^8	5.00×10^{-9}
4[a]	20	2.77×10^8	3.62×10^{-9}	3.72×10^8	2.69×10^{-9}
5[a]	3.7	7.77×10^6	0.13×10^{-6}	8.38×10^6	0.12×10^{-6}
1[b]	15	2.97×10^8	3.37×10^{-9}	3.43×10^8	2.93×10^{-9}
2[b]	15	2.93×10^8	3.41×10^{-9}	3.46×10^8	2.89×10^{-9}
3[b]	15	2.92×10^8	3.42×10^{-9}	3.47×10^8	2.89×10^{-9}
4[b]	20	5.22×10^8	1.92×10^{-9}	1.08×10^9	0.92×10^{-9}
5[b]	3.7	1.47×10^7	0.68×10^{-7}	1.62×10^7	0.62×10^{-7}
1[c]	15	1.04×10^7	0.96×10^{-7}	1.18×10^7	0.85×10^{-7}
2[c]	15	1.03×10^7	0.97×10^{-7}	1.19×10^7	0.84×10^{-7}
3[c]	15	1.03×10^7	0.97×10^{-7}	1.19×10^7	0.84×10^{-7}
4[c]	20	1.84×10^7	0.55×10^{-7}	2.14×10^7	0.47×10^{-7}
5[c]	3.7	0.51×10^6	1.96×10^{-6}	0.55×10^6	1.82×10^{-6}

[a]Extended-extended states transitions [possibility (i)].
[b]Extended-tail or tail-extended states transitions [possibilities (ii) – (iii)].
[c]Tail-tail states transitions [possibility (iv)].

or tail states? In order to determine the origin of PL, one also needs to know the Stokes shift in the PL spectra. For example, if the excitonic PL is occurring through radiative transitions from extended-to-extended states, then the Stokes shift observed in the PL spectra should only be equal to the exciton binding energy. For this reason, it is important to determine the exciton binding energy corresponding to all the four transition possibilities.

The ground-state singlet exciton binding energy E_s in a-Si:H is obtained as [6.5, 6.10]:

$$E_s = \frac{9\mu_{ex}e^4}{20(4\pi\varepsilon_0)^2\varepsilon^2\hbar^2} \tag{6.60}$$

This gives $E_s \sim 16\,\text{meV}$ for possibility (i), 47 meV for possibilities (ii) and (iii), and 0.33 eV for possibility (iv). The known optical gap for a-Si:H is 1.8 eV [6.2] and E_0 estimated from experiments is about 1.44 eV at 15 K [6.11], 1.45 eV at 20 K [6.9], and 1.36 eV at 3.7 K [6.50] (see Table 6.1). Considering that the PL in a-Si:H originates from excitonic states [possibility (i)], we find a Stokes shift of 0.36 eV, 0.35 eV, and 0.44 eV at temperatures 15 K, 20 K, and 3.7 K, respectively. Such a large Stokes shift is not possible due only to the exciton-binding energy, which at most is in the meV range for possibilities (i)–(iii) as given above. As the nonradiative relaxation of excitons is very fast, not much PL may occur through possibility (i) even at a time delay of 500 ps measured by Wilson et al. [6.11] and Stearns [6.9]. It may possibly occur from transitions only through possibilities (ii)–(iv), which means at least one of the charge carriers has relaxed to the tail states before recombining radiatively. This can be easily explained from the following: As stated above, the charge carrier–lattice interaction is much stronger in a-Si:H than in crystalline Si [6.5]. As a result, it is well established [6.4] that an excited hole gets self-trapped very fast in the tail states in a-Si:H. Thus a Stokes shift of about 0.4 eV observed experimentally at 3.7 K, 15 K, and 20 K is due to the relaxation of holes in excitons to the tail states plus the excitonic binding energy, which is in the meV range and hence plays a very insignificant role. The radiative lifetimes of such transitions calculated from the present theory and listed in Table 6.2 fall in the ns time range at temperatures 15–20 K, which agrees very well with those observed by Wilson et al. [6.8, 6.11] and Stearns [6.9]. At 3.7 K, the calculated radiative lifetime is found to be in the μs range, which also agrees very well with Aoki et al.'s experimental results of the singlet exciton's radiative lifetime [6.50] (see Figures 5.13, 5.15, and 5.16).

(iii) Possibility (iv) – Tail-to-tail states recombinations

For possibility (iv), as a type II singlet geminate pair, when both e and h are localized in their tail states, we get $\mu_x = 3.55\,m_e$, $a_{ex} = 0.223\,\text{nm}$, and $t'_e = 1.29 \times 10^{10}\,\text{m}^{-1}$. Using these in Equation (6.55), with $i = 1$, one gets $R_{spt1} = 1.03 \times 10^7\,\text{s}^{-1}$, which gives $\tau_{r1} = 0.97 \times 10^{-7}\,\text{s}$ at 15 K, $R_{spt1} = 1.84 \times 10^7\,\text{s}^{-1}$, which gives $\tau_{r1} = 0.55 \times 10^{-7}\,\text{s}$ at 20 K, and $R_{spt1} = 0.51 \times 10^6\,\text{s}^{-1}$ with the corresponding $\tau_{r1} = 2.0\,\mu\text{s}$ at 3.7 K.

Using Equation (6.55) with $i = 2$, we get $R_{spt2} = 1.18 \times 10^7\,\text{s}^{-1}$ which gives $\tau_{r2} = 0.85 \times 10^{-7}\,\text{s}$ at 15 K, and $R_{spt2} = 2.14 \times 10^7\,\text{s}^{-1}$, which gives $\tau_{r2} = 0.47 \times 10^{-7}\,\text{s}$ at 20 K. The corresponding results at 3.7 K are obtained as $R_{spt2} = 0.55 \times 10^6\,\text{s}^{-1}$ and $\tau_{r2} = 2.0\,\mu\text{s}$. Wilson et al. [6.11] have also observed the lifetime in the μs time range, which may be attributed to transitions through possibility (iv). Aoki et al. [6.50] have observed another PL peak at 3.7 K (τ_G, Figure 5.16) with a radiative lifetime in the μs range but slightly longer than the singlet exciton radiative lifetime (τ_S) at 3.7 K. According to the radiative lifetime calculated here (Table 6.2), the PL peak at τ_G may be attributed to transitions from the tail-to-tail states from a type II singlet pair [possibility (iv)] because the corresponding radiative time is about $2\,\mu$s, longer than the radiative lifetime for possibilities (i)–(iii).

A geminate pair of type I may be expected to have a larger separation and hence longer radiative lifetime. Indeed, in Figure 5.15 a peak, denoted by τ_D, appears at much larger lifetime of ms and Aoki et al. have attributed this peak to the distant or nongeminate pairs. According to the present theory, this peak may be attributed to the type I geminate pairs or distant pairs, but it is difficult to distinguish between the two (see below).

For possibility (iv), we have calculated the rates using the same value of E_0 as for possibilities (ii) and (iii), which may not be correct. However, as E_0 cannot be measured experimentally it is difficult to find a correct value for it for possibility (iv). One way to find a value of E_0 for possibility (iv) is from the expected corresponding Stokes shift, when both the charge carriers have relaxed to their tail states. This can be done as follows [6.32]: The measured Stokes shift in a-Si:H is found to be 0.440 at 3.7 K [6.50] and 0.350 eV at 20 K [6.9] for the possibilities (ii)–(iii). Here we consider only these two different experimental values as examples. For possibility (iv), when both e and h are in their tail states, a double Stokes shift can be assumed. Accordingly, one gets the Stokes shift of 0.880 eV at 3.7 K and 0.700 eV at 20 K. Then, considering $E_C = 1.8$ eV in a-Si:H, the PL peak is expected to occur at $E_{mx} = 1.8 - 0.880 = 0.920$ eV at 3.7 K and 1.1 eV at 20 K, which give from Equations (6.58) and (6.59) $E_0 = 0.918$ eV at 3.7 K and 1.094 eV at 20 K. Using these values of E_0, the maximum value of rates at $\hbar\omega = E_{mx}$ are calculated and the inverse of these reveals the corresponding radiative lifetimes ($\tau_r = 1/R_{sp}$). For type I geminate pairs, the lifetime is calculated for e–h separation, $r_{eh} = 2L$ and $3L$, where $L = 0.235$ nm, the average interatomic separation in a-Si:H; and for the type II geminate pairs, $r_{eh} = a_{ex}$ is used.

The radiative lifetimes thus calculated in a-Si:H using the experimental value of E_{mx} at 3.7 K from Aoki et al.'s experiment [6.50] and at 20 K from Stearns' experiment [6.9] are listed in Table 6.3. At 3.7 K, for the type II geminate pairs, the radiative lifetime is obtained in the μs time range, which agrees with the lifetime τ_G observed by Aoki et al. At 20 K the

Table 6.3 Rates of spontaneous emission and corresponding radiative lifetimes calculated from Equation (6.55) (with $i = 2$) for geminate pairs of types I and II in a-Si:H. Rates for type I are calculated for the average e–h separation, $r_{e-h} = 2L$ and $3L$ ($L = 0.235$ nm) at the same E_0 and E_{mx} values

E_0 (eV)	E_{mx} (eV)	T (K)	Radiative lifetime (τ_r)		
			Geminate I		Geminate II
			$r_{eh} = 2L$	$r_{eh} = 3L$	$r_{eh} = a_{ex}$
0.918	0.920	3.7	1.10 ms	0.15 s	15.00 μs
1.094	1.100	20	6.20 μs	0.87 ms	88 ns

same lifetime is in the ns time range, which agrees with the lifetime observed by Stearns. Here it is important to note that the input values used to calculate the rate of spontaneous emission at 3.7 K and 20 K are obtained from two different experiments, namely references [6.50] and [6.9], and therefore are different. Aoki et al. have not observed any radiative lifetime in the ns time range at any temperature, including 20 K as observed by Stearns. The value of E_{mx} is not available to the author at 3.7 K and 20 K from the same experiment.

The results for the type I geminate pairs calculated at $r_{eh} = 2L$ and $3L$ are in the ms to s time range, which agree with the radiative lifetimes, τ_D, measured by Aoki et al. [6.49] but attributed to nongeminate pairs by them (See Figure 5.15). It is not possible to distinguish between the lifetime of type I geminate pairs and nongeminate pairs unless one can distinguish the corresponding separation between e and h in each. Therefore, it is possible that the measured lifetime is actually due to type I geminate pairs because the type II geminate pairs were not known then [6.51]. In this view, the lifetime τ_G measured in the μs time range can be associated to the type II geminate pairs, τ_D in the ms to s range to type I geminate pairs, and any other longer lifetimes to nongeminate pairs.

We have discussed here only the type II singlet geminate pairs, because without consideration of the spin–orbit coupling the transition matrix element [Equation (6.11)] vanishes for a triplet spin configuration. As the magnitude of the spin–orbit coupling is usually smaller than that of $\hat{H}_{xp}$ in Equation (6.3), the radiative lifetime of triplets is expected to be longer. The excitonic Bohr radius of triplet excitons and triplet type II geminate pair may not be very different and therefore the lifetime τ_T (see Figure 5.15), may be attributed to both.

As the rates of recombination of nongeminate pairs or distant pairs have the same expression as in Equation (6.55), it is not possible to distinguish between nongeminate and geminate pairs of the type I. The only difference can be the average separation between carriers, as it is well established that nongeminate pairs are usually formed from those geminate pairs excited by higher-energy photons, the separation between e and h in nongeminate pairs being larger than the critical separation given by $r_c \leq n^{-1/3} \leq 2r_c$ for small generation rate [6.30, 6.31]. According to the present theory a larger separation will yield an exponentially smaller recombination rate and hence larger radiative lifetime in the time range of seconds [6.51] (see Figure 5.15). The radiative lifetime for $r_{eh} \geq 3L$ indeed produces such longer lifetimes at 3.7 K. One may therefore conclude that the excited pairs with a separation of $3L$ or more may be classified as nongeminate pairs.

The longer lifetime obtained for possibility (iv) is because of the exponential factor $\exp(-2t'_e |r_{e-h}|)$ appearing in the rate of recombination in Equation (6.55) through Equation (6.48), which is proportional to the probability of quantum tunneling a barrier of height E_c for a distance $|r_{e-h}|$. Therefore, the expression also implies that the rate of spontaneous emission reduces and hence the radiative lifetime is prolonged in possibility (iv) due to the quantum tunneling through a distance of $|r_{e-h}|$ before the radiative recombination.

In evaluating the transition matrix element of the exciton–photon interaction operator [Equation (6.3)], which depends only on the relative momentum between e and h in an exciton, the wave functions of e and h are used instead of the wave function of an exciton [6.10, 6.42]; for example, in the form of $\Psi(R,r) = \exp\left(i\dfrac{\mathbf{P} \cdot \mathbf{R}}{\hbar}\right)\phi(r)$. From this point of view the theory presented here appears to address the recombination between an excited pair of an electron and a hole, not that between an electron and a hole in an exciton. The fact that the interaction operator is independent of the center of mass momentum $\mathbf{P}$ means that the integral $Z_{lm\lambda}$ [Equation (6.10)] always gives the average value of the separation between e and h. In an exciton, the separation is the excitonic Bohr radius and in an e-and-h pair it would be their average separation. As far as the recombination in amorphous solids is concerned, therefore, this is the only difference between an excitonic recombination and free e and h recombination.

(iv) Results of two-level approximation

As we have determined the energy, E_{mx}, at which maxima of PL peaks have been observed in the five samples of a-Si:H, we can calculate R_{sp12} in Equation (6.32) and the corresponding radiative lifetime at these photon energies for a comparison. Here we will use $|r_{e-h}| = a_{ex}$ and $\omega = E_{mx}/\hbar$ in Equation (6.32). For extended-to-extended state recombination, $a_{ex} = 4.67$ nm and, using $E_{mx} = 1.4$ eV for Wilson et al.'s sample, we get $\omega = 2.1 \times 10^{15}$ Hz, $R_{sp12} = 7.4 \times 10^{10}$ s^{-1}, and $\tau_r = 1/R_{sp12} = 13$ ps. For Stearn's sample, with $E_{mx} = 1.45$ eV leading to $\omega = 2.2 \times 10^{15}$ Hz, we get $R_{sp12} = 8.5 \times 10^{10}$ s^{-1} and the corresponding $\tau_r = 12$ ps. For Aoki's sample with $E_{mx} = 1.36$ eV and the corresponding $\omega = 2.0 \times 10^{15}$ Hz, we get $R_{sp12} = 6.2 \times 10^{10}$ s^{-1} and $\tau_r = 16$ ps. Thus, within the two-level approximation, one cannot get any results for the radiative lifetime longer than ps, which does not agree with the experimental results. This is a further clear indication of the fact that the two-level approximation is not valid for condensed-matter systems.

6.2.4 Temperature dependence of PL

The maximum rates calculated from Equations (6.54) and (6.55) at the emission energy E_{mx} increase exponentially as the temperature increases. This agrees very well with the model used by Wilson et al. [6.11] to interpret the observed temperature dependence in their PL intensity. Such temperature dependence is independent of the origin of the PL, whether the recombination originates from the extended or tail states. However, the maximum rates obtained in Equations (6.54) and (6.55) vanish at $T = 0$ because x_{01} [Equation (6.56)] and x_{02} [Equation (6.57)] become zeros as $T \to 0$, which is in contradiction with the observed rates [6.9, 6.11]. The reason for this is that one cannot derive the rate at 0 K from Equations (6.54) and (6.55). At 0 K the, the electron-and-hole-probability distribution functions become unity; $f_c \to 1$ for $E_e < E_{Fn}$ and $f_v \to 1$ for $E_h > E_{Fp}$ as $T \to 0$ and using this the rates at 0 K become equal to the pre-exponential factors as obtained in Equations (6.46)–(6.48). Thus, the temperature dependence of the rates obtained here can be expressed as:

$$R_{spi}(T) = R_{0i}\{[1 - \Theta_1(T)] + \exp[-(\hbar\omega - E_0)/\kappa_B T]\Theta_2(\hbar\omega - E_0)\}, \quad i = 1, 2 \qquad (6.61)$$

where R_{0i} are the pre-exponential factors of Equations (6.54) and (6.55) corresponding to the four possibilities. The first step function, $\Theta_1(T)$, is used to indicate that the first term of Equation (6.61) vanishes for $T > 0$ and the second step function $\Theta_2(T)$ shows that there is no spontaneous emission for $\hbar\omega - E_0 < 0$. For an estimate of R_{0i} we have used the values of E_{mx} and E_0 obtained at $T = 3.7$ K, which gives $R_{01} = 7.77 \times 10^6$ s^{-1}, 1.47×10^7 s^{-1}, and 0.51×10^6 s^{-1} for extended–extended, extended–tail, and tail–tail states, respectively, and the corresponding values for R_{02} are obtained as 8.38×10^6 s^{-1}, 1.62×10^7 s^{-1} and 0.55×10^6 s^{-1}.

Wilson et al. [6.11] have fitted their observed rates to the following model:

$$\frac{1}{\tau_r} = v_1 + v_0 \exp(T/T_0) \qquad (6.62)$$

and the best fit has been obtained with $v_1 = 10^8 \, \text{s}^{-1}$, $v_0 = 0.27 \times 10^6 \, \text{s}^{-1}$, and $T_0 = 95 \, \text{K}$. Stearns [6.9] has obtained a best fit to his data with $v_1 = 2.7 \times 10^8 \, \text{s}^{-1}$, $v_0 = 6.0 \times 10^6 \, \text{s}^{-1}$, and $T_0 = 24.5 \, \text{K}$. According to Equation (6.62), $\tau_r^{-1} = v_1$ at $T = 0$. Comparing these results with Equation (6.61), one should have $R_{0i} = v_1$, but R_{0i} is obtained an order of magnitude smaller than v_1. Also in Equation (6.62), $v_1 > v_0$ whereas according to Equation (6.61) $v_1 = v_0 = R_{0i}$. These discrepancies may be attributed to the fact that Equation (6.62) is obtained by fitting to the experimental data and not by any rigorous theory. As a result, both Equations (6.61) and (6.62) produce similar results for the radiative lifetime, but individual terms on the right-hand sides contribute differently. This is also apparent from the fact that according to Equation (6.62), v_1 contributes to the rates also at nonzero temperatures but, according to Equation (6.61), R_{0i} contributes only at $T = 0 \, \text{K}$. Stearns [6.9] has fitted the observed rate of emission at the emission energy of 1.43 eV to Equation (6.62). We have plotted the calculated rates in Equation (6.54) as a function of the temperature for possibility (ii) at the same emission energy (1.43 eV) and our results are shown in Figures 6.2(a) and (b), respectively.

Aoki et al. [6.50] have measured the QFRS spectra of a-Si:H and a-Ge:H at various temperatures (see Figure 5.16), where all PL peaks appearing at τ_S (μs), τ_G (μs), τ_T (ms), and

(a)

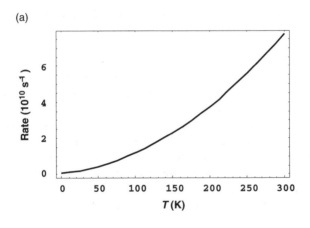

(b)

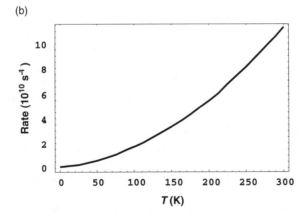

Figure 6.2 Rates plotted as a function of temperature at an emission energy of 1.43 eV and calculated for possibility (ii) from Equation (6.54) in (a) $i = 1$ and (b) $i = 2$

τ_D (0.1 s) show a shift toward a shorter time scale as the temperature increases. This behavior is quite consistent with the temperature dependence obtained from the present theory as well.

6.2.5 Excitonic concept

Another question may arise here is: Why is the concept of excitons necessary to explain the radiative lifetime when the excited charge carriers can recombine without forming excitons? This can be answered as follows: Instead of using the reduced mass of excitons and the excitonic Bohr radius, if one uses the effective masses of e and h and the distance between them, one can calculate the radiative lifetime of a free excited electron–hole pair (not an exciton) from Equations (6.54) and (6.55). The radiative lifetime thus obtained for such a recombination is also in the ns range for possibilities (i)–(iii). For possibility (iv), of course, as the separation between e and h is expected to be larger one may get smaller rates and hence a longer radiative lifetime. However, there are two important observations which support the formation of excitons. 1) A free e–h pair will relax down to the tail states in a ps time scale, resulting in PL only from possibility (iv), which is not supported by the observed PL (see Figures 5.13–5.15). 2) PL has a double-peak structure appearing in the time-resolved spectra corresponding to the singlet and triplet excitons. No such spin-correlated peaks can be obtained with the recombination of free electron–hole pairs.

The theory presented here for the radiative recombination from the tail states is different from the radiative tunneling model [6.2], where the rate of transition is given by: $R_{spt} \sim R_0$ $\exp(-2d/d_0)$, d being the separation between e and h, d_0 is the larger of the extents of the electron and hole wave functions, usually considered to be 10–12 Å, and R_0 is the limiting radiative rate expected to be about 10^8–$10^9 \, s^{-1}$. Here d, being the distance between a pair of excited charge carriers (not an exciton), is not a fixed distance, like a_{ex}; it can be of any value. As a result, the radiative tunneling model cannot explain the appearance of the double-peak structure in PL originating from the tail states. Although the rate of recombination from the tail states derived in Equation (6.55) also has a similar exponential dependence as in the radiative tunneling model, it depends on the excitonic Bohr radius, which is a constant, and hence the exponential factor becomes a constant. This explains very well the peak structure in PL.

Kivelson and Gelatt [6.7] have developed a comprehensive theory for PL in amorphous semiconductors based on a trapped-exciton model. A trapped exciton considered by them is one in which the hole is trapped in a localized gap state and the electron is bound to the hole by their mutual Coulomb interaction. The model is thus similar to possibility (ii) considered here, but possibilities (i) and (iii) have not been considered by them. The square of the dipole transition matrix element is estimated to be $\sim(2a_h)^5/(a_B)^3$ by them. This is probably one of the main differences between the present theory and that of Kivelson and Gelatt. The other important difference between the present theory and that of Kivelson and Gelatt is in possibility (iv) involving recombinations in the tail states. For this possibility, Kivelson and Gelatt have followed the radiative tunneling model and used the exponential factor as $\exp(-2d/a_B^*)$. As discussed above such an exponential dependence on varying d cannot explain the peak structures observed in PL spectra.

It is quite clear from the expressions derived in Equations (6.54) and (6.55) that rates do not depend on the excitation density. This agrees very well with the measured radiative lifetime by Aoki et al. [6.49], who have found that the radiative lifetime of singlet and triplet excitons is independent of the generation rate (see Chapter 5). This provides additional support for the PL observed in a-Si:H to be due to excitons. One of the speculations, as described above, has been that the shorter PL time in the ns range observed by Wilson et al. may be due to the high excitation density used in the TRS measurements. The present theory clarifies this point very successfully that it is due not to the excitation density but to the combined effect of temperature and fast nonradiative relaxation of charge carriers to lower energy states that one does not observe any peak in the ns time range at 3.7 K.

Considering distant-pair recombinations, Levin et al. [6.31] have studied PL in amorphous silicon at low temperatures by computer simulation. They have also used the radiative tunneling model for recombinations in the tail states, but they have not considered radiative recombination of excitons. Therefore, results of their work also cannot explain the occurrence of the double-peak structure in PL.

6.3 PHOTOINDUCED CHANGES IN AMORPHOUS CHALCOGENIDES

One of the photoinduced changes observed in amorphous semiconductors, particularly in amorphous chalcogenides (a-Chs), is the reduction in the bandgap by illumination and it is commonly known as photodarkening (PD) [6.5, 6.25, 6.52, 6.53]. In some cases, PD is also accompanied by photoinduced volume expansion (PVE) in the material (see Chapter 7). There are two kinds of PD observed in a-Chs, one that disappears by annealing but remains even if the illumination is stopped [6.27] and the other that disappears once the illumination is stopped [6.28]. The former is called metastable PD and the latter transient PD, which has been observed only recently. Much effort has been devoted to understanding the phenomena of PD and PVE in the last two decades [6.5, 6.54–6.57]; however, no model has been successful in resolving all issues observed. One of the recent models is repulsive electronic interaction [6.58]. As chalcogenides, like $As_2Se(S)_3$, have layered structures, the charge carriers excited by illumination in these materials move in these layers. The hole mobility is larger than the electron mobility in these materials and hence photo-generated electrons reside mostly in the conduction-tail states while holes diffuse away faster in the un-illuminated region through valence-extended and -tail states. Therefore, the layers that absorb photons during the illumination become negatively charged, giving rise to a repulsive inter-layer Coulomb interaction, which increases the inter-layer separation and causes PVE. The same force is assumed to induce an in-plane slip motion that increases the LP–LP interaction and causes PD. However, the repulsive Coulomb force model is qualitative and more recent calculations [6.59] indicate that the force is too weak to cause PVE. More recently [6.60] a unified description of the photo-induced volume changes in chalcogenides based on tight-binding (TB) molecular dynamics (MD) simulations of amorphous selenium has been putforward (see Chapter 7). For causing movements in an atomic network, one requires the involvement of lattice vibrations, of which (surprisingly) none of the models of PD and VE has given any account. However, the involvement of lattice vibrations has recently been considered [6.61] in inducing photostructural changes in glassy semiconductors.

In 1988, Singh discovered [6.62] a drastic reduction in the bandgap due to pairing of charge carriers in excitons and exciton–lattice interaction in nonmetallic crystalline solids. The magnitude of reduction in the bandgap varies with the magnitude of the exciton–phonon interaction, which is different in different materials; the softer the structure is, the stronger is the carrier–phonon interaction. It has been established that due to their planar structure [6.5], the carrier–phonon interaction in a-Chs is stronger than in a-Si:H, which satisfies the condition for the occurrence of Anderson's negative-U [6.63] in a-Chs but not in a-Si:H. This is the basis of the hole-pairing model for the creation of light-induced defects in a-Chs [6.5].

In this section, based on Holstein's approach [6.64], the energy eigenvalues of positively and negatively charged polarons and paired charge carriers, created by excitations due to illumination, are calculated. It is found that the energies of the excited electron (negative charge) and hole (positive charge) polarons becomes lowered due to the carrier–phonon interaction [6.65]. Also the like excited charge carriers can become paired because of the negative-U effect [6.63] caused by the strong carrier–lattice interaction, and the energy of such paired states is also lowered. Thus, the hole polaronic state and paired-hole states overlap with the lone-pair and tail states in a-Chs, which expands the valence band and reduces the bandgap energy and hence causes PD. Formation of polarons, as well as pairing of holes, increases the bond length on which such localizations occur, which causes VE. The energy eigenvalues of such polarons and pairing of charge carriers have been calculated.

6.3.1 Effect of photo-excitation and phonon interaction

Based on Holstein's approach [6.64], we consider here a model amorphous solid in the form of a linear chain of atoms. The electronic Hamiltonian in such a chain can be written as a sum of charge carrier ($\hat{H}_{el}$), phonon ($\hat{H}_{ph}$), and charge carrier–phonon interaction ($\hat{H}_I$) energy operators in the real coordinate space as [6.5, 6.65]:

$$\hat{H} = \hat{H}_{el} + \hat{H}_{ph} + \hat{H}_I \tag{6.63}$$

In a-Chs, as the lone-pair orbitals overlap with the valence band, the combined valence band becomes much wider than the conduction band. As the effective mass of charge carriers is inversely proportional to the corresponding band width [6.5] [see Equations (3.17) and (3.20)], the hole's effective mass becomes smaller than the electron's effective mass in a-Chs. As a result, holes move faster than electrons in these materials. However, in the tail states the charge carriers become localized.

In such a system, we consider that a photon is absorbed to excite an electron (e) in the conduction state and hole (h) in the valence state. Considering the effect of strong carrier–phonon interaction, the excited hole may become a positive-charge polaron and an electron may become a negative-charge polaron. Furthermore, two excited electrons and two excited holes can become paired and localized on a bond due to the effect of negative-U, a fact which has already been established [6.5, 6.63]. Pairing of holes on a bond breaks the bond and creates a pair of dangling bonds, which is used to explain the creation of light-induced defects in a-Ch [6.5].

Let us first consider the case of an excited hole in the valence band. The eigenvector of such an excited hole can be written as:

$$|h,0> = \sum_l C_l d_{0l}^+ |0> \tag{6.64}$$

where h denotes the hole and 0 denotes valence band; the conduction band is denoted by 1. C_l represents the probability amplitude coefficient, d_{0l}^+ ($= a_{0l}$) is the creation operator of a hole, a_{0l} is the annihilation operator of an electron in the valence states on site l, and $|0>$ represents the vacuum state with all valence states completely filled and all conduction states completely empty. Using Equations (6.63) and (6.64), one can solve the Schrödinger equation, $\hat{H} \, | h,0 > \, = W_h| \, h,0 >$, to get a secular equation such as:

$$W_h C_l = \left(-\sum_n \frac{\hbar^2 \nabla_n^2}{2M} + \frac{1}{2} \sum_{m,n} M \omega_m \omega_n x_m x_n - A_l^h x_l - E_h^0 \right) C_l + T_h \left(C_{l-1} + C_{l+1} \right) \tag{6.65}$$

where W_h is the energy eigenvalue of the hole, the first two terms within the parentheses correspond to the kinetic and potential energies of nuclear vibrations, x_n is the nth bond length of a diatomic molecule in the chain vibrating with a frequency ω_n, A_l^h is the hole–phonon coupling coefficient, $E_h^0 = E_h^l$, a constant of energy of a hole localized at site l and hence site independent, and T_h is the hole-transfer energy between nearest neighbours from l to $l \pm 1$ in the chain. Although the vibration frequency ω_m has a subscript m, being the intramolecular vibrational frequency of identical molecules, it is site independent. Considering only the diagonal terms in the vibrating potential, multiplying Equation (6.65) by C_l^* and then summing over all l, we get:

$$W_h = -\sum_n \frac{\hbar^2 \nabla_n^2}{2M} + \frac{1}{2} \sum_m M \omega_m^2 x_m^2 - \sum_l A_l x_l \, |C_l^2| - E_h^0 + T_h \left(C_{l-1} + C_{l+1} \right) C_l^* \tag{6.66}$$

where $\sum_l C_l^* C_l = \sum_l |C_l|^2 = 1$ is used. Setting $\dfrac{\partial W_h}{\partial x_q} = 0$, the energy eigenvalue, W_h, can be minimized with respect to the bond length, x_q. This gives the bond length at the minimum energy as:

$$x_q^{(0)} = \frac{A_q^j \, |C_q^{(0)}|^2}{M \omega^2} \tag{6.67}$$

where the superscript (0) denotes the value of quantities at the minimum energy and subscript q denotes the qth bond at which the hole is localized. The subscript q has been dropped from the frequency ω as it is independent of bond sites. The bond length increases by $x_q^{(0)}$ due to the localization of the hole on the bond. Substituting Equation (6.67) into Equation (6.66), we calculate the minimum energy of the hole, denoted by W_h^0, as:

$$W_h^0 = -E_h^0 - 2T_h - E_{hq} \tag{6.68}$$

where E_{hq} represents the hole–polaron binding energy obtained as:

$$E_{hq} = \frac{\left(A_q^{h^2}/M\omega^2\right)^2}{48T_h} \tag{6.69}$$

This is the energy by which the energy of an excited hole is lowered from the free-hole-state energy, which is at $-E_h^0 - 2T_h$. This means that the hole-energy state moves upward in the energy gap by releasing an energy E_{hq} to phonons.

6.3.2 Excitation of a single electron–hole pair

The generation of excited electrons and holes occurs in pairs through photoexcitations. The excited electrons may also form polarons in the conduction states and their energy will also be lowered. For a single pair of excited charge carriers on our linear chain, the secular equation can be written as:

$$W_{eh}C_l = \left[(E_e^{l+X} - E_h^l) - (A_{l+X}^e x_{l+X} - A_l^h x_l) + \frac{1}{2}\sum_m M\omega^2 x_m^2 - U_{eh} \right] C_l$$
$$- T_e (C_{l+X+1} + C_{l+X-1}) + T_h (C_{l+1} + C_{l-1}) \tag{6.70}$$

where W_{eh} is the energy eigenvalue of the excited pair, $E_e^l = E_e^0$ represent the energy of an electron at site l and is also site independent, U_{eh} is the Coulomb interaction energy between the excited pair, T_e is the energy of transfer of electron between nearest neighbors. M is the atomic mass and ω is the frequency of vibration between nearest-neighbor atoms. X is the distance between the excited pair, and x_l is the bond length between nearest neighbors of site l. Minimizing the energy with respect x, we obtain from Equation (6.70) as reported [6.5]:

$$W_{eh} = E_e^0 - E_h^0 - U_{12} - 2(T_e - T_h) - E_{eq} - E_{hq} \tag{6.71}$$

where E_e^0 and E_h^0 are the site-independent energies of the excited electron and hole, respectively, without the lattice and Coulomb interactions between the excited charge carriers, and E_{eq} is the electronic polaron-binding energy, which can be obtained from Equation (6.69) by replacing the subscript h by e. The energy of an excited pair of charge carriers without the lattice interaction is given by:

$$W_{eh}^0 = E_e^0 - E_h^0 - U_{eh} - 2(T_e - T_h) \tag{6.72}$$

In the excitation of a free pair of charge carriers, one usually neglects U_{eh} and then the energy W_{eh}^0 is close to the optical energy gap in most materials. Subtracting Equation (6.72) from Equation (6.71), we get the reduction in the optical gap due to formation of excited pair of polarons as:

$$\Delta W = W_{eh} - W_{eh}^0 = -E_{eq} - E_{hq} \tag{6.73}$$

The lowering of an excited pair's energy also means that the bond on which such polaronic formation occurs will be stretched by $x_q^{(0)}$ and the bond may break. A bond breaking due to such single excitation has been recently demonstrated by numerical simulation [6.66; see also Chapter 7] as well. This can also be seen from Equation (6.67) that the bond becomes stretched due to an excitation in which a hole is localized on the bond and hence the bonding becomes weaker.

6.3.3 Pairing of like excited charge carriers

Here we consider that two pairs of electrons and holes are excited in the chain. Assuming that the separation between electron and hole in one excited pair is x and that in the other is x' and that the separation between two holes is X, the secular equation analogous to Equation (6.70) can be written as [6.5, 6.65]:

$$W(x,x',X)C_l = [(E_e^{l+x} - E_h^l) + (E_e^{l-X+x'} - E_h^l) - (A_{l+x}^e x_{l+x} - A_l^h x_l)$$
$$- (A_{l-X+x'}^e x_{l-X+x'} - A_{l-X}^h x_{l-x}) + \frac{1}{2}\sum_m M\omega_m^2 x_m^2 + U_{12}']C_l$$
$$- T_e(C_{l+x+1} + C_{l+x-1} + C_{l-X+x'+1} + C_{l-X+x'-1})$$
$$+ T_h(C_{l+1} + C_{l-1} + C_{l-X+1} + C_{l-X-1}) \qquad (6.74)$$

where $W(x, x', X)$ is the energy eigenvalue of the two excited pairs of electrons and holes, and U_{12}' is the total Coulomb interaction between the two pairs of excited charge carriers. It is well established that Se-based chalcogenides have linear chain-like structures and hence are flexible, and in such structures the carrier–phonon interaction is considered to be very strong. Such a strong carrier–phonon interaction can induce pairing of like excited charge carriers on a bond due to Anderson's −U effect [6.63], known for valence-band electrons. In most solids, usually one of the two interactions, electron–phonon or hole–phonon, is stronger than the other. Therefore, here we first consider the case that the hole–phonon interaction is stronger and therefore that two holes can be paired on a bond and the two excited electrons will form two polarons on the chain. The pairing of electrons can also be studied in an analogous way as described below.

In solving Equation (6.74), it is assumed that the energy eigenvalues of two electronic polarons are already derived in an analogous way as for the hole polaron [Equation (6.68)]. Our interest here is to calculate the energy eigenvalue of the excited state with two holes localized on the same bond, i.e., $X = 0$. In this case the secular equation [Equation (6.74)] reduces to:

$$W_{2h(x,x',0)}C_l = \left[2W_e^0 - 2E_h^l + 2A_l^h x_l + \frac{1}{2}\sum_m M\omega_m^2 x_m^2 + U_{12}'\right]C_l + 2T_h(C_{l+1} + C_{l-1}) \qquad (6.75)$$

where the subscript $2h$ on $W_{2h}(x, x', 0)$ denotes the localization of the two holes and W_e^0 is the energy of an electronic polaron derived in an analogous way as in Equation (6.68) for a hole polaron and is obtained as:

$$W_e^0 = E_e^0 - 2T_e - E_{eq} \qquad (6.76)$$

Following the steps used to derive Equations (6.65) and (6.66), here again we can minimize the energy with respect x_l to get:

$$W_{2h}^0 = 2W_e^0 - 2E_h^0 + 4T_h + U_h - E_{hh} = 2(E_e^0 - E_h^0) - 4(T_e - T_h) + U_{12}' - 2E_{eq} - E_{hh} \quad (6.77)$$

where

$$E_{hh} = \frac{1}{6T_h}\left(\frac{(A_q^h)^2}{M\omega^2}\right)^2 \quad (6.78)$$

The bond length of a bond on which such a pairing occurs becomes twice as large as when only a single-hole polaron is formed and is given by:

$$x_q^{hh} = \frac{2A_q^h C_q^* C_q}{M\omega^2} = 2x_q^0 \quad (6.79)$$

The energy of two excitations without the charge carrier–phonon interaction can be written as:

$$W_{2h}^{00} = 2(E_e^0 - E_h^0) - 4(T_e - T_h) + U_{12}' \quad (6.80)$$

Thus, the energy of a pair of excitations with two holes localized on a bond is lowered by ΔE given as:

$$\Delta E = W_{2h}^{00} - W_{2h}^0 = 2E_{eq} + E_{hh} \quad (6.81)$$

In an analogous way one can derive the energy eigenvalue, W_{2e}^0, of a pair of excited electrons localized on an antibonding orbital of a bond and two hole polarons localized separately elsewhere as:

$$W_{2e}^0 = 2(E_e^0 - E_h^0) - 4(T_e - T_h) + U_{12}' - 2E_{hq} - E_{ee} \quad (6.82)$$

where E_{ee} can be obtained from Equation (6.78) by replacing the subscript h by e. It may be noted here that unlike the case of pairing of holes on a bond, pairing of electrons on a bond does not break the bond. In this case, the two-hole polaron's energy states move upward in the bandgap and the paired-electron energy states in the conduction band move downward, resulting in a narrowing of the bandgap.

We have shown above that the localization of an excited hole on a bond increases its bond length and the bond can break. Such a localization occurs by the formation of a hole polaron due to strong interaction between the hole and lattice vibrations. The hole-polaron state has an energy lower than the free-hole state and moves upward, mixing with the lone-pair orbitals in chalcogenides that widens the valence band and narrows the bandgap. Such a strong charge carrier–phonon interaction is possible in a-Chs because of their linear flexible structure and weak coordination, which can also induce pairing of excited holes on a bond. In this case the bond length of a bond increases twice as much as in the case of polaron formation and their binding energy is eight times larger than the polaron binding energy. Such paired-hole states contribute significantly to both PD and PVE. PD is caused by the lowering of the paired-hole state energy and such a state widens the valence band even further and hence reduces the bandgap. As the bond length expands twice as much as in the case of a single-hole polaron, this contributes significantly to volume expansion as well. Moreover, pairing of holes on a bond breaks the bond because of the removal of covalent electrons and causes photo-induced bond breaking in a-Chs, as has already been established.

Let us arrive at some estimates of the possible reduction in the bandgap due to formation of polarons and bipolarons (paired holes) on a bond. Depending on the material, ΔE [Equation (6.81)] can be in the region of a fraction of an eV. We have estimated ΔE in As_2S_3 as follows. The energy of lattice vibration of a bond can be written as [6.5]:

$$E(q) = E_0 + \frac{1}{2} M\omega^2 (q - q_0)^2 \tag{6.83}$$

where E_0 and q_0 are the energy and the interaction coordinate, respectively, at the minimum of the vibrational energy. The vibrational force along the interaction coordinate can be obtained as:

$$A = \left(\frac{\partial E}{\partial q}\right)_{q=0} = -M\omega^2 q_0 \tag{6.84}$$

Using this in Equations (6.69) and (6.78), we obtain:

$$\frac{E_{hh}}{T_h} = 8\frac{E_{hp}}{T_h} = \frac{1}{6}\left(\frac{M\omega^2 q_0^2}{T_h}\right)^2 \tag{6.85}$$

For As_2S_3, the phonon energy of the symmetric mode is $344\,cm^{-1}$ [6.4]. Using this and applying Toyozawa's criterion [6.67] of strong carrier–phonon interaction as $E_{hp} \geq T_h$, we get $T_h = 12.33\,meV$ from Equation (6.85), which gives $E_{hp} = 12.33\,meV$, $E_{hh} = 98.40\,meV$, and $\Delta E = 0.123\,eV$. This agrees very well with the observed reduction in the bandgap in As_2S_3 of about $0.16\,eV$ [6.5, 6.68]. A similar narrowing in the bandgap is expected when two excited electrons become paired on an anti-bonding orbital and two hole polarons are formed elsewhere. In this case, however, as no bond breaking may occur, the substance will go back to the original state after the exciting energy source (illumination) is stopped. In materials where the excited pairs of charge carriers form only a pair of polarons, without any pairing of like-charge carriers, the reduction in bandgap will be equal to $E_{ep} + E_{hp} \sim 2E_{hp} = 25\,meV$. Reductions in the bandgap in various materials, estimated from experimental data [6.68], are listed in Table 4; accordingly, most observed reductions are found to be in the range of 0.02–$0.17\,eV$.

Thus, it is shown above that in materials with strong carrier–lattice interaction, the excited charge carriers can form polarons as well as like-excited-charge carriers can become paired as self-trapped bipolarons on a bond because energetically a paired like-charge carrier excited state is more stable. The planar structure of chalcogenides with weak coordination enables these materials to be more flexible and hence possess stronger carrier–phonon interaction. The energy states of both polarons and bipolarons are lower than that of excited free-charge carriers. Thus the energy of a hole polaron and bipolaron (paired holes) moves up further in the lone-pair orbitals and -tail states, which expands the valence band. Likewise, the energy

Table 6.4 Bandgap reduction, ΔE estimated from the observed data [6.68] in various amorphous materials

Amorphous material	ΔE (eV)
As_2S_3	0.16
GeS_2	0.17
S	0.12
As_2Se_3	0.07
Se	0.06
$GeSe_2$	0.03
As_2Te_3	0.06
Sb_2S_3	0.02

states of electron polaron and bipolaron (paired electrons) in the anti-bonding orbitals lower the conduction mobility edge down. These effects together are responsible for the reduction in the bandgap and hence PD. It should be noted that all the above possibilities might not occur together in the same material. In materials where the electron–lattice interaction is larger than the hole–lattice interaction, the formation of electron polaron and bipolaron will be more efficient and materials with stronger hole–lattice interaction will have hole polaron and bipolaron formation more efficient. Accordingly, a varying degree of PD is expected to occur in different materials and this is quite in agreement with the results listed in Table 4.

It has been established [6.5] that pairing of holes on a bond breaks the bond as soon as two excited holes become localized on it. It is also possible for a bond to be broken due to localization of a single hole on a bond. The bond breaks due to the removal of covalent electrons, and a pair of dangling bonds is created. This is the essence of the pairing-hole theory of creating light-induced defects in a-Chs. Such light-induced defects are reversible by annealing. However, pairing of excited electrons does not break a bond, it only reduces the bandgap, and such an excited state will revert to the original material after the illumination has been switched off. This concept can be applied to explain both metastable and transient PD. The former occurs due to either formation of hole polarons or pairing of holes or both such that bonds are broken, and which cannot be recovered by stopping the illumination. It remains metastable and the material reverses back to its original form by thermal annealing. The latter occurs due to pairing of electrons and/or formation of polarons without any bond breaking and then the material reverses back to its original form after the illumination is stopped. Usually, transient PD is observed more than is metastable PD. This is because there are three processes contributing to the transient PD; pairing of electrons, formation of positive-charge polarons (without bond breaking), and negative-charge polarons, in comparison with only two possible channels of formation of positive-charge polarons (with bond breaking) and pairing holes contributing to metastable PD.

6.4 CONCLUSIONS

It is demonstrated that the effective-mass approach can be applied to amorphous structures to understand many electronic and optical properties that are based on the free-carrier concept. Using the effective-mass approximation, it is shown that the excitation-density-independent PL observed in a-semiconductors arises from the radiative recombination of excitons, types I and II geminate pairs, and nongeminate pairs. Both the Stokes shift and radiative lifetimes should be taken into account in determining the PL electronic states. Although the radiative lifetime for possibilities (i)–(iii) are of the same order of magnitude, the Stokes shift observed in PL suggest that these recombinations occur from extended-to-tail states [(possibility (ii) in a-Si:H. The singlet radiative lifetime is found to be in the ns time range at temperatures >15 K and is μs at 3.7 K, and triplet lifetime in the ms range. In possibility (iv), as carriers have to tunnel to a distance equal to the excitonic Bohr radius, the radiative lifetime is prolonged. The PL from possibility (iv) can arise from two types of geminate pairs, excitonic (type II) and nonexcitonic (type I), and nongeminate or distant pairs.

It is also shown that the effective mass of a charge carrier changes in amorphous semiconductors as it crosses its mobility edge. This also influences the radiative lifetime of an exciton as it crosses the mobility edges. A large Stokes shift implies a strong carrier–lattice interaction in a-Si:H and therefore PL occurs in thermal equilibrium. Results of two-band approximation and nonequilibrium are therefore not applicable for a-Si:H.

Finally, using the effective mass-approximation, it is shown quantum mechanically that it is the strong carrier–lattice interaction in a-Chs that causes all three phenomena; creation of the light-induced defects, PD, and VE. The results of the present theory agree qualitatively with those obtained from the molecular dynamics simulation (see Chapter 7).

ACKNOWLEDGEMENTS

The author has benefited very much from discussions with Professors T. Aoki, K. Shimakawa, Keiji Tanaka, and Sandor Kugler during the course of this work. The work is supported by the Australian Research Council's large grants (2000–2003) and IREX (2001–2003) schemes, and a bilateral exchange grant (2005/06) by the Australian Academy of Science and the Japan Society for the Promotion of Science (JSPS).

REFERENCES

[6.1] N.F. Mott and E.A. Davis, *Electronic Processes in Non-crystalline Materials* (Clarendon Press, Oxford, 1979).

[6.2] R.A. Street, *Hydrogenated Amorphous Silicon* (Cambridge University Press, Cambridge, 1991).

[6.3] D. Redfield and R.H. Bube, *Photoinduced Defects in Semiconductors* (Cambridge University Press, Cambridge, 1996).

[6.4] K. Morigaki, *Physics of Amorphous Semiconductors* (World Scientific, London, 1999).

[6.5] J. Singh and K. Shimakawa, *Advances in Amorphous Semiconductors* (Taylor and Francis, London, 2003).

[6.6] R.A. Street, *Adv. Phys.*, **30**, 593 (1981).

[6.7] S. Kivelson and C.D. Gelatt, Jr., *Phys. Rev. B*, **26**, 4646 (1982).

[6.8] B.A. Wilson, P. Hu, T.M. Jedju, and J.P. Harbison, *Phys. Rev. B*, **28**, 5901 (1983).

[6.9] D.G. Stearns, *Phys. Rev. B*, **30**, 6000 (1984).

[6.10] J. Singh, T. Aoki, and K. Shimakawa, *Philos. Mag. B*, **82**, 855 (2002).

[6.11] B.A. Wilson, P. Hu, J.P. Harbison, and T.M. Jedju, *Phys. Rev. Lett.*, **50**, 1490 (1983).

[6.12] T. Aoki, S. Komedoori, S. Kobayashi, C. Fujihashi, A. Ganjoo, and K. Shimakawa, *J. Non-Cryst. Solids*, **299–302**, 642 (2002). The group has recently notified that an error was made in measuring the E_{mx} in this paper and that it has been corrected in ref. [6.20].

[6.13] S.P. Deppina and D.J. Dunstan, *Philos. Mag. B*, **50**, 579 (1984).

[6.14] F. Boulitrop and D.J. Dunstan, *J. Non-Cryst. Solids*, **77–78**, 663 (1985).

[6.15] R. Stachowitz, M. Schubert, and W. Fuhs, *J. Non-Cryst. Solids*, **227–230**, 190 (1998).

[6.16] S.P. Deppina and D.J. Dunstan, *Philos. Mag. B*, **50**, 579 (1984).

[6.17] S. Ishii, M. Kurihara, T. Aoki, K. Shimakawa, and J. Singh, *J. Non-Cryst. Solids*, **266–269**, 721 (1999).

[6.18] T. Aoki, S. Komedoori, S. Kobayashi, T. Shimizu, A. Ganjoo, and K. Shimakawa, *Nonlinear Optics*, **29**, 273 (2002).

[6.19] T. Aoki, *J. Mat. Sci. – Mater. Eng.*, **14**, 697 (2003).

[6.20] J. Singh and I.-K. Oh, *J. Appl. Phys.*, **97**, 063516 (2005).

[6.21] E.A. Davis, in *Amorphous Semiconductors*, Edited by M.H. Brodsky, 2nd Edition (Springer, Berlin, 1985), Ch. 3, p. 41.

[6.22] D.L. Staebler and C.R. Wronski, *Appl. Phys. Lett.*, **31**, 527 (1977).

[6.23] M. Stutzmann, *Philos. Mag. B*, **56**, 63 (1987).

[6.24] H.M. Branz, *Phys. Rev. B*, **59**, 5498 (1999).

[6.25] K. Shimakawa, A. Kolobov, and S.R. Elliott, *Adv. Phys.*, **44**, 475 (1995).

[6.26] H. Hisakuni and K. Tanaka, *Appl. Phys. Lett.*, **56**, 2925 (1994).

[6.27] K. Shimakawa, N. Yoshida, A. Ganjoo, Y. Kuzukawa, and J. Singh, *Philos. Mag. Lett.*, **77**, 153 (1998).

[6.28] A. Ganjoo, K. Shimakawa, K. Kitano, and E.A. Davis, *J. Non-Cryst. Solids*, **299–302**, 917 (2002).

[6.29] D.J. Dunstan and F. Boulitrop, *Phys. Rev. B*, **30**, 5945 (1984).

[6.30] B.I. Shklovskii, H. Fritzsche, and S.D. Baranovskii, *Phys. Rev. Lett.*, **62**, 2989 (1989).

[6.31] E.I. Levin, S. Marianer, and B.I. Shklovskii, *Phys. Rev. B*, **45**, 5906 (1992).

[6.32] J. Singh, presented at the 21st International Conference on Amorphous and Nanocrystalline Semiconductors, ICANS21, Lisbon, Portugal, 4–9 September 2005, to be published in *J. Non-Cryst. Solids* (2006, in press).

[6.33] J. Singh, *J. Non-Cryst. Solids*, **299–302**, 444 (2002).

[6.34] J. Singh, *Nonlinear Optics*, **29**, 111 (2002).

[6.35] J. Singh, *J. Mater. Sci.*, **14**, 171 (2003).

[6.36] G.D. Cody, in *Semiconductors and Semimetals*, vol. 21, part B (1984), p. 11.

[6.37] P.K. Basu, *Theory of Optical Processes in Semiconductors* (Clarendon Press, Oxford, 1997).

[6.38] H.T. Grahn, *Introduction to Semiconductor Physics* (World Scientific, Singapore, 1999).

[6.39] W. Dumke, *Phys. Rev.*, **105**, 139 (1957).

[6.40] G. Lasher and F. Stern, *Phys. Rev. A*, **133**, 553 (1964).

[6.41] H.B. Beb and E.W. Williams, in *Semiconductors and Semimetals*, Edited by R.K. Willardson and A.C. Beer (Academic Press, London, 1972), Vol. 8, p. 18.

[6.42] J. Singh, *Excitation Energy Transfer Processes in Condensed Matter* (Plenum, New York, 1994).

[6.43] W. Van Roosbroeck and W. Shockley, *Phys. Rev.*, **94**, 1558 (1954).

[6.44] S.R. Elliott, *The Physics and Chemistry of Solids* (John Wiley & Sons Ltd, Chichester, 1998).

[6.45] J. Shah and R.C.C. Leite, *Phys. Rev. Lett.*, **22**, 1304 (1969).

[6.46] S. Kivelson and C.D. Gelatt, Jr., *Phys. Rev. B*, **19**, 5160 (1979).

[6.47] L. Ley, in: *The Physics of Hydrogenated Amorphous Silicon II*, Edited by J.D. Joanpoulos and G. Lucovsky (Springer-Verlag, Berlin, 1984), p. 61.

[6.48] W.E. Spear, in *Amorphous Silicon and Related Materials*, Edited by H. Fritzsche (World Scientific, Singapore, 1988).

[6.49] T. Aoki, T. Shimizu, S. Komedoori, S. Kobayashi, and K. Shimakawa, *J. Non-Cryst. Solids* (2006) in press.

[6.50] T. Aoki, T. Shimizu, D. Saito, and K. Ikeda, *J. Optoelectron. Adv. Mat.*, **7**, 137 (2005).

[6.51] T. Aoki, presented at the 21st International Conference on Amorphous and Nanocrystalline Semiconductors, ICANS21, Lisbon, Portugal, 4–9 September 2005, to be published in *J. Non-Cryst. Solids* (2006, in press).

[6.52] K. Tanaka, *Phys. Rev. B*, **57**, 5163 (1998).

[6.53] K. Tanaka, *J. Non-Cryst. Solids*, **35–36**, 1023 (1980).

[6.54] K. Tanaka, *Rev. Solid State Sci.*, **4**, 641 (1990).

[6.55] K. Tanaka, in *Handbook of Advanced Electronic and Photonic Materials*, Edited by H.S. Nalwa (Academic Press, San Diego, 2001), p. 119.

[6.56] S.R. Elliott, *J. Non-Cryst. Solids*, **81**, 71 (2001).

[6.57] A.V. Kolobov, H. Oyanagi, K. Tanaka, and K. Tanaka, *Phys. Rev. B*, **55**, 726 (1997).

[6.58] K. Shimakawa, *J. Optoelectron. Adv. Mater*, **7**, 145 (2005).

[6.59] E.V. Emelianova, G.J. Adriaenssens, and V.I. Arkhipov, *Philos. Mag. Lett.*, **84**, 47 (2004).

[6.60] J. Hegedüs, K. Kohary, D.G. Pettifor, K. Shimakawa, and S. Kugler, *Phys. Rev. Lett.*, **95**, 206803 (2005).

[6.61] M.I. Klinger, V. Halpern, and F. Bass, *Phys. Status Solids. B*, **230**, 39 (2002).

[6.62] J. Singh, *Chem. Phys. Lett.*, **149**, 447 (1988).

[6.63] P.W. Anderson, *Phys. Rev. Lett.*, **34**, 953 (1975).

[6.64] T. Holstein, *Ann. Phys.*, **8**, 325 (1959).

[6.65] J. Singh and I.-K. Oh, *J. Non-Cryst. Solids*, **351**, 1582 (2005).

[6.66] J. Hegedüs, K. Kohary, and S. Kugler, *J. Opto-electro. Adv. Mater.*, **7**, 59 (2005).

[6.67] Y. Toyozawa, *J. Phys. Soc. Jpn.*, **50**, 1861 (1981).

[6.68] K. Tanaka, *J. Non-Cryst. Solids*, **59–60**, 925 (1983).

7 Light-induced Volume Changes in Chalcogenide Glasses

S. Kugler[a,b], J. Hegedüs[a,c], and K. Kohary[d]

[a]*Department of Theoretical Physics, Budapest University of Technology and Economics, H-1521 Budapest, Hungary*
[b]*Department of Electronics and Computer Engineering, Tokyo Polytechnic University, Japan*
[c]*Department of Theoretical Physics and Material Sciences Center, Philipps University Marburg, Renthof 5 D-35032 Marburg, Germany*
[d]*Department of Materials, University of Oxford, Parks Road, Oxford OX1 3PH, UK*
Corresponding author's e-mail: kugler@eik.bme.hu

7.1. INTRODUCTION

Chalcogenide glasses are disordered solids, which contain a considerable amount of chalcogen atoms (S, Se, and Te). In the stable condensed phases chalcogenide elements form covalent bonds with two nearest neighbors in accordance with the $8 - N$ rule. These atoms have six electrons in the outermost shell with a configuration of s^2p^4. Electrons in s states do not participate in bonding since these states have energies well below those of the p states. Two covalent bonds are formed between chalcogen atoms by two p electrons, but the other electron pair – also called a lone pair (LP) – remains unbonded.

The structure of pure amorphous selenium (a-Se) – a representative chalcogenide material – can be described as a random mixture of rings and helical chains accompanied by coordination defects. The trigonal phase consists of infinite helical chains, and the monoclinic phase's main unit is an eight-membered selenium ring. In these structures the bond length is close to 2.38 Å, the bond angle 100°, and the dihedral angle 102°, which remain mainly preserved also in the amorphous phase. Beyond the disorder introduced by the random arrangement of chains and rings in a-Se the one- and three-fold coordination defects (C1 and C3) also play a decisive role. The average coordination number, however, remains still close to two. According to the Phillips–Thorpe model a-Se is considered to have an underconstrained network due to its average coordination number being two. In an amorphous structure where coordination defects modify the continuous random network the presence of dangling bonds is expected due to the existing unpaired electrons. However, no electron-spin resonance (ESR) signal has been observed in a-Se. A possible explanation of this phenomenon is that the defects are not neutral: one-fold coordinated atoms are negatively charged (C1) and three-fold coordinated atoms are positively charged (C3), which are called valence alternation pairs (VAPs).

Chalcogenide glasses exhibit various changes in structural and electronic properties during bandgap illumination, like photo-induced volume change, photodarkening, and photo-induced change in the phase state (photo-crystallization and photo-amorphization). A size effect can be observed: photodarkening cannot be induced in As_2S_3 films which are thinner than 50 nm [7.1]. These phenomena do not occur in the crystalline chalcogenides nor in any other amorphous semiconductors. The microscopic structural changes are facilitated by two factors common to chalcogenide glasses: the low average coordination number and the structural freedom of the noncrystalline state. During the illumination some of the films can expand (a-As_2S_3, a-As_2Se_3, etc.), and some shrink (a-GeS_2, a-$GeSe_2$, etc.) [7.2]. Several investigations have been carried out in order to provide an explanation of these phenomena [7.3–7.8]. It has been established that there is a configurational rearrangement with changes in atomic coordination in the vicinity of the excitation [7.5–7.7]. In a simplistic model such changes in the local bonding environment were explained by the formation of VAPs [7.5]. In amorphous selenium, the model material of chalcogenide glasses, the formation of new inter-chain bonds has also been suggested [7.6]. However, an atomistic study of amorphous selenium has revealed that the structural rearrangements are less local than used in such simple models and has given evidence that further possible bond formations and bond breakings are responsible for photo-induced effects [7.7]. We proposed a simple, unified description of the photo-induced volume changes in chalcogenides based on tight-binding (TB) molecular dynamics (MD) simulations of amorphous selenium [7.8]. We have found that the microscopic rearrangements in the structure (like bond breaking and bond formation) are responsible for the macroscopic volume change under illumination. The first *in situ* surface-height measurement [7.9] on amorphous selenium was carried out recently and supports our proposed mechanism. Recently, two excellent books appeared on this topic [7.10, 7.11].

The layout of this chapter is as follows: Section 2 gives an overview on our MD computer code. The details of high-quality, void-free sample preparation using our code can be found in Section 3. The two subsequent sections contain our microscopic and macroscopic descriptions of photo-induced volume changes.

7.2 SIMULATION METHOD

We have recently developed a molecular dynamics (MD) computer code (ATOMDEP program package) to simulate the preparation procedures of real amorphous structures (growth by atom-by-atom deposition on a substrate and rapid quenching) [7.12–7.16]. A standard velocity Verlet algorithm was applied in our MD simulations in order to follow the atomic-scale motions. To control the temperature we applied the velocity-rescaling method. We chose $\Delta t = 1$ fs or 2 fs for the time step, depending on the temperature. In our works the growth of amorphous carbon [7.12], silicon [7.13, 7.14], and selenium [7.15, 7.16] films were simulated by this MD method. This computer code is convenient for investigating photo-induced volume changes as well if the built-in atomic interaction can handle the photoexcitation.

For calculating the interatomic forces in a-Se we used tight-binding (TB) [7.16, 7.17] and self-consistent field tight-binding (SCF-TB) [7.16, 7.18] models. TB parameterization [7.17] has been introduced for selenium following the techniques developed by Goodwin et al. [7.19]. It was thoroughly tested by MD calculations in liquid and amorphous phases and the results were compared to experiments and to *ab initio* calculations. The agreement with experiments and *ab initio* calculations is rather good apart from the fact that the number of coordination defects in the solid and liquid phases is higher than the experimentally measured values. The authors have improved their TB Hamiltonian by including the Hubbard correction [7.18]. This implies that either the algorithm has to be made self-consistent or the perturbation theory must be applied. Our choice was the first alternative. For a reasonable accuracy only a few SCF and MD steps were needed. Convergence criteria were considered to be satisfied if the deviation of atomic charges between the actual and the previous iterations was less than 0.01 electron/atom. First, we constructed a tight-binding Hamiltonian, and then diagonalized it. After obtaining a solution, we added the Hubbard terms and recalculated the Hamiltonian matrix. The procedure was repeated until the necessary convergence was reached. The Hamiltonian matrix changed only slightly in one MD step, due to the small atomic movements. Therefore, we could use the eigenvectors from the previous MD step as the starting point in the self-consistency cycle. A convergence problem occurred when we tried to use the SCF method. The solution oscillated, and did not converge at the optimized value of 0.875 eV for the Hubbard parameter. To handle this issue we introduced a damping in the SCF cycles by linearly combining the new solution with the previous one. This method slowed down the convergence speed, but eliminated the oscillations as well.

To test our computer code we performed two test runs [7.20]. Crystalline forms of selenium consist of chains and eight-membered rings. It is very likely that these local arrangements can be found in noncrystalline forms of selenium as well. The initial configuration of the eight-membered ring in our simulation had bond lengths of 2.38 Å and bond angles of 102°. Dihedral angles were 100°. For the eighteen-membered selenium chain (with 1-dimensional periodic boundary condition) these values were: 2.36 Å, 100°, and 98°, respectively. Every Se atom had two first-neighbors, i.e. there was no coordination defect. Before illumination (photoexcitation) the individual ring and the chain were relaxed for 4 ps at $T = 500$ K. During this period the structures were stable. When a photon was absorbed an electron from the highest occupied molecular orbital (HOMO) was transferred to the lowest unoccupied molecular orbital (LUMO). This is a simple model of photoexcitation when an electron is shifted from the valence band to the conduction band (electron–hole pair creation). After excitation one bond length in the ring started to increase and bond-breaking occurred. A similar result was published [7.21] by a Japanese group for S_8. They performed MD simulation within the framework of density functional theory in the local density approximation. In our second MD simulation we investigated the linear chain structure. The same procedure was performed to model the excitation. A very similar result was obtained; a bond inside the chain was broken immediately after a HOMO electron was excited. Two snapshots of this process can be seen in Figure 7.1.

7.3 SAMPLE PREPARATION

To mimic the thin-film structures, we fabricated glassy networks, for which we applied periodic boundary conditions (PBC) in two dimensions [x,y]. The samples were open in the z direction. When we illuminated the cell, it could expand or shrink into the open direction. The volume changes in the sample can be derived by measuring the distance between atoms at the two open ends. The initial simulation-cell geometry was a rectangular box of size $12.78 \times 12.96 \times 29.69$ (*xyz* in Å). The 162-atom sample had an initial density of $4.33\,g/cm^3$.

We prepared our samples from the liquid phase by rapid quenching. Our 'cook and quench' sample preparation procedure was as follows: first we set the temperature of the system to 5000 K for the first 300 MD steps. During the following 2200 MD steps, we decreased linearly the temperature from 700 K to 250 K, driving the sample through the glass transition and reaching the condensed phase. Then we set the final temperature to 20 K and relaxed the sample for 500 (1 ps) MD steps. The closed box was opened in the z-direction at the 3000th MD step. We thus obtained two surfaces with an increased number of one-fold coordinated atoms and increased potential energy. This final topology corresponded to a thin-film structure. One problem remained: the localized vibration modes at the surface were excited by the opening procedure. This caused an inhomogeneity in the temperature distribution: the sample had higher temperatures at the ends. Therefore, we homogenously redistributed the atomic kinetic energies according to the Maxwell–Boltzmann distribution, to speed up the thermalization process. We did this three times at the 4000th, 6000th, and 7000th MD steps. The Hubbard parameter, U, was also changed

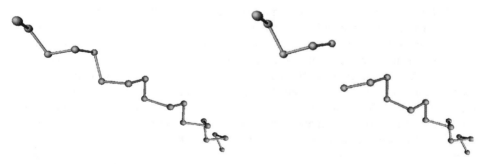

Figure 7.1 Left panel: a snapshot of an 18-atom linear chain at room temperature. Right panel: the same linear chain after excitation

during quenching. By increasing U, one can expect a greater tendency to form a nearly fully two-fold coordinated structure, which is claimed to be the situation for selenium. Therefore, we set the Hubbard parameter at 5 eV during the quenching in the first 4000 MD steps. After the opening procedure was completed, during the relaxation phase at the 4000th MD step we restored U from 5 eV to 0.875 eV, which is the optimized value according to Lomba et al. [7.18]. The system was relaxed for at total of 40 000 MD steps (80 ps) at 20 K.

The most important steps of the preparation procedure can be seen in Figures 7.2–7.5. The initial structure was a bulk glassy selenium network. As in the first phase (0–0.6 ps) the temperature was set at 5000 K in the first MD step, and the potential energy abruptly increased from −540 eV to −480 eV (Figure 7.3). This means that the initial bonding

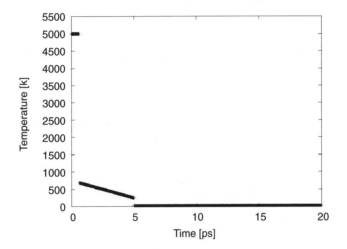

Figure 7.2   Temperature as a function of time during the sample preparation

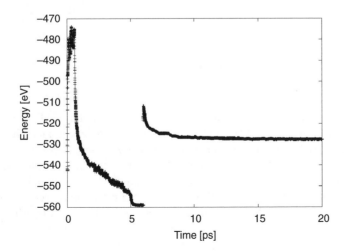

Figure 7.3   A typical curve of the potential energy of the sample vs time during preparation

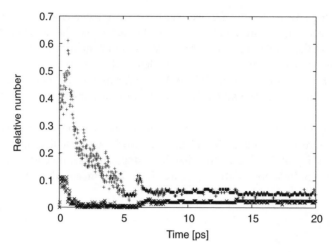

Figure 7.4   Time development of the relative numbers of 3-fold (×) and 1-fold (+) coordinated atoms

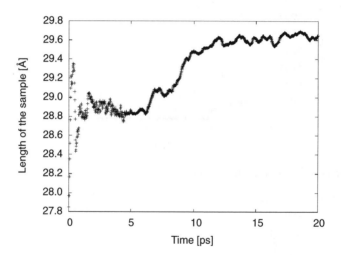

Figure 7.5   Length of the sample during preparation

topology was completely destroyed since at high temperatures all the bonds break immediately. Here, two atoms were considered bonded when the bond length between them was less than 2.7 Å. The relative number of one-fold coordinated atoms was about 40% and that of the three-fold coordinated atoms was 10% during the first 0.6 picoseconds. At 0.6 picoseconds, when the temperature was set to be about 700 K, this situation changed drastically, the percentage of one-fold and three-fold coordinated atoms decreased to 20–25% and 1–2%, respectively. At this point the potential energy decreased by 60 eV within less than 0.1 ps. In the second phase (0.6–5 ps), as we reduced the temperature further down from 700 K to 250 K the number of one-fold coordinated atoms decreased from 20% to 5% but the number of three-fold coordinated atoms did not change (See Figure 7.4).

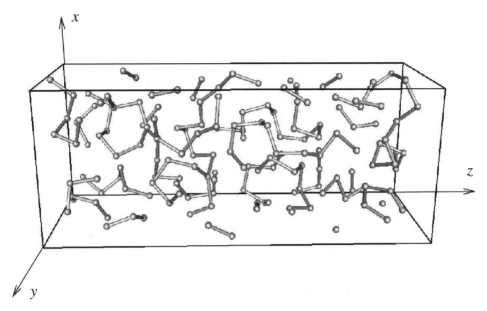

Figure 7.6 Snapshot of a final glassy selenium network. The sample can expand in the z direction, since at the ends we do not apply periodic boundary conditions

The cooling rate during this phase was about 10^{14} K/s. At 5 ps, another sudden decrease by 5 eV in the potential energy (Figure 7.3) was observed when the temperature was reduced from 250 K to 20 K. After this the system was equilibrated for another 1 ps and then the network was opened by releasing the PBC in the z-direction. Examining the potential energy (Figure 7.3), an immediate increase from −560 eV to −510 eV can be seen. This corresponds to breaking approximately 10–20 bonds at the surface. Subsequently, a quick recovery occurred, involving about a 5–10 eV decrease in the potential energy in 0.5 picoseconds and then another 5–10 eV decrease on a longer time scale (10–100 picoseconds). The first corresponds to the formation of new bonds, as one sees in the change of coordination numbers (Figure 7.4), while the second corresponds to a large-scale structural change, i.e. volume expansion (Figure 7.5). During the last five picoseconds the sample became stable and the volume no longer changed significantly. If we changed the initial velocities of the atoms only during the first MD step we fabricated topologically different glassy networks under similar physical conditions. For a 162-atom system, one SCF step on a 1200 MHz computer took about 2–3 seconds. The visualization of the structures and of the time development of the system was carried out by self-written Java software, JGLMOL, which is available free of charge [7.22].

Samples prepared at 20 K had densities from 3.95 to 4.19 g/cm^3. The number of coordination defects ranged from 3 to 12%. Most of these defects were located on the surfaces. The structure consisted mainly of branching chains, but some rings could also be found. The samples were accepted if the volume fluctuation was less than 0.5% in 60 picoseconds. We prepared altogether 30 samples, and 17 were considered to be stable and useful for further studies. One of them is displayed in Figure 7.6. A radial distribution function of a representative sample can be seen in Fig 7.7.

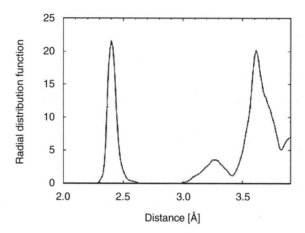

Figure 7.7 Radial distribution function of an amorphous selenium network as a function of atomic distances. The model was prepared by MD simulation at 20 K

7.4 LIGHT-INDUCED PHENOMENA

In amorphous selenium immediately after the absorption of a photon, an electron–hole pair became separated in space [7.8]. Therefore, they can be treated independently, i.e. we can investigate the roles of excited electrons and holes separately. We ran two sets of computer simulations: first, to model the excited electron creation we put an extra electron into the LUMO; and secondly, we annihilated an electron in HOMO (hole creation).

7.4.1 Electron excitation

A covalent bond between two-fold and three-fold coordinated atoms was broken (C2 + C3 ⇒ C1 + C2) in the majority of cases when an additional electron was put into the LUMO, as seen in Figure 7.8.

The bond-breaking significantly affects the bond lengths, alternations between shrinkage and elongation in the vicinity of the broken bond being displayed in Figure 7.8. Our localization analysis revealed that the LUMO was localized at this site before the bond-breaking as it can be seen in Figure 7.9. A release of excitation restores all bond lengths to their original value.

The time development of photo-induced bond-breaking due to an added electron and the corresponding volume expansion in one of our amorphous selenium sample is shown in Figure 7.10. We have selected one sample (Figure 7.10) from our simulations, which seems to be a typical run. Similar changes were observed in each amorphous selenium network. Before the excitation at 5 ps the bond length was about 2.55 Å. In this particular case bond-breaking occurred at a weaker bond due to the C3 site, which had a larger interatomic separation than the majority of the nearest-neighbor bonds of 2.4 Å. During the illumination, this weaker bond (2.55 Å) increased by 10–20% (in this example to 3 Å) and it decreases to its original value after the de-excitation. (Arrows show the excitation and de-excitations in Figure 7.10.) The volume change follows the bond-breaking and it shows damped oscillations on the picosecond time scale.

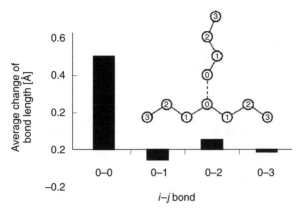

Figure 7.8   Bond breaking and bond-length changes due to electron addition in the LUMO of an amorphous selenium network. Here i and j refer to the numbered atoms in the Figure. The dark rectangles represent the average changes in bond lengths between i–j atoms

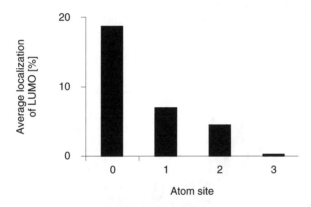

Figure 7.9   Average localizations of LUMO at the atoms 0, 1, 2, 3. See Figure 7.8 for notations

7.4.2 Hole creation

We observed that inter-chain bonds were formed after creation of a hole and they cause contraction of the sample (Figure 7.11). This contraction always appears near to atoms where the HOMO is localized. Since the HOMO is usually localized in the vicinity of a one-fold coordinated atom, the inter-chain bond formation often takes place between a one-fold coordinated atom and a two-fold coordinated atom ($C\{1,0\} + C\{2,0\} \Rightarrow C\{1,1\} + C\{2,1\}$, where the second number means the number of inter-chain bonds). However, sometimes we also observed the formation of inter-chain bonds between two two-fold C2 coordinated atoms ($C\{2,0\} + C\{2,0\} \Rightarrow C\{2,1\} + C\{2,1\}$).

In order to model the collective effect of photo-induced changes in amorphous selenium, we also performed simulations with five excited electron creations and five hole creations. We put five excited electrons from the five highest occupied energy levels (one electron from one level) to the five lowest unoccupied energy levels (again, one electron to each

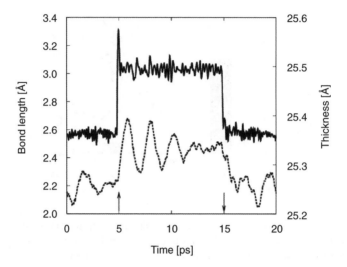

Figure 7.10 Bond length of broken bond (solid line) and thickness of sample (dotted line) during photoexcitation plotted as a function of time. Excitation starts at 5 ps and ends at 15 ps (arrows) [Reproduced from J. Hegedüs et al., *Phys. Rev. Lett.*, **95**, 206803. Copyright (2005) by the American Physical Society]

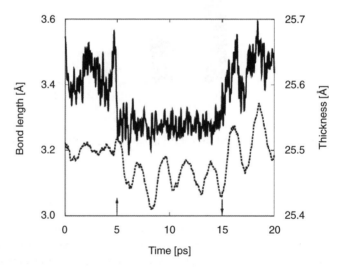

Figure 7.11 Photo-induced local contraction due to the addition of a hole in a-Se as a function of time. Distance between two atoms in different Se chains at sites where HOMO is localized (solid line). Thickness of sample is denoted with dashed line. Hole is created at 5 ps and annihilated at 15 ps [Reproduced from J. Hegedüs et al., *Phys. Rev. Lett.*, **95**, 206803. Copyright (2005) by the American Physical Society]

level). We found similar effects as described above for single electron/hole creation: bond-breakings and inter-chain bond formations have similar characteristics. Nevertheless, in the five-excited-electron creation case, further bond-breaking occurred not only at the C3 sites, but as well at some C2 sites. In the case of the five-hole creation, we observed that inter-chain bonds were formed between C1 and C2 sites and also between C2 and C2 sites.

The bond-breaking and inter-chain bond formation can be understood in terms of a change in the bond strength before and during the excitations. We calculated the bond energies [7.8, 7.23] within the TB representation. We obtained a decrease in the bonding energy of 0.24 eV after a bond-breaking for a typical case. In contrast, a hole addition leads to an increase in the inter-chain bond energy of 0.042 eV.

7.5 MACROSCOPIC MODELS

7.5.1 Ideal, reversible case (a-Se)

The light-induced volume expansion and volume shrinkage in amorphous selenium occur simultaneously and these are additive quantities as our molecular dynamics simulations have confirmed. The expansion of the thickness d_e is proportional to the number of excited electrons n_e ($d_e = \beta_e n_e$), while the shrinkage d_h is proportional to the number of created holes n_h ($d_h = \beta_h n_h$), where the parameters β_e and β_h are the average thickness changes caused by an excited electron and a hole, respectively. The time-dependent equation of thickness change can then be written as:

$$\Delta(t) = d_e(t) - d_h(t) = \beta_e n_e(t) - \beta_h n_h(t) \tag{7.1}$$

Assuming $n_e(t) = n_h(t) = n(t)$ we get

$$\Delta(t) = (\beta_e - \beta_h)n(t) = \beta n(t) \tag{7.2}$$

where β is a characteristic constant of different chalcogenide glasses related to photo-induced volume (thickness) change. The sign of this parameter governs whether the material shrinks or expands. The number of electrons excited and holes created is proportional to the duration time of illumination. Their generation rate G depends on the number of incoming photons and on the photon absorption coefficient. After the photon absorption, the separated excited electrons and holes migrate within the amorphous sample and then eventually recombine. A phenomenological equation for this dominant process can be written as:

$$dn_e(t)/dt = G - Cn_e(t)n_h(t) \tag{7.3}$$

where C is a constant. Using $n_e(t) = n_h(t) = n(t)$ and $\Delta(t) = \beta n(t)$, we obtain a fundamental equation for the time-dependent volume change, namely,

$$d\Delta(t)/dt = G\beta - (C/\beta)\Delta^2(t) \tag{7.4}$$

Solution of this nonlinear differential equation is obtained as:

$$\Delta(t) = \beta(G/C)^{1/2}\tanh\left[(GC)^{1/2}t\right] \tag{7.5}$$

Recently, the photo-induced expansion of amorphous selenium films was measured *in situ* for the first time using optoelectronic interference and enhanced by image processing [7.9]. Figure 7.12 shows the measured time evolution of the surface height over the interval of 0–300 s together with its best fit.

After the light is turned off ($G = 0$), Equation (7.4) can be written as:

$$d\Delta(t)/dt = -(C/\beta)\Delta^2(t) \tag{7.6}$$

with the solution

$$\Delta(t) = a/[a(C/\beta)t + 1] \tag{7.7}$$

Figure 7.13 displays the measured volume change and the fitted theoretical curve to the measured data. Light was switched off at $t = 800$ s.

7.5.2 Nonideal, irreversible case (a-As$_2$Se$_3$)

In the ideal case we assumed that each local structure variation was reversible and the original local structure was reconstructed after the electron–hole recombination. However, the

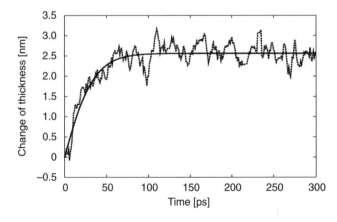

Figure 7.12   Time development of volume expansion in amorphous selenium (dashed line) and its theoretical best fit (solid line)

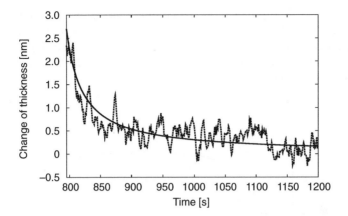

Figure 7.13   The measured shrinkage of a-Se (dashed line) and the fitted curve (solid line) after stopping the illumination

result of a measured volume change on flatly deposited a-As$_2$Se$_3$ film is quite different from the ideal selenium case (see Figure 2b in Ref. [7.9]). To explain the difference we must take into account a large number of irreversible changes in the local atomic arrangement, i.e. after turning off the light the local configuration remains the same and there is no electron–hole recombination. The total volume change includes both the reversible and irreversible changes and it can be written as:

$$\Delta_{\text{total}}(t) = \Delta_{\text{rev}}(t) + \Delta_{\text{irr}}(t) \tag{7.8}$$

The reversible part follows Equations (7.4) and (7.6) during and after the illumination, respectively, with the corresponding solutions given in Equations (7.5) and (7.7).

We now consider the irreversible component. During the illumination, the generation rate of irreversible microscopic change is time dependent. Let's consider that an upper limit exists for the maximum number of electrons and holes causing irreversible changes and let these denoted by $n_{\text{e,irr,max}}$ and $n_{\text{h,irr,max}}$, respectively. To simplify the derivation let us assume that $n_{\text{e,irr,max}} = n_{\text{h,irr,max}}$. In this case, one can write the electron generation rate as:

$$G_{\text{e}}(t) = C_{\text{e}}\left[n_{\text{e,irr,max}} - n_e(t)\right] \tag{7.9}$$

Note that there is no recombination term in Equation (7.9). Following Equation (7.4), we obtain that the irreversible expansion is governed by:

$$d\Delta_{\text{irr}}(t)/dt = G_{\text{irr}} - C_{\text{irr}}\Delta(t) \tag{7.10}$$

Equation (7.10) then leads to the solution:

$$\Delta_{\text{irr}}(t) = (G_{\text{irr}}/C_{\text{irr}})(1 - \exp\{-C_{\text{irr}}t\}) \tag{7.11}$$

Using Equation (7.11) in Equation (7.8), the best fit of the volume expansion [$\Delta_{total}(t)$], and that of the reversible and irreversible parts [$\Delta_{rev}(t)$, and $\Delta_{irr}(t)$] are displayed in Figure 7.14.

After illumination there is no volume change caused by the irreversible microscopic effects. Figure 7.15 shows the shrinkage after switching off the illumination, which is the fit as obtained in the reversible case.

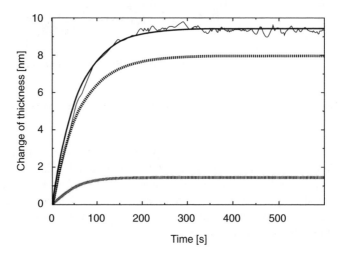

Figure 7.14 Time development of volume expansion of a-As$_2$Se$_3$. Thin solid line is the measured curve, thick solid line is the fitted line [$\Delta_{total}(t) = \Delta_{rev}(t) + \Delta_{irr}(t)$]. Lower dashed curve is the best fit of $\Delta_{rev}(t)$, while the upper one is that of the irreversible part $\Delta_{irr}(t)$

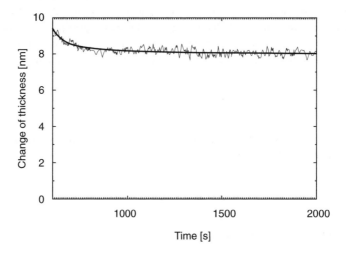

Figure 7.15 The measured decay (thin solid line) and the fitted theoretical curves (thick solid line) for the shrinkage as a function of time after stopping the illumination

7.6 CONCLUSIONS

We have proposed a new explanation for the photo-induced volume changes in chalcogenide glasses. We have found that the covalent bond-breaking occurs in these glasses with excited electrons, whereas holes contribute to the formation of inter-chain bonds. In the ideal situation both processes are reversible. The interplay between photo-induced bond-breaking and inter-chain bond formation leads to either volume expansion or shrinkage. In the nonideal case, only a part of the processes is irreversible and the total expansion includes the reversible and irreversible changes. Our microscopic explanation of the macroscopic photo-induced volume change is consistent with the first *in situ* surface-height measurements.

ACKNOWLEDGEMENTS

This work has been supported by the OTKA Fund (Grants No. T043231, T048699) and by Hungarian-British and Hungarian-Japanese intergovernmental Science and Technology Programmes (No. GB-17/03 and No. JAP-8/02). Simulations have been carried out using computer resources provided to us by the Tokyo Polytechnic University. We are indebted to Prof. Takeshi Aoki for this possibility. We would also like to acknowledge many valuable discussions with Prof. David Pettifor at the University of Oxford, Prof. Koichi Shimakawa (Gifu University. Japan), and Prof. Peter Thomas (Philipps University, Marburg, Germany).

REFERENCES

[7.1] K. Tanaka, N. Kyoya, and A. Odajima, *Thin Solid Films*, **111**, 195 (1984).

[7.2] Y. Kuzukawa, A. Ganjoo, and K. Shimakawa, *J. Non-Cryst. Solids*, **227–230**, 715 (1998); *Philos. Mag. B*, **79**, 249 (1999).

[7.3] K. Shimakawa, A. Kolobov, and S.R. Elliott, *Adv. Phys.*, **44**, 475 (1995).

[7.4] K. Shimakawa, N. Yoshida. A. Gahjoo, A. Kuzukawa, and J. Singh, *Philos. Mag. Lett.*, **77**, 153 (1998).

[7.5] H. Fritzsche, *Solid-State Commun.*, **99**, 153 (1996).

[7.6] A.V. Kolobov, H. Oyanagi, K. Tanaka, and K. Tanaka, *Phys. Rev. B*, **55**, 726 (1997).

[7.7] X. Zhang and D.A. Drabold, *Phys. Rev. Lett.*, **83**, 5042 (1999).

[7.8] J. Hegedüs, K. Kohary, D.G. Pettifor, K. Shimakawa, and S. Kugler, *Phys. Rev. Lett.*, **95**, 206803 (2005).

[7.9] Y. Ikeda and K. Shimakawa, *J. Non-Cryst. Solids*, **338**, 539 (2004).

[7.10] J. Singh and K. Shimakawa, *Advances in Amorphous Semiconductors* (Taylor and Francis, London and New York, 2003).

[7.11] *Photo-Induced Metastability in Amorphous Semiconductors*, Edited by A.V. Kolobov (Wiley-VCH, 2003).

[7.12] K. Kohary and S. Kugler, *Phys. Rev. B*, **63**, 193404 (2001).

[7.13] S. Kugler, K. Kohary , K. Kádas, and L. Pusztai, *Solid-State Commun.*, **127**, 305 (2003).

[7.14] K. Kohary and S. Kugler, *Mol. Simul.*, **30**, 17 (2004).

[7.15] J. Hegedüs, K. Kohary, and S. Kugler, *J. Non-Cryst. Solids*, **338**, 283 (2004).

[7.16] J. Hegedüs and S. Kugler, *J. Phys.: Condens. Matter,* **17,** 6459 (2005).

[7.17] D. Molina and E. Lomba, *Phys. Rev. B*, **60**, 6372 (1999).

[7.18] E. Lomba, D. Molina, and M. Alvarez, *Phys. Rev. B*, **61**, 9314 (2000).

[7.19] L. Goodwin, A.J. Skinner, and D.G. Pettifor, *Europhys. Lett.*, **9**, 701 (1989).

[7.20] J. Hegedüs, K. Kohary, S. Kugler, and K. Shimakawa, *J. Non-Cryst. Solids*, **338**, 557 (2004).

[7.21] F. Shimojo, K. Hoshino, and Y. Zempo, *J. Phys.: Condens. Matter*, **10**, L177 (1998).

[7.22] http://www.physik.uni-marburg.de/~joco/JGLMOL.html

[7.23] D. Pettifor, *Bonding and Structure of Molecules and Solids* (Clarendon Press, Oxford, 1995).

[7.24] J. Hegedüs, K. Kohary, and S. Kugler, *J. Non-Cryst. Solids*, **352**, 1587 (2006).

8 Optical Properties of Glasses

A. Edgar

School of Chemical and Physical Sciences, Victoria University, Box 600, Wellington,
New Zealand
e-mail: Andy.Edgar@vuw.ac.nz
Telephone 0064 4 463 5949
Fax 0064 4 463 5237

Optical Properties of Condensed Matter and Applications   Edited by J. Singh
© 2006 John Wiley & Sons, Ltd

8.1 INTRODUCTION

Historically, glass was first valued for jewellery and decoration, but as the last millenium developed it became apparent that glass had two key attributes which made it an especially valuable material: it could be worked into a variety of shapes by processes such as casting, drawing, moulding, polishing, and blowing, and it was transparent to visible radiation. This combination permitted the development of key technologies: manufacturing of containers with transparent walls, building and vehicle windows, and optical instruments and appliances such as spectacles, telescopes, cameras, and microscopes. It is difficult to appreciate from the context of the 21st century what a difference having a transparent, weatherproof, durable window material must have made to daily living in the middle ages, or to the ability to correct long- and short-sightedness with spectacles. In the 20th century, the discovery of methods of making low-loss silica optical fibre has revolutionized communications and computing technologies. In the 21st century, we can expect further developments based on these two key attributes of glass as a material: the optical transparency [which can be extended from the visible to the ultraviolet (UV) and the infrared (IR)], and the ability to shape or engineer a glass into an arbitrary geometry or structure. This chapter first outlines the scientific and technical background to optical glass and its applications before presenting a discussion of some topics of current interest. Sections 2–5 discuss the complex refractive index, its dispersion, and factors which affect it such as temperature, stress, chemical composition, and magnetic field. In sections 6 and 7 we describe the origins of color in glass, and the effects of doping with open-shell ions from the 3d and 4f series of the periodic table, before outlining the fluorescence properties of rare-earth-doped glasses. In section 8 we outline the optical fibre application of glass. Finally, in sections 9–12 we review current work on refractive index microengineering in glasses and transparent glass ceramics.

8.2 THE REFRACTIVE INDEX

Well annealed glass is generally an optically homogeneous (on the scale of the wavelength of light), nonmagnetic, and isotropic material. The behavior of an electromagnetic wave propagating through the glass can be described by a complex refractive index $n - ik$, or equivalently a complex dielectric constant $\varepsilon' - i\varepsilon''$, which are related through

$$(n - ik)^2 = \varepsilon' - i\varepsilon'' \tag{8.1}$$

or equivalently

$$\varepsilon' = n^2 - k^2 \tag{8.2}$$

and

$$\varepsilon'' = 2nk \tag{8.3}$$

In Figure 8.1 we show how the refractive index for a representative glass, silica, varies with wavelength, with data taken from a handbook of optical constants [8.1]. The key features

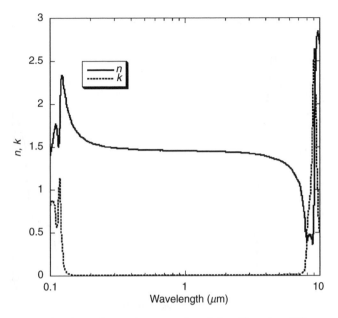

Figure 8.1   Complex refractive index (n, k) vs wavelength for silica glass SiO_2

are a broad region where n is almost (but not quite) wavelength independent, (the transmittance 'window' of the glass), bounded by a sharp rise in k at short wavelengths (the 'UV edge'), and a sharp rise in k at longer wavelengths (the IR edge).

It is shown in many texts [8.2] that for a plane monochromatic electromagnetic wave of frequency ω, polarized in the x direction and propagating through nonconducting glass in the z direction, the electric field strength E_x is given by

$$E_x(z) = E(0)\exp(-kz\omega/c)\exp[i\omega(t - nz/c)] \tag{8.4}$$

The wavelength in the glass is given by $\lambda = \lambda_0/n$, where λ_0 is the vacuum wavelength, and the amplitude of the wave decreases exponentially in the propagation direction, with an intensity decaying as $\exp(-\alpha z)$, where $\alpha = 2\omega k/c$ is called the absorption coefficient. In Figure 8.2, we show the absorption coefficient in silica glass as a function of wavelength.

Clearly, the abrupt rise in k at the two edges gives rise through $\alpha = 2\omega k/c$ to strong absorption (note the logarithmic scale in Figure 8.2) marking the perimeter of the transmission window. We also note that the well known absorption features at 1.39 μm and 2.7 μm due to hydroxyl impurities in silica glass are too small to appear in Figure 8.2, the 1.39 μm absorption peak being of the order of $10^{-5}\,cm^{-1}$ in high-purity SiO_2. Although these bands are very weak, over a long pathlength in the order of kilometers they and other effects (section 8) have a dominant influence on the choice of wavelength used for fibre-optic communications within the broad silica window. The sharp rise in k also contributes to strong reflections from glass surfaces at the window boundaries. For normal incidence, the intensity reflection coefficient for an air–glass interface is given by:

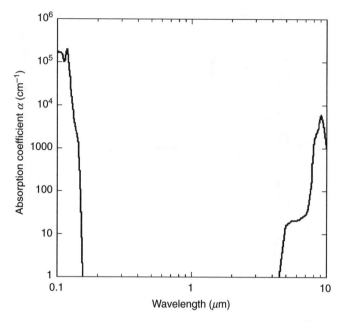

Figure 8.2   Absorption coefficient plotted as a function of wavelength for silica SiO_2

$$R = \left| \frac{(n-1)^2 + k^2}{(n+1)^2 + k^2} \right| \tag{8.5}$$

and in Figure 8.3 we show the effect of the edges on the reflection coefficient.

8.3 GLASS INTERFACES

The complex refractive index of glasses is dominated in the window of transparency by the real part. In that case, the reflection and transmission coefficients for normal incidence at the interface between two glass media are readily calculated from Equation (8.5) using the relative refractive index $m = n_2/n_1$ for the refractive indices for light in the medium of incidence (n_1) and transmission (n_2); the result is shown in Figure 8.4.

For light, which makes an angle of incidence θ_i with the normal to the interface, the reflection and transmission coefficients depend upon the polarization of the incident light. Light with its electric field vector polarized in the plane defined by the incident, and reflected ray, and the normal is referred to as p-polarized; the perpendicular polarization is called s-polarized. The reflection and transmission coefficients for intensity follow from Fresnel's equations and are given by

$$R_s = \left[\frac{-\cos\theta_i + m\cos\theta_t}{\cos\theta_i + m\cos\theta_t} \right]^2 , \quad T_s = 1 - R_s \tag{8.6}$$

$$R_p = \left[\frac{m\cos\theta_i - \cos\theta_t}{m\cos\theta_i + \cos\theta_t} \right]^2 , \quad T_p = 1 - R_p \tag{8.7}$$

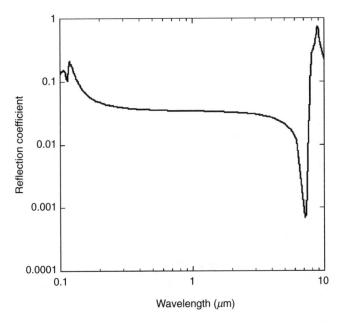

Figure 8.3 Reflection coefficient at normal incidence vs wavelength for an air/silica glass interface

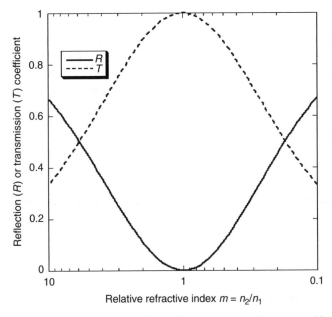

Figure 8.4 Reflection (R) and transmission (T) coefficients for intensity at normal incidence for light incident from medium 1 onto medium 2 plotted as a function of the relative refractive index

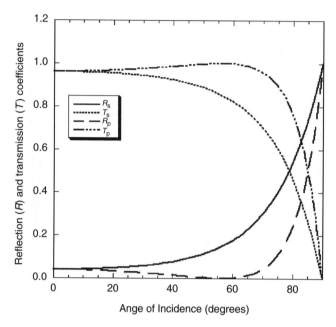

Figure 8.5   Plot of reflection (R) and transmission (T) coefficients vs the angle of incidence in degrees for light incident from medium 1 onto medium 2 for a relative refractive index $m = 1.5$

where m is the relative refractive index (assumed real) defined earlier and θ_t is the angle the transmitted ray makes with the normal, related to θ_t through Snell's law. Equations (8.6) and (8.7) reduce to Equation (8.5) for normal incidence, $n_1 = 1$, and $k = 0$.

The reflection and transmission coefficients are shown in Figure 8.5 for the case when the relative refractive index is 1.5. The reflection coefficient (R_p) for the p-polarized wave goes to zero at the Brewster angle θ_B, which from Equation (8.7) and Snell's law can be calculated to be

$$\tan \theta_B = m \tag{8.8}$$

If the refractive index on the incident side is greater than that on the 'transmitted' side of the interface, then the incident beam will be fully reflected from the interface if the angle of incidence exceeds the critical angle θ_c where

$$\sin \theta_c = m \tag{8.9}$$

In Figure 8.6 we show the Brewster and critical angles as a function of the relative refractive index.

In many applications of glasses involving interfaces, these two angles play a crucial role. For example, most gas lasers use windows cut at the Brewster angle to minimize reflection losses for one polarization. The critical angle is the defining parameter for the propagation of guided waves down fibre optic cables.

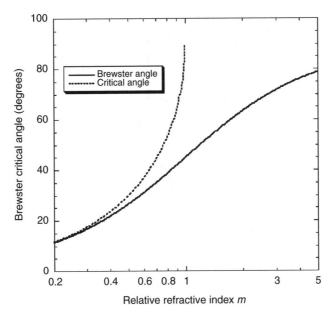

Figure 8.6 Brewster and critical angles as a function of relative refractive index m

8.4 DISPERSION

The dispersion is the derivative of the refractive index with respect to wavelength, $dn/d\lambda$, and plays a critical role in the design of many optical instruments. A change in refractive index results in a change of focal length in lenses or deviation in prisms, and so a nonzero dispersion gives rise to chromatic aberrations (e.g., colored fringes of images) in optical systems. To describe the dispersion quantitatively, a figure of merit known as the Abbe number is generally used and is given by:

$$v_d = \frac{n_d - 1}{n_F - n_C} \qquad (8.10)$$

where n_d is the refractive index at the helium d-line (587.6 nm), n_F at the blue line of hydrogen (486.13 nm), and n_C (656.27 nm) at the red line of hydrogen. We note that sometimes the sodium D-line at 589.6 nm is substituted for the helium line, in which case the Abbe number is denoted by v_D. An alternative figure of merit, v_e, is based on the e-line of mercury (546.07 nm), and F' and C' lines of cadmium (479.99 nm, 643.85 nm, respectively).

The fraction [Equation (8.10)] expresses the refracting power of the glass relative to the dispersion. The Abbe number is a basic parameter for optical instrument designers who have to deal with problems of chromatic aberration, and is commonly presented as an 'Abbe diagram' where the Abbe number is plotted as a function of refractive index for different glass families. An Abbe diagram for oxide glasses, based on data for the Schott range [8.3], is shown in Figure 8.7. Usually, glasses with a large refractive index have high dispersion; it is the latter which dominates in Equation (8.10), and so such glasses have a low Abbe number and appear in Figure 8.7 to the upper right. Glasses with small refractive indexes

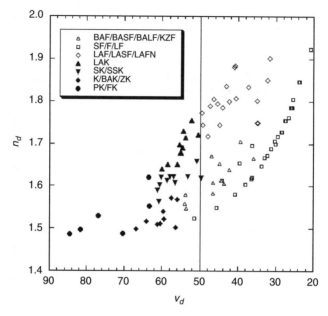

Figure 8.7   Abbe diagram for common optical glasses. Note that the Abbe number increases from right to left. The boundary at $v_d = 50$ divides crown glasses (left) from flint glasses (right).

usually have low dispersion and high Abbe numbers, and so appear in the lower left of Figure 8.7. For achromatic lens design, a second parameter called the relative partial dispersion $P_{g,F}$, defined as $(n_g - n_F)/(n_F - n_C)$, is also used to achieve a more comprehensive match of glass dispersion [8.3] for different optical elements. The refractive index n_g is that at the blue line of mercury (435.83 nm). The Abbe number is an indirect measure of the first derivative of the refractive index, and the partial dispersion is an indirect measure of the second derivative.

The refractive index n_d, Abbe number v_d, and partial dispersion $P_{g,F}$ have been historically used as a 3-parameter summary of the refractive index and its dispersion, and are still widely used for comparing different glasses, but for analytical work it is advantageous to return to the refractive index itself. Glass manufacturers generally measure the refractive index at a variety of wavelengths corresponding to atomic emission lines and fit the data to a curve, which represents the sum of classical damped oscillator responses to an electromagnetic field. Assuming several such oscillators are responsible for the absorption process near the absorption edge of a material, the refractive index is often written as:

$$n^2 - 1 = \sum_j \frac{B_j \lambda^2}{(\lambda^2 - \lambda_j^2)} \tag{8.11}$$

with the expectation that the formula (known as the Sellmaier formula) would only be a good representation for n for wavelengths well away from the resonance wavelengths λ_j. In that case Equation (8.11) can also be expressed as an (even) series in inverse powers of λ by expanding the functions:

$$\lambda^2/(\lambda^2 - \lambda_j^2) = 1 + (\lambda_j/\lambda)^2 - (\lambda_j/\lambda)^4 + \dots \tag{8.12}$$

so that

$$n^2 - 1 = A + A_1\lambda^2 + A_2\lambda^{-2} + A_3\lambda^{-4} + A_4\lambda^{-6} + A_5\lambda^{-8} \tag{8.13}$$

Here the notation follows that used by Schott [8.3], and the λ^2 term (with a negative coefficient) is added to improve the fit towards the IR region of the spectrum. Manufacturers will generally quote the coefficients A_k, or B_j and $C_j = (\lambda_j)^2$, so that the refractive index can be computed for any wavelength, or they may simply quote the values of n at several wavelengths. The accuracy of Equation (8.11) for interpolating the refractive index in the visible region is good to within about 1 part in 10^5 for a six-parameter fit.

Clearly the Abbe number, being defined in terms of visible wavelengths, is of less relevance in the IR range, and so equivalent alternatives are used for IR-transmitting glasses.

The refractive index and Abbe number are used in a common commercial numeric designation of optical glasses, which summarizes their properties in a six- or nine-digit code. For example, a glass described as 805 254 would have a refractive index n_d of 1.805 and an Abbe number of 25.4. Schott append the density in $g\,cm^{-3}$ to give a nine-digit code. The chemical composition is indicated by a letter code, which commonly contains K for crown glasses or F for flint glasses. Historically, crown glasses with a low dispersion were based on SiO_2–CaO–Na_2O, whilst flint glasses were based on SiO_2–PbO. The current spectrum of glasses contains many other combinations of elements, but they are still labeled as low-dispersion crown glasses if the Abbe number is greater than 50, and high-dispersion flint glasses if the Abbe number is less than 50. There are few systematics to the rest of the letter labeling system, and individual manufacturers have their own systems. But as an example, the common Schott BK-7 borosilicate glass is a crown K glass, the B indicates the boron content, whilst the 7 is a manufacturer's running number.

8.5 SENSITIVITY OF THE REFRACTIVE INDEX

8.5.1 Temperature dependence

The refractive index is a function of other variables than wavelength, notably temperature. From the Sellmaier formula, assuming only a single resonance in the UV region, the refractive-index variation with temperature may be parameterized as:

$$\frac{dn}{dt} = \left(\frac{n(\lambda,T_0)^2 - 1}{2n(\lambda,T_0)}\right)\left(D_0 + 2D_1\Delta T + 3D_2\Delta T^2 + \frac{E_0 + 2E_1\Delta T}{\lambda^2 - \lambda_0^2}\right) \tag{8.14}$$

where T_0 is a reference temperature, and the Ds and Es and λ_0 are experimentally determined parameters.

The significance of the temperature dependence of the refractive index lies in the implications for precision refractive optics. It is also important in high-powered laser applications where defocusing can occur in glass materials which have a significant absorption coefficient at the laser wavelength, and so show substantial localized heating in the focal zone. It is also important for applications such as glass etalons where temperature-independent conditions for interference are desirable.

The refractive index in a sample of bulk glass must be homogeneous for many optical applications, but inhomogeneities known as Striae can arise during the casting process, arising from convective flow and differential cooling rates. Entrapped gas bubbles, undissolved crystalline inclusions, platinum particles from the melt crucible, and crystals arising from mild devitrification all give rise to light scattering, which is also detrimental to glass performance. The glass manufacturer's catalogues can be consulted for specifications describing Striae and scattering.

8.5.2 Stress dependence

The refractive index is a function of the stress on a glass sample. The stress may be of internal origin, as, for example, after a poorly executed annealing process, or external. Sometimes applied external stress is an undesired consequence of an experimental situation, as in windows for cryogenic apparatus, which are subject to stress from thermal contraction of the support structure. The key attribute of stress is that it makes the material optically birefringent; i.e., the refractive index depends upon the polarization of the optical EM field. Since a stress field is described by a tensor, the birefringence is a tensorial quantity, but in the simple case of uniaxial stress σ the refractive indices for light polarized parallel and perpendicular to the stress direction may be written as:

$$n_{//} = n + \frac{dn_{//}}{d\sigma}\sigma \tag{8.15}$$

and

$$n_{\perp} = n + \frac{dn_{\perp}}{d\sigma}\sigma \tag{8.16}$$

The derivatives in Equations (8.15) and (8.16) are often referred to as the photoelastic coefficients. An alternative pair of birefringence parameters can be defined in terms of strain and related to those above [8.3] through the theory of elasticity.

Birefringence polarization modulators based on exciting a block of silica glass into longitudinal resonant oscillation at ultrasonic frequencies are available commercially. A three-component resonator system which uses a feedback loop to keep the oscillation amplitude constant, based on the design first used for internal friction measurements by Robinson and Edgar [8.4], is used commercially in ellipsometers.

8.5.3 Magnetic field dependence – the Faraday effect

In the presence of a magnetic field, normally optically isotropic materials such as glasses become birefringent. The refractive index depends upon the relative orientation of the magnetic field, the propagation direction for the light, and the polarization. The simplest geometry to analyse and the most important in practical applications is that where the propagation direction of linearly polarized light is parallel with the magnetic field, and in this case the manifestation of the effect is that the plane of polarization is rotated through an angle θ, proportional to the field strength H, after traversing a path length s through the glass. This effect is known as the Faraday effect after its discoverer, Michael Faraday. The proportionality constant is called the Verdet constant V and is defined through:

$$\theta = VsH \tag{8.17}$$

Plane-polarized light may be decomposed into two counter-rotating circularly polarized waves travelling in the same direction. The Faraday effect arises through a difference in the refractive indices n_+, n_- for the right-hand and left-hand rotating components, respectively. The relationship in Equation (8.17) follows from the more fundamental definition,

$$n_- - n_+ = \left(\frac{2c}{\omega}\right)VH \tag{8.18}$$

The Faraday effect has both diamagnetic and paramagnetic contributions; the former is present in all glasses, but it is small, almost temperature independent, and characterized by a positive value for V, whilst the latter only occurs when the glass contains paramagnetic ions, is relatively large, negative, and has a large temperature coefficient. Both effects are strongly wavelength dependent. The diamagnetic effect arises from the Zeeman splitting of excited states. For example, the splitting of an excited state which is two-fold degenerate results in a relative shift of the two associated transitions from the ground state; each transition is allowed only for one or other polarization. From Equation (8.11) the value of λ_j for each transition is different and so are the refractive indices n_+, n_-. The general form of the wavelength and temperature dependence of the Verdet constant is generally written [8.5] in terms of the frequency of the transition v, the number of ions per unit volume N, and the absolute temperature T, as:

$$V = 4\pi N v^2 \sum_{n,g} \left| \frac{A(n,g)}{\left(v^2 - v_{n,g}^2\right)^2} + \frac{B(n,g)}{\left(v^2 - v_{n,g}^2\right)} + \frac{C(n,g)}{T\left(v^2 - v_{n,g}^2\right)} \right| \tag{8.19}$$

where the sum runs over all the states g of the ground multiplet and excited state levels n between which transitions are possible. The T^{-1} dependence for the C term reflects the temperature dependence of the population in the ground multiplet, which mirrors that of the Curie law for magnetic susceptibility, and is a reasonable approximation near room temperature. However, deviations from Curie's law, arising from crystal-field splittings in the ground multiplets are often observed for rare-earth ions [8.6], and the same is also true for Verdet constants. In Figure 8.8, we show the (reciprocal) Verdet constant for a phosphate

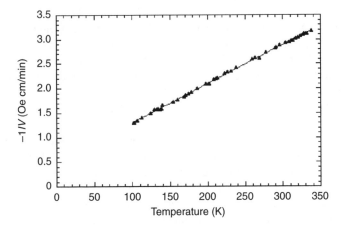

Figure 8.8 eciprocal of the Verdet constant for Ce^{3+} ions in a phosphate glass using data taken from Borelli [8.5]. The line is a least-squares fit to the Curie–Weiss law, which gives a variation as $(T + 65K)^{-1}$

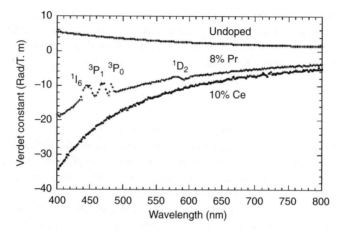

Figure 8.9   Verdet constants for undoped and rare-earth-doped fluorozirconate glass recorded at room temperature as a function of wavelength [Reproduced from A. Edgar et al., *J. Non-Cryst Solids*, **231**, 257 (1998) by permission of Elsevier]

glass [8.5] plotted as a function of temperature; the relationship may be fitted with the Curie–Weiss law which shows that V varies with temperature T as $(T + 65K)^{-1}$, not as T^{-1}. Van Vleck and Hebb [8.7] proposed that Faraday rotation should be proportional to magnetic susceptibility, implying that their temperature dependencies should be the same even if they are not Curie-like, but a test of this hypothesis by Edgar et al. [8.8] in the case of Ce^{3+} ions in fluoride glasses found significant deviations which were explained in terms of an inappropriate assumption in Van Vleck and Hebb's analysis – that the excited state splitting of the Ce^{3+} ions was relatively small. Nonetheless, the temperature dependence of the magnetic susceptibility is a good indicator of what to expect for the Faraday rotation.

The experimental wavelength dependence of the Faraday rotation for undoped and rare-earth-doped ZBLAN fluorozirconate glass [8.8] is shown in Figure 8.9 and illustrates several aspects of Equation (8.19). First, the glass itself has a small positive Verdet constant arising from optical transitions which make up the UV edge near 200 nm. For Ce^{3+} doping, the Verdet constant is large and negative and shows the smooth wavelength dependence predicted by the C term in Equation (8.19) which arises from the intense 4d–5f Ce^{3+} transitions in the near-UV around 270–350 nm. Finally, for Pr^{3+} the main effect is again from the same 4d–5f transitions, which are further towards the UV region, but there are also weaker 4f–4f transitions in the visible region which cause deviations from Equation (8.19). Both the deviation and the absorption at those wavelengths must be considered in any application.

Yamani and Asahari [8.9] give a useful compilation of Verdet constants for diamagnetic and paramagnetic glasses. Diamagnetic glasses generally have smaller Verdet constants than paramagnetic ones, but they offer the advantage for practical applications that they have no absorption lines across the visible region of interest, and the Faraday rotation is nearly temperature independent. Paramagnetic glasses can have much larger Verdet constants, but they have a significant temperature dependence and there can be absorption lines or bands in the area of interest. Glasses doped with 3d ions show strong broad absorption so are of little interest as Faraday rotation materials. For the rare-earths, trivalent cerium and divalent europium do not generally have any absorption in the visible region, and have intense 4d–5f transitions in the near-UV region, and so give a large Verdet constant. Terbium has a large ground-state g-value, resulting in large ground-state splittings, and so also has a large Verdet constant, but with the complication of weak 4f absorption lines in the visible region.

The Faraday effect finds applications in optical isolators, modulators, and in electric-current sensing. In the latter application, a fibre optic sensor is typically used to measure the large current flowing in overhead power lines. A few turns of the fibre are wrapped around the cable, and the current detected through the surface magnetic field it generates; the advantage over competing electrical means of current sensing is the optical isolation from the high voltage provided by the fibre optic cable.

8.5.4 Chemical perturbations – molar refractivity

It has been known for centuries that the refractive index of a glass increases when heavy elements are added to it; heavy in the sense of those having a high atomic number such as lead, lanthanum, or barium. These elements are commonly found in flint glasses. The underlying relationship between the bulk property and the atomic character can be discussed using the Clausius–Mossotti (or Lorentz–Lorenz) relationship, which is derived in most standard texts on electromagnetism (e.g., Griffiths [8.10])

$$\frac{n^2-1}{n^2+2} = \frac{1}{3\varepsilon_0} \sum_i N_i \alpha_i \qquad (8.20)$$

where n is the refractive index, ε_0 is the permittivity of free space, and N_i is the density of atoms or ions of atomic polarizability α_i. If the glass is composed of compounds whose refractive indices can be measured separately, then the Clausius–Mossotti relation implies that the glass refractive index can be computed from the 'molar refractivities' defined as,

$$MR_j = \left(\frac{n_j^2-1}{n_j^2+2}\right)\frac{MW_j}{\rho_j} \qquad (8.21)$$

where MW_j is the molecular weight of the compound and ρ_j is the density, so that the glass refractive index can be computed as,

$$\frac{n^2-1}{n^2+2} = \frac{\rho_{glass}}{MW_{glass}} \sum_j f_j MR_j \qquad (8.22)$$

where f_j is the mole fraction of compound j in the glass. In practice, the ions or atoms in a glass do not have a unique polarizability; for example, for oxygen ions the polarizability depends on whether they are in bridging or non-bridging positions. Nonetheless Equation (8.22) does give a useful first-order guide as to the effect of different chemical perturbations on the refractive index.

8.6 GLASS COLOR

Glasses may appear colored in transmitted light due to a variety of mechanisms, including absorption by transition metal ions, band-edge cut-off, and colloidal precipitates. Band-edge coloration simply implies that the fundamental edge for absorption has moved into the visible region of the spectrum, and many glasses based on anions other than oxygen, such as sulfur, selenium, and tellurium, qualify in this regard. Transition metal ion doping by elements from the 3d series with their open valence-shell electronic structure results in absorption bands in the visible and IR regions arising from 3d–3d electronic transitions. In principle, rare-earth-ion doping can also give rise to coloration, but the absorption per ion for the usual 4f–4f transitions is much weaker and narrower than for 3d–3d transitions.

8.6.1 Coloration by colloidal metals and semiconductors

It has long been known that the addition of minute quantities of metals such as Cu, Ag, and Au to a glass can under certain preparation conditions give rise to a strong coloration. The classic example is 'ruby glass', which is a silicate glass containing colloidal gold. The effect arises from the strong perturbation that small particles of metal make to the propagation of EM waves through the glass, due to the huge difference in the dielectric constants of the glass and metal. The effect is to scatter and absorb the light, with the balance between these two effects, and the wavelength and angular dependence (for scattering), depending on the particle size and the wavelength. The theoretical basis for discussing the subject is Mie scattering theory, as described for example by Van der Hulst [8.11], Kerker [8.12], or Bohren and Huffman [8.13]. The latter authors, for example, show that the absorption and scattering cross-sections are, respectively, given for small spherical particles of radius a by,

$$C_{abs} = \frac{kV}{3} \left[\frac{27}{(\varepsilon' + 2)^2 + \varepsilon''^2} \right] \varepsilon'' \tag{8.23}$$

and

$$C_{scat} = \frac{k^4 V^2}{18\pi} |\varepsilon - 1|^2 \left[\frac{27}{(\varepsilon' + 2)^2 + \varepsilon''^2} \right] \tag{8.24}$$

where k is the wavenumber in the glass, $\varepsilon = \varepsilon' + j\varepsilon''$ is the dielectric constant of the particle relative to the glass, and $V (= \frac{4}{3} \pi a^3)$ is the particle volume. The extinction coefficient is given by,

$$\alpha = N(C_{scat} + C_{abs}) \tag{8.25}$$

where N is the particle concentration, and the light intensity I after traversing a thickness z of glass is,

$$I = I_0 \exp(-\alpha z) \tag{8.26}$$

The relationships in Equation (8.23), and (8.24) are only valid for particles that are small relative to the wavelength in the sense that $|\varepsilon ka| \ll 1$. The interesting implications are that whilst the absorption rises as the volume fraction of particles increases (i.e., as NV increases), the scattering rises as $(NV)a^3$ increases, and so the balance between absorption and scattering effects changes with the particle size. For small particles, absorption dominates, but, with increasing particle size, scattering becomes more important in determining the extinction [8.14]. The other interesting effect is the possibility of a plasma resonance: for metals the real part of the dielectric constant is negative, and so it is possible that the factor ($\varepsilon' + 2$) goes to zero for some particular wavelength, giving rise to a peak in the scattering and absorption. The sharpness of the peak depends on the value of ε'' at that wavelength; the lower the value the sharper the resonance. The red color of ruby glass, for example, is due to a plasma resonance in gold in the blue-green. Stookey et al. [8.15] have shown how a nonspherical Ag particle precipitated on an NaF needle in a glass results in a split plasma resonance, which permits the tuning of the absorption band across the visible spectrum.

Glasses may also be produced which contain semiconducting particles such as CdS, CdTe, and CdSe, and alloys of these compounds. In this case the major perturbation to glass transparency arises near the wavelength appropriate to the bandgap of the semiconductor, where the real and imaginary refractive indices change markedly and the absorption rises dramatically. When the bandgap lies in a transparent region for the host glass, the effect is an abrupt rise in the absorption coefficient of the composite glass ceramic at that wavelength. By engineering the concentration and alloying nature and size of the semiconducting particles, a range of glasses can be produced with sharp cut-off characteristics in their transparency at a range of wavelengths, and these have practical applications in fabricating the well known colored edge-filters used in optical spectroscopy.

8.6.2 Optical absorption in rare-earth-doped glass

The rare-earth (or lanthanide) series comprises the 14 elements from cerium to lutetium. They usually occur in the trivalent form in glasses, although divalent europium, samarium, and dysprosium are not uncommon. It is possible to dope rare-earth ions in concentrations of up to tens of percent or more in some glasses (in which case they are really a basic glass constituent rather than a dopant), but even so the resulting glasses do not strongly absorb light and are not as strongly colored as glasses containing 3d transition metal series ions. The basic reason is related to different electronic structures. In the 4f series, the electronic configuration comprises an inner unfilled 4f shell and outer filled $5s^2$, $5p^6$ shells. The effect of the filled outer shells is to partially screen the 4f valence electrons from both the static and dynamic effect of the ligands and glass environment, and so the crystal or ligand field is a relatively small perturbation on the free-ion Hamiltonian. In contrast, for the 3d series, the valence electrons are outer electrons and are unshielded from the environment. The electronic energy-level structure for the rare-earth ions in solids has been described in many texts (e.g., Dieke [8.16]). In summary, the ground configuration $4f^n$ ($n = 1, 13$) is split by the intra-atomic electron–electron interaction forces into 'terms' [S, L] which are labeled by the total orbital angular momentum, L, and total spin angular momentum, S. [We ignore the case of lutetium ($n = 14$) here, since it is not spectroscopically active in its normal valence state.] Spin–orbit coupling splits the terms further into multiplets of degeneracy $2J + 1$ which are labeled by the total angular momentum $J = |L + S|, |L + S - 1|. . . .|L - S|$. The resulting energy-level schemes are well known, and are reproduced in several texts as 'Dieke' diagrams [8.16]. In Figure 8.10, we show the particular case for Pr^{3+} ($4f^2$). The diagram also indicates schematically that there are non-4f states, principally those of the excited $5d^1 4f^{n-1}$ configuration and ion–ligand charge-transfer states, which lie at higher energies, or sometimes overlap with the highest $4f^n$ states.

The effect of the glass environment on the n electrons in the 4f shell can be described by an interaction of the form,

$$H = \sum_{i=1,n} \sum_{\substack{l=1,\infty \\ m=-l,l}} B_m^l C_m^l (\theta_i, \varphi_i)$$ (8.27)

where the C_m^l are spherical tensor operators in the electron coordinates θ, ϕ, and the B_m^l are crystal field parameters. The form of Equation (8.27) was originally developed as a spherical harmonic expansion of an assumed electrostatic interaction between the electrons of the transition metal or rare-earth ion and the point charges or point dipoles of the surrounding ions. However, it was soon found that the magnitude of the B parameters was much larger than that calculated from an electrostatic model, and it is now known that although Equation (8.27) provides a good parametric description of the splitting patterns for the multiplets, the actual parameters contain substantial contributions from effects such as wavefunction overlap and covalency. They are therefore generally referred to as ligand field parameters rather than crystal field parameters [8.17].

Terms with $l > 6$ have matrix elements that are identically zero within the $4f^n$ configuration, and so they are usually omitted from Equation (8.27). For crystals, some of the remaining B_m^l parameters are necessarily zero as a consequence of the crystal symmetry, but for glasses there is no such constraint and furthermore the ligand field varies from site to site. The ligand field is a small perturbation compared with the spin orbit or electrosta-

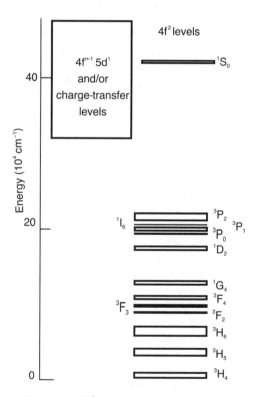

Figure 8.10   Energy level diagram for Pr^{3+} ions

tic energies, and this together with the site-to-site distribution in ligand fields means that resolved ligand field levels within an $[S, L, J]$ multiplet are not generally observed in glasses, unlike in crystals. The effect is generally just to broaden the transitions associated with each $[S, L, J]$ free-ion level, albeit with some unresolved structure.

Thus, the absorption spectrum of any particular rare-earth ion does not vary substantially from one glass to another, because it is the unresolved ligand field splittings that are glass-specific. The absorption is also quite weak. This is also a direct result of the shielding effect of the outer $5s^2$, $5p^6$ shells, which has two consequences with regard to transition intensities. Firstly, electric dipole transitions within a pure 4f configuration are strictly forbidden by Laporte's parity rule [8.16, 8.17]. However, the odd-parity (l odd) components in Equation (8.27) have a special role in breaking this selection rule. They have no matrix elements within states drawn from a 4f configuration, and so do not give rise in the first order to any crystal field splitting, but they do admix states from excited configurations with the opposite-parity into the 4f states. To the extent that the states now contain a small admixture of opposite-parity states, electric dipole transitions between the 'impure' 4f states are now permitted. Thus electric dipole transitions are weakly allowed through configuration mixing. (We note that magnetic dipole transitions are allowed within a 4f configuration, but they are approximately six orders of magnitude weaker than allowed electric dipole transitions, and so the dominant effect is usually the electric dipole transitions induced by configuration mixing, as discussed here.)

The second key effect is that the electron-vibrational interaction between the 4f electronic states and the dynamic glass environment is also weak due to the outer shell shielding. In the case of the 3d ions, the electron–phonon coupling to odd-parity vibrations results in allowed 'vibronic' transitions, as explained in the next section, giving rise to broad vibronic sidebands that are more intense than the purely 'zero-phonon' electronic transitions. For the rare earths, this effect is much weaker and consequently the optical absorption is much less than for the same concentration of a 3d transition metal ion. In Figure 8.11, we show the experimental optical absorption spectrum for Pr^{3+} ions in ZBLA glass, which can be compared with the energy-level scheme shown in Figure 8.10. The transitions from the $^3H_4$

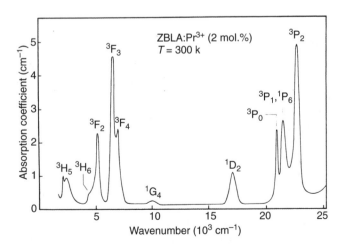

Figure 8.11 Optical absorption spectrum of Pr^{3+} ions in ZBLA glass [Reproduced from J.L. Adam and W.A. Sibley, *J. Non-Cryst. Solids*, **76**, 267 (1985) by permission of Elsevier]

Table 8.1   Electronic configurations for ions of the 3d series

d^1	d^2	d^3	d^4	d^5	d^6	d^7	d^8	d^9
		V^{2+}	Cr^{2+}	Mn^{2+}	Fe^{2+}	Co^{2+}	Ni^{2+}	Cu^{2+}
Ti^{3+}	V^{3+}	Cr^{3+}	Mn^{3+}	Fe^{3+}				
		Mn^{4+}						

ground state are labeled by the final state. It is evident that the ligand field splittings of the multiplets cannot be resolved. The combination of the transitions to the $^1D_2$ state in the yellow, and the ^3P states in the blue, results in the glass having a light green tinge.

8.6.3 Absorption by 3d metal ions

The 3d series of metal ions are characterized by progressive filling of the 3d shell as shown in Table 8.1.

The unshielded outer-shell electrons interact strongly with the ligands and the glass environment, unlike the case for the rare-earth ions. The effect of the crystal field is greater than that of spin–orbit coupling for the 3d ions, and so we consider the effects of the crystal field first. In a crystal, the ligand field has a well defined symmetry that markedly reduces the number of parameters in the Hamiltonian in Equation (8.27) (in addition, the summation only contains terms up to and including $l = 4$ since the matrix elements are identically zero for $l > 4$ within the states of a 3d configuration), and the powerful formalism of point group theory may be brought to bear on the problem. For glasses, there is no such symmetry, but nonetheless we find that dopant ions are typically found in environments that are a distribution about an 'average' environment. For the purpose of discussion, let us suppose that the average nearest-neighbor environment is octahedral with six identical ligands. The symmetry of a perfect octahedron then limits the ligand field to have the form:

$$H = B_4 \left[C_0^4 + \sqrt{\frac{5}{14}} \left(C_4^4 + C_{-4}^4 \right) \right] \tag{8.28}$$

so that only one parameter is required. Tanabe and Sugano [8.19] have calculated the eigenvalues of the Hamiltonian [Equation (8.28)] for the electrostatic and ligand field terms for this case, and for the similar cases of tetrahedral and cubic coordination, assuming particular values for the intra-shell coulombic interaction specified by the Racah parameters B and C [8.17]. Their plots of energy vs cubic crystal field are a standard aid in understanding $3d^n$ absorption spectra. A modern account is given by Figgis and Hitchman [8.17]. Ligand field theory abounds in different normalizations, and Tanabe and Sugano [8.19] use an equivalent but proportional parameter Dq to parameterize the ligand field rather than B_4. In Figure 8.12, we show the Tanabe–Sugano diagram for the d^7 configuration in an octahedral field for $C/B = 4.63$.

The predicted absorption spectra, ignoring the effects of non-octahedral ligand field perturbations, spin–orbit coupling, and vibrational interactions, may be read off the diagram with the known typical Dq and B values for different ligand coordinations. The implicit assumption here is that the crystal field is dominated by the ligands, but this should be a better approximation for glasses than for crystals due to the lack of long-range order in

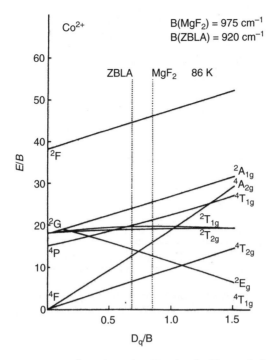

Figure 8.12 Energy levels for the d^7 configuration [Reprinted with permission from Y. Suzuki et al., *Phys. Rev. B*, **35**, 4472. Copyright (1987) the American Physical Society]

glasses. Fuxi [8.21] gives a table of estimated values of Dq and B for different ions in different glasses. It should be noted that the observed spectra are not the sharp lines that might be expected by running a vertical line up Figure 8.12 at the expected value of Dq/B. At the very least, the distribution in crystal fields is about the average, and the spin–orbit coupling splits and shifts term sub-levels, resulting in substantial inhomogeneous broadening of the order of hundreds of cm^{-1}.

An additional and substantial homogeneous broadening effect for the 3d ions arises from the effect of strong electron–phonon coupling. The broadening effect is best explained through the 'configuration coordinate' model (e.g., see Figgis and Hitchman [8.17]), which accounts for transitions that involve both electronic transitions and phonon excitations. This means that substantial absorption occurs on the high-energy side of the purely electronic transition, and in fact the model shows that the most probable transition is typically not the 'zero-phonon' transition but one which corresponds to the creation of several phonons. The overall result is that the absorption band is approximately a broad gaussian envelope, comprising many individual electron–phonon excitations, whose center is displaced to the high-energy side of the purely electronic transition. (The reverse occurs for fluorescence, the emission band lying on the low-energy side of the zero-phonon line.)

The strength of the optical absorption for 3d ions is greater than that for 4f ions since the configuration admixture effect is larger (because the 3d outer shell is unshielded); in addition odd-parity vibrations enhance the transition probability also through configuration admixture effects. A selection rule important in determining the relative strengths of electronic transitions for the 3d series is the spin selection rule that $\Delta S = 0$, which arises because the electric dipole operator in the transition-probability expression does not involve spin.

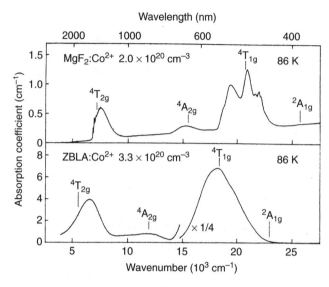

Figure 8.13   Optical absorption spectrum of Co^{2+} in ZBLA glass and MgF_2. [Reprinted with permission from Y. Suzuki et al., *Phys. Rev. B*, **35**, 4472. Copyright (1987) the American Physical Society]

However, spin–orbit mixing of pure *LS* terms means that such spin-forbidden transitions can occur, albeit with a reduced intensity. In Figure 8.13, we show the observed absorption spectrum for Co^{2+} ions in a ZBLA glass and MgF_2. The major spin-allowed transitions can be directly interpreted with the aid of Figure 8.12. The strong absorption in the red-green region leaves most Co-doped glasses with a deep blue coloration. Figures 8.11 and 8.13 show the typical behavior that rare-earth ions show narrower sets of absorption lines than do the 3d transition metal ions. Figure 8.13 shows resolved fine structure within the $^4T_{1g}$ multiplet; such structure is frequently resolved in crystals but rarely in glasses.

In summary, transition metal ions give rise to strong broad absorption bands in glasses. The best guide to the interpretation and prediction of spectra is a knowledge of the average local coordination together with the energy-level diagrams of Tanabe and Sugano [8.19].

8.7 FLUORESCENCE IN RARE-EARTH-DOPED GLASS

Fluorescence in rare-earth-doped glass is the basis for many applications such as neodymium-doped glass lasers, up-converters, erbium-doped fibre amplifiers for telecommunications, and fibre lasers. Fluorescence can be excited by optical pumping via the parity-allowed transition into the $4f^{n-1}5d$ configuration or charge-transfer excited states lying in the UV region indicated in Figure 8.10. From there, relaxation to the excited sub-states of the $4f^n$ configuration either via radiative or nonradiative (phonon-assisted) mechanisms can occur, followed by radiative (fluorescent) decay to the ground state either directly, or via a cascade through intermediate states. Alternatively, the fluorescence can be excited by pumping into the higher-lying states of the $4f^n$ configuration directly, but this is a weaker process as it is parity forbidden. As with the absorption, the fine structure due to crystal field splittings of the free-ion multiplets is not resolved; the glassy environment again gives rise to a distribution of crystal fields and this, together with the small magnitude of the crystal field, gives rise to an inhomogeneous broadening of the lines rather than a splitting, although unresolved structure can sometimes be observed. The unresolved structure can be probed further using techniques such as fluorescent-line narrowing (FLN), which rely on the power and narrow linewidth of modern lasers to isolate transitions due to a particular glass site from within a broad envelope.

Since the ligand field splittings are rarely resolved for rare-earth ions in glasses, it is not usually possible to parameterize the effects with a small set of ligand field parameters as is typically done for rare-earth ions in crystalline solids. However, there is great interest in the relative magnitudes of the transition probabilities for the various possible fluorescent transitions, since these are essential inputs to any consideration of practical device performance. Judd [8.22] and Ofelt [8.23] have shown that it is possible to predict the intensity of any allowed transition from just three parameters (Ω_2 Ω_4, Ω_6) that summarize the effect of admixtures of excited-state configurations into the ground state, induced by odd-parity components of the ligand field. The electric dipole oscillator strength for absorption or emission is given by Quimby [8.24]:

$$f_{ed}(a,b) = \frac{8\pi^2 m\nu\chi_{ed}}{3h(2J+1)n^2} \sum_{t=2,4,6} \Omega_t |\langle a\|U^t\|b\rangle|^2 \qquad (8.29)$$

where J is the total angular momentum of the initial level, n is the refractive index of the glass, ν is the frequency of the transition, and $\chi_{ed} = n(n^2 + 2)^2/9$ is a local field correction for electric dipole transitions. The quantities U^2, U^4, U^6 are double-reduced matrix elements of the unit tensor operator U^t, which are tabulated, for example, by Carnall et al. [8.25]. From the oscillator strengths, the radiative lifetime can be calculated as the probability per unit time for the decay as:

$$A = \frac{8\pi^2\nu^2 e^2 n^2}{mc^3} f_{ed}(a,b) \qquad (8.30)$$

The Judd–Ofelt parameters Ω_t thus serve as a three-parameter base from which all the electric dipole transition probabilities $[L, S, J] \Rightarrow [L', S', J']$ within a given configuration may be calculated, alabeit for the entire multiplet rather than for individual ligand-field levels, but since these are usually unresolved this is not an important limitation. The parameters Ω_t are specific to a given configuration and given glass host, since they parameterize the electron–ligand interaction. They can themselves be estimated from the absorption spectrum; in principle, measurements of the absorption cross-section for just three transitions suffice, but in practice a least-squares fitting to measurements from several lines is usually performed. The measurements yield the absorption cross-section σ_{abs}, which is related to the absorption oscillator strength by:

$$f_{abs} = \frac{mc}{\pi e^2} \int \sigma_{abs}(\nu)\,d\nu \qquad (8.31)$$

For transitions that have a significant magnetic dipole character, the oscillator strength for magnetic-dipole-allowed transitions is given by:

$$f_{md} = \frac{h\nu\chi_{md}}{6(2J+1)n^2 mc^2} |\langle a\|L + 2S\|b\rangle|^2 \qquad (8.32)$$

where $\chi_{md} = n^3$ is the local field correction for magnetic dipole transitions. The magnetic dipole oscillator strength should first be subtracted from the observed oscillator strength f_{abs} to yield $f_{ed} = f_{abs} - f_{md}$ for use in determining the Judd–Ofelt parameters from Equation (8.29).

Tables of Judd–Ofelt parameters for rare-earth ions in heavy-metal fluoride glasses have been given by Quimby [8.24], and for Nd^{3+} and Er^{3+} ions in a variety of glasses by Fuxi [8.21].

Whilst this discussion has focused on the radiative lifetime, one should be aware that there are parallel nonradiative processes involving energy migration between the rare-earth ions (in sufficiently concentrated systems), and ubiquitous phonon-assisted decay, which act to reduce the lifetime estimated from purely radiative considerations.

In Figure 8.14, we show the fluorescence from Pr^{3+} ions in ZBLA glass. This spectrum was generated by pumping from the ground state into the $^3P_2$ state. It is noticeable that the low-temperature spectrum shows slightly narrower bands, but the residual inhomogeneous ligand field broadening still prevents resolution of any ligand field splittings.

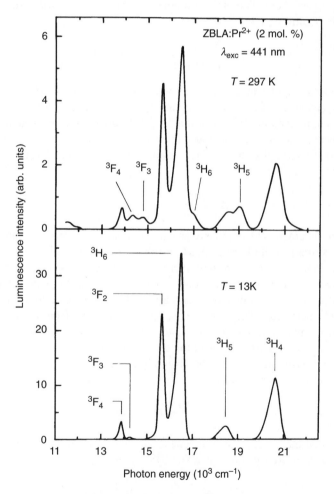

Figure 8.14 Luminescence spectrum for Pr^{3+} ions in ZBLA glass [Reproduced from J.L. Adam and W.A. Sibley, *J. Non-Cryst. Solids*, **76**, 267 (1985) by permission from Elsevier]

Fluorescence from 3d ions is infrequently observed in glasses, with the notable exception of the $3d^3$ (e.g., Cr^{3+}) and $3d^5$ (e.g., Mn^{2+}) ions.

8.8 GLASSES FOR FIBRE OPTICS

The glasses used to make fibre optics for long-distance communications are necessarily characterized by a low extinction coefficient in the transmission window, so low that in the case of silica it is too small to appear in Figure 8.1. As a practical engineering matter, the attenuation is measured in units of *dB/km* rather than as an extinction coefficient in cm^{-1} with the relation between the two being:

$$\text{attenuation}\Big/ \text{unit length} (\text{dB}/\text{km}) = \frac{1}{L}\log_{10}\left(\frac{P(L)}{P(0)}\right) = 43\,429\,\alpha \qquad (8.33)$$

where $P(x)$ is the power at distance x from the origin of a fibre optic, and the numerical value is for the attenuation coefficient α in units of cm^{-1}. Practical values of the minimum attenuation for silica are in the region of 0.2 *dB/km* or $5 \times 10^{-6}\,cm^{-1}$. This implies that the intensity is reduced by a factor of ten after about 50 km of cable. The attenuation in the window region for fibre optic glasses contains contributions from several effects that are shown schematically in Figure 8.15.

The so-called 'V-curve' [8.26] is defined by electronic edge absorption and Rayleigh scattering on the short-wavelength side of the minimum, and multiphonon absorption on the long-wavelength side; these effects can be described semi-empirically by [8.27]:

$$\alpha = A\exp(-a/\lambda) + B/\lambda^4 + C\exp(c/\lambda) \qquad (8.34)$$

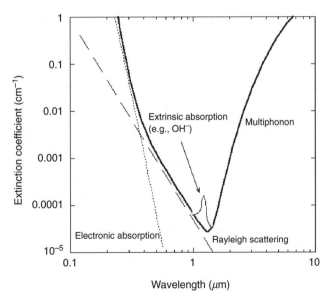

Figure 8.15 Schematic diagram for the spectral dependence of extinction coefficient for typical glasses

The first term of Equation (8.34) describes the absorption associated with the band edge. In crystalline solids, the absorption associated with electronic bandgap transition should in principle commence abruptly when the photon energy first exceeds the bandgap energy. However, there is a weak tail (the 'Urbach tail') extending into the bandgap zone which is thought to arise principally from phonon-assisted electronic transitions, and from localized defect-related electronic states. The involvement of phonons relaxes the electron wavevector selection rules which otherwise confine transitions to be vertical on the usual energy-band diagram. In glasses, the structural disorder means that the concept of an abrupt onset of electronic states must be abandoned, and there is a continuous transition from high-mobility bulk states, which lie well within the effective band, to localized low-mobility states, which lie in the nominal bandgap. It is these latter states, together with phonon-assisted transitions, which give rise to the Urbach tail in glasses. However, this contribution is usually dominated, at least in wide-gap materials, closer to the V-curve minimum by Rayleigh scattering arising from chemical and structural perturbations on a molecular scale in the glass. We can expect from Equation (8.24) [specifically in the form of Equation (8.40)] that this effect will scale as the inverse of the fourth power of the wavelength, as the square of the refractive index deviation from the average, and as the volume of the micro-regions which show refractive index deviation.

On the long-wavelength side of the V-curve, the attenuation is determined by multiple phonon absorption. In a crystal, such transitions are permitted through anharmonicity of the lattice vibrational potential energy, giving rise to a long-wavelength tail in the absorption associated with the IR absorption bands. For different materials, the position of the IR bands shifts depending upon the atomic masses involved. In a simple estimation, the frequency of the IR bands is given by:

$$\omega_0^2 = K/\mu \tag{8.35}$$

where K is the force constant and μ is the reduced mass for the vibration. Thus for glasses made from higher-mass atoms and/or with weaker bonds, we expect that the IR bands will move to longer wavelengths (lower frequencies), and that the multi-phonon edge will move with them. Thus the V curve for heavy-atom glasses moves to longer wavelengths along with the IR bands, and so we observe minima at longer wavelengths for the IR-transmitting heavy-metal fluoride and chalcogenide glasses. Since the Rayleigh scattering contribution is not directly dependent on the atomic mass, the effect is that the V-curve minima are expected to be deeper for these glasses than for silica.

So far this potential for lower fibre-optic attenuation than for silica has remained just that, due to extrinsic scattering by impurities and inclusions. For the heavy-metal fluoride glasses, residual transition metal ion concentrations, particularly of Fe^{2+}, Co^{2+}, and Cu^{2+}, give rise to electronic transitions near the V-curve minima, and there are also wavelength-independent scattering contributions from metallic inclusions such as Pt particles originating from the crucible used for glass melting, and from inadvertent crystallization in the fibre-drawing process. All of these dominate the contributions in Equation (8.34) near the V-curve minimum, and so silica retains its position as the fibre optic glass of choice. Even for silica, there is the well known hydroxide impurity vibrational overtone near 1.39 μm (with weaker overtones at 1.25 and 0.95 μm) shown schematically in Figure 8.15, which lies very close to the V-curve minimum at 1.55 μm, so that in practice it is the two minima on either side of this impurity absorption that are of practical use, resulting in the two telecommunications windows 1.2–1.3 μm and 1.55–1.6 μm.

Whilst operation of a fibre optic cable at the wavelength of minimum attenuation is clearly desirable for long-distance communication, it is also important that the dispersion is minimal since otherwise distortion of a pulse shape occurs in the time domain, limiting the maximum bit rate. Since the physical factors, which determine the position of the V-curve minimum and the wavelength for zero dispersion, are quite different, there is no reason to expect the wavelengths of minimum extinction and zero dispersion to be the same. If we use the Sellmaier formula with just two terms to express the refractive index due to one electronic resonance in the UV region and a vibrational resonance in the IR as,

$$n^2 - 1 = \frac{B_e \lambda^2}{(\lambda^2 - \lambda_e^2)} - \frac{B_v \lambda^2}{(\lambda^2 - \lambda_v^2)} \tag{8.36}$$

then we find that by computing the dispersion $dn/d\lambda$ that there is a zero in the dispersion at, approximately,

$$\lambda = \left(\frac{B_e \lambda_e^2 \lambda_v^2}{B_v} \right)^{1/4} \tag{8.37}$$

In the case of silica fibre, this zero is at $1.27\,\mu m$ compared with the extinction minimum at $1.55\,\mu m$.

8.9 REFRACTIVE INDEX ENGINEERING

By systematically manipulating the refractive index within a volume of glass it is possible to fabricate devices that show useful optical properties. The classic example is the optical fibre with a cladding, which has a lower refractive index than the core, giving a structure that guides light rays within the core by total internal reflection. In this case the refractive index variation is achieved, in concept at least, by simple fusion of a cylinder of core material and a pipe of cladding material, followed by stretching. The success of fibre optics has generated interest in extending the manipulation of refractive indices to other situations.

GRadient INdex (GRIN) fibres, where the refractive index is engineered to vary radially with a gradual rather than a step transition, have improved transmission properties compared with the simple step fibres. The gradation is on the scale of microns, but it is also possible to produce index gradients over a larger scale of millimetres, giving rise to a family of GRIN optical elements [8.28]. For example, the simplest GRIN device is the rod lens, essentially a scaled-up version of the graded index fibre, in which a radial index gradient results in the periodic focusing of light along the axis of the rod. GRIN lenses are typically a few mm in diameter, with focal lengths which can be of the same scale, and are used as relay lenses or in endoscopes. One advantage of GRIN over conventional lenses is that they have planar faces, which assists fabrication and systems integration. Other applications include fibre optic couplers and imaging, where arrays of GRIN micro-lenses can be used to image large objects at short object–detector distances [8.9].

The index profile can be produced in a variety of ways. In the ion-exchange method, a borosilicate glass containing a highly polarizable ion such as thallium or cesium is immersed in a bath of a molten salt containing an alkali ion of lower polarizability, for example sodium or potassium [8.9]. Ion exchange results in a change in the refractive index of the glass, which can be understood through the Clausius–Mossotti relation [Equation (8.20)] as a consequence of the replacement of an ion of one polarizability with another. The radial variation in refractive index directly reflects the ionic diffusion profile. Thallium salts are toxic, and so thallium-free processes (e.g., Hornschuh et al. [8.29]) based on exchanges between the ions Li, K, Na, and Ag have been developed. The ion-exchange process can also be used to generate thin planar waveguide structures on glass surfaces, which are more suited to opto-electronic integration than are conventional fibres. Photolithography can be used to define waveguide geometry. GRIN structures have also been produced in polymers, with the advantage that larger diameters are possible; for example, Koike et al. [8.30] have made GRIN contact lens structures from polymer GRIN material.

Spatial variations of refractive index on a much smaller scale are generated in the inscription of Bragg gratings [8.31] into optical fibres. The motivation here is to produce a periodic variation in refractive index, i.e., a grating, along the fibre length which can act as an interference filter at wavelengths related in the usual way to the grating period. In Figure 8.16, we show the reflectivity of a Bragg grating inscribed in silica.

Such filters find widespread use in wavelength-dependent multiplexing in fibre optic communications, and in fibre optic sensors [8.33]. In the limiting case, the grating acts as a mirror, and so can be used as an *in situ* reflector for fibre lasers. The grating is usually generated by an interference pattern from a high-intensity UV laser (typically at 244 nm from a CW frequency-doubled 488 nm argon ion laser), the interference pattern producing a permanent or temporary refractive-index change in the glass at the positions of the interference antinodes. The interference patterns can be generated using a phase mask [8.32], or by an arrangement such as Lloyd's mirror configuration. In the common silica-based fibre, the photosensitivity is established by germanium doping and enhanced by hydrogen loading [8.31]. Refractive index changes of the order of 10^{-4}–10^{-2} can be produced. The germanium doping introduces an absorption band at 242 nm, and it is clear that the basic interaction with the UV light is through resulting Ge–O bond breaking and reconstruction. But the exact mechanism of refractive index change is still a matter of discussion, with two interpreta-

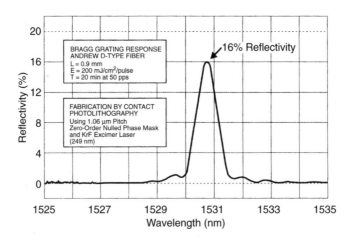

Figure 8.16   Reflectivity of a Bragg grating inscribed in a silica optical fibre [Reproduced from K.O. Hill et al., *Appl. Phys. Lett.*, **62**, 1035 (1993), by permission of the American Institute of Physics]

tions. One is that a periodic densification effect is responsible, with the refractive index being periodically modulated either directly due to the changed particle density [Equation (8.20)], or indirectly, for example by the resulting periodic core-cladding stress modulating the refractive index through the photoelastic effect [8.34]. The other model involves microscopic defects with changed atomic polarizability such as color centers [8.35]. Bragg gratings have been observed in other glasses (e.g., Ce-doped ZBLAN), but the mechanisms can be expected to be quite different from those applied for silica.

However, the best prospects for refractive index engineering come with the micromachining technique [8.36]. In this method, a high energy and high repetition rate femtosecond laser (e.g., Ti: Sapphire) or oscillator is focused via a microscope objective onto a micrometer-scaled volume inside a material that is normally transparent at the laser wavelength. Within the focal volume, the intensity is so high that electron–hole pairs are generated through nonlinear interactions, and the resulting bond disruptions can lead to permanent refractive index changes in the order 10^{-3} of through mechanisms that are probably similar to those applied to Bragg gratings. (Although we note that color centers have been discounted by Streltsov and Borelli [8.37] in the case of silica and borosilicate glass.) By scanning the focused spot parallel to the surface, and also normal to the surface (to an extent limited by the focal length of the microscope objective), 3-dimensional structures can be written in the glass. If the change in refractive index is positive, these structures act as light waveguides.

If the energy dumped into the focal spot is increased, the concentration of electron–hole pairs multiplies in an avalanche process. Their energy is transferred to the lattice through the electron–phonon interaction, and rapid, highly localized heating occurs. The material within the focal volume is essentially rendered into a plasma, which first expands and then cools, leaving behind a permanent structural modification of approximately spherical geometry in which the refractive index varies radially [8.38]. The refractive index is smaller at the core than at the outer perimeter, which has been radially compressed, and so is just the opposite to that described earlier. In front of and beyond the focal volume, the electric field strength is inadequate to trigger the effect. By tuning the energy, focusing, and repetition rate of the laser or oscillator, the diameter of the microsphere and the refractive index can be adjusted. Again, by scanning the material perpendicular to the beam with an XY stage, or along the axis of the beam, or all of these, two- or three-dimensional light-guiding structures can be machined into the glass. In Figure 8.17, we show how the refractive index varies over the width of a simple linear waveguide that has a tubular structure. The light guiding is achieved within the walls of the tubular structure.

The technique is now gaining commercial interest for micromachining in a range of transparent solids; multiplexers, Bragg gratings, and filters have all been fabricated using this technique. The technique can also be used for write-once mass storage [8.40]. Schaffer et al. [8.39] have reviewed the physical processes in femtosecond micromachining.

8.10 TRANSPARENT GLASS CERAMICS

8.10.1 Introduction

Glass ceramics comprise glassy matrices containing crystallites, which may range in size from nanometers to microns, and in volume fractions up to several tens of percent. They

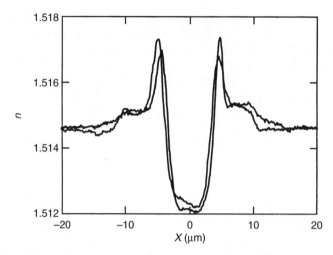

Figure 8.17 Variation of refractive index across the width of light waveguide produced by laser micromachining [Reproduced with permission from Professor Mazur from http://mazur-www.harvard.edu/publications.pjp?function=display&rowid=541]

are produced by thermally controlled nucleation and growth of crystals, often with the aid of a nucleating agent, from a prequenched glass. The crystals can impart improved mechanical properties to the host glass through their inhibition of crack propagation. In addition, the balance between the thermal expansion coefficients of the crystals (sometimes negative) and the host glass permits adjustment of the composite expansion coefficient, and it is possible to design materials such as the Schott product Zerodur® which has zero expansion coefficient at room temperature. The outstanding mechanical and thermal properties of these glass ceramics find applications in areas as widespread as artificial joints, machineable ceramics, magnetic disk substrates, and telescope mirror blanks.

However, the major interest in the present context is in 'nanophase' or 'nanocrystalline' glass ceramics where crystals of size ≤50 nm are uniformly dispersed in a transparent host glass, resulting (for favorable conditions, discussed below) in a transparent glass ceramic. Beall and Pinckney [8.41] have published a recent review of these materials. Generally the crystals are themselves transparent, but it is worth pointing out that the first optical application for nanophase glass ceramics was based on metallic crystallites. Stookey et al. [8.15] pioneered the work on photochromic glasses where the reversible darkening and/or coloration in silver- and copper-doped borosilicate glasses is based on photochemical reactions involving silver and copper ions, and colloidal silver particles [8.3]. In this case the optical effects are due to Mie scattering from the metallic particles described earlier (section 8.6.1). The first major application of glass ceramics containing transparent crystals evolved from the earlier development of low thermal-expansion-coefficient glass ceramics containing 'stuffed' β-quartz when it was found that these could be made in transparent form if the particle size was sufficiently small. The term 'stuffed' here refers to β-quartz where some of the Si^{4+} ions have been replaced by Al^{3+} ions and monovalent cations such as Li^+; in the limit this is the crystal β-eucryptite $LiAlSiO_4$. This immediately gives rise to the possibility of a transparent material able to survive high temperatures without cracking, and this type of transparent glass ceramic now finds applications in fire-doors, stove tops, and cookware.

However, in the past decade, there has been growing interest in transparent glass ceramics for photonic applications. These materials combine the ease of formation and fabrication of a glass with the desirable properties of a crystalline environment for photo-active dopant ions, such as those from the rare-earth series. Critical to most applications is the transparency of the glass, since in general each crystallite will act as a scattering center, and in applications such a lasers and optical amplifiers it is absolutely critical that any mechanism (scattering, absorption, etc.), that removes light intensity from the beam is minimal. In 1993, Wang and Ohwaki [8.42] triggered the current interest when they reported that a glass ceramic comprising Er^{3+}- and Yb^{3+}-doped fluoride crystals in an oxyfluoride glass was as transparent as the precursor glass and showed 100 times the up-conversion efficiency when pumped at 0.97 μm, a result comparable with the best achieved in single crystals. The potential advantages of this new class of photonic material were appreciated by Tick et al. [8.43], who adapted the material for Pr^{3+} amplifier operation at 1.31 μm by replacing Yb with Y and Zn fluorides, and the Er^{3+} with Pr^{3+}. They showed that the glass ceramic had a greater quantum efficiency than the competing ZBLAN fluoride glass, and remarkably low scattering, raising the potential of this material as a fibre amplifier. In a later paper, Tick [8.44] went further to suggest that, under appropriate conditions of particle size and volume fraction, glass ceramics could be ultratransparent – that is, have a transparency similar to that of pure glass. These three papers acted as the catalyst for the substantial interest over the past decade in transparent glass ceramics for photonics applications.

8.10.2 Theoretical basis for transparency

An understanding of light-scattering processes in glass ceramics is clearly important to optimize the applications. We begin by examining the regime in which the particles act as independent scatterers. Mie first developed a theory of independent scattering from spherical particles, which reduces to the Rayleigh theory for particles much smaller than the wavelength of light. The Mie theory is exact in the sense that it is based on solutions of Maxwell's equations resulting from the matching of electromagnetic waves at the surface of the sphere, but those solutions are in the form of approximations to the resulting infinite series. There is also the Rayleigh–Gans (or Rayleigh–Debye) theory, which can be applied to particles somewhat larger than those for which Rayleigh theory is appropriate, and for nonspherical geometries; it works by dividing a volume into Rayleigh scattering micro-volumes, and then adding the scattering from each micro-volume, taking into account position-dependent phase shifts.

The scattering cross-section for extinction is defined as the area which, if it were completely 'black', i.e., completely absorbing, would remove an equivalent intensity from the beam per scattering center. For a spherical particle of refractive index n_1 and radius a, immersed in a medium of refractive index n_m, the scattering cross-section C_{sca} for light of vacuum wavelength λ is

$$C_{sca} = \frac{24\pi^3 V^2}{(\lambda/n_m)^4}\left(\frac{n^2-1}{n^2+2}\right)^2 \qquad (8.38)$$

where V is the particle volume and n is the relative refractive index n_1/n_m. This is actually Equation (8.24) expressed for a real dielectric constant and in terms of refractive indices. A light beam of initial intensity I_0 which passes through a thickness z of scattering particles with concentration of N m^{-3} will emerge with intensity I where,

$$I = I_0 \exp(-NC_{sca}z) = I_0 \exp(-\alpha z) \tag{8.39}$$

If the difference in refractive index $|n_1 - n_m|$ is much less than the average $\bar{n} = (n_1 + n_m)/2$, then Equation (8.38) can be approximated in the expression for the extinction coefficient $\alpha = NC_{sca}$, by:

$$NC_{sca} = \left(\frac{128}{9}\right)(NV)\frac{\pi^4 a^3}{(\lambda)^4}(\Delta n\bar{n})^2 \tag{8.40}$$

which illustrates some of the key dependencies. The extinction coefficient depends linearly on the volume fraction, on the cube of the particle size, and on the square of the difference in refractive indices. Thus the designer of transparent glass ceramics should seek to minimize all of these quantities, particularly the particle size and the refractive index mismatch, to achieve transparency in the Rayleigh regime.

Tick et al. [8.45] pointed out that for the estimated volume fraction of particles (~30%) present in the oxyfluoride glass ceramics, the Rayleigh theory for dilute and weak scatterers give results for the extinction coefficient that are an order of magnitude larger than those observed. Since a theory for nucleated glass ceramics was not available, Tick et al. [8.45] turned to the work of Hopper [8.46, 8.47] for light scattering in glasses that showed spinodal decomposition. Hopper showed that the extinction coefficient should be,

$$\alpha = (6.3 \times 10^{-4})k_0^4 \theta^3 (\Delta n\bar{n})^2 \tag{8.41}$$

where θ is the mean phase width and k_0 is the wavevector in free space. We note first that the Rayleigh-like wavelength dependence of the extinction coefficient as the inverse fourth power is preserved in this theory. Making the large conceptual step of applying the theory to nucleated crystals, we can equate θ to $(a + W/2)$ where a is the particle radius and W the interparticle spacing. If we take a value of 50% for the volume fraction, Hopper's theory predicts an extinction coefficient about two orders of magnitude lower than that obtained from the Rayleigh theory [8.45]. However, the validity of this theory applied to a glass showing nucleated rather than spinodal decomposition remains questionable.

Hendy [8.48] has presented a more recent analysis of light scattering for the late-stage phase-separated materials and has deduced an extinction coefficient given by,

$$\alpha = \frac{14}{15\pi}\langle\varphi\rangle\langle1-\varphi\rangle\left(\frac{\Delta n}{n}\right)^2 k^8 a^7 \tag{8.42}$$

where φ is the volume fraction of the phase, $k = 2\pi/\lambda$, and the other terms have their usual meaning. This formula has a very different dependence on wavelength and particle size to that of the Rayleigh and Hopper models; in particular note the very high orders for the wavelength and particle-size dependence. As far as the author is aware, there have been no reports of a $1/\lambda^8$ wavelength dependence for the extinction coefficient observed experimentally. This would be expected to give rise to visible coloration effects, more pronounced than those for Rayleigh scattering. In fact there have been very few attempts to investigate the wavelength dependence of the scattering and confront it with the theories, but this is difficult in the case of rare-earth or other doping in glass ceramics of practical interest, since electronic absorption can dominate the extinction coefficient in the UV region. We conclude that the theoretical work supports the idea that cooperative scattering effects reduce the overall scattering, but this still awaits rigorous experimental confirmation. It would be interesting to attempt a direct numerical simulation of the scattering problem for fixed particle size as a function of concentration, perhaps with the aid of the discrete dipole approximation [8.49] used in many similar scattering problems.

In the absence of an established theory, qualitative ideas have guided the development of transparent glass ceramics. The ideas of effective media suggest that if the size scale of the refractive index perturbations are much smaller than the wavelength of light in the medium, then the material should be described by a single effective dielectric constant (given, for example, by the Maxwell–Garnett theory) and so there should be no scattering. Tick [8.44] has suggested the following four empirical criteria for a transparent glass ceramic:

1) The particle size must be less than 15 nm.
2) The interparticle spacing must be comparable with the crystal size.
3) Particle size distribution must be narrow.
4) There cannot be any clustering of the crystals.

One might also add that the refractive index difference must be small. If there is no refractive index mismatch, then the material is of course fully transparent. It is worth pointing out that this can only occur for cubic (i.e., nonbirefringent) crystals, but can explain why some glass ceramics involving a single crystalline phase can be transparent for large crystallite sizes.

Experimentally, particle sizes are often measured from the X-ray diffraction pattern line widths using the Scherrer formula [8.50]. This is necessarily a mean-size estimate, and gives no information on the size distribution, and is limited by the instrumental resolution to particle sizes less than about 100 nm. Transmission electron microscopy (TEM) measurements are better in this regard, and can give particle-size distributions, mean spacings, and volume fractions, but are much more time-consuming. Estimates of volume fraction from the starting compositions are likely to be overestimates because of loss of volatile components during the melting process. For example, the oxyfluoride glass ceramics are susceptible to loss of fluoride through SiF_4 volatilization.

As an example of the insight that can be gained from TEM work, Dejneka [8.51] has found agreement with Rayleigh scattering for 7% volume fraction LaF_3 nanocrystals of average size 15 nm, with a refractive index mismatch of 0.07 to an oxyfluoride host glass. However, a sample that contained 300 nm particles when quenched, and that was quite opaque, became more transparent on annealing; TEM studies showed that the effect of annealing was simply to fill in the spaces between the large crystals with small 20–30 nm crystals. This cannot be explained by Rayleigh scattering, and suggests that a cooperative effect is indeed involved in the scattering process, as the theory predicts.

8.10.3 Rare-earth-doped transparent glass ceramics for active photonics

Rare-earth-doped transparent glass ceramics have been an area of substantial activity during the past decade, primarily driven by telecommunication applications. Dejneka [8.52] and Goncalves et al. [8.53] have presented reviews of photonic applications of rare-earth-doped transparent glass ceramics. These materials offer the following advantages for applications such as fibre amplifiers and up-conversion:

1) They can be almost as transparent as simple glasses.
2) The active rare-earth ions may selectively partition into the crystals.
3) The rare-earth ions are in a crystalline environment where their radiative decay characteristics are superior to those typically found in a glassy environment. This is especially the case for heavy-metal halides such as LaF_3 and $Pb_xCd_{1-x}F_2$.
4) The glass ceramics include silicate-based varieties that are compatible with bonding to silica fibre.

In Table 8.2, we present a representative sample of rare-earth-doped transparent glass ceramics, which have been studied during the past decade. A great deal of work has been done on the oxyfluorides containing $Pb_xCd_{1-x}F_2$ and LaF_3. In the latter case, it is perhaps not surprising that the nanocrystalline phase readily accommodates a variety of trivalent rare-earth ions since there is excellent charge and ionic radius match. However, Dejneka has [8.51] made an interesting observation that oxyfluorides doped with GdF_3 rather than LaF_3 shows crystallization in a hexagonal tysonite phase that has not been previously reported; similarly YF_3 and TbF_3 are also found in the hexagonal phase although this is not the room-temperature-stable phase for these compounds. We have also found unreported phases, and phases unstable at room temperature and pressure, in ZBLAN glasses containing chloride and bromide crystals. It seems likely that the increased role of surface as against volume

Table 8.2 A selection of rare-earth-doped transparent glass ceramics, which have been investigated for photonics applications. The 'principal' components are those which are present in the melt in molar percentages greater than ~10%

Glass type and principal components	Secondary components	Crystal	Dopant	Crystal size (nm)	Applications	References
Oxyfluoride SiO_2, Al_2O_3, CaF_2		CaF_2	Er^{3+}	3–12 nm	Up-conversion	Qiao et al. [8.57]
Oxyfluoride SiO_2, Al_2O_3, CaF_2	CaO	CaF_2	Eu^{2+}	11–18 nm	Scintillation	Fu et al. [8.58]
Oxyfluoride SiO_2, CdF_2	Al_2O_3, ZnF_2	β-PbF_2	Er^{3+}	2.5–12 nm	Fibre amplifier	Tikhomirov et al. [8.54, 8.59, 8.60]
Oxyfluoride SiO_2, Al_2O_3, CdF_2, PbF_2	ZnF_2, YF_3, YbF_3	Cd/PbF_2	Pr^{3+}	9–18 nm	Fibre amplifier	Tick et al. [8.45]

Table 8.2 *Continued*

Glass type and principal components	Secondary components	Crystal	Dopant	Crystal size (nm)	Applications	References
Oxyfluoride SiO_2, Al_2O_3, CdF_2, PbF_2	ZnF_2	PbF_2	Tm^{3+}	7–18 nm		Mattarelli et al. [8.61]
Oxyfluoride SiO_2, Al_2O_3, CdF_2, PbF_2	GdF_3	$Pb_xCd_{1-x}F_2$	Yb^{3+}, Nd^{3+}, Tm^{3+}	18 nm	Up-conversion	Qiu et al. [8.62]
Oxyfluoride (germanate) GeO_2, PbF_2, PbO		PbF_2	Er^{3+}	10 nm	Fibre amplifier	Mortier and Auzel [8.63]
Oxyfluoride (germanate) GeO_2, PbF_2, PbO		β-PbF_2	Er^{3+}	8–30 nm		Mortier et al. [8.55]
Oxyfluoride (germanate) $PbGeO_3$, PbF_2, CdF_2		β-PbF_2	Er^{3+}, Eu^{3+}	5–10 nm		Bueno et al. [8.64]
Oxyfluoride (germanate) $PbGeO_3$, PbF_2, CdF_2		β-PbF_2	Er^{3+}	5–10 nm	Up-conversion	Gouveia-Netro et al. [8.65]
Oxyfluoride SiO_2, CdF_2, PbF_2	Al_2O_3, ZnF_2, YF_3	$Pb_xCd_{1-x}F_2$	Pr^{3+}	8–14 nm	Fibre amplifiers	Braglia et al. [8.66]
Oxyfluoride (sol–gel) SiO_2	LaF_3	LaF_3	Eu^{3+}, Tm^{3+}	<10 nm	Waveguide	Ribeiro et al. [8.67]
Oxyfluoride SiO_2, Al_2O_3, Na_2O	RE fluorides	LaF_3	Eu^{3+}	10–20 nm		Dejneka [8.52]
Oxyfluoride (sol–gel) SiO_2	LaF_3	LaF_3	Er^{3+}, Yb^{3+}	10–20 nm	Up-conversion	Biswas et al. [8.68]
Oxyfluoride SiO_2, Al_2O_3, Na_2O, LaF_3		LaF_3	Er^{3+}	6–9 nm	Up-conversion	Wang et al. [8.69]
Fluorozirconates ZrF_4, LaF_3	AlF_3, GaF_3	LaF_3	Er^{3+}			Mortier et al. [8.70]
Fluorozirconates ZrF_4, $BaBr_2$, NaF	AlF_3, LaF_3	$BaBr_2$	Eu^{2+}	<20 nm	Storage phosphors	Secu et al. [8.71]
Fluorozirconates ZrF_4, $BaCl_2$, NaF	AlF_3, LaF_3	$BaCl_2$	Eu^{2+}	<10 nm	Storage phosphors	Schweizer et al. [8.72]
Fluorozirconates ZrF_4, $BaBr_2$, RbF, LiF	AlF_3, LaF_3	Rb_2BaBr_4	Eu^{2+}	<32 nm	Storage phosphors	Edgar et al. [8.73]

energies for nanoscale crystallites stabilizes phases that would normally be unstable at STP, and that one can in general 'expect the unexpected' in nanoscale glass ceramics.

For the case of $Pb_xCd_{1-x}F_2$, the nanocrystalline phase was identified from the XRD pattern as a solid solution of CdF_2 and PbF_2. It is surprising that trivalent rare-earth ions selectively partition into these nano-crystals, given the mismatch of charge and ionic radius. Tikhomirov et al. [8.54] proposed that in erbium-doped oxyfluorides, the Er^{3+} ions are involved in nucleating centers that comprise mixed-phase orthorhombic α-PbF_2:ErF_3 nucleation centers for the subsequent growth of β-PbF_2 nano-crystals. This mechanism can explain the selective uptake of rare-earth ions, and has considerable implications for the quantum efficiencies, since one can expect at least a region of high rare-earth concentration near the core of the nano-crystal, which may be subject to concentration quenching. One would therefore expect a marked dependence of the optical performance of the material on the nucleation and growth conditions. Just this effect was observed by Mortier et al. [8.55] in Er^{3+}-doped germanate glasses, with clear evidence in the $^4I_{11/2}$–$^4I_{13/2}$ fluorescent transition for concentration-quenching effects. In contrast to these results, for sol–gel-derived oxyfluoride glasses, the rare-earth ion seems to limit nucleation and growth of the crystals [8.56], and so the role of the rare-earth ion is clearly process dependent.

It is noticeable from Table 8.2 that one common factor for all the listed transparent glass ceramics is simply particle size, which suggests that this factor alone may dominate the issue of transparency.

8.10.4 Ferroelectric transparent glass ceramics

Ferroelectric crystals from the ABO_3 family such as $LiNbO_3$ are widely used in nonlinear electro-optics. But the expense of single-crystal production has prompted a search for glass ceramic materials containing ferroelectric crystals that may be more cost effective, and which also offer the usual glass advantages of fabrication versatility. Such materials can be used directly as an optically isotropic medium for use in Kerr birefringence modulators, or can be electrically poled at an elevated temperature, resulting in a linear electro-optic effect as required for a Pockels modulator and for second-harmonic generation. A number of transparent glass ceramics containing ferroelectric crystals have recently been reported [8.74–8.77], and in a niobium–lithium–silicate glass containing sodium niobate crystals, a Kerr effect comparable with nitrobenzene, a standard Kerr cell liquid, has been observed [8.78]. Nonlinear properties of glasses are the subject of Chapter 10.

8.10.5 Transparent glass ceramics for X-ray storage phosphors

X-ray storage phosphor (XRSP) imaging plates [8.79] are solid-state replacements for photographic film used in medical, dental, and industrial radiography. The currently favored material used in imaging plates is powdered crystalline BaFBr doped with ~1000 ppm divalent europium. The image is stored as a spatial variation of radiation-induced trapped electrons and holes, and is read out by stimulating their recombination with a raster-scanned laser beam. The recombination energy is transferred to the europium ions and appears as photo-stimulated luminescence (PSL). Although the imaging plate technology offers many advantages over the traditional photographic film process such as reuseability, wider dynamic range, freedom from chemical developers, and direct digital image read-out, the spatial resolution is not as good as that of fine-grained photographic film. The reason is that the focused laser beam used in the read-out process is scattered by the powder grains, so that the PSL occurs not only from the region of the focusing spot, but also from the surrounding material, limiting the resolution to about 100 μm. While this is adequate for most applications, such as chest X-rays, it is inadequate for mammography and for crack detection in materials testing. It may be thought that the problem could be overcome by embedding the material in a refractive-index-matching binder, but BaFBr is birefringent, so matching is not possible.

Consequently, the idea of a transparent XRSP is very attractive for minimizing optical scattering and improving the spatial resolution in imaging-plate radiography, but unfortunately it seems that the density of stable trapping centers in simple glasses examined so far is insufficient for a viable XRSP. However, it has been shown [8.71, 8.80, 8.81] that fluorozirconate glass ceramics containing europium-doped $BaBr_2$, Rb_2BaBr_4, $RbBa_2Br_5$ or $BaCl_2$ nano-crystals show a significant XRSP effect, in the latter case up to 80% of that for BaFBr:Eu. The magnitude of the effect depends on the crystallite size, and it was suggested by Secu et al. [8.71] that there was a surface layer on the $BaCl_2$ crystallites of about 7 nm in thickness which was not PSL active. Large crystallites of size >~30 nm are therefore optimal for a large PSL effect. However, an increasing crystallite size also results in more light scattering, and so there is a trade-off between PSL efficiency and transparency. For samples, which are transparent to the eye (in 3-mm-thick slices), the relative efficiency falls to about 10%. Nonetheless, good X-ray images can be recorded, as shown in Figure 8.18, and the spatial resolution of the glass ceramic, specified by the modulation transfer function, is at least a factor of two better than for BaFBr.

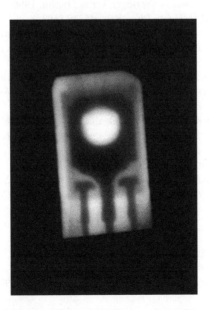

Figure 8.18 X-ray image of the internal structure of a power transistor recorded in a ZBLAN transparent glass ceramic

8.11 CONCLUSIONS

The optical properties of glasses is a subject that has both mature and developing aspects; the historical use of glass in optical instruments means that a great deal of information has been gathered and analysed regarding the bulk optical properties of traditional glass in the visible region. Yet there are new and fascinating applications that have emerged in the past thirty years, dominated by fibre optics and the telecommunications industries. The revolution in human communications generated by the internet and the resources that it delivers are, of course, based on high-speed, high-capacity optical links provided by optical pulse propagation on glassy silica fibres. And within the more recent time frame of the past decade, there have been new and fascinating developments: novel glass families optimized for optical performance in the IR and UV spectral regions, glass fibre lasers and amplifiers, a whole new class of glassy materials developing in the form of transparent glass ceramics, new physical phenomena such as photo-stimulated luminescence, and new technologies for refractive index engineering. The next few years promise to deliver equally intriguing developments in optical glass science.

REFERENCES

[8.1] *Handbook of Optical Constants of Solids*, edited by E.D. Palik (Academic Press, Orlando, 1985).

[8.2] M. Garbuny, *Optical Physics* (Academic Press, New York, 1965).

[8.3] *The Properties of Optical Glass*, edited by H. Bach and N. Neuroth (Springer-Verlag, Berlin, 1995).

[8.4] W.H. Robinson and A. Edgar, *IEEE Trans. Sonics Ultrasonics SU*, **21**, 98 (1974).

[8.5] N.F. Borelli, *J. Chem. Phys.*, **41**, 3289 (1964).

[8.6] R.L. Carlin, *Magnetochemistry* (Springer-Verlag, Berlin, 1986).

[8.7] J.H. Van Vleck and M.H. Hebb, *Phys. Rev.*, **46**, 17 (1934).

[8.8] A. Edgar, D. Giltrap, and D.R. MacFarlane, *J. Non-Cryst. Solids*, **231**, 257 (1998).

[8.9] M. Yamane and Y. Asahara, *Glasses for Photonics* (Cambridge University Press, Cambridge, 2000).

[8.10] D.J. Griffiths, *Introduction to Electrodynamics* (Prentice Hall, Upper Saddle River, 1999).

[8.11] H.C. Van der Hulst, *Light Scattering by Small Particles* (John Wiley & Sons, Inc., New York, 1957).

[8.12] M. Kerker, *The Scattering of Light* (Academic Press, New York, 1969).

[8.13] C.F. Bohren and D.R. Huffman, *Absorption and Scattering of Light by Small Particles* (John Wiley & Sons, Inc., New York, 1983).

[8.14] A. Edgar, *J. Non-Cryst. Solids*, **220**, 78 (1997).

[8.15] D.S. Stookey, G.B. Beall, and J.E. Pierson, *J. Appl. Phys.*, **49**, 5114 (1978).

[8.16] G.H. Dieke, *Spectra and Energy Levels of Rare Earth Ions in Crystals* (Interscience Publishers, New York, 1968).

[8.17] B.N. Figgis and M.A. Hitchman, *Ligand Field Theory and its Applications* (Wiley-VCH, Weinheim, 2000).

[8.18] J.L. Adam and W.A. Sibley, *J. Non-Cryst. Solids*, **76**, 267 (1985).

[8.19] Y. Tanabe and S. Sugano, *J. Phys. Soc. Jpn*, **9**, 766 (1954).

[8.20] Y. Suzuki, W.A. Sibley, O.H. El Bayoumi, T.M. Roberts, and B. Bendow, *Phys. Rev. B*, **35**, 4472 (1987).

[8.21] G. Fuxi, *Optical and Spectroscopic Properties of Glass* (Springer-Verlag, Berlin, 1992).

[8.22] B.R. Judd, *Phys. Rev.*, **127**, 750 (1962).

[8.23] G.S. Ofelt, *J. Chem. Phys.*, **37**, 511 (1962).

[8.24] R.S. Quimby, *Active phenomena in doped halide glasses*, in *Fluoride Glass Fibre Optics*, edited by I.D. Aggarwal and G. Lu (Academic Press, Boston, 1991).

[8.25] W.T. Carnall, H. Crosswhite, and H.M. Crosswhite, *Energy level structure and transition probabilities of the trivalent lanthanides in LaF3*, Argonne National Laboratory Report, Argonne, IL (1977).

[8.26] B. Bendow, in *Fluoride Glass Fiber Optics*, edited by I.D. Aggarwal and G. Lu (Academic Press, Boston, 1991).

[8.27] J.S. Sanghera and I.D. Aggarwal, *Infrared Fiber Optics* (CRC Press, Boca Raton, 1998).

[8.28] D.T. Moore, *Appl. Optics*, **19**, 1035 (1980).

[8.29] S. Hornschuh, B. Messerschmidt, T. Possner, U. Possner, and C. Russel, *J. Non-Cryst. Solids*, **347**, 121 (2004).

[8.30] Y. Koike, A. Asakawa, S.P. Wu, and E. Nihei, *Appl. Optics*, **34**, 4669 (1995).

[8.31] J. Albert, *Mater. Res. Bull.*, **23**, 36 (1998).

[8.32] K.O. Hill, B. Malo, F. Bilodeau, D.C. Johnson, and J. Albert, *Appl. Phys. Lett.*, **62**, 1035 (1993).

[8.33] A. Othonos and K. Kalli, *Fiber Bragg Gratings: Fundamentals and Applications in Telecommunications and Sensing* (Artech House Books, London, 1999).

[8.34] H.G. Limberger, P.-Y. Fonjallaz, R.P. Salathe, and F. Cochet, *Appl. Phys. Lett.*, **68**, 3069 (1996).

[8.35] M. Kristensen, *Phys. Rev.*, **64**, 144201/1 (2001).

[8.36] K. Hirao and K. Miura, *J. Non-Cryst. Solids*, **239**, 91 (1998).

[8.37] A.M. Streltsov and N.F. Borelli, *J. Opt. Soc. Am. B*, **19**, 2496 (2002).

[8.38] J.N. Glezer and E. Mazur, *Appl. Phys. Lett.*, **71**, 882 (1997).

[8.39] C.B. Schaffer, A. Brodeur, E. Mazur, *Meas. Sci. Technol.* **12**, 1784 (2001).

[8.40] J.N. Glezer, M. Milosavljevic, R.J. Finlay, T.-H. Her, J.P. Callan, and E. Mazur, *Optics Lett.*, **21**, 2023 (1996).

[8.41] G.H. Beall and L.R. Pinckney, *J. Am. Ceram. Soc.*, **82**, 5 (1999).

[8.42] Y. Wang and J. Ohwaki, *Appl. Phys. Lett.*, **63**, 3268 (1993).

[8.43] P.A. Tick, N.F. Borelli, and I.M. Reaney, *Opt. Mater. (Amsterdam)*, **15**, 81 (2000).

[8.44] P.A. Tick, *Opt. Lett.*, **23**, 1904 (1998).

[8.45] P.A. Tick, N.F. Borelii, L.K. Cornelius, and M.A. Newhouse, *J. Appl. Phys.*, **78**, 6367 (1995).

[8.46] R.W. Hopper, *J. Non-Cryst. Solids*, **70**, 111 (1985).

[8.47] R.W. Hopper, *J. Non-Cryst. Solids*, **49**, 263 (1982).

[8.48] S.C. Hendy, *Appl. Phys. Lett.*, **81**, 1171 (2002).

[8.49] B.T. Draine and P.J. Flateau, *J. Opt. Soc. Am. A*, **11**, 1491 (1994).

[8.50] H.P. Klug and L.E. Alexander, *X-ray Diffraction Procedures*, 2nd edition (John Wiley & Sons, Inc., New York, 1974).

[8.51] M. Dejneka, *J. Non-Cryst. Solids*, **239**, 149 (1998).

[8.52] M. Dejneka, *MRS Bull.*, **23**, 57 (1998).

[8.53] M. Clara Goncalves, L.F. Santos, and R.M. Almeida, *C. R. Chim.*, **5**, 845 (2002).

[8.54] V.K. Tikhomirov, D. Furniss, A.B. Seddon, I.M. Reaney, M. Beggoria, M. Ferrari, M. Montagna, and R. Rolli, *Appl. Phys. Lett.*, **81**, 1937 (2002).

[8.55] M. Mortier, P. Goldner, C. Chateau, and M. Genotelle, *J. Alloys Compd.*, **323–324**, 245 (2001).

[8.56] W. Luo, Y. Wang, F. Bao, L. Zhou, and X.J. Wang, *J. Non-Cryst. Solids*, **347**, 31 (2004).

[8.57] J. Qiao, X. Fan, J. Wang, and M. Wang, *J. Non-Cryst. Solids*, **351**, 357 (2005).

[8.58] J. Fu, J.M. Parker, P.S. Flower, and R.M. Brown, *Mater. Res. Bull.*, **37**, 1843 (2002).

[8.59] V.K. Tikhomirov, D. Furniss, and A.B. Seddon, *J. Mater. Sci. Lett.*, **21**, 293 (2002).

[8.60] V.K. Tikhomirov, A.B. Seddon, M. Ferrari, M. Montagna, L.F. Santos, and R.M. Almeida, *Europhys. Lett.*, **64**, 529 (2003).

[8.61] M. Mattarelli, V.K. Tikhomirov, M. Montagna, E. Moser, A. Chiasera, S. Chaussedent, G. Nunzi Conti, S. Pelli, G.C. Righini, L. Zampredi, and M. Ferrari, *J. Non-Cryst. Solids*, **345–346**, 354 (2004).

[8.62] J. Qiu, A. Mukai, A. Makishima, and Y. Kawamoto, *J. Phys.: Condens. Matter*, **14**, 13827 (2002).

[8.63] M. Mortier and F. Auzel, *J. Non-Cryst. Solids*, **256–257**, 361 (1999).

[8.64] L.A. Bueno, P. Melnikov, Y. Messadeq, and S.J.L. Ribeiro, *J. Non-Cryst. Solids*, **247**, 87 (1999).

[8.65] A.S. Gouveia-Netro, E.B. da Costa, L.A. Bueno, and S.J.L. Ribeiro, *J. Alloys Compd.*, **375**, 224 (2004).

[8.66] M. Braglia, C. Bruschi, G. Dai, J. Kraus, S. Mosso, M. Baricco, L. Battezzati, and F. Rossi, *J. Non-Cryst. Solids*, **256–257**, 170 (1999).

[8.67] S.J.L. Ribeiro, C.C. Araujo, L.A. Bueno, R.R. Goncalves, and Y. Messadeq, *J. Non-Cryst. Solids*, **348**, 180 (2004).

[8.68] A. Biswas, G.S. Maciel, C.S. Friend, and P.N. Prasad, *J. Non-Cryst. Solids*, **316**, 393 (2003).

[8.69] J. Wang, J. Qiao, X. Fan, and M. Wang, *Physica B (Amsterdam)*, **353**, 242 (2004).

[8.70] M. Mortier, A. Monteville, G. Patriarche, G. Maze, and F. Auzel, *Opt. Mater.*, **16**, 255 (2001).

[8.71] M. Secu, S. Schweizer, J.-M. Spaeth, A. Edgar, G.V.M. Williams, and U. Rieser, *J. Phys.: Condens. Matter*, **15**, 1097 (2003).

[8.72] S. Schweizer, L.W. Hobbs, M. Secu, J.-M. Spaeth, A. Edgar, G.V.M. Williams, and J. Hamlin, *J. Appl. Phys.*, **97**, 083522–1 (2005).

[8.73] A. Edgar, G.V.M. Williams, P.K.D. Sagar, M. Secu, S. Schweizer, J.-M. Spaeth, X. Hu, P.J. Newman, and D.R. MacFarlane, *J. Non-Cryst. Solids*, **326–327**, 489 (2003).

[8.74] Y. Takahashi, Y. Benino, T. Fujiwara, and T. Komatsu, *J. Non-Cryst. Solids*, **316**, 320 (2003).

[8.75] A.A. Zhilin, G.T. Petrovsky, V.V. Goulbkov, A.A. Lipovskii, D.K. Tagantsev, B.V. Tatarintsev, and M.P. Shepilov, *J. Non-Cryst. Solids*, **345–346**, 182 (2004).

[8.76] K. Hirano, Y. Benino, and T. Komatsu, *J. Phys. Chem. Solids*, **62**, 2075 (2001).

[8.77] Y. Benino, Y. Takahashi, T. Fujiwara, and T. Komatsu, *J. Non-Cryst. Solids*, **345–346**, 422 (2004).

[8.78] A.A. Lipovskii, D.K. Tagantsev, B.V. Tatarintsev, and A.A. Vetrov, *J. Non-Cryst. Solids*, **318**, 268 (2003).

[8.79] J.A. Rowlands, *Phys. Med. Biol.*, **47**, R123 (2002).

[8.80] S. Schweizer, L. Hobbs, M. Secu, J.-M. Spaeth, A. Edgar, and G.V.M. Williams, *J. Appl. Phys.*, **97**, 0835522/1 (2005).

[8.81] A. Edgar, G.V.M. Williams, and G.A. Appleby, *J. Lumine.*, **108/1–4**, 19 (2004).

9 Properties and Applications of Photonic Crystals

H.E. Ruda and N. Matsuura

Centre for Nanotechnology
University of Toronto, 170 College Street, Toronto, Ontario M5S 3E4, Canada

9.1 INTRODUCTION

Photonic crystals (PCs) are periodic, dielectric, composite structures in which the interfaces between the dielectric media behave as light-scattering centers. PCs consist of at least two component materials having different refractive indices, which scatter light due to their refractive index contrast. The one-, two-, or three-dimensional (1D, 2D, or 3D) periodic arrangement of the scattering interfaces may, under certain conditions, prevent light with wavelengths comparable to the periodicity dimension of the PC from propagating through the structure. The band of forbidden wavelengths is commonly referred to as a 'photonic

Optical Properties of Condensed Matter and Applications Edited by J. Singh
© 2006 John Wiley & Sons, Ltd

bandgap' (PBG). Thus, PCs are also commonly referred to as photonic bandgap (PBG) structures.

PCs have great potential for providing new types of photonic devices. The continuing demand for photonic devices in the areas of communications, computing, and signal processing, using photons as information carriers, has made research into PCs an emerging field with considerable resources allocated to their technological development. PCs have been proposed to offer a means for controlling light propagation in small, sub-micron-scale volumes – the photon-based equivalent of a semiconductor chip – comprising optical devices integrated together onto a single compact circuit. Proposed applications of PCs for the telecommunication sector include optical cavities, high-Q filters, mirrors, channel add/drop filters, superprisms, and compact waveguides for use in so-called planar lightwave circuits (PLCs).

Practical applications of PCs generally require man-made structures, as photonic devices are designed primarily for light frequencies ranging from the UV to the near-IR regime (i.e., ~100 nm to ~2 μm, respectively) and PCs having these corresponding periodicities are not readily available in nature. 1D PCs in this wavelength range may be easily fabricated using standard thin-film deposition processes. However, 2D and 3D PC structures are significantly more difficult to fabricate and remain among the more challenging nanometer-scale architectures to realize with cost-effective and flexible patterning using traditional fabrication methodologies. Recently, there has been considerable interest in PC-based devices that has driven advanced fabrication technologies to the point where techniques are now available to fabricate such complex structures reliably on the laboratory scale. In addition to traditional semiconductor nanostructure patterning methods based on advanced patterning/ etching techniques developed by the semiconductor industry, novel synthesis methods have been identified for 2D and 3D periodic nanostructured PC arrays. There are several excellent reports reviewing these fabrication techniques in the literature [9.1–9.4] and this growing field has already been the subject of numerous recent reviews, special issues, and books in the area of theoretical calculations (both bandstructure and application simulations), 2D PC structures, 3D PC structures, and opal-based structures [9.5–9.7].

Recently, there has been great interest in exploring the use of PCs for applications in the active field of telecommunications, such as in the area of PLCs (e.g., for optical switching). In such applications, the PC properties should be adjustable to create 'tunable' photonic bandgaps. This development increases the functionality of all present applications of PCs by allowing the devices in such applications to be adjustable, or tunable. We review here recent developments in the engineering of tunable nanometer-scale architectures in 2D and 3D. This review aims to organize this ever-changing volume of information such that interested theorists can design structures that may be easily fabricated with certain materials, and such that technologists can try to meet existing fabrication 'gaps' and issues with current systems.

9.2 PC OVERVIEW

9.2.1 Introduction to PCs

The simplest PC structure is a multilayer film, periodic in 1D, consisting of alternating layers of material with different refractive indices (Figure 9.1). Theoretically, this 1D PC can act

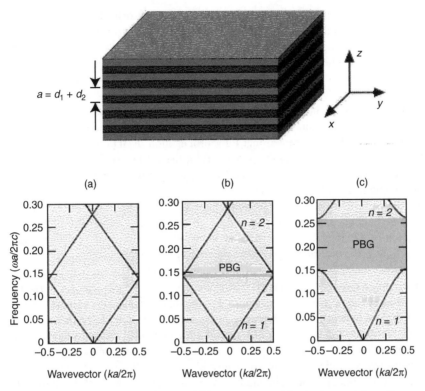

Figure 9.1   Example of a 1D photonic crystal. For a structure with a basis consisting of a green and blue layer, having respective thicknesses of d_1 and d_2, and lattice constant of $a = d_1 + d_2$. In the band diagrams shown in (a) through (c), we assume that $d_1 = d_2 = d = a/2$. Band diagrams are calculated for different dielectric constants ε: (a) GaAs bulk ($\varepsilon = 13$); (b) GaAs/GaAlAs multilayer ($\varepsilon_1 = 13$, $\varepsilon_2 = 12$); and (c) GaAs/air multilayer ($\varepsilon_1 = 13$, $\varepsilon_2 = 1$) [Reproduced from J.D. Joannopoulos et al., *PCs: Moulding the Flow of Light*, 1995 by permission of Princeton University Press]

as a perfect mirror for reflecting light with wavelengths within its photonic bandgap, and for light incident normal to the multilayer surface. 1D PCs are found in nature, as seen for example in the iridescent colors of abalone shells, butterfly wings, and some crystalline minerals [9.4], and in man-made 1D PCs (i.e., also known as Bragg gratings). The latter are widely used in a variety of optical devices, including dielectric mirrors, optical filters, and in optical fiber technology.

The center frequency and size (i.e., frequency band) or so-called stop band of the PBG depends on the refractive index contrast (i.e., n_1/n_2, where n_1 and n_2 represent the refractive indices of the first and second materials, respectively) of the component materials in the system. Figure 9.1 shows an example of a 1D PC, with a periodic arrangement of low-loss dielectric materials. This multilayer film is periodic in the z-direction and extends to infinity in the x- and y-direction. In 1D, a photonic bandgap occurs between every set of bands, at either the edge or at the center of the Brillouin zone – photonic bandgaps will appear whenever n_1/n_2 is not equal to unity [9.7]. For such multilayer structures, corresponding photonic bandgap diagrams show that the smaller the contrast, the smaller the bandgaps [9.7]. In 1D PCs, if light is not incident normal to the film surface, no photonic bandgaps will exist. It is also important to note that at long wavelengths (i.e., at wavelengths much larger than the periodicity of the PC), the electromagnetic wave does not probe the fine structure of the crystal lattice and effectively sees the structure as a homogeneous dielectric medium.

The phenomena of light waves traveling in 1D periodic media was generalized for light propagating in any direction in a crystal, periodic in all 3 dimensions, in 1987, when two independent researchers suggested that light propagation in 3D could be controlled using 3D PCs [9.8, 9.9]. By extending the periodicity of the 1D PC to 2D and 3D, light within a defined frequency range may be reflected from any angle in a plane in 2D PBG structures, or at any angle in 3D PBG structures.

Since the periodicity of the PCs prevents light of specific wavelengths (i.e., those within the photonic bandgap) from propagating through them in a given direction, the intentional introduction of 'defects' in these structures allows PCs to control and confine light. Propagation of light with wavelengths that were previously forbidden can now occur through such 'defect states' located within the photonic bandgap. Defects in such PCs are defined as regions having a different geometry (i.e., spacing and/or symmetry) and/or refractive index contrast from that of the periodic structure. For example, in a 2D PC comprising a periodic array of dielectric columns separated by air spaces, a possible defect would include the removal of a series of columns in a line. Specific wavelengths of light forbidden from propagating through defect-free regions would then be able to propagate through the line defect, but not elsewhere. Indeed, by appropriately eliminating further columns, light may be directed to form optical devices, including, for example, a low-loss 90° bend in a 2D waveguide, as shown in the theoretical simulation [9.7] (Figure 9.2).

Clearly, photons controlled and confined in small structures, of a size in the order of the wavelength of light, and using the extremely tight bend-radii offered by PCs would facilitate miniaturization and the fabrication of PLCs [9.7]. In addition, since the periodicity of the PC gives rise to the existence of bandgaps that change the dispersion characteristics of light at given frequencies, defect-free PCs give rise to other interesting phenomena, including highly dispersive elements, through the so-called superprism effect [9.10]. Possible designs of PC-based optical devices have been extensively explored using such properties [9.1, 9.11], generating much excitement in the optical telecommunication field [9.6].

9.2.2 Nano-engineering of PC architectures

Most of the promising applications of 2D and 3D PCs depend on the center frequency and frequency range for the photonic bandgaps. A so-called 'complete', 'full', or 'true', PBG is defined as one that extends throughout the entire Brillouin zone in the photonic band structure – that is, for all directions of light propagation for photons of appropriate frequency. An incomplete bandgap is commonly termed a 'pseudo-gap' or a 'stop band', because it only occurs in the reflection/transmission along a particular propagation direction. A complete gap occurs when a stop band's frequencies overlap in all directions in 3D. The center frequencies and stop band locations of the PBGs critically depend on the

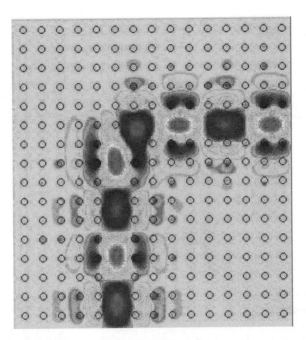

Figure 9.2   Theoretical simulation of a low-loss 90° bending of light in a 2D waveguide [Reproduced from J.D. Joannopoulos, et al., *PCs: Moulding the Flow of Light*, 1995 by permission of Princeton University Press]

unit-cell structure [9.7–9.9, 9.12, 9.13]. In particular, the PC properties depend on the symmetry of the structure (i.e., the unit-cell arrangements), the scattering-element shape within the unit cell, the fill factor (i.e., the relative volume occupied by each material), the topology, and the refractive index contrast (Figure 9.3).

It has been shown that a triangular lattice with circular cross-section scattering elements in 2D, or a face-centered cubic/diamond lattice with spherical scattering elements in 3D, tends to produce larger PBGs [9.7]. Also, as discussed above, the dielectric contrast is an important determining factor with 2D and 3D PC structures. The lower the structure dimension, the more readily are PBGs manifested since an overlap of the PBGs in different directions is more likely (i.e., something which is a certainty in 1D structures). In 3D structures, calculations have determined that the minimum dielectric contrast (n_1/n_2) required to obtain a full PBG is about two [9.2, 9.12, 9.14]. For full PBGs, the ideal structure typically consists of a dielectric–air combination, to obtain both the greatest dielectric contrast as well as reduced losses associated with light propagation in optical materials other than air [9.7].

The relationship between the fill factor, structure symmetry, and scattering-element shape on the size and location of the PBG is complex [9.7, 9.15–9.18] and will not be discussed in detail here. However, it is clear that the ability to adjust, or tune, one or more of these parameters and thus tune the PC properties is a very exciting development for future PC device applications.

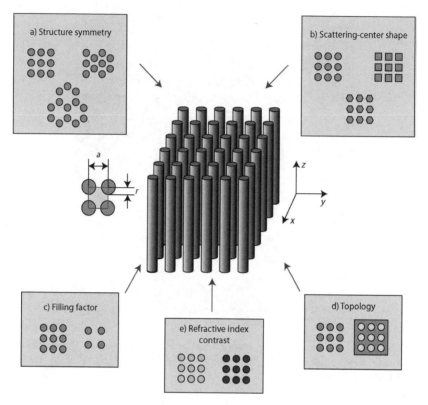

Figure 9.3 PBG parameters that affect the frequency and associated stop band. In particular, key PBG parameters for the 2D periodic arrangement of one material (i.e., the rods, in grey) embedded in a second material (i.e., air), periodic in the *x–y* plane (shown in central figure) include the structure symmetry, scattering-center shape, fill (filling) factor, refractive index contrast, and topology

9.2.3 Materials selection for PCs

The physical architecture of a given PC is just one of the design considerations – other important ones are the optical properties of the PC materials. In particular, the refractive index of a given material and its electronic bandgap determine the performance and appropriate range of frequencies of PC devices fabricated from such a material.

For large PBGs, component materials need to satisfy two criteria; first, having a high refractive index contrast, and secondly, having a high transparency in the frequency range of interest. It may be a challenge to satisfy both these criteria at optical wavelengths. Suitable classes of PC materials include conventional semiconductors and ceramics, since at wavelengths longer than their absorption edge (or electronic bandgap) they can have both high refractive indices and low absorption coefficients. In addition, these materials have other very useful electronic and optical properties, which may complement the functions served by the presence of the PBG.

Another, often overlooked, consideration for PC material selection is the ability to translate the desired 'bulk', single-crystal properties of well known materials into nanometer-scale PC properties. Such structures typically have extremely large surface areas, and the microstructure of the constituent PC elements must be controlled during fabrication. Consequently, the final properties of the PC elements often vary from those of the bulk properties. This is extremely relevant when considering the functionality of a final PC device, such as those relying on electronic properties (e.g., lasing).

Finally, it is interesting to note that there is a strong correlation between fabrication methods selected and the component materials. For example, 'top-down' dry-etching of semiconductors (e.g., Si, GaAs, InP, etc.) into nanostructured 2D arrays is well characterized and relatively commonplace, whereas the same cannot be said for ceramic materials. The opposite is generally true for chemical or 'bottom-up' synthesis using sol–gel technology, in which ceramic PCs are relatively easy to fabricate when sol–gel-based infiltration techniques are used (e.g., in 3D inverse opal fabrication). It is also clear that the fabrication technique primarily determines the PC structure that can be fabricated, which in turn, determines the PC properties.

9.3 TUNABLE PCs

Although conventional PCs offer an ability to control light propagation or confinement through the introduction of defects, once such defects are introduced the propagation or confinement of light in these structures is not controllable. Thus, discretionary switching of light, for example, or re-routing of optical signals, is not available with fixed defects in PCs. There are two approaches that have been pursued to tune the properties of PCs: these are by tuning the refractive index of the constituent materials, or by altering the physical structure of the PC. In the latter case, the emphasis has been principally on changing the lattice constant, although other approaches are also relevant (i.e., including for example, the fill factor, structure symmetry, and scattering-element shape). In the text below we discuss the state of the art in both of these areas.

9.3.1 Tuning PC response by changing the refractive index of constituent materials

We discuss recent progress in using four approaches within this category – namely, tuning the PC response by using (a) light, (b) applied electric fields, (c) temperature or electrical field, and (d) changing the concentration of free carriers (using electric field or temperature) in semiconductor-based PCs.

(a) PC refractive index tuning using light

One approach to modifying the behavior of PCs is to use intense illumination of the PC by one beam of light to change the optical properties of the crystal in a nonlinear fashion; this, in turn, can thus control the properties of the PC for another beam of light. An example of a nonlinear effect includes the flattening of the photon dispersion relation near a PBG, which relaxes the constraints on phase matching for second- and third-harmonic generation [9.19]. In addition, light location near defects can enhance a variety of third-order nonlinear processes in 1D PCs [9.20].

A new approach for PBG tuning has been proposed based on photo-reversible control over molecular aggregation, using the photochromic effect in dyes [9.21]. This can cause a reversible change over the photonic stop band. Structures that were studied included opal films formed from 275 nm diameter silica spheres, infiltrated with photochromic dye: two dyes were considered – namely, 1,3-dihydro-1,3,3-trimethylspiro-(2*H*-indole-2,3′-[3*H*]naphth[2,1-*b*][1,4]oxazine)   (SP)   and   *cis*-1,2-dicyano-1,2-bis(2,4,5-trimethyl-3-thienyl)ethene (CMTE). For the SP dye-opal, a reversible 15 nm shift in the reflectance spectrum was observed following UV irradiation, which was ascribed to changes in refractive index due to resonant absorption near the stop band. Smaller shifts (of about 3 nm) were observed for the CMTE dye-opal. The recovery process on cessation of UV illumination was quite slow, taking about 38 s for the SP dye.

Nonlinear changes in the refractive index have also been studied in PCs comprising 220 nm diameter SA polystyrene spheres infiltrated with water [9.22]. In these studies, the optical Kerr effect was used to shift the PBG. 40 GW/cm^2 of peak pump power at 1.06 µm (35 ps pulses at 10 Hz repetition rate) was used to shift the PBG 13 nm. The large optical nonlinearity originates from the delocalization of conjugated π-electrons along the polymer chains, leading to a large third-order nonlinear optical susceptibility. Time response was measured as a function of the delay time between pump and probe, and confirmed a response time of several picoseconds.

(b) PC refractive index tuning using an applied electric field

There have been a number of studies focusing on using ferroelectric materials to form PCs [9.23–9.26]. The application of an electric field to such materials can be used to change the refractive index and tune the PC optical response. For example, lead lanthanum zirconate titanate (PLZT) inverse opal structures have been fabricated by infiltration using 350 nm diameter polystyrene sphere templates and annealing at 750 °C [9.25]. The films were formed on indium tin oxide (ITO)-coated glass to enable electric-field-induced changes in reflectivity to be measured due to the electro-optical effect. Applying voltages of up to about 700 V across films of thickness of about 50 mm only achieved a few nm of peak shift, attributed to the very modest changes in refractive index from the applied field (i.e., from 2.405 to 2.435, as a result of the bias). It should be noted, however, that the intrinsic response of the electro-optical effect in these materials is known to be in the GHz range and hence highly suitable for rapid tuning. Other reports include the formation of inverse opal barium strontium titanate (BST) PCs using infiltration of polystyrene opals [9.23, 9.24]. BST is most interesting in that it provides a high-refractive-index material and a factor of at least two times higher breakdown field strength than PLZT, and hence offers a much wider range of applied field for tuning. Reports of using other ferroelectric materials include a high-temperature infiltration process for the ferroelectric copolymer, poly(vinylidene difluoride–trifluoroethene), infiltrated into 3D silica opals with sphere diameters of 180, 225, and 300 nm [9.26].

(c) Refractive index tuning of infiltrated PCs

In this case, the approach has been to consider modulation of the refractive index of a PC infiltrated with a tunable medium – in particular, the most popular approach has been to use liquid-crystals to infiltrate porous 2D and 3D PC structures. Such liquid crystals can behave as ferroelectrics whose refractive index may be tuned either using an applied electric field, or by thermal tuning. Reported results for this approach, using 3D inverse opal structures, have been restricted to infiltration of a ferroelectric-liquid crystal material into a silicon/air PC [9.27]. Reported changes in the refractive index of the liquid crystal are from 1.4 to 1.6 under applied field [9.27]. However, since the ferroelectric liquid has a higher refractive index than air, the infiltration results in a significant decrease in the refractive index contrast. This means that the original full photonic bandgap of the inverse opal silicon PC no longer exists, and the practical utility of the silicon structure as a PC is effectively lost. Theoretical simulations have shown that partial surface wetting of the internal inverse opal surface can retain the full photonic band in silicon [9.27], but it is questionable whether such a complex structure can ever be practically fabricated. Finally, the presence of a ferroelectric liquid crystal surrenders one of the main advantages of the original concept for PC structures: that is, permitting light propagation in air [9.28]. The temperature tuning of the liquid crystal material was shown to result in very small changes in refractive index (change in $n < 0.01$ over a 70 °C change) and thus only provide minimal shifts in the transmittance through the PC over a large temperature range [9.29].

(d) PC refractive index tuning by altering the concentration of free carriers (using electric field or temperature) in semiconductor-based PCs

An elegant way to rapidly tune the PBG of semiconductor-based PCs is to adjust the refractive index by modulating the free-carrier concentration using an ultra-fast optical pulse [9.30]. Using this approach, reflectivity of a two-dimensional silicon-based honeycomb PCs with 412 nm air holes (100 µm in length) in a 500 nm periodic array, was studied with a pump-probe approach [9.30]. By varying the delay between the pump and probe beams, the speed of PBG tuning was measured to be about 0.5 ps. The reflectance relaxation (corresponding to return of the PBG to its original position) occurred on a time scale of 10 to 100 ns, characteristic of recombination of excess of electrons and holes. Although these results are very encouraging, this approach cannot suppress or correct for, light-scattering losses caused by structural imperfections, which remain important considerations for currently fabricated PCs.

9.3.2 Tuning PC response by altering the physical structure of the PC

The second approach that we discuss for tuning the response of a PC is based on changes to the physical structure of the PC. We discuss the following approaches for tuning using this approach: using (a) temperature, (b) an applied magnetic field, (c) strain/deformation, (d) piezoelectric effects, and (e) micro-electro-mechanical systems (MEMS) [i.e., actuation].

(a) Tuning PC response using temperature

An example of this approach is a study of the temperature tuning of PCs fabricated from self-assembled polystyrene beads [9.31]. The PBG in these structures was fine-tuned by annealing samples at temperatures from 20 to 100 °C, resulting in a continuous blue shift in the stop-band wavelength from 576 to 548 nm. New stop-bands appeared in the UV transmission spectra when the sample was annealed above about 93 °C – the glass transition temperature of the polystyrene beads.

(b) Tuning PC response using magnetism

An example of this approach is the use of an external applied magnetic field to adjust the spatial orientation of a PC [9.32]. This can find application in fabricating photonic devices such as tunable mirrors and diffractive display devices. These authors fabricated magnetic PCs by using monodispersed polystyrene beads self-assembled into a ferromagnetic fluid comprising magnetite particles, with particle sizes smaller than 15 nm [9.32]. On evaporation of the solvent, a cubic PC lattice was formed with the nano-particles precipitating out into the interstices between the spherical polystyrene colloids. These authors then showed how the template could be selectively removed by calcination or wet etching, to reveal an inverse opal of magnetite – such structures being proposed as being suitable for developing magnetically tunable PCs.

(c) Tuning PC response using strain

The concept in this case is quite straightforward – deforming or straining the PC changes the lattice constant or arrangement of dielectric elements in the PC with concomitant change in the photonic-band structure. Polymeric materials would appear to be most suited to this methodology, owing to their ability to sustain considerable strains. However, concerns of reversibility and speed of tuning clearly would need to be addressed. Theoretical predictions for the influence of deformation on such systems include a report on a new class of PC based on self-assembling cholesteric elastomers [9.33]. These elastomers are highly deformable when subjected to external stress. The high sensitivity of the photonic-band structure to strain, and the opening of new PBGs are discussed in this paper [9.34]. Charged colloidal crystals were also fixed in a poly(acrylamide) hydrogel matrix to fabricate PCs whose diffraction peaks were tuned by applying mechanical stress [9.35]. The PBG shifted linearly and reversibly over almost the entire visible spectral region (from 460 nm to 810 nm).

Modeling of the photonic band structure of 2D silicon-based triangular PCs under mechanical deformation was also reported [9.35]. The structures considered comprised a silicon matrix with air columns. The authors showed that while a 3% applied shear strain provides only minor modifications to the PBG, uniaxial tension can produce a considerable shift. Other modeling includes a study of how strain can be used to tune the anisotropic optical response of 2D PCs in the long-wavelength limit [9.36]. Their calculations showed that the decrease in dielectric constant per unit strain is larger in the direction of the strain than normal to it. Indeed, the calculated birefringence is larger than that of quartz. They suggest that strain tuning of this birefringence has attractive application in polarization-based optical devices.

To appreciate the sensitivity of such structures to mechanical tuning, it is instructive to refer to some recent work on PMMA inverse opal PC structures that were fabricated using silica opal templates [9.37]. Under the application of uniaxial deformation of these PCs, the authors found a blue shift of the stop-band in the transmission spectrum – the peak wavelength of the stop-band shifted from about 545 nm in the undeformed material to about 470 nm under a stretch ratio of about 1 to 6.

Another practical approach that has been applied to physically tuning PC structures is that of thermal annealing. One such study showed how the optical properties of colloidal PCs comprising silica spheres can be tuned through thermal treatment [9.38]. This was attributed to both the structural and physiochemical modifications of the material on annealing. A shift in the minimum transmission from about 1000 nm (unannealed) to about 850 nm (annealed at about 1000 °C) was demonstrated, or about a maximum shift in Bragg wavelength of ~11%.

A quite novel application of strain tuning was recently reported [9.39]. The authors studied 2D PCs comprising arrays of coupled optical micro-cavities fabricated from vertical-cavity-surface-emitting laser structures. The influence of strain, as manifested by shifts in the positions of neighboring rows of micro-cavities with respect to each other, corresponded to alternating square or quasi-hexagonal shear-strain patterns. For strains below a critical threshold value, the lasing photon-mode locked to the corresponding mode in the unstrained PC. At the critical strain, switching occurred between square and hexagonal lattice modes.

Finally, there has been a proposal for using strains in a PC to tune the splitting of a degenerate photon-state within the PBG, suitable for implementing tunable PC circuits [9.40]. The principle they applied is analogous to the static Jahn–Teller effect in solids. These authors showed that this effect is tunable by using the symmetry and magnitude of the lattice distortion. Using this effect they discussed the design of an optical valve that controls the resonant coupling of photon modes at the corner of a T-junction waveguide structure.

(d) Tuning PC response using piezoelectric effects

In this section we discuss using piezoelectric effects to physically change PC structures and hence tune them. A proposal was made for using the piezoelectric effect to distort the original symmetry of a 2D PC from a regular hexagonal lattice to a quasi-hexagonal lattice under applied electric field [9.41]. The original bands decomposed into several strained bands, dependent on the magnitude and direction of the applied field. In the proposed structures, the application of ~3% shear strain is shown to be suitable for shifting 73% of the original PBG, which they refer to as the tunable bandgap regime. An advantage of such an approach is that such structures are suggested to be capable of operation at speeds approaching MHz. Another report [9.42] discusses the design and implementation of a tunable silicon-based PBG micro-cavity in an optical waveguide, where tuning is accomplished using the piezoelectric effect to strain the PC – this was carried out using integrated piezoelectric micro-actuators. These authors report on a 1.54 nm shift in the cavity resonance at 1.56 μm, for an applied strain of 0.04%.

There have also been reports of coupling piezoelectric based actuators to PCs. One such report [9.43] discusses a poly(2-methoxyethyl acrylate)-based PC composite directly coupled to a piezoelectric actuator to study static and dynamic stop-band tuning characteristics – the stop-band of this device could be tuned through a 172 nm tuning range, and could be modulated at up to 200 Hz.

(e) Tuning PC response using micro-electro-mechanical systems (MEMS) actuation

There have been a number of interesting developments in this field including PC-based devices comprising suspended 1D PC mirrors separated by a Fabry–Perot cavity (gap) [9.44]. When such structures are mechanically perturbed, there can be a substantial shift in the PBG due to strain. The authors discuss how a suite of spectrally tunable devices can be envisioned based on such structures – these include modulators, optical filters, optical switches, wavelength-division multipliers (WDM), optical logical circuits, variable attenuators, power splitters, and isolators. Indeed, the generalization of these concepts beyond 1D was discussed in a recent patent [9.45] and covers tunable PC structures. This report [9.45], as well as others [9.46], considered an extension of these ideas to form families of micromachined devices. The latter authors [9.46] modeled and implemented a set of micromachined vertical-resonator structures for 1.55 μm filters comprising two PC (distributed Bragg reflector [DBR]) mirrors separated by either an air gap or semiconductor heterostructures. Electromechanical tuning was used to adjust the separation between the mirrors and hence fine-tune the transmission spectrum. The mirror structures were implemented using strong index-contrast InP/air DBRs giving an index contrast of 2.17, and weak-contrast (i.e., index contrast of 0.5) silicon nitride/silicon dioxide DBRs. In the former case, a tuning range of over 8% of the absolute wavelength was achieved – varying the inter-membrane voltage by up to 5 V gave a tuning range of over 110 nm. Similar planar structures are discussed in other papers [9.47] except for where the mirrors are formed from two slabs of PC separated by an adjustable air gap (also see reference [9.45]). These structures are shown to be able to perform as either flat-top reflection or all-pass transmission filters, by varying the distance between the slabs, for normally incident light. Unlike all previously reported all-pass reflection filters, based on Gires–Tournois interferometers using multiple dielectric stacks, their structure generates an all-pass transmission spectrum, significantly simplifying signal extraction and optical alignment – also the spectral response is polarization independent owing to the 90° rotational symmetry of their structure.

9.4 SELECTED APPLICATIONS OF PC

In this section a selection of some interesting application areas for PCs are discussed, focusing on integrated optics or PLCs, with notable omission of other important areas such as microwave-based PCs and sensing applications (some of which importantly cover the field of biotechnology). Regretfully, these omissions are necessary, as covering any one of the many other fields would inevitably entail another such special review to do the subject justice.

To appreciate the applications possibilities within this limited scope, recall that some of the key attractive properties that PCs possess for PLCs include, in particular, an ability to strongly confine light [9.1], as well as unique dispersive properties [9.10]. The property of confinement may be exploited to realize compact channel waveguides, sharp bends, and also to achieve high isolation between adjacent channels (i.e., so-called low cross-talk) [9.7]. Dispersive properties, on the other hand, readily lend themselves to optical functions including wavelength separation (e.g., the so-called superprism effect) as well as pulse-shape modification (e.g., pulse compression) [9.1]. It should be noted that, as far as the maturity of technology is concerned, current PLCs are very simple systems. Typically, they comprise components having sizes from mm to cm, permitting a very limited number of different functions to be integrated together on a chip, very much reminiscent of the state of electronic integrated circuits (ICs) in the early 1960s.

9.4.1 Waveguide devices

As discussed above there are a number of different types of waveguide devices that have been developed in PCs. These include both channel waveguides and coupled resonant optical waveguide (CROW) structures [9.48]. Practical implementation of these structures relies on achieving low losses. Planar PCs are able to achieve ~4–10 dB/unit cell and so ~30 dB can be achieved within a few cells. There are three key loss mechanisms to consider in these structures: (a) scattering losses at structural imperfections in the PCs, (b) out-of-plane losses, and (c) losses caused by transverse electric/transverse magnetic (TE/TM) coupling.

When considering scattering losses at imperfections it should be recognized that, intrinsically, PC waveguides should not suffer any in-plane losses as such structures are designed to forbid all propagating modes at the operating wavelength. One therefore expects that scattering from such imperfections should have a much lower impact in PC structures than for ridge waveguides having similar dimensions. On the other hand, in multimode waveguides, this type of scattering can excite higher-order modes and therefore become more significant. PLCs are ideally made as two-dimensional devices, with the third dimension assumed theoretically to be infinite. However, in real finite PLC structures, out-of-plane losses from diffraction and scattering provide the dominant loss mechanism in PC waveguides. In the case of structures formed by etching hole arrays into a solid semiconductor structure, losses originate from poor guiding by holes and their depth being insufficiently long. This causes part of the waveguide mode to be scattered away. Such losses can be minimized by increasing the aspect ratio of the etched holes. TE/TM coupling losses can occur in cases such as when the PC surrounding a channel waveguide provides a bandgap for only the TE polarization. In this case, once the confined TE mode is converted to TM, it no longer experiences the photonic bandgap and is therefore free to leak out of the channel waveguide and propagate through the crystal.

CROWs are quite distinct from conventional waveguides and have no real analogy in traditional guided-wave devices. The waveguide is formed through an array of coupled defects in the PC [9.48]. In some sense these structures are analogous to quantum electronic structures such as resonant tunneling structures in which extended states are coupled to standing-wave-type cavity states to control the transmission flux through the structure. In the case of CROWs, by varying the type of defects and their spacing, one can control the propagation of light from defect to defect, and this allows for the tailoring of the wavelength response of the structure as well as its group velocity [9.48]. Some impressive experimental results have been demonstrated for light propagation in such structures. Importantly, such structures overcome some of the problems of the more conventional PC-based waveguiding devices, showing very little propagation loss as well as providing a means to guide light around sharp bends for ultra-compact optical device and systems applications. Such structures have been successfully demonstrated in the microwave regime where CROWs are real competitors for conventional PC channel waveguides [9.49].

Although somewhat unglamorous, the problem of coupling fiber to PCs presents one of the biggest engineering challenges for implementing PC-based PLCs in practice. The problem originates from the fact that fibers possess a circular cross-section of ~5–7 μm diameter as distinct from the PC waveguides having cross-sections of only several 100s of nm. This leads to severe mode mismatch between the fiber and PC, and is responsible for insertion losses that can be ~20–30 dB. Various approaches to solving this problem have been discussed in the literature.

9.4.2 Dispersive devices

Light guiding in PC-based structures can be supplemented by unique dispersive devices using the same systems. This again provides a distinct advantage for PCs over traditional alternatives for realizing compact, highly functional optical systems for a variety of applications. One of the most important aspects of this is waveguide dispersion using the superprism effect. That is for performing beam steering and multiplexing and de-multiplexing (MUX/DEMUX) for dense wavelength-division multiplexing (DWDM), for example. Such dispersion effects in periodic structures have been known for some time [9.50] and were rediscovered in PCs for application in WDM systems. The superprism effect, as described by Kawakami and co-workers [9.10], uses the band structural asymmetry near the Γ-point, where the wavevector changes much more dramatically with change in frequency in the Γ–K direction as compared with the Γ–M direction. These workers show that an angular dispersion of 50° can be observed for a change in input wavelength from 900 nm to 1000 nm. As discussed above, PC systems are, in principle, scalable to any wavelength range. Thus, when such a system is scaled to the 1.55 μm regime for telecommunication applications, this dispersive behavior would correspond to ~2° for the 50 GHz channel spacing in a typical DWDM system. If one assumes a series of output waveguides that are laterally separated by 5 μm, a distance of ~150 μm would be required to separate the channels. This is an impressively small distance compared with current phased-array waveguides where dimensions are typically on the mm to cm scale.

9.4.3 Add/Drop multiplexing devices

Some of the key components in DWDM systems are so-called add/drop devices that allow for the selective removal or addition of a particular wavelength channel to an optical data stream, thereby allowing all the channels to be fully utilized. The first PC-based add/drop device was proposed in 1998 [9.51] and was based on the resonance created by two defects in intersecting PC waveguide devices, enabling coupling between two channel waveguides at resonance frequency. It should be noted that such devices are closely related to CROW structures discussed above. By analogy to quantum electronics, whereas CROWS involve a series of coupled cavities, these devices appear as the equivalent of a single isolated quantum well/dot structure coupled to their waveguides. Unfortunately, experimental realization of this approach has proven to be particularly challenging. The main reason for this is that there is a strong dependence of the resonance condition on the size of the defect, making it impractical to control the dimensions of the defect to the required tolerance.

9.4.4 Applications of PCs for LEDs and lasers

The concept of using micro-structured mirrors for developing semiconductor laser structures is fairly mature. Periodic micro-structures have been commonly used in distributed feedback (DFB) and distributed Bragg reflector (DBR) lasers, as well as for vertical-cavity structures (e.g., vertical-cavity-surface-emitting lasers [VCSEL] using Bragg stacks). However, the dielectric contrast in such structures is typically less than 1% as compared with almost 4 : 1 in PC-based structures. This is particularly significant as it can lead to much shorter interaction lengths of ~1 μm as compared with hundreds of μm for conventional DFB/DBR structures. Importantly, this reveals the possibility of creating edge-emitting laser elements with very small optical volume, as is discussed below. Our discussions begin by considering how PCs can be fruitfully applied to improve the performance of LEDs.

The principal active material for light-emitting diodes (LEDs) remains semiconductors with reported internal quantum efficiencies as high as 99.7% [9.52]. The utility of these materials then depends on how efficiently this generated light can be extracted. As a consequence of Snell's Law, only the radiation falling within a small cone (i.e., ~16° for GaAs) can escape – everything else is totally internally reflected. A simple calculation shows that the extraction efficiency scales as $1/4n^2$ where n is the semiconductor refractive index. Thus, the small cone of light allowed by Snell's Law represents only a few percent (e.g., about ~2%) of the total available solid angle for GaAs, explaining the low observed external quantum efficiency of ~2–4% in standard commercial LEDs. High-brightness LEDs have recently come to market and are finding applications in display and lighting applications owing to their superior brightness and lifetime compared with incandescent sources. These devices make use of light extracted from more than one facet. Another approach is to place the active layer on a low-index substrate and then roughen it. The resulting surface scatters a large fraction of the light out of the material and leads to measured efficiencies of as high as ~30% [9.52]. However, one of the most promising approaches to developing LEDs is based on PCs, as these structures offer the potential to improve both the extraction efficiency of the light as well as control over the direction of light emission from the structures. As regards the latter, PCs enable one to suppress emission in unwanted directions, and to enhance it in desired directions. Such structures have been predicted to be suitable for producing light emitters with external quantum efficiencies exceeding 50% [9.1]. Two main factors for controlling the emission are: (a) increased extraction and photon recycling, and (b) altering the fundamental emission process.

One of the key motivations for studying PCs is their promise for strong spontaneous emission enhancement – i.e., owing to the fact that the spontaneous emission rate of an excited atom can be increased by placing it in a micro-cavity. In particular, such a micro-cavity can be formed by creating a defect within a PC structure as discussed above. The maximum enhancement factor is achieved when the radiating dipole is so oriented as to experience the maximum interaction with the cavity mode. The cavity enhancement factor f_P or Purcell factor is given by $(3/4\pi^2)(Q/V)(\lambda/n)^3$, where Q is the quality factor of the cavity, and V is the modal volume, n is the refractive index of the cavity material, and λ is the wavelength. PC micro-cavities offer high-Q micro-cavities with exceptionally low mode volume suitable for very high spontaneous-emission-enhancement factors. Recently, it was shown that one could indeed control timing of light emission from CdSe nano-crystals (i.e., quantum dots) embedded in a titania inverse opal PC. Londahl et al. [9.53] found that by changing the PC lattice constant one could accelerate/decelerate the rate of spontaneous emission from the quantum dots by as much as a factor of two. Speeding up of the spontaneous emission could lead to more efficient light sources such as LEDs, while slowing it down could help create more efficient solar cells.

The application of PCs for developing lasers also appears to be very promising. Typical semiconductor-based lasers typically have between 10^4 to 10^5 modes of which only one is the desired lasing mode. Clearly, therefore, a lot of light is wasted before stimulated emission occurs. In addition, since all of these modes are still there after the onset of lasing, they contribute to noise. PC structures offer a means for reducing the number of available modes and thus reducing both the lasing threshold condition and also reducing dramatically the laser noise characteristics. This may be described by reference to the so-called β-factor, where $\beta = \Gamma\lambda^4/8\pi vn^3\Delta\lambda$. β Quantifies the mode ratio of the desired mode to all the other modes and determines the laser threshold and Γ is the confinement factor (the ratio of the gain volume to cavity volume). Defects can be used in the PCs to place modes in an otherwise forbidden spectral regime – for example, one can artificially create a mode within the forbidden bandgap by using a defect and the β-factor will then assume a value of unity because the single allowed mode becomes the lasing mode. The principle of realizing a laser using a PC structure is quite straightforward then – one needs to design the PC such that the defect frequency coincides sufficiently closely with the gain peak of the respective emitter material.

One other interesting development in PCs is related to harnessing surface plasmons. This may be applied to develop sensors, light emitters, or light-detecting structures. In particular, we focus on the application of plasmonic structures for light emitters. Plasmonics rely on plasmon-related effects in metallic structures, which can offer exciting phenomena when applied to light-emitting devices. Surface plasmons, quanta of electron oscillations at metal/dielectric interfaces, have been used to explain the super-transmission effect where a transmission of ~4% was measured for a thin metal film perforated with holes comprising ~2% area fill-factor and size well below the cut-off wavelength (150 nm for 1.55 μm light emission) [9.54]. The best explanation to date for this is that incident light excites the top-surface plasmons, which then couple to the surface plasmon on the other side of the metal film. This latter surface-plasmon mode couples to radiation modes, making the whole process appear as if the incident light were directly transmitted through the metal film. Surface-plasmon modes also have been observed in small metal particles with the possibility of enhancing interactions between light and other structures such as for photodetectors using such an approach. The concept of developing PCs from such plasmonic systems has obvious attractiveness as a means to use such resonances to transfer light from an active semiconductor structure to the outside world, improving light-extraction efficiency – one of the key problems for light-source design. The design of such systems would rely on matching the plasma resonance frequency to that of the emission, and by using control over the PC structure, control the coupling.

9.5 CONCLUSIONS

This paper discussed the principles and properties of PCs, both passive and tunable, with emphasis on application of such structures for PLCs. New phenomena in these systems are offering a glimpse of the far-reaching prospects of developing photonic devices which can, in a discretionary fashion, control the propagation of modes of light in an analogous fashion to how nanostructures have been harnessed to control electron-based phenomena. Analogous to the evolution of electronic systems, one can anticipate a path toward development of compact, active, integrated photonic systems, as envisioned based on the technology outlined in this review. We discuss proposals and indeed demonstrations of a wide variety of crystal-based photonic devices with applications in areas including communications, computing, and sensing, for example. In such applications, photonic crystals can offer both a unique performance advantage, as well as a potential for substantial miniaturization of photonic systems.

ACKNOWLEDGEMENTS

The authors gratefully acknowledge support from NSERC, CSA, CIPI, OCE, and AFOSR.

REFERENCES

[9.1] T.F. Krauss and R.M. de la Rue, *Prog. Quant. Electron.*, **23**, 51 (1999).

[9.2] V. Berger, *Opt. Mater.*, **11**, 131 (1999).

[9.3] D.J. Norris and Y.A. Vlasov, *Adv. Mater.*, **13**, 371 (2001).

[9.4] V. Mizeikis, S. Juodkazis, A. Marcinkevicius, S. Matsuo, and H. Misawa, *J. Photochem. Photobiol. C*, **2**, 35 (2001).

[9.5] *Photon. Nanostruct.*, **1**, (2003); the whole first issue is dedicated to fundamentals and applications of photonic crystals.

[9.6] S.G. Johnson and J.D. Joannopoulos, Photonic Crystals: Road from Theory to Practice (Kluwer Academic Publishers, Borton, MA, 2002).

[9.7] J.D. Joannopoulos, R.D. Meade, and J.N. Winn, PCs: Moulding the Flow of Light, (Princeton University Press, Princeton, NJ, 1995).

[9.8] S. John, *Phys. Rev. Lett.*, **58**, 2486 (1987).

[9.9] E. Yablonovitch, *Phys. Rev. Lett.*, **58**, 2059 (1987).

[9.10] H. Kosaka, T. Kawashima, A. Tomita, M. Notomi, T. Tamamura, T. Sato, and S. Kawakami, *Phys. Rev. B*, **58**, R10096 (1998).

[9.11] V. Berger, *Curr. Opin. Solid State Mater. Sci.*, **4**, 209 (1999).

[9.12] K.M. Ho, C.T. Chan, and C.M. Soukoulis, *Phys. Rev. Lett.*, **65**, 3152 (1990).

[9.13] S. Satpathy, Z. Zhang, and M.R. Salehpour, *Phys. Rev. Lett.*, **64**, 1239 (1990).

[9.14] E. Yablonovitch, *J. Opt. Soc. Am. B*, **10**, 283 (1993).

[9.15] Y. Xia (editor), Special Issue on Photonic Crystals, *Adv. Mater.*, **13**, (2001).

[9.16] C. Anderson and K. Giapis, *Phys. Rev. Lett.*, **77**, 2949 (1996).

[9.17] R.D. Meade, A.M. Rappe, K.D. Brommer, and J.D. Joannopoulos, *J. Opt. Soc. Am. B*, **10**, 328 (1993).

[9.18] R.D. Meade, A.M. Rappe, K.D. Brommer, J.D. Joannopoulos, and O.L. Alerhand, *Phys. Rev. B*, **48**, 8434 (1993).

[9.19] J. Martorell, R. Vilaseca, and R. Corbalan, *Appl. Phys. Lett.*, **70**, 702 (1997).

[9.20] H. Inouye and Y. Kanemitsu, *Appl. Phys. Lett.*, **82**, 1155 (2003).

[9.21] Z.-Z. Gu, T. Iyoda, A. Fujishima, and O. Sato, *Adv. Mater.*, **13**, 1295 (2001).

[9.22] X. Hu, Q. Zhang, Y. Liu, B. Cheng, and D. Zhang, *Appl. Phys. Lett.*, **83**, 2518 (2003).

[9.23] I. Soten, H. Miguez, S.M. Yang, S. Petrov, N. Coombs, N. Tetreault, N. Matsuura, H.E. Ruda, and G.A. Ozin, *Adv. Funct. Mater.*, **12**, 71 (2002).

[9.24] N. Matsuura, S. Yang, P. Sun, and H.E. Ruda, *Appl. Phys. A*, **81**, 379 (2005).

[9.25] B. Li, J. Zou, X.J. Wang, X.H. Liu, and J. Zi, *Appl. Phys. Lett.*, **83**, 4704 (2003).

[9.26] T.B. Xu, Z.Y. Cheng, Q.M. Zhang, R.H. Baughman, C. Cui, A.A. Zakhidov, and J. Su, *J. Appl. Phys.*, **88**, 405 (2000).

[9.27] S. John and K. Busch, *J. Lightwave Technol.*, **17**, 1931 (1999).

[9.28] J.D. Joannopoulos, *Braz. J. Phys.*, **26**, 53 (1996).

[9.29] K. Yoshino, Y. Kawagishi, M. Ozaki, and A. Kose, *Jpn. J. Appl. Phys.*, **38** Part 2, L786 (1999).

[9.30] S.W. Leonard, H.M. van Driel, J. Schilling, and R.B. Wehrspohn, *Phys. Rev. B*, **66**, 161102-1 (2002).

[9.31] B. Gates, S.H. Park, and Y. Xia, *Adv. Mater.*, **12**, 653 (2000).

[9.32] B. Gates and Y. Xia, *Adv. Mater.*, **13**, 1605 (2001).

[9.33] P.A. Bermel and M. Warner, *Phys. Rev. E*, **65**, 010702(R)-1 (2001).

[9.34] Y. Iwayama, J. Yamanaka, Y. Takiguchi, M. Takasaka, K. Ito, T. Shinohara, T. Sawada, and M. Yonese, *Langmuir*, **19**, 977 (2003).

[9.35] S. Jun and Y.-S. Cho, *Opt. Express*, **11**, 2769 (2003).

[9.36] C.-S. Kee, K. Kim, and H. Lim, *Physica B*, **338**, 153 (2003).

[9.37] K. Sumioka, H. Kayashima, and T. Tsutsui, *Adv. Mater.*, **14**, 1284 (2002).

[9.38] H. Miguez, F. Meseguer, C. Lopez, A. Blanco, J.S. Moya, J. Requena, A. Mifsud, and V. Fornes, *Adv. Mater.*, **10**, 480 (1998).

[9.39] H. Pier, E. Kapon, and M. Moser, *Nature*, **407**, 880 (2000).

[9.40] N. Malkova and V. Gopalan, *Phys. Rev. B*, **68**, 245115-1 (2003).

[9.41] S. Kim and V. Gopalan, *Appl. Phys. Lett.*, **78**, 3015 (2001).

[9.42] C.W. Wong, P.T. Rakich, S.G. Johnson, M. Qi, H.I. Smith, E.P. Ippen, L.C. Kimmerling, Y. Jeon, G. Barbastathis, and S.-G. Kim, *Appl. Phys. Lett.*, **84**, 1242 (2004).

[9.43] S.H. Foulger, P. Jiang, A. Lattam, D.W. Smith, J. Ballato, D.E. Dausch, S. Grego, and B.R. Stoner, *Adv. Mater.*, **15**, 685 (2003).

[9.44] S. Rajic, J.L. Corbeil, and P.G. Datskos, *Ultramicroscopy*, **97**, 473 (2003).

[9.45] N. Matsuura, H.E. Ruda, and B.G. Yacobi, *Configurable Photonic Device*, US patent 6 961 501 (2005).

[9.46] H. Hillmer, J. Daleiden, C. Prott, F. Römer, S. Irmer, V. Rangelov, A. Tarraf, S. Schüler, and M. Strassner, *Appl. Phys. B*, **75**, 3 (2002).

[9.47] W. Suh and S. Fan, *Opt. Lett.*, **28**, 1763 (2003).

[9.48] A. Yariv, Y. Xu, R.K. Lee, and A. Scherer, *Opt. Lett.*, **24**, 711 (1999).

[9.49] M. Bayindir, B. Temelkuran, and E. Ozbay, *Phys. Rev. B*, **61**, R11855 (2000).

[9.50] P.S.J. Russell, *Phys. Rev. A*, **33**, 3232 (1986).

[9.51] S. Fan, P.R. Villeneuve, J.D. Joannopoulos, and H.A. Haus, *Opt. Express*, **3**, 4 (1998).

[9.52] I. Schnitzer, E. Yablonovitch, C. Caneau, and T.J. Gmitter, *Appl. Phys. Lett.*, **62**, 131 (1993).

[9.53] P. Londahl, A.F van Driel, I.S. Nikolaev, A. Irman, K. Overgaag, D. Vanmaekelbergh, and W.L. Vos, *Nature*, **430**, 654 (2004).

[9.54] T.W. Ebbesen, H.J. Lezec, H.F. Ghaemi, T. Thio, and P.A. Wolff, *Nature*, **391**, 667 (1998).

10 Nonlinear Optical Properties of Photonic Glasses

K. Tanaka

*Department of Applied Physics, Graduate School of Engineering, Hokkaido University,
Sapporo 060-8628, JAPAN
e-mail: keiji@eng.hokudai.ac.jp*

10.1 INTRODUCTION

In this article, we define photonic glass as a high-purity inorganic glass, which can be applied to photonics. There are many kinds of applications which utilize linear and nonlinear optical properties. The most important one is undoubtedly the optical fibre, which has densely surrounded the earth for optical communications. The optical fibre is also used in functional devices such as Bragg reflectors, optical amplifiers, and lasers [10.1]. In addition, integrated glass waveguides are currently being developed for optical signal processing such as wavelength division [10.2, 10.3]. In these applications, not only linear but also nonlinear optical properties play important roles. On the one hand, large nonlinearities are needed for all-optical switches, power stabilizers, soliton fibres, super-continuum generators, etc. [10.4, 10.5]. Nonlinear optical excitations are useful also for three-dimensional memories and photo-electronic microfabrication [10.4]. On the other hand, we need small nonlinearity in high-intensity glass lasers and nondispersive optical fibres.

Optical Properties of Condensed Matter and Applications   Edited by J. Singh
© 2006 John Wiley & Sons, Ltd

Despite of the wide range of applications, glass science still remains at a preliminary level in comparison with crystal science [10.6–10.10]. This is because of three essential features. First, owing to the lack in long-range structural periodicity, X-ray diffraction patterns are less informative in glasses, and explicit atomic structures cannot be determined. Second, also owing to the lack of periodicity, Bloch wave functions cannot be postulated, in principle, for electrons and atomic vibration. The wave number is no more a good quantum number, and only the density-of-states has some physical meaning. The conventional one-electron approximation is largely limited. Under these circumstances, we may analyse a kind of many-body problem for electrons and atoms using computer simulations. However, simulated results cannot necessarily provide universal insights, because of restricted calculating conditions such as enormously fast quenching speeds. Third, glass is thermodynamically quasi-stable, having glass-transition temperatures that vary with preparation procedures and post-preparation treatments. The quasi-stability also causes variations in macroscopic properties. We cannot envisage an ideal glass structure, even in a statistical sense, which is in marked contrast to the situation in an ideal crystal and an ideal gas, which are perfectly periodic and completely random [10.10], respectively. These three features pose many unresolved problems. As for linear optical properties, for instance, we cannot yet give definite interpretations to the origins of Tauc gaps [10.9, 10.11], minimal Urbach-edge energy of ~50 meV [10.12, 10.13], and so forth.

Optical nonlinearity in photonic glasses also needs to be studied further. Experimentally, most previous work has investigated nonlinear refractivity [10.5, 10.14], probably because of its wider relevance to all-optical switches, etc. Nonlinear absorption is comparatively less well studied. In addition, in many cases, nonlinear properties have been evaluated at fixed laser wavelengths, and spectral studies are largely limited. Theoretically, in conventional approaches, nonlinear refractive glasses have been regarded as dielectrics, where photoelectronic excitations are treated as virtual [10.4]. Accordingly, the relationship between absorption and refractivity is not clear. On the other hand, interpretations of the nonlinear absorption in glasses have been naive, most taking only the density-of-states into account, neglecting the chemical origin of electronic wave functions [10.15]. However, as described in Section 10.3.1, the transition amplitudes of one- and two-photon absorptions have contrasting forms, so that parallel studies on these absorption processes may add new insights.

In the present article, we will study two topics on the optical nonlinearity in photonic glasses. After a brief review of the photonic glass in Section 10.2, Section 10.3 focuses on a unified understanding of the third-order nonlinearity. The optical nonlinearity in glasses is analysed in Section 10.3 using semiconductor terminology, which may be complementary to the dielectric approach mentioned above. In the present method, as illustrated in Figure 10.1, the relationship among atomic structures (atom and bonding), electronic structures, optical absorption spectra (α, E_g, and β), and refractivity spectra (n_0 and n_2) becomes clearer. Spectral dependence is obtained in a straightforward way. In Section 10.4 we will consider the role of nonlinear excitations in photoinduced phenomena. It is often asserted that, when a photoinduced phenomenon is induced by light of photon energy less than the bandgap ($\hbar\omega < E_g$), the nonlinear excitation is held responsible. However, this may be a hasty conclusion, because it neglects mid-gap states, whose existence is inherent to the glass. We will see that photo-excitation mechanisms in glasses are not as simple as those in crystals.

For other nonlinear properties, such as the second-order nonlinearity in poled glasses [10.2, 10.16] and large nonlinearity in nano-particle dispersed glasses [10.2, 10.17], the reader may refer to a recent review article [10.18].

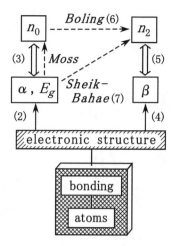

Figure 10.1 Relations between atomic structure (atoms and bonding), electronic structure, absorption spectra (α and β), and refractive-index spectra (n_0 and n_2). n_0, α and E_g are linear properties, and n_2 and β are nonlinear. Solid single arrows represent Equations (10.2) and (10.4), thick double-ended arrows represent linear and nonlinear Kramers–Krönig relations, Equations (10.3) and (10.5), and dashed arrows show the Moss rule ($E_g \sim n_0$), Bolings' relation [Equation (10.6)], and Sheik-Bahaes' relation [Equation (10.7)]

10.2 PHOTONIC GLASS

Several kinds of photonic glasses are available at present, which include oxides [10.6, 10.7], chalcogenides [10.6, 10.8], and halides [10.7, 10.19]. As is well known, studies on nonlinear effects in halide glasses have been limited due to their smallness [10.2, 10.7, 10.19], and accordingly these will not be included in this article. On the other hand, both oxide and chalcogenide glasses consist of the group VIb atoms (O, S, Se, and Te), which are characterized by s^2p^4 valence-electron configurations, so that their atomic and electronic structures can be considered in a unified way [10.10].

Figure 10.2 shows schematic atomic networks in SiO_2 and As_2S_3. The structure of such simple glasses can be grasped at three atomic scales. First, it is demonstrated that the so-called short-range structure, which includes the atomic coordination number, bond length, and bond angle, is similar to that in the corresponding crystal [10.6, 10.8, 10.9]. Second, the medium-range order, which encompasses atomic structures of 0.5–3 nm, is believed to exist, but the actual structures have not been elucidated, because experimental methods are limited. Small rings, such as 3-membered Si–O rings [Figure 10.2(a)], probably exist in SiO_2 [10.6, 10.20], and distorted layer structures [Figure 10.2(b)] are proposed for $As_2S(Se)_3$ and $GeS(Se)_2$ [10.8]. Third, there are point-like defects such as E′ centres [Figure 10.2(a)] in SiO_2 and D^0 in GeS_2, which are neutral dangling bonds with typical densities at the ppm level so that their features can be analysed through electron-spin signals [10.6, 10.9]. In covalent stoichiometric glasses such as As_2S_3, Raman-scattering spectra show unambiguously wrong bonds [Figure 10.2(b)], i.e., homopolar bonds in stoichiometric glasses, which exist with typical density of ~1 at.% [10.6, 10.8, 10.13, 10.21]. Many other defects such as

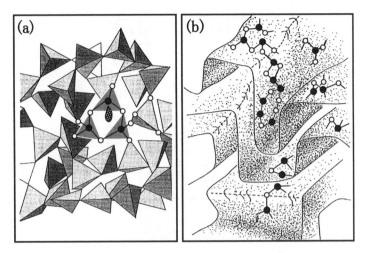

Figure 10.2   Atomic structures of (a) SiO$_2$ and (b) As$_2$S$_3$ glasses. In (a), Si and O are shown as a solid and an open circle with 4- and 2-fold coordination. In (b), As and S are shown as a solid and an open circle with 3- and 2-fold coordination. The bond lengths are 0.16 nm in (a) SiO$_2$ and 0.23 nm in (b) As$_2$S$_3$, so that side lengths of these illustrations are 2–3 nm. Note that, near the centers, (a) includes a small ring and an E′ centre and (b) contains wrong bonds (As–As and S–S). In PbO–SiO$_2$ glass, Si atoms in SiO$_2$ (a) are replaced by Pb atoms with some changes in their atomic coordination

charged defects (D$^+$ and D$^-$) have been proposed, but it is difficult to experimentally confirm their existence.

Constituent atoms and an amorphous structure determine the electronic structure in three levels [10.6, 10.8–10.10]. First, the short-range structure governs the bandgap energy E_g, which ranges over 4–10 eV for oxides and 1–3 eV for chalcogenides (see Figure 10.3). Accordingly, the latter are regarded as a kind of amorphous semiconductor. For simple stoichiometric glasses such as SiO$_2$ and As$_2$S$_3$, the origins of the valence and conduction bands are known, as will be described in Section 10.3.2. The bandgap energy governs the refractive index n_0 of a material, which is ~1.7 in oxides and ~2.5 in chalcogenides, being consistent with the Moss rule, $n_0^4 E_g \approx 77$ [10.22]. Second, the medium-range structure is assumed to affect densities-of-states at band edges, which may govern the steepness of exponential Urbach edges [10.20, 10.23]. Third, the defects such as dangling bonds and wrong bonds are likely to produce gap states, which produce residual optical absorption [10.10, 10.13].

It seems that the defect-induced absorption is inherent to glasses [10.10]. For an ideal insulator crystal, which is conceptually obtained through infinitesimally slow cooling of a melt to 0 K, we can envisage complete transparency arising from zero gap-states. By contrast, the glass is prepared through freezing at around a finite temperature, i.e., the glass-transition temperature, so that it necessarily contains structural disorders such as strained bonds and defects. And, the defective bonds are very likely to produce gap states, which give rise to the mid-gap absorption and/or weak optical absorption tails. Interestingly, the tail in chalcogenide glasses cannot be detected by photoconductive measurements [10.12, 10.21].

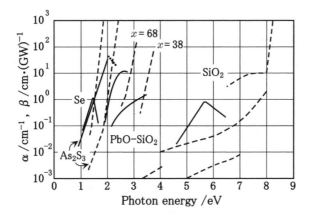

Figure 10.3 Spectral dependence of the one- and two-photon absorption coefficients, α (dashed lines) and β (solid lines), respectively, in SiO_2, As_2S_3, Se, and two xPbO–$(100 - x)$SiO$_2$ glasses with $x = 38$ and 68. α in SiO$_2$ at $\hbar\omega \leq 8\,eV$ may be influenced by light scattering, where some different spectra have been reported, so that α can be read as attenuation

10.3 NONLINEAR ABSORPTION AND REFRACTIVITY

10.3.1 Fundamentals

The polarization P induced by an electric field E in a material can be written in cgs units as [10.4, 10.5]:

$$P = \chi^{(1)}{:}E + \chi^{(2)}{:}E{\cdot}E + \chi^{(3)}{:}E{\cdot}E{\cdot}E... \tag{10.1}$$

Here, the first term $\chi^{(1)}{:}E$ depicts the conventional linear response, and other terms give nonlinear responses. The overall feature is illustrated on a frequency axis in Figure 10.4. $\chi^{(1)}$ is related with the linear refractive index n_0 and absorption coefficient α [Figure 10.5(a), (a′)]. Here, we should note that the glass is macroscopically isotropic, and accordingly it cannot provide even-order nonlinear effects arising from $\chi^{(2)}{:}E{\cdot}E$, with notable exceptions in poled glasses [10.4, 10.5, 10.16, 10.18]. In such cases, then the third-order nonlinearity $\chi^{(3)}{:}E{\cdot}E$ $\cdot E$ becomes the most important nonlinearity, and in some cases the fifth-order effects may appear as well [10.24, 10.25].

The third term provides at least three effects [10.4, 10.5]. First is the third-harmonic generation (see Figure 10.4). That is, if the frequency of E is ω, E^3 contains a term of 3ω, and the component may be detected if a phase-matching condition for the fundamental and the overtone light could be satisfied. However, in isotropic materials such as a glass, the condition is in general difficult to be satisfied [10.26], resulting into low generation efficiency. In addition, if ω is in the near-IR, 3ω may be located in the UV, at which frequency a glass may not be transparent. Accordingly, such overtone generations are rather limited. Second is the so-called optical Kerr effect, including self-(de)focusing, which arises from the intensity-dependent refractive-index change n_2I. Such a change appears because the third term of Equation (10.1), $\chi^{(3)}{:}E{\cdot}E{\cdot}E$, can be rewritten as $n_2I\,E$, in which n_2 is related to $\chi^{(3)}$. Third

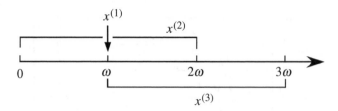

Figure 10.4 Roles of electric susceptibilities, $\chi^{(1)}$, $\chi^{(2)}$, and $\chi^{(3)}$, in frequency scale upon excitation with light of frequency ω

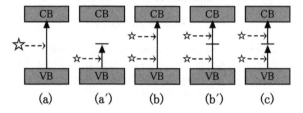

Figure 10.5 Schematic illustrations of (a) one-photon, (b) two-photon, (b′) resonant two-photon, and (c) two-step absorptions from valence (VB) to conduction (CB) bands in a semiconductor. (a′) shows a mid-gap absorption. Note that the resonant two-photon absorption (b′) is a one-step process, which occurs resonantly with mid-gap states, while the two-step absorption (c) consists of two successive one-photon processes

is the two-photon absorption, which is schematically illustrated in Figure 10.5(b) and (b′). As is well known, the absorbed light intensity is proportional to a time average of $E \cdot dP/dt$, which provides a term of βI^2 ($\propto E^4$), i.e., two-photon absorption, where β is related to the imaginary part of $\chi^{(3)}$. These two quantities, $n_2 I$ and βI^2 with the proportionality factors of n_2 and β, respectively, are very useful in the control of light at optical communication frequencies.

Quantitatively, however, $\chi^{(3)}$ is very small. As a result, the nonlinear effects can be detected only when the power density of incident light is comparable to typical electric-field strength in atoms, which is roughly 500 MW/cm² [10.4]. Despite of the small magnitude of $\chi^{(3)}$, substantial third-order signals may appear if pulsed light is focused onto a small spot and is propagated in long (or thick) samples, such as optical fibres.

We can also delineate optical responses in terms of quantum mechanics. A conventional one-photon absorption coefficient $\alpha(\hbar\omega)$ for amorphous materials is written as [10.6, 10.9]:

$$\alpha(\hbar\omega) \propto |\langle \varphi_f | H | \varphi_i \rangle|^2 \int D_f(E + \hbar\omega) D_i(E) \, dE \tag{10.2}$$

where H is the photon–electron interaction Hamiltonian, the simplest form being proportional to r, $\varphi(r)$ is the electron wave function, $D(E)$ is the density-of-states, and the subscripts i and f represent an initial and a final state. Here, the momentum conservation is neglected because of nonextended wave functions, and the transition amplitude $\langle \varphi_f | H | \varphi_i \rangle$ is assumed to be independent of $\hbar\omega$ for simplicity. The refractive-index spectrum $n_0(\hbar\omega)$ can then be calculated from $\alpha(\hbar\omega)$ using the conventional Kramers–Krönig relation [10.27]:

$$n_0(\omega) - 1 = (c/\pi)\,\wp\!\int \{\alpha(\Omega)/(\Omega^2 - \omega^2)\}\,d\Omega \qquad (10.3)$$

where $\wp$ denotes the principal value of the integral. Since the absorption spectrum $\alpha(\Omega)$ is governed by E_g, the Moss rule ($n_0{}^4 E_g \approx 77$) [10.22], which was derived for semiconductors with a bandgap of E_g, can be regarded as a simplified representation of Equation (10.3).

When photons with $\hbar\omega \approx E_g/2$ are absorbed, we may envisage two-photon absorption. Following the same approach used in Equation (10.2), the two-photon absorption coefficient, $\beta(\hbar\omega)$, can be written as [10.28]:

$$\beta(\hbar\omega) \propto \left|\sum_n \langle \varphi_f | H | \varphi_n \rangle \langle \varphi_n | H | \varphi_i \rangle / (E_{ni} - \hbar\omega)\right|^2 \int D_f(E + 2\hbar\omega)\, D_i(E)\, dE \qquad (10.4)$$

for degenerate cases, i.e., when the two photons have the same energy. Here, the subscript n represents an intermediate state, and $E_{ni} = E_n - E_i$, where E_n is the energy of the n-th electronic state. The two-photon absorption governs the intensity-dependent refractive index n_2, which can be calculated using a nonlinear Kramers–Krönig relation for a nondegenerate case [10.27, 10.29]. In the present context of degenerate cases, we can express it as a rough approximation in the form [10.4]:

$$n_2(\omega) \approx (c/\pi)\,\wp\!\int \{\beta(\Omega)/(\Omega^2 - \omega^2)\}\,d\Omega \qquad (10.5)$$

which appears to be practically useful if $\beta(\Omega)$ is located in a narrow region at sufficiently higher frequencies than calculated $n_2(\omega)$, i.e., $\Omega \gg \omega$, as demonstrated recently [10.30].

Figure 10.6 shows typical linear and nonlinear spectra of absorption and refractivity for a direct-gap semiconductor with bandgap energy of E_g [10.4, 10.28, 10.29]. The absorption

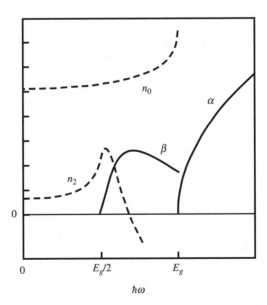

Figure 10.6 Schematic illustrations of absorption spectra, α and β, and refractivity spectra, n_0 and n_2, governed by one- (α and n_0) and two-photon (β and n_2) processes in an ideal crystal

edge of multi-photon absorption processes is located at E_g/m, where m is the number of photons simultaneously absorbed. Accordingly, the two-photon absorption starts at $E_g/2$ as shown in Figure 10.6. Note that, in ideal crystalline semiconductors, all the electronic states in the vicinity of band edges are extended, so that a simple band theory can be applied. Such a situation applies also to indirect-gap semiconductors although spectral shapes change considerably [10.31].

However, the situation changes substantially in a disordered material where mid-gap states exist, which influence the absorption in two ways. One is resonant two-photon absorption. If a gap state which satisfies $E_{ni} - \hbar\omega = 0$ in Equation (10.4) exists, the two-photon absorption at around the gap state becomes resonantly enhanced [see Figure 10.5(b′)]. The other is, as illustrated in Figure 10.5(c), a mid-gap state may cause two-step absorption, which is a successive one-photon absorption process [10.32]. The two-step absorption is detected, e.g., at the Urbach-edge region in As_2S_3 [10.33]. It is needless to say that not only two-, but multi-photon absorptions can be envisaged to occur in this way. However, to the author's knowledge, theoretical considerations on these processes in glasses, including mid-gap states, still remain to be studied.

10.3.2 Two-photon absorption

Figure 10.3 shows one- and two-photon absorption (attenuation) spectra of SiO_2, As_2S_3 [10.33] and Se. We see that these three glasses possess qualitatively the same features. In the one-photon spectrum, the so-called Urbach edge appears at $\hbar\omega \geq 8.0\,eV$, 1.9 eV, and 1.4 eV in SiO_2, As_2S_3 and Se, which are consistent with the Tauc gaps in these materials as being $E_g \approx 10\,eV$ [10.19, 10.27], 2.4 eV, and 2.0 eV [10.6], respectively. Below the edge, residual absorption (attenuation) seems to exist. However, the spectral shapes are different among several studies, probably reflecting absorption due to defects and impurities and also light scattering due to structural fluctuations. On the other hand, the two-photon absorption appears at $\hbar\omega \geq E_g/2$, which may have peaks at ~5.8 eV, ~2.0 eV, and ~1.5 eV, respectively, in these three glasses. It may be mentioned here that two-photon photocurrents have been detected in Se [10.34].

The spectral shape of the two-photon absorption shown in Figure 10.3 may resemble that of the theoretical curve shown in Figure 10.6. Nevertheless, the correspondence is not accurate. Specifically, in As_2S_3, the decrease in β at $\hbar\omega \geq 2.0\,eV$ is affected by two-step absorption [10.32, 10.33], and accordingly, the peak may be deceptive. In addition, the observed peaks in SiO_2 and Se are substantially sharper than the corresponding theoretical ones shown in Figure 10.6, where the vertical axis is plotted on a linear scale. Therefore, further studies are needed for interpreting the experimental absorption shape.

The spectral features pointed out above can be understood as follows [10.28]: In SiO_2, as illustrated in Figure 10.7, the top of the valence band is composed of lone-pair p-electron states of O atoms and the bottom of the conduction band is governed by anti-bonding character states of Si 3p and 3d states [10.10]. The one-photon absorption occurs from the valence band to the conduction band. The fact that the two-photon spectrum lies at an energy around half of the optical gap (~10 eV) suggests that the two-photon transition also occurs from the valence to the conduction band. Similar interpretations can apply to As_2S_3 and Se as well.

Figure 10.3 shows also an interesting correlation in the spectra of As_2S_3. That is, the exponential β spectrum below the peak is nearly parallel to the weak absorption tail in α at $\hbar\omega$

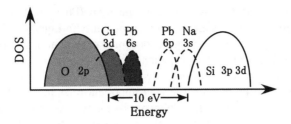

Figure 10.7   Density-of-states (DOS) of the valence and conduction bands in SiO_2. Also added are Cu, Pb, and Na states in SiO_2. Occupied states are shaded

$\approx 1.5\,eV$. In other words, E_β ($\approx 150\,meV$) in $\beta \propto \exp(\hbar\omega/E_\beta)$ is roughly the same with the characteristic energy E_W (≈ 200–$300\,meV$) of the one-photon weak-absorption tail, which has a form of $\alpha \propto \exp(\hbar\omega/E_W)$ at $\hbar\omega \approx 1.5\,eV$ [10.6, 10.12]. Note that this comparison is meaningful, since all the data of As_2S_3 in the Figure have been obtained using a single high-purity ingot. Such a correlation may exist also in Se and SiO_2, but this is not clear due to limited experimental results.

The nearly parallel exponential spectra, i.e., $E_\beta \approx E_W$, are ascribable to gap states. This coincidence implies that the two-photon absorption process occurs resonantly with the gap states which give rise to the weak absorption tail. This can be explained from Equation (10.4), where if $1/(E_{ni} - \hbar\omega)$ can be approximated as $\delta(E_{ni} - \hbar\omega)$, $\Sigma_n 1/(E_{ni} - \hbar\omega)$ behaves as a density-of-states at $\hbar\omega \approx 1.5\,eV$. The gap state also behaves as D_f at $\hbar\omega \approx 1.5\,eV$ in the linear absorption [Equation (10.2)], so that $E_\beta \approx E_W$ can appear.

Figure 10.3 also includes the spectra for PbO–SiO_2 glasses, which present some unique features [10.35]. $\alpha(\hbar\omega)$ shows an exponential Urbach edge at around $\hbar\omega \approx 3\,eV$ and $\alpha \approx 1 \sim 100\,cm^{-1}$, which shifts nearly in parallel to lower energies with the PbO content. On the other hand, $\beta(\hbar\omega)$ appears to be nonexponential. There exists a rise at $\sim 2.0\,eV$, which seems to be common to all the compositions [10.35]. This fairly sharp threshold suggests that two-photon transitions in these glasses occur between the states which are separated in energy by more than $\sim 4\,eV$, which is substantially greater than the energy of the Urbach edge at ~ 3 eV. Another notable feature in β in PbO–SiO_2 glasses is an increase in the maximal value by an order from 1 to $10\,cm/GW$ with nearly twice an increase in the PbO content from 38 to 68 at.%.

These contrasting features in $\alpha(\hbar\omega)$ and $\beta(\hbar\omega)$ spectra in PbO–SiO_2 glasses can be interpreted as follows [10.28]: The electronic structure of PbO–SiO_2 glass is shown in Figure 10.7, which shows that the one-photon absorption edge is governed by intra-atomic 6s → 6p transitions in Pb. This interpretation can explain the red-shift in one-photon absorption edges, since the 6s and 6p bands arising from Pb atoms probably become wider with increasing PbO content, and hence an increase in the interatomic interaction. However, such a transition cannot be assumed for the two-photon absorption, because it cannot occur between s and p states in a single atom [see Equation (10.4) and note that $H \propto r$]. Then, a possible two-photon transition with small photon energy can be O (2p) → Pb (6p), which is consistent with the observation of different thresholds in $\alpha(\hbar\omega)$ and $\beta(2\hbar\omega)$ at ~ 3 and $\sim 4\,eV$, as the initial states (Pb 6s and O 2p) are different. This interpretation is also consistent with the dramatic increase observed in β with the increase in PbO content, which is ascribable to resonant two-photon absorption. This means that the two-photon absorption can be resonantly induced by the Pb (6s) states, as the term $\langle$Pb 6p|H|Pb 6s$\rangle\langle$Pb 6s|H|O 2p$\rangle$ /$(E_{\text{Pb6s,O2p}} - \hbar\omega)$ possibly governs the transition probability in Equation (10.4). Note that, as Pb (6s) is an occupied state, like the highest occupied molecular orbital (HOMO), the two-step absorption [10.32] [Figure 10.5(c)] cannot occur, which has been confirmed experimentally [10.35]. In short, in contrast to the simple glass such as SiO_2, the three atomic levels (O 2p, Pb 6s, and Pb 6p) seem to play important roles in $PbO-SiO_2$ glasses. Comparatively, defective structures are less important in this system. Similar interpretations may be applicable to other heavy-metal oxide glasses containing Bi_2O_3 [10.15, 10.36].

10.3.3 Nonlinear refractivity

Since measurements of n_2 are relatively difficult, several relations for evaluations of n_2 using linear optical properties have been proposed. These relations can be grouped into two kinds. One of these is specifically applicable to transparent materials, and the most famous in this kind is probably Bolings' relation [10.4] given by:

$$n_2 (10^{-13}\,\text{esu}) \approx 391(n_{\text{d}} - 1)/v_{\text{d}}^{5/4} \tag{10.6}$$

where n_{d} is the linear refractive index at the d-line ($\lambda = 588\,\text{nm}$) and v_{d} is the Abbe number. This relation is known to provide satisfactory agreements in materials with small n_{d} (≤ 1.7). In the other kind, which applies to semiconductors, the optical properties can be connected with the bandgap energy E_{g}. For instance, Sheik-Bahae et al. [10.29] have shown for many ($\sim$30) crystals with $E_{\text{g}} \approx 1 - 10\,\text{eV}$ that the relation given by:

$$n_2 n_0 = KG(\hbar\omega/E_g)/E_g^4 \tag{10.7}$$

presents a good approximation, where K is a material-independent constant and $G(\hbar\omega/E_g)$ represents a universal spectral dependence. As shown by the solid line in Figure 10.8, this relation gives satisfactory fits also to many data on oxide and chalcogenide glasses [10.28]. Here, for E_{g}, we take the Tauc gaps or the photon energies at $\alpha = 10^3\,\text{cm}^{-1}$. This satisfactory fitting suggests that, irrespective of the crystalline or noncrystalline nature of the material, n_2 is governed by E_{g}. We can assume that, with a scale of optical wavelengths, there is no marked difference between crystals and glasses as far as the electronic wave functions of conduction and valence bands are concerned. The structural differences between crystals and glasses have secondary effects on n_2, which may explain a large deviation, e.g., in $Ag_{20}As_{32}Se_{48}$, from the line of Equation (10.7). Moreover, as illustrated in Figure 10.8, the Moss rule ($n_0^4 E_g \approx 77$) also applies to these glasses, which further enforces the importance of E_{g} in these glasses.

It is interesting to examine if we can relate $n_2(\hbar\omega)$ with $\beta(\hbar\omega)$. Since the $\beta(\hbar\omega)$ data are limited to only some photon-energy ranges [10.28], we may approximate tentatively the spectra in As_2S_3 and SiO_2 by Gaussian curves, which are shown by dashed lines in Figure 10.9. Then, $n_2(\hbar\omega)$ can be calculated using the nonlinear Kramers–Krönig relation [Equation (10.5)] as shown by solid lines in Figure 10.9. It can be seen that the calculated $n_2(\hbar\omega)$ shapes [Figure 10.9] resemble the theoretical curves for a crystalline semiconductor shown in Figure 10.6. We also see in Figure 10.9 that the agreements between the calculated $n_2(\hbar\omega)$ and the experimental n_2, plotted by open symbols [10.5], are surprisingly good. This agree-

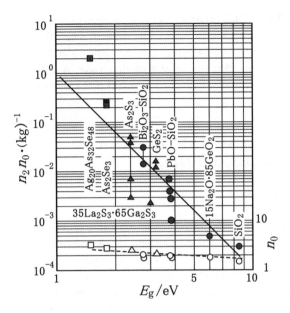

Figure 10.8 Linear and nonlinear refractivity, n_0 (open symbols) and n_2 (solid symbols), in some oxide (circles), sulfide (triangles), and selenide (squares) glasses as a function of the optical gap E_g. The solid and dashed lines depict Sheik-Bahaes' relationship [Equation (10.7)] and the Moss relation $n_0^4 E_g \approx 77$

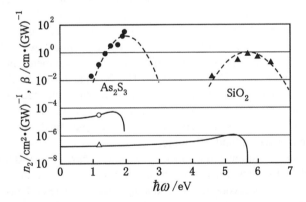

Figure 10.9 Two-photon absorption spectra $\beta(\hbar\omega)$ in As_2S_3 (solid circles) and SiO_2 (solid triangles), fitted Gaussian profiles (dashed lines), calculated $n_2(\hbar\omega)$ (solid lines), and experimental n_2 (open circle and triangle). $\beta(\hbar\omega)$ data are the same as those shown in Figure 10.3

ment suggests that Equation (10.5) can practically be employed for estimating $n_2(\hbar\omega)$, and that n_2 in the near-IR region is governed by the two-photon absorption, which has been confirmed also in crystalline semiconductors [10.5]. Similar observations have been made in PbO–SiO$_2$ glasses [10.30].

The above result also provides an answer to the frequently asked question, 'Why does As$_2$S$_3$ show higher n_2 than does SiO$_2$ at optical communication wavelengths ($\hbar\omega \approx 1\,\text{eV}$)

[10.2, 10.3, 10.5, 10.14]?' This is because a higher n_2 arises from the greater β (~10 cm/GW) at a lower-photon-energy region in As_2S_3. The greater β can then be connected with the smaller E_g, because Equation (10.4) gives $\beta \propto E_g^{-2}$, provided that the denominator $(E_{ni} - \hbar\omega)^2$ governs gross features. A higher n_2 at a smaller value of E_g can also be obtained from Equation (10.7), which presents, $n_2 \propto E_g^{-4}$. In short, the energy gap E_g appears to play a decisive role [10.18].

10.4 NONLINEAR EXCITATION-INDUCED STRUCTURAL CHANGES

10.4.1 Fundamentals

Oxide and chalcogenide glasses are known to exhibit a variety of photoinduced phenomena [10.8–10.10, 10.37, 10.38]. The phenomena appear to be related with metastable structural changes, while it is difficult to identify real atomic changes in disordered glass structures. Some of the photoinduced changes can be recovered by thermal annealing.

It should be noted that, in comparison with the photoinduced phenomena in crystals such as F-center formation in alkali halides, two features add complexities to the glass. One is that the photo-induced change in a glass depends upon many factors such as photon energy and (peak) intensity of illuminating light, cw or pulse (pulse duration and repetition), light-spot size (volume and interfacial stress effects), light polarization, illuminating temperature, and illumination atmosphere (air, vacuum, etc.) [10.39]. For instance, many photo-induced changes become smaller at higher T_i, ultimately disappearing at $T_i \approx T_g$ [10.10, 10.21], where T_i is the temperature at which illumination is provided and T_g is the glass-transition temperature. The other is that a glass property changes from sample to sample, reflecting the quasi-stability. For instance, SiO_2, which is one of the simplest oxide glasses, cannot be uniquely defined due to small, but crucial, differences in stoichiometry and impurities such as H, O_2, OH, and Cl [10.20, 10.37, 10.40, 10.41]. In a chalcogenide, As_2S_3, it is known that evaporated and annealed films undergo different photo-induced changes of photo-polymerization and increases in disorder, respectively [10.8, 10.21]. In short, we are confronted with a variety of exposure conditions and samples, and as a result it is more difficult to obtain a universal understanding.

Nonlinear optical excitations induced by photons with $\hbar\omega/E_g \approx 0.1$–0.8 seem to provide notable photoinduced effects. Such low-energy photons can penetrate into bulk samples more than ~1 cm, which is favorable for producing volume-related effects. For instance, in Ge-doped SiO_2 fibres with a diameter of ~100 μm, Bragg reflectors are inscribed using excimer lasers [10.34]. In silica glasses, three-dimensional optical fabrications have been explored using intense fs–ps Ti:sapphire lasers [10.42, 10.43]. In chalcogenides, several studies report refractive-index changes upon exposures to near-IR pulses [10.14, 10.21, 10.44, 10.45]. At these photon energies, linear absorption is relatively small, and, accordingly, nonlinear excitations can be responsible for photo-electronic excitations.

However, fundamental mechanisms are still vague. In general, a photo-induced phenomenon occurs through an energy transfer from photons to electrons and hence to atomic structures. Accordingly, the process may be divided into two parts, photo-electronic excitation and electro-structural (vibrational) change, and the latter may be regarded as either athermal or thermal. (In athermal processes the temperature rise in a material due to illumination can be neglected but it cannot be neglected in thermal processes). In this section, we will study comparatively the oxide SiO_2 and the chalcogenide As_2S_3 for athermal photo-induced phenomena induced by one-photon and multi-photon processes.

Nevertheless, it is necessary to estimate a temperature rise induced by the light illumination. The rise ΔT can be evaluated very roughly for the two extreme cases: Upon pulsed irradiation with a peak intensity I and a pulse duration τ, the temperature rise can be estimated as $\Delta T \approx Q/cV$, where Q is the absorbed light energy, c the specific heat in the units of J/K·cm^3, and V the related volume. V can be evaluated from the light-spot size, the sample thickness, or light-penetration length such as α^{-1} or $(\beta I)^{-1}$ and the thermal diffusion length $(k\tau/c)^{1/2}$, where k is the thermal conductivity. For instance, when a SiO_2 bulk sample with $k \approx 10^{-2}$ W/cm·K and $c \approx 2$ J/cm^3·K is exposed to a 1 ns pulse with a peak intensity of 1 kW and spot diameter of 10 μm (≈ 1 GW/cm^2) and $\hbar\omega = 8$ eV ($\alpha^{-1} \approx 1$ mm), one gets $\Delta T \approx 5$ K. For pulses of duration shorter than 1 ns, the thermal-diffusion length $(k\tau/c)^{1/2}$ becomes shorter than ~25 nm, and it can practically be neglected. On the other hand, a steady-state temperature rise in an absorbing sample exposed to cw or high-repetition pulses with a time-averaged intensity I can be estimated from $\Delta T \approx Q/(2\pi kr)$ [10.46], where Q ($\approx \alpha IV$) is the absorbed light power and r is the radius of the light spot. If $Q = 1$ mW and $r = 10$ μm, this gives $\Delta T \approx 10$ K for SiO_2.

10.4.2 Oxides

Structural responses of SiO_2 to bandgap and mid-gap excitations are summarized in Figure 10.10(a). There are three kinds of experiments performed to study the photo-structural changes in SiO_2: In one type of experiment, type (i), Ti:sapphire laser pulses of sub-ps duration and energy of $\hbar\omega \approx 1$ eV ($\hbar\omega/E_g \approx 0.1$) [10.47–10.50] are employed, and, in types (ii) and (iii), excimer laser pulese of ~10 ns and $\hbar\omega \approx 5$–8 eV ($\hbar\omega/E_g \approx 0.5$) [10.20, 10.51, 10.52] and (super-) bandgap excitations ($\hbar\omega/E_g \geq 1$) [10.20, 10.53, 10.54] are used, respectively. All the exposures in Figure 10.10(a) are provided and measured at room temperature, so that $T_i/T_g \approx 300/1500 = 0.2$.

Studies using bandgap illumination are limited, probably because of the large bandgap energy of $E_g \approx 10$ eV. Takigawa et al. [10.53] report for synthetic SiO_2 windows that a single shot of 9.8 eV laser light (Ar excimer laser, 5 ns) with 280 mJ/cm^2 ($= 56$ MW/cm^2) power causes topological surface damage and crystalline Si formation. They also mention that lower energy shots with $\hbar\omega \approx 5$–8.5 eV do not produce crystalline Si, despite the presence of marked surface damage. Awazu has demonstrated that, in thermally grown amorphous SiO_2 films, super-bandgap excitations with energy 14 and 18 eV, which can be regarded as cw, produce small Si–O rings [10.20]. Akazawa [10.54] reports that, under super-bandgap excitations (synchrotron radiation of 100–300 eV), amorphous SiO_2 films evaporate while accumulating Si–Si wrong bonds. In short, a (super-)bandgap excitation tends to produce Si wrong bonds and evaporation. These phenomena in SiO_2 resemble photo-enhanced vaporization, which is more efficiently induced in heated As_2S_3 when exposed to (super-)bandgap illumination [10.55]. It is not yet well established whether or not photodarkening (PD) (see Section 10.4.3) exists in SiO_2 [10.20], partly because accurate measurements of the absorption edge at ~9 eV are not easy.

On the other hand, many studies have demonstrated some changes at small-photon-energy excitations with $\hbar\omega/E_g \approx 0.1$–0.8. When the incident peak power is greater than the order of J/cm^2 levels, or GW/cm^2 for 1 ns pulses, both surface and/or internal damage occurs with ablation [10.20, 10.47, 10.49]. When the peak intensity is around 1 MW/cm^2, three kinds of phenomena appear, which are: the creation of defective structures such as Si–Si bonds [10.56] and small rings [10.20], volume changes [10.37, 10.51], and refractive-index changes [10.37, 10.48]. The volume and the refractive-index change have been correlated through the Lorentz–Lorenz relation [10.37]. However, the role of the defect creation in these macroscopic changes awaits further studies [10.52].

Is the multi-photon excitation really responsible for these photo-structural changes? It is not yet unambiguously understood whether the excitation is triggered by one-photon or by multi-photon processes. For a two-photon excitation to be dominant, at least, a condition is that $\beta I \gg \alpha$ must be satisfied. For instance, at $\hbar\omega \approx 5$ eV, since $\alpha = 10^{-2}$ cm^{-1} and $\beta = 10^{-1}$ cm/GW (see Figure 10.3), the light intensity must be greater than ~100 MW/cm^2. This intensity level is located in Figure 10.10 at the border between the areas of structural damage

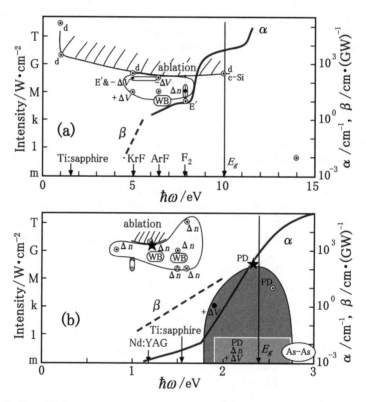

Figure 10.10   Photo-induced phenomena scaled by the excitation photon energy $\hbar\omega$ and the (peak) light intensity in (a) SiO$_2$ and (b) As$_2$S$_3$. T, G, M, etc. on the left-hand-side vertical axis represent TW, GW, MW, kW, W, mW. Also shown are the one- and two-photon absorption spectra, α (solid lines) and β (dashed lines), respectively. Their scales are shown on the right vertical axis. Photon energies of related laser light are indicated on the horizontal axis as Nd:YAG, Ti:sapphire, KrF, ArF, and F$_2$. E_g represents the bandgap energy. In the abbreviations, E$'$ indicates an E$'$-center formation, $+\Delta V$ a volume expansion, $-\Delta V$ a volume contraction, Δn a refractive-index change, WB a wrong-bond formation, PD a photodarkening, d damage. In the shaded region in (b), photodarkening, refractive-index increase, and volume expansion are shown to occur

and structural modifications. Here, we should note that α is measured using a weak probe light, but the condition of $\beta I \gg \alpha$ should be satisfied under intense illumination that may produce transitory mid-gap absorption at defective sites [10.57], which will further enhance the one-photon absorption coefficient α. Such defect effects may cause different mid-gap absorptions as shown in Figure 10.3 for SiO_2. The intense illumination also gives rise to a temperature increase, which is likely to add band-edge absorption [10.9]. Under these circumstances, distinctions among various excitation processes become more-or-less difficult. For instance, Kajihara et al. [10.58] have proposed that the exposure to 7.9 eV photons causes the formation of E′-centers in SiO_2 as a result of a two-step absorption process rather than a two-photon absorption process (see Figure 10.5).

10.4.3 Chalcogenides

The chalcogenide glasses have an optical gap in the range of 1–3 eV, so that several kinds of prominent changes are induced by exposure to visible light [10.3, 10.8–10.10, 10.21]. Among these, PD (red-shift of the optical absorption edge) and a related refractive-index increase have attracted substantial interest, because these two phenomena are simple, athermal, inherent to covalent chalcogenide glasses such as Se and As_2S_3, and have promise for optical applications.

For understanding the fundamental mechanism of PD and related phenomena, spectral studies are valuable [10.8, 10.21]. For instance, in As_2S_3 with $E_g \approx 2.4$ eV, it is known that the bandgap illumination at room temperature, which corresponds to $T_i/T_g \approx 300/450 \approx 0.7$, gives a PD of ~50 meV and a refractive-index increase of ~0.02 (see Table 10.1). The two changes are quantitatively related through the linear Kramers–Krönig relation [Equation (10.3)]. It has also been demonstrated that even an intense sub-bandgap illumination can produce notable changes [10.21]. In As_2S_3, continuous light of energy 2.0 eV ($\hbar\omega/E_g \approx 0.8$) and intensity greater than 100 W/cm² can produce PD that is comparable to that induced by bandgap illumination, and a more prominent volume expansion. Pulsed light of energy $\hbar\omega$ = 1.65 eV ($\hbar\omega/E_g \approx 0.7$) and peak intensity of 100 MW/cm² can also induce the volume expansion. Tanaka has proposed that these sub-gap illumination effects are triggered by complete photo-excitation of the band-tail states (Chapter 5 in reference [10.21]).

Table 10.1 Photodarkening ΔE and photoinduced refractive-index change Δn induced in As_2S_3 glass by pulsed light of energies 1.17 eV and 2.33 eV and also by continuous-wave (cw) bandgap light of ~2.4 eV for comparison. Related parameters (α, β, and excitation light intensity) are also listed. β at 2.33 eV cannot be evaluated due to two-step absorption [10.33]. In the pulse experiments, ΔE and Δn are evaluated at a fixed absorbed photon number of 10^{23}–10^{24}/cm³, which may not provide saturated changes

$\hbar\omega$ (eV)	ΔE (meV)	Δn	α (cm^{-1})	β (cm/W)	Intensity (W/cm²)
1.17	0 ± 5	0.005	10^{-3}	10^{-10}	10^9
2.33	20	0.003	300	?	10^7
cw	50	0.02			~0.05

However, mid-gap illumination ($\hbar\omega \approx E_g/2$) provides different features [10.39]. When the illumination is weak, no detectable changes appear within experimental time scales. When it is intense and the condition, $\beta I \gg \alpha$, is satisfied, as listed in Table 10.1, the excitation gives rise to a refractive-index change, which is consistent with previous studies [10.21, 10.44, 10.45] *but no discernible PD*, 0 ± 5 meV, is observed. Raman scattering spectroscopy shows that the mid-gap excitation causes a density increase (a few percent) in wrong bonds, As–As and S–S, which is in contrast with undetectable changes upon sub-bandgap excitation. The temperature rise induced by the mid-gap excitation is estimated to be ~10 K, which can be practically neglected in comparison to the glass-transition temperature of ~450 K. That is, the process appears to be athermal.

To understand the above two-photon excitation effect, we must answer at least two questions: 1) Why does the excitation produce the wrong bonds? And 2) why does the refractive index increase?

For answering the first question, we may follow the idea that the two-photon absorption can be resonantly enhanced by mid-gap states if the states satisfy $E_{ni} - \hbar\omega \approx 0$ in Equation (10.4), The mid-gap states are ascribable to anti-bonding states of As–As bonds [10.33]. In such cases, the resonant two-photon absorption can occur spatially selectively at around the defective sites, and, as a result, As clusters may grow through some mechanism at present unspecified. On the other hand, bandgap illumination probably causes defect-unrelated excitations resulting in different structural changes, such as bond twisting [10.8, 10.21], which probably broaden the valence band. In short, as illustrated in Figure 10.11, reflecting the different transition probabilities between the two-photon and linear excitations, different structural changes could follow.

Question 2) may be addressed as follows. Taking the linear Kramers–Krönig relation [Equation (10.3)] into account, we assume that the refractive-index increase should be accompanied by some absorption increase such as the PD (reduction in bandgap), but experimentally no PD has been detected. A notion can be put forward that the inhomogeneous structures containing As and S clusters produce some stresses, which cause the refractive-index increase through photo-elastic effects. Otherwise, we may envisage that it occurs due to an emergence of some mid-gap absorption processes. Further studies, such as composition dependence of the absorption spectra, will be valuable in order to examine these notions.

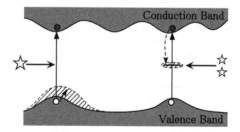

Figure 10.11 Schematic illustrations of the bandgap excitations with one-photon (left) and resonant two-photon (right) processes. One-photon excitations may cause modifications in the medium-range structures, which broaden the valence band, and as a result, the PD appears. Two-photon excitations may produce localized defect centers such as small As clusters

10.5 CONCLUSIONS

We have highlighted two topics of nonlinear optical properties in oxide and chalcogenide glasses. One is a unified understanding of the absorptive and dispersive third-order nonlinearity, and the other is the role of nonlinear excitation in photo-induced phenomena.

The optical nonlinearity in glasses is governed by the bandgap energy and is modified by detailed electronic structures. The role of the bandgap energy is similar to that in crystals, since the short-range structures in a glass are similar to those in its crystalline counterpart. However, the nonlinearity in glasses is affected by the band-tail and mid-gap states, having localized atomic wave functions, which could provide spatially selected resonant absorption.

Different photo-structural changes appear in these materials depending upon the photon energy and light intensity. When light intensity is weak and $\hbar\omega \approx E_g/2$, one-photon absorption by mid-gap states can be held responsible for these photostructural changes. If the intensity is high, we should also take two-step and (resonant) two-photon processes into account. If $\hbar\omega \approx E_g$, the band-to-band excitations become dominant. These photoelectronic excitations seem to provide different photo-structural changes. In SiO_2, the mid-gap excitation creates defective structures, which produce volume contraction and refractive-index increase, while bandgap illumination tends to produce Si–Si homopolar bonds. In As_2S_3, two-photon excitations produce wrong bonds (As–As and S–S), while bandgap illumination gives rise to the photodarkening and photoexpansion.

Finally, it should be emphasized that, in contrast to crystals, glasses have a wider variety of microscopic and macroscopic controllability. Microscopic structures can be modified easily by illumination. For instance, a glass can be modified to become anisotropic by some poling techniques [10.16, 10.18] or producing crystallites [10.18, 10.50]. Macroscopic shapes can arbitrarily be modified by mechanical and thermal treatments. Long and wide homogeneous samples, including fibres and film forms, can be prepared, in which small optical nonlinearity is seemingly enhanced by the geometrical factors. These controllable characteristics provide superior features in glasses as opposed to crystals, which may have larger nonlinearity, while the atomic structure is uniquely determined and sample scales are restricted.

ACKNOWLEDGEMENTS

The author would like to thank to his students, K. Sugawara, N. Terakado, and N. Minamikawa, for illustrations and comments.

REFERENCES

[10.1] G.P. Agrawal, *Nonlinear Fiber Optics*. 3rd Edition (Academic Press, San Diego, 2001).

[10.2] E.M. Vogel, M.J. Weber, and D.M. Krol, *Phys. Chem. Glasses*, **32**, 231 (1991).

[10.3] A. Zakery and S.R. Elliott, *J. Non-Cryst. Solids*, **330**, 1 (2003).

[10.4] R.W. Boyd, *Nonlinear Optics*. 2nd Edition (Academic Press, Boston, 2003).

[10.5] R.L. Sutherland, *Handbook of Nonlinear Optics*. 2nd Edition (Marcel Dekker, New York, 2003).

[10.6] S.R. Elliott, *Physics of Amorphous Materials*. 2nd Edition (Longman Scientific & Technical, Essex, 1990).

[10.7] G. Fuxi, *Optical and Spectroscopic Properties of Glass* (Springer-Verlag, Berlin, 1992).

[10.8] M.A. Popescu, *Non-Crystalline Chalcogenides* (Kluwer Academic Publishers, Dordrecht, 2001).

[10.9] J. Singh and K. Shimakawa, *Advances in Amorphous Semiconductors* (Taylor & Francis, London, 2003).

[10.10] K. Tanaka, Group VIb Amorphous Materials and Applications, in *Non-Crystalline Materials for Optoelectronics*, edited by G. Lucovsky and M. Popescu, Optoelectronic Materials and Devices (INOE, Bucharest, 2004), Vol. 1, Ch. 3.

[10.11] S.M. Malik and S.K. O'Leary, *J. Non-Cryst. Solids*, **336**, 64 (2004).

[10.12] K. Tanaka and S. Nakayama, *Jpn. J. Appl. Phys.*, **38**, 3986 (1999).

[10.13] K. Tanaka, *J. Opt. Adv. Mater.*, **4**, 505 (2002).

[10.14] A. Jha, X. Liu, A.K. Kar, and H.T. Bookey, *Curr. Opin. Solid State Mater. Sci.*, **5**, 475 (2001).

[10.15] K. Imanishi, Y. Watanabe, T. Watanabe, and T. Tsuchiya, *J. Non-Cryst. Solids*, **259**, 139 (1999).

[10.16] Y. Quiquempois, P. Niay, M. Douay, and B. Poumellec, *Curr. Opin. Solid State Mater. Sci.*, **7**, 89 (2003).

[10.17] S. Vijayalakshmi and H. Grebel, Nonlinear Optical Properties of Nanostructures, in *Handbook of Nanostructured Materials and Nanotechnology*, edited by H.S. Nalwa (Academic Press, San Diego, 2000), Vol. 4, Ch. 8.

[10.18] K. Tanaka, *J. Mater. Sci. Mater. Electron.*, **16**, 633 (2005).

[10.19] J. Lucas, *Curr. Opin. Solid State Mater. Sci.*, **2**, 405 (1997).

[10.20] K. Awazu and H. Kawazoe, *J. Appl. Phys.*, **94**, 6243 (2003).

[10.21] *Photo-Induced Metastability in Amorphous Semiconductors*, edited by A.V. Kolobov (Wiley-VCH, Weinheim, 2003).

[10.22] T.S. Moss, *Optical Properties of Semi-conductors* (Butterworths Science Publisher, London 1959).

[10.23] K. Tanaka, *Phys. Rev. B*, **36**, 9746 (1987).

[10.24] K. Ogusu, J. Yamasaki, S. Maeda, M. Kitao, and M. Minakata, *Opt. Lett.*, **29**, 265 (2004).

[10.25] G. Boudebs, S. Cherukaulappurath, M. Guignard, J. Troles, F. Smektala, and F. Sanchez, *Opt. Commun.*, **232**, 417 (2004).

[10.26] M. Baudrier-Raybaut, R. Haidar, Ph. Kupecek, Ph. Lemasson, and E. Rosencher, *Nature*, **432**, 374 (2004).

[10.27] D.C. Hutchings, M. Sheik-Bahae, D.J. Hagan, and E.W. Van Stryland, *Opt. Quantum Electron.*, **24**, 1 (1992).

[10.28] K. Tanaka, *J. Non-Cryst. Solids*, **338–340**, 534 (2004).

[10.29] M. Sheik-Bahae and E.W. Van Stryland, Optical Nonlinerities in the Transparency Region of Bulk Semiconductors, in *Semiconductors and Semimetals*, edited by E. Garmire and A. Kost (Academic Press, San Diego, 1999), Vol. 58, Ch. 4.

[10.30] K. Tanaka and N. Minamikawa, *Appl. Phys. Lett.*, **86**, 121112 (2005).

[10.31] M. Dinu, *IEEE Quant. Electron.*, **39**, 1498 (2003).

[10.32] R.A. Baltraneyunas, Yu.Yu. Vaitkus, and V.I. Gavryushin, *Sov. Phys. JETP*, **60**, 43 (1984).

[10.33] K. Tanaka, *Appl. Phys. Lett.*, **80**, 177 (2002).

[10.34] R.C. Enck, *Phys. Rev. Lett.*, **31**, 220 (1973).

[10.35] K. Tanaka, N. Yamada, and M. Oto, *Appl. Phys. Lett.*, **83**, 3012 (2003).

[10.36] N. Sugimoto, H. Kanbara, S. Fujiwara, K. Tanaka, Y. Shimizugawa, and K. Hirao, *J. Opt. Soc. Am. B*, **16**, 1904 (1999).

[10.37] *Defects in SiO$_2$ and Related Dielectrics: Science and Technology*, edited by G. Pacchioni, L. Skuja, and D.L. Griscom (Kluwer Academic Publishers, Dordrecht, 2000).

[10.38] *Handbook of Advanced Electronic and Photonic Materials and Devices*, edited by H.S. Nalwa (Academic Press, San Diego, 2001), Vol. 5, Chs. 2, 3, and 4.

[10.39] K. Tanaka, *Philos. Mag. Lett.*, **84**, 601 (2004).

[10.40] K. Saito, M. Ito, A.J. Ikushima, S. Funahashi, and K. Imamura, *J. Non-Cryst. Solids*, **347**, 289 (2004).

[10.41] T. Bakos, S.N. Rashkeev, and S.T. Pantelides, *Phys. Rev. B*, **69**, 195206 (2004).

[10.42] S. Juodkazis, H. Misawa, and I. Maksimov, *Appl. Phys. Lett.*, **85**, 5239 (2004).

[10.43] D. Ehrt, T. Kittel, M. Will, S. Nolte, and A. Tünnermann, *J. Non-Cryst. Solids*, **345/346**, 332 (2004).

[10.44] M. Asobe, *Opt. Fiber Technol.*, **3**, 142 (1997).

[10.45] N. Ho, J.M. Laniel, R. Vallee, and A. Villeneuve, *Opt. Lett.*, **28**, 965 (2003).

[10.46] A. Arun and A.G. Vedeshwar, *Physica B: Condens. Matter (Amsterdam)* **229**, 409 (1997).

[10.47] L. Luo, C.L.D. Wang, H. Yang, H. Jiang, and Q. Gong, *Appl. Phys. A*, **74**, 497 (2002).

[10.48] A. Saliminia, N.T. Nguyen, M.-C. Nadeau, S. Petit, S.L. Chin, and R. Vallée, *J. Appl. Phys.*, **93**, 3724 (2003).

[10.49] T.Q. Jia, Z.Z. Xu, X.X. Li, R.X. Li, B. Shuai, and F.L. Zhao, *Appl. Phys. Lett.*, **82**, 4382 (2003).

[10.50] Y. Shimotsuma, P.G. Kazansky, J.R. Qiu, and K. Hirao, *Phys. Rev. Lett.*, **91**, 247405 (2003).

[10.51] M. Takahashi, T. Uchino, and T. Yoko, *J. Am. Ceram. Soc.*, **85**, 1089 (2002).

[10.52] K. Kajihara, L. Skuja, M. Hirano, and H. Hosono, *J. Non-Cryst. Solids*, **345/346**, 219 (2004).

[10.53] Y. Takigawa, K. Kurosawa, W. Sasaki, K. Yoshida, E. Fujiwara, and Y. Kato, *J. Non-Cryst. Solids*, **116**, 293 (1990).

[10.54] H. Akazawa, *J. Vac. Sci. Technol.*, *B*, **19**, 649 (2001).

[10.55] K. Tanaka, *Philos. Mag. Lett.*, **79**, 25 (1999).

[10.56] Y. Ikuta, S. Kikugawa, M. Hirano, and H. Hosono, *J. Vac. Sci. Technol.*, *B*, **18**, 2891 (2000).

[10.57] N. Fukata, Y. Yamamoto, K. Murakami, M. Hase, and M. Kitajima, *Appl. Phys. A – Mater. Sci. Process.*, **79**, 1425 (2004).

[10.58] K. Kajihara, Y. Ikuta, M. Hirano, and H. Hosono, *J. Non-Cryst. Solids*, **322**, 73 (2003).

11 Optical Properties of Organic Semiconductors and Applications

T. Kobayashi and H. Naito

Department of Physics and Electronics, Osaka Prefecture University, 1-1 Gakuen-cho, Sakai, Osaka 599-8531, Japan
e-mail: tkobaya@pe.osakafu-u.ac.jp

11.1 INTRODUCTION

Organic materials have fascinating optical properties and have been extensively investigated in many research fields associated with light. Compared with inorganic materials, a great advantage of organic materials is their variety. In inorganic materials, repeating units consist either of the same atoms or a few different at the most, so there is not much room to control one property of the material while keeping the rest unchanged. In organic molecules, there are an almost infinite number of combinations of atoms and therefore it is possible to design organic materials to have desirable properties. In addition, the molecular arrangement and its dimensionality can be also controlled. Therefore, organic materials have received considerable research attention as model systems for understanding the interaction between light and materials. From the application viewpoint for fabricating light-emitting devices and nonlinear optical devices, π-conjugated polymers are the most promising materials because of their good fluorescence yield, large optical nonlinearity, and ultrafast class

Optical Properties of Condensed Matter and Applications   Edited by J. Singh
© 2006 John Wiley & Sons, Ltd

relaxation time. In this chapter, we will investigate the fundamental optical properties of π-conjugated polymers.

11.2 MOLECULAR STRUCTURE OF π-CONJUGATED POLYMERS

Although π-conjugated polymers are composed mainly of only carbon and hydrogen atoms, there are many kinds of polymer backbones. A few representative polymer backbones are shown in Figure 11.1. The most important feature of π-conjugated polymers is that unsaturated carbons lie on a chain, over which π electrons are delocalized one-dimensionally. The side chains provide the thermal stability, solubility, hydrophilicity, and many other properties to the polymer backbones, but the fundamental optical (and electrical) properties are determined by the π electrons. In this section, we will first review the chemical structure of π-conjugated polymers, and then introduce some theoretical approaches in the next subsection.

Carbon consists of an atomic nucleus and 6 electrons, and their electronic configuration is $1s^2 2s^2 2p^2$ in the ground state. To form four bonding orbitals, the carbon is first promoted to the configuration of $1s^2 2s 2p^3$, and then 1 electron in the 2s orbital and some in 2p orbitals are hybridized. For instance, in diamond, all of the 4 electrons in the L-shell form four identical sp^3-hybridized orbitals to bond with 4 carbon atoms. In π-conjugated polymers, instead of forming the sp^3-hybridized orbitals, 1 electron in the 2s orbital and 2 electrons in 2p ortibals are hybridized to three identical sp^2 orbitals, which are in the same plane and are arranged at an angle of about 120 degrees [see Figure 11.2 (a)]. Since they do not have angular momentum in the bonding direction, they are called σ orbitals. Each carbon atom is combined with other carbon atoms and hydrogen atoms by the σ orbitals to form the backbone of π-conjugated polymers [see Figure 11.2 (b)]. These σ orbitals have maximum amplitude in the bonding direction so that overlap of the σ orbitals of adjacent carbon atoms is expected to be large. In fact, the σ bonds are strong and can be excited only by photons

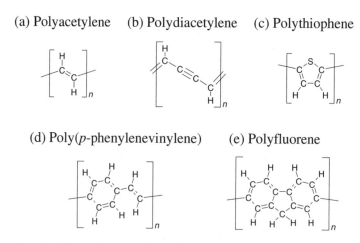

(a) Polyacetylene   (b) Polydiacetylene   (c) Polythiophene

(d) Poly(p-phenylenevinylene)   (e) Polyfluorene

Figure 11.1   Some backbone structures of π-conjugated polymers

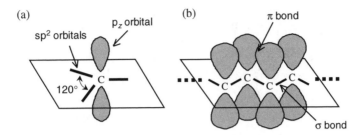

Figure 11.2   (a) sp^2-hybridized orbitals and p$_z$ orbital. (b) σ and π bonds in a single chain of poly-acetylene (in this figure hydrogen atoms and σ bonds connecting carbon and hydrogen atoms are omitted)

in the ultraviolet (UV) energy range. On the other hand, the remaining p$_z$ orbital of the carbon atom is called a π orbital, and does not have its maximum amplitudes in the bonding direction so that the π orbital overlaps only slightly with other π orbitals of the neighboring carbon atoms, above and below the plane containing the σ bonds, to form π bond(s). The π bonds are not strong and it is possible to excite the electronic state of an electron in the π bond by photons in the visible range. In other words, electrons in the π bonds determine the optical properties of π-conjugated polymers in the visible range, and electrons in the σ bonds are usually not taken into consideration. An electron in the linked π orbitals (the π bond) and one in the σ orbitals (the σ bond) are sometimes called a π electron and a σ electron, respectively.

In the electronic structure of π-conjugated polymers, the linked π orbitals form the π band. Each orbital has two states, i.e. spin-up and -down states, but supplies only one electron. Thus, one may expect that the π band is a half-filled band and therefore π-conjugated polymers should show efficient conductivity like metals. However, in the actual polymers, intervals between carbon atoms are alternately modified to form bond alternation in order to reduce the total energy (see Figure 11.3). This alternation splits the π band into two equal half bands, separated by a bandgap; the lower half is fully filled and the upper one is completely empty. Although this bond alternation increases the elastic energy, the energy of electrons is lowered enough to compensate for the increase. Consequently, π-conjugated polymers show the electronic properties of semiconductors or insulators. It may be noted that, in the polymer backbone, purely single bonds do not appear although the bond alternation is depicted using a single line as well as a double line. The interatomic distances between carbon atoms are modified only slightly by the alternation. Several electron-diffraction measurements have revealed that the bond lengths are modified by less than 5% from the mean length [11.1]. Therefore, it is reasonable to consider that π electrons are delocalized over the whole polymer backbone.

11.3   THEORETICAL MODELS

Although a single chain of a π-conjugated polymer can be considered as a one-dimensional semiconductor, its theoretical treatment is not simple because of some interactions. In inorganic materials, it is usually assumed that nuclei do not move even after excited states are

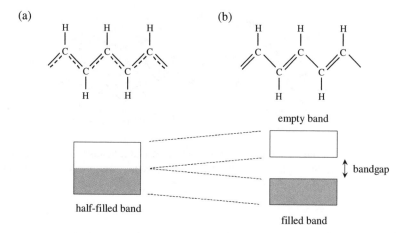

Figure 11.3 (a) In the case where carbon atoms are placed with the same intervals, the linked π orbitals form a half-filled band. In the above chemical structure, the solid lines are the σ bonds and dashed lines indicate the linked π orbitals. (The dotted lines are *not* π bonds because each carbon atom has only 4 bonding orbitals.) (b) In the case where the intervals between carbon atoms are alternately modulated, the half-filled band splits into a filled band and an empty band separated by a bandgap

created. However, in π-conjugated polymers it is essential to take into consideration coupling between electronic states and bond orders when their optical properties are discussed. To do this, the tight-binding Hamiltonian including the electron–phonon (or electron–lattice) interaction is often used as [11.2]:

$$H = \frac{1}{2}\sum_n K(u_{n+1}-u_n)^2 - \sum_n t_{n+1,n}\left(C_{n+1}^+ C_n + C_n^+ C_{n+1}\right) \tag{11.1}$$

where K is the elastic constant, u_n is the displacement of the carbon atom at site n, and C_n^+ and C_n are electron creation and annihilation operators, respectively. The second term in Equation (11.1) has the same form as in the tight-binding model but the transfer integral, $t_{n+1,n}$, is defined to be proportional to the interval between the carbon atoms at positions $n + 1$ and n using the following relation:

$$t_{n+1,n} = t_0 - \alpha(u_{n+1}-u_n) \tag{11.2}$$

where α is a parameter indicating coupling strength between an electron and nucleus, and t_0 is a constant. This model can be applied to polymers with degenerate ground states like polyacetylene: in polyacetylene, *cis* and *trans* forms have the same ground-state energy [see Figure 11.4 (a)]. However, in many other π-conjugated polymers, the ground state of geometrical isomers is not degenerate [see Figure 11.4 (b)]. In order to include this nondegeneracy effect in such polymers, the following transfer integral is alternatively used:

$$t_{n+1,n} = \left[1+(-1)^n\,\delta_0\right]t_0 - \alpha(u_{n+1}-u_n) \tag{11.3}$$

(a)   (b)

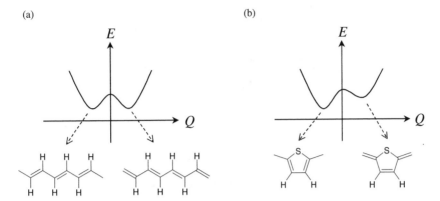

Figure 11.4 Examples of π-conjugated polymers with (a) degenerate and (b) nondegenerate ground states

where δ_0 is a dimensionless bond-alternation parameter and is introduced here to solve the nondegenerate ground states. Calculations using these Hamiltonians have succeeded in explaining topological charged excited states, such as solitons, polarons, bipolarons, and molecular deformations due to photoexcitations, which are discussed in later sections. However, recently it has been recognized that the electron–electron interaction is also essential in understanding the electronic and optical properties of π-conjugated polymers and it is taken into account using the following Hubbard–Peierls Hamiltonian [11.3]:

$$H = -\sum_{n,\sigma} t_{n+1,n} \left(C^+_{n+1,\sigma}C_{n,\sigma} + C^+_{n,\sigma}C_{n+1,\sigma} \right) + \sum_{n} U\rho_{n\downarrow}\rho_{n\uparrow} + \frac{1}{2}\sum_{n\neq m}\sum_{\sigma,\sigma'} V_{m,n}\rho_{n,\sigma}\rho_{m,\sigma'} \quad (11.4)$$

where U is the on-site Hubbard repulsion (the nearest-neighbor hopping integral), V is the nearest-neighbor charge density–charge density interaction, $\rho_{n,\sigma} = C^+_{n,\sigma}C_{n,\sigma}$, and σ indicates spin (up or down). We do not solve this Hamiltonian in this book because it is still impossible to solve it for a realistic π-conjugated polymers consisting of more than a few hundred sites and containing significant structural disorder. Here, we would only like to stress that electron–phonon and electron–electron interactions play an essential role in the electronic structure of π-conjugated polymers and that all of their optical and electronic properties cannot be described within the framework of either band theory or effective-mass approximation, which are very efficient theoretical approaches to our understanding basic optical and electronic properties of inorganic semiconductors.

However, it is also true that many similarities exist between inorganic and organic materials. For instance, according to the one-dimensional exciton theory developed for inorganic semiconductors, discrete exciton energy levels appear below the continuum state and most of the oscillator strength concentrates on the lowest exciton level. These features are observed in π-conjugated polymers as well (see Figure 11.5). Therefore, more familiar and intuitive theories for inorganic materials can be approximately used for organic materials as long as their applicable range is paid attention to. The electron–phonon interaction is taken into account throughout this chapter, but the effect of electron–electron interaction is briefly reviewed only in Section 11.7.

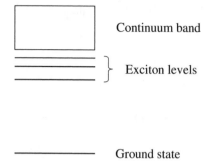

Figure 11.5   Schematic energy-band structure in π-conjugated polymers

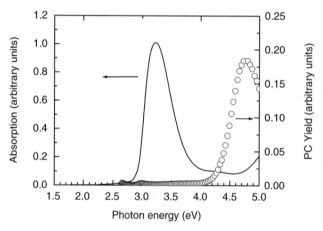

Figure 11.6   Absorption and photoconductivity (PC)-yield spectra of a spin-coated film of polyflu-orene. Exciton binding energy can be estimated from the difference between the onsets of absorption and PC yield spectra

11.4 ABSORPTION SPECTRUM

Absorption measurement is the most fundamental spectroscopy and is very helpful in under-standing the characteristic features of π-conjugated polymers, including large electron–phonon interaction. Figure 11.6 shows the absorption spectrum of a spin-coated film of polyfluorene, where a broad and featureless band is observed at 3.2 eV. This band corresponds to the transition from the ground state to the lowest excited state. In spin-coated films, the polymer forms an amorphous phase, and the broad width of the band results from significant inhomogeneous broadening. In π-conjugated polymers, ideally π electrons are delocalized over the whole polymer chain but structural disorders, such as bending or twist-ing around a bond between fluorene units, limit the delocalization of π electrons and increases the resonance energy of the polymer. Since actual polymers have a large distrib-ution of delocalization lengths of π electrons, especially in amorphous films and in solution, such a broad and featureless absorption spectrum is often observed. For analysing the absorption spectrum in π-conjugated polymers, the simple band picture without Coulomb interaction (electron–electron interaction) is not valid; this, in fact, can be confirmed from a comparison between absorption and photoconductivity (PC)-yield spectra (see Figure 11.6). PC yield indicates a probability that an absorbed photon generates a pair of charged carriers. If the Coulomb interaction is negligible in a system, a photoexcitation always produces a pair of oppositely charged carriers, which will contribute to a photocurrent in the system. On the other hand, in a system with electron–electron interaction, the excited pairs of charge carriers can form excitons due to their Coulomb interaction and then an excess of energy will be necessary to separate such a pair from each other. This required excess of energy can be estimated from the photon-energy difference between the onsets of absorption and PC-yield spectra and is called the 'exciton binding energy.' The exciton binding energy of the spin-coated film of polyfluorene is estimated to be about 0.1 eV from a comparison between absorption and PC-yield spectra shown in Figure 11.6. In many other π conjugated polymers, similar exciton binding energies have been reported [11.4]. Although the band in the PC-yield spectrum in Figure 11.6 corresponds to a continuum state in Figure 11.5, it is difficult to recognize the band in the absorption spectrum. This is because most of the oscillator strength concentrates on the lowest exciton level, which is a major feature of one-dimensional systems.

Some π-conjugated polymers have a tendency to align in a regular manner and to form ordered films, where structural disorders are much suppressed. Figure 11.7 shows an absorption spectrum of a spin-coated film of polythiophene, where a vibronic structure consisting of several bands with the same energy interval is clearly observed. In order to explain this vibronic structure, we have schematically illustrated the energy-potential curves in Figure 11.8, where a few vibrational energy levels of a phonon mode are indicated by the horizontal lines. If this system has electron–phonon interaction, transitions from the zero vibrational level in the ground state to the excited vibrational levels in the excited state are allowed, and discrete transition bands appear in the absorption spectrum. In Figure 11.8,

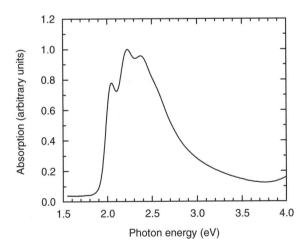

Figure 11.7 Absorption spectrum of a spin-coated film of polythiophene

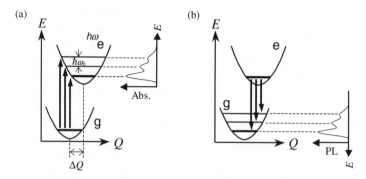

Figure 11.8 Schematic illustration of vibronic potential-energy curves for (a) absorption and (b) photoluminescence (PL) processes. In this Figure, e and g indicate electronic excited and ground states. $\hbar\omega_0$ is the phonon energy of the associated mode, and ΔQ is the difference between the potential minima of the g and e curves. If ΔQ is zero, i.e. the case of no electron–phonon interaction, the transitions to the higher vibrational levels are forbidden

the potential minimum of the excited state is slightly shifted from that of the ground state. The magnitude of this shift, ΔQ, represents the strength of electron–phonon interaction in the system. The transition to the zero vibrational level in the excited state is called '0–0 transition' and corresponds to purely an electronic transition. On the other hand, other transitions are called '0–1 transition,' '0–2 transition,' etc. After a photoexcitation corresponding to a 0–n transition, an electronic excited state is created and n phonons are emitted. In many π-conjugated polymers the associated phonon mode is a C=C stretching mode with phonon energy of 0.18 eV. Such a vibronic structure is always observed in their ordered films. Although π-conjugated polymers have many phonon modes, most of their phonon energies are much less than 0.18 eV and hence their contributions can be included in inhomogeneous broadening.

The vibronic structure can be simulated by taking into account the electron–phonon interaction and the associated phonon mode [11.5]. In a system without electron–phonon interaction, the transition matrix element can be calculated simply from:

$$m = \int \Psi_e^0(r)^*(-er)\Psi_g^0(r)\,dr \tag{11.5}$$

where Ψ_g^0 and Ψ_e^0 are electronic wave functions for the ground and excited states, and r is the electronic coordinate. However, in a system with a significant electron–phonon interaction, the electronic state is influenced by the displacement of atoms, which can be taken into account by adding the following term as a perturbation in the Hamiltonian:

$$H_{\text{int}}(r,Q) = -u(r)Q \tag{11.6}$$

where Q is the generalized coordinate for the associated phonon mode. In Equation (11.6) we neglect the higher term of Q for simplicity. The perturbed wavefunction is then written as:

$$\Phi_{in}(r,Q) = \Psi_i(r,Q)\cdot\xi_{in}(Q) \tag{11.7}$$

where i and n indicate the electronic state and vibrational level, respectively, and ξ_{in} describes the vibrational wave function of atoms. This perturbed wave function is referred to as the 'vibronic wave function.' Using this vibronic wave function, the transition matrix element of transitions from the ground vibrational level of the ground state to the nth vibrational level of the excited state results in:

$$
\begin{aligned}
M_{0n} &= \int \Phi_{g0}(r,Q)^* (-er) \Phi_{en}(r,Q-\Delta Q)\,drdQ \\
&= \int \psi_g(r)^* (-er)\psi_e(r)\,dr \times \int \xi_{g0}(Q)^* \xi_{gn}(Q-\Delta Q)\,dQ \\
&= m \times \int \xi_{g0}(Q)^* \xi_{gn}(Q-\Delta Q)\,dQ
\end{aligned}
\tag{11.8}
$$

where we focus on transition only from zero vibrational level in the ground state. In Equation (11.8), the first factor is the same as given in Equation (11.5), which involves transitions only between electronic states. The second factor as an integral in Equation (11.8), dependent on the vibrational levels, determines the vibronic structure. Since the absorption (and PL) intensity is proportional to the square of the transition matrix element, the following formula is more practical:

$$
F_{0n} = \left| \int \xi_{g0}(Q)^* \xi_{gn}(Q-\Delta Q)\,dQ \right|^2
\tag{11.9}
$$

Equation (11.9) is referred to as the Franck–Condon factor and is denoted by F_{0n}.

Using the harmonic oscillator approximation, Equation (11.9) can be simplified into the following form:

$$
F_{0n}(S) = \frac{e^{-S} S^n}{n!}
\tag{11.10}
$$

where Huang–Rhys parameter, S, represents the strength of electron–phonon interaction for an associated phonon mode, and is given by:

$$
S = \frac{\Delta Q^2}{2}
\tag{11.11}
$$

In Figure 11.9 we show some vibronic structures calculated using Equation (11.10) for several Huang–Rhys parameters. Equation (11.10) is the same as the Poisson distribution, and always gives the maximum intensity for the 0–S transition. For instance, in a case where the Huang–Rhys parameter, S, is 2, the 0–2 transition has the maximum intensity. However, it is still difficult to reproduce perfectly the observed absorption spectrum of π-conjugated polymers using Equation (11.10). This is because the spectrum has large inhomogeneous broadening even in ordered films and contributions from higher excited states.

11.5 PHOTOLUMINESCENCE

Many π-conjugated polymers show photoluminescence (PL) in the visible spectral range. For instance, polyfluorene, poly(p-phenylenevinylene) (PPV), and polythiophene show blue, green, and red emissions, respectively. Some of their derivatives have good fluores-

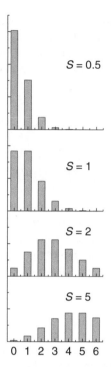

Figure 11.9 Examples of vibronic structure calculated using Equation (11.10) for several Huang–Rhys parameters

cence efficiency and are expected to be used in fabricating light-emitting devices. In these materials, after excited states are created by photoexcitation, they immediately relax into the lowest excited state (Kasha's rule) and then emit light. Therefore, their PL straightforwardly reflects the nature of the lowest excited state. In the PL spectrum, vibronic structure also appears due to transitions from the lowest vibrational level in the excited state to vibrational levels in the ground state [see Figure 11.8 (b)]. This process can also be described using Equation (11.10) and then only the Huang–Rhys parameter determines the vibronic structure under the harmonic-oscillator approximation. Therefore, symmetrical absorption and PL spectra are expected to be observed as shown in Figure 11.10; the upper figure represents the case of small S and the lower one that of large S. The differences between the absorption and PL maxima is called the Stokes shift, which results from many relaxation mechanisms of the polymer chain occurring after the photoexcitation. However, as shown in the lower figure of Figure 11.10, a vibronic structure with large Huang–Rhys parameter could be the main reason of the larger Stokes shift. In this case, the Stokes shift is roughly estimated to be $2\omega_0 S$, where ω_0 is the associated phonon energy.

In π-conjugated polymers, the bond alternation becomes modified after the photoexcitation to reduce the total energy of the excited state by forming a self-trapped exciton, whose structure for polythiophene is depicted in Figure 11.11. This self-trapping process is induced right after the photoexcitation and is complete within 10 ps [11.6]. Therefore, most of the photoexcitations are emitted from the self-trapped state, and this reduction in energy also contributes to the observed Stokes shift.

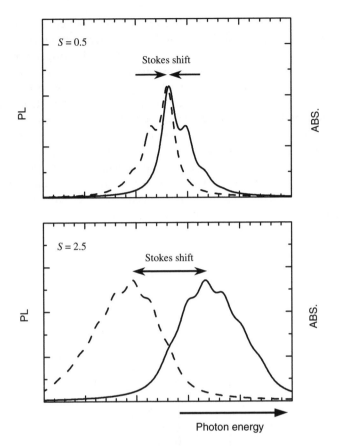

Figure 11.10 Mirror image of absorption and PL spectra. The difference between their maxima is called the Stokes shift. These are ideal cases where only one excited state appears in the observed spectral range. In actual π-conjugated films, the observed absorption spectrum does not entirely agree with that expected from the PL spectrum because the observed absorption spectrum contains inhomogeneous broadening and contributions from higher excited states

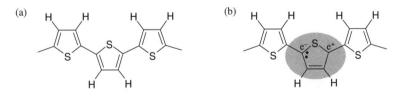

Figure 11.11 Bond arrangements (a) in the ground state and (b) in self-trapped excited state for polythiophene

In disordered films of π-conjugated polymers, π electrons are not fully delocalized: torsion and bending of the polymer backbone limit delocalization of the π electrons. In such cases, films should be considered as an ensemble of π conjugation of various lengths. The emitted energy from a segment of the backbone becomes red-shifted as its delocalization increases. Therefore, after the photoexcitation, excited states prefer to migrate from shorter segments to longer ones and exclusively emit luminescence from longer conjugation segments. This migration process can be one of the reasons for the observation of Stokes shift in disordered films. However, PL does not usually originate from the longest segment in disordered films. As an excited state migrates to longer segments, it becomes more difficult to find further longer segments nearby. Within its lifetime, an excited state can only migrate to segments with a certain length of π conjugation. When a disordered film is excited by photons of high enough energy, the PL spectrum is independent of the excitation photon energy. On the other hand, when the photon energy is less than the threshold, PL shows dependence on the photon energy. In this condition, those segments to which an excited state cannot migrate from shorter segments can be directly excited and PL from these segments can be observed. This measurement is called 'site-selective fluorescence,' and we show an example of this measurement in Figure 11.12 [11.7].

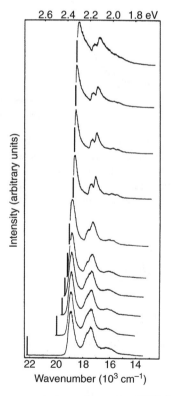

Figure 11.12   Site-selective fluorescence measurements on PPV films. The different spectra were obtained by varying the excitation energy (indicated by the vertical lines), starting at the bottom far from resonance and moving into resonance going up the figure [Reproduced from S. Heun et al., *J. Phys. Condens. Matter*, **5**, 247 (1993) by permission of IOP Publishing Ltd]

11.6 NONEMISSIVE EXCITED STATES

As in inorganic semiconductors, it is possible to generate charge carriers in π-conjugated polymers by doping. In π-conjugated polymers, these carriers are called soliton, polaron, or bipolaron, where bond alternation is modified and charge is delocalized within 10–30 carbon atoms [11.8] as illustrated in Figure 11.13. A soliton is formed in polyacetylenes, and polarons and bipolarons are formed in the π-conjugated polymers, where the ground state of geometrical isomers is not degenerate. These carriers govern conductivity in doped π-conjugated polymers. In undoped π-conjugated polymers, solitons or polarons can be created by photoexcitations but they decay nonradiatively within a lifetime of μs–ms. Thus, the formation of such charged excitations serves as one of the nonradiative decay channels in some π-conjugated polymers. The long-lived photoexcitations in π-conjugated polymers can be observed by cw photo-induced absorption (PA) measurements. In PA measurements, a pump beam creates photoexcitatioins and a probe beam detects transitions of photoexcitations to higher excited states. CW PA is usually conducted using a lock-in amplifier with mechanical chopper, so that ultrafast photoexcitations contribute little to the cw PA signal and only long-lived photoexcitations appear in the observed spectrum. In Figure 11.14 we show cw PA spectra of disordered and highly ordered films of poly(3-hexylthiophene) (P3HT) in companion with PA-detected magnetic resonance (PADMR) spectra [11.3], from which it is possible to know the spin number of each PA band. In a highly ordered film of P3HT [Figure 11.14 (b)], several PA bands with spin 1/2 are observed. In a one-dimensional model, polarons have two localized states in the gap between the HOMO and LUMO as shown in Figure 11.15. The higher state is empty and the lower one is occupied by an electron in the case of a polaron with a positive charge. Thus, the P2 band observed in Figure 11.14 (b) can be assigned to the P2 transition from the lower localized state to the higher one in Figure 11.15, and the P1 band observed in Figure 11.14 (b) can be assigned to the P1 transition from the ground state to the lower localized state in Figure 11.15. Since these

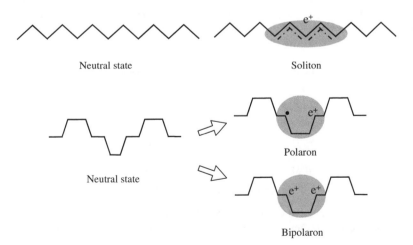

Figure 11.13 Charged excited states in π-conjugated polymers. In π-conjugated polymers, charge induces modification of the bond alternation to form stable excited states

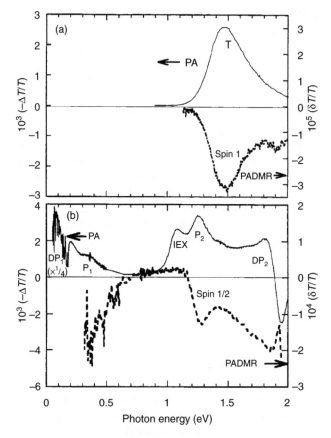

Figure 11.14   (upper) PA and (lower) λ-PADMR spectra of (a) disordered and (b) ordered poly-thiophene films. In the upper figures, the PA bands correspond to transitions of the long-lived excited states to higher excited states. The lower figures show the spin number of each PA band [Reproduced with permission from O.J. Korovyanko et al., *Phys. Rev. B*, **64**, 235112. Copyright (2001) by the American Physical Society]

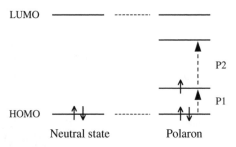

Figure 11.15   Energy structure of polarons. Solid arrows indicate electrons (spin direction) and broken arrows indicate transitions appearing in the bandgap after the polaron formation

bands (P1 and P2) are also observed in isolated polythiophene chains in an inert polystyrene matrix, they can be attributed to one-dimensional polarons, which are located in the disordered portions of the polythiophene film. A highly ordered film contains the crystallized portions of the chains as well, where interchain interaction is expected to shift the energy level of the PA bands slightly. Thus, the DP1 and DP2 bands in Figure 11.14 (b) are assigned to delocalized polarons [11.9]. In addition, a PA band with spin zero (denoted by IEX) is observed at 1.07 eV. In polythiophene films, intrachain neutral excitations, i.e., self-trapped excitons, decay into the ground state within a lifetime of the order of 1 ns; such short-lived excitations cannot be observed in cw PA measurements. Thus, the PA band with spin zero is attributed to the interchain neutral excitations, which are pairs of polarons that have opposite charges and are located on neighboring chains. On the other hand, only a broad PA band with spin 1 is observed in Figure 11.14 (a). This band corresponds to a triplet–triplet transition, which indicates efficient intersystem crossing in disorder portions of the polymer. Thus, in π-conjugated polymers after photoexcitation, many kinds of excited states have to be considered.

11.7 ELECTRON–ELECTRON INTERACTION

From symmetry considerations, energy levels in one-dimensional π-conjugated polymers are classified into odd-parity (B_u) or even-parity (A_g) states. Since the ground state is an A_g state, the B_u state is the one-photon-transition-allowed excited state and it appears in the absorption spectrum. On the other hand, an A_g state is a one-photon-transition-forbidden state and has no transition dipole moment. Although the latter state is absent in the absorption spectrum, both A_g as well as B_u states play important roles in PL and nonlinear optical processes. For instance, according to Kasha's rule, photoexcitations in higher excited states immediately relax nonradiatively to the lowest excited state. If the lowest excited state has B_u symmetry, then a PL with radiative lifetime in the ps–ns range would be observed. However, if the lowest excited state is an A_g state, radiative lifetime increases dramatically and most of the photoexcitations preferably decay nonradiatively. Therefore, the symmetry of the lowest excited state is one of the most important factors in determining fluorescence yield in π-conjugated polymers. In nonlinear optical processes, both A_g and B_u states work as resonance energy states: when incident photon frequency resonates with the energy interval between the A_g and B_u states, an enhancement in the nonlinear optical response is obtained. In particular, in one-dimensional π-conjugated polymers A_g states are essential to our understanding of their nonlinear optical properties because the transition dipole moments between the lowest B_u state and higher A_g states are significantly large [11.10].

Energy levels of A_g symmetry are strongly dependent on the electron–electron interaction, and even the Hartree–Fock framework, without any configuration interaction, is not enough to describe an A_g state in π-conjugated polymers. In some π-conjugated polymers the lowest excited state is indeed an A_g state. The appearance of an A_g state between the ground and the lowest B_u states cannot be explained using any theories without electron–electron interaction. Although several theoretical models, including the electron–electron interaction, have been developed recently, it is still difficult to apply these models to π-conjugated polymers, which consist of a huge number of sites and have significant structural disorders. Here, we introduce a basic concept of group theory applicable to π-conjugated polymers and some clear experimental results.

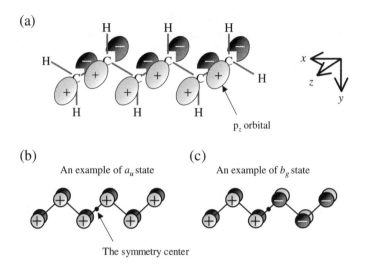

Figure 11.16 Symmetry of electronic states in $C_{2n}H_{2n+2}$ (in the case of $n = 3$). (a) The molecular structure of $C_{2n}H_{2n+2}$ with the emphasized p_z orbitals. The shape of the p_z orbital is expressed by two touching spheres. In this figure, the touching points are in the plane containing the carbon atoms, and since the two spheres of each p_z orbital above and below the plane have opposite signs, they are distinguished by '+' and '−'. (b) and (c) show two examples of combinations of $2n$ p_z orbitals

As a simple model of π-conjugated polymers, we consider the symmetry of a polyene, $C_{2n}H_{2n+2}$, in Figure 11.16 (a). This molecular structure possesses C_{2h} group symmetry and is invariant under the operation of space inversion at the symmetry center or rotation about the symmetry axis though 180 degrees. In group theory, wave functions whose sign changes or remains unchanged after the rotation are labeled by 'b' and 'a', respectively. Furthermore, wave functions whose sign changes or remains unchanged after the space inversion are labeled by 'u' and 'g', respectively. A p_z orbital has a shape represented by two touching spheres with opposite signs [see Figure 11.16 (a)], and thus a sign of the p_z orbital reverses after space inversion. Considering this fact, a sign of π electrons consisting of $2n$ p_z orbitals can be easily determined. Figures 11.16 (b) and (c) are two examples of combinations of $2n$ p_z orbitals having different symmetry. In Figure 11.16 (b), all the '+' spheres are located in the $+z$ direction. In this case, a sign of this π-electron level does not change after the rotation but changes after the space inversion. Thus, this π-electron level has a_u symmetry. Similarly, the π-electron level shown in Figure 11.16 (c) has b_g symmetry. When all combinations of six p_z orbitals are examined, you will find all π-electron levels in the molecule can be classified into 'a_u' or 'b_g' states. As illustrated in Figure 11.17, this molecule has $2n$ energy levels of π electrons, and therefore a_u and b_g states appear alternately from bottom to top. Since each level has spin-up and -down states, this molecule has $4n$ states for $2n$ π electrons. Thus, the electronic configuration of this molecule is determined by the way that $2n$ π electrons are arranged to occupy $4n$ states. In Figure 11.17, three examples of electronic configuration are illustrated. The symmetry of a configuration can be calculated by multiplying the symmetry of $2n$ π electrons using the following relations:

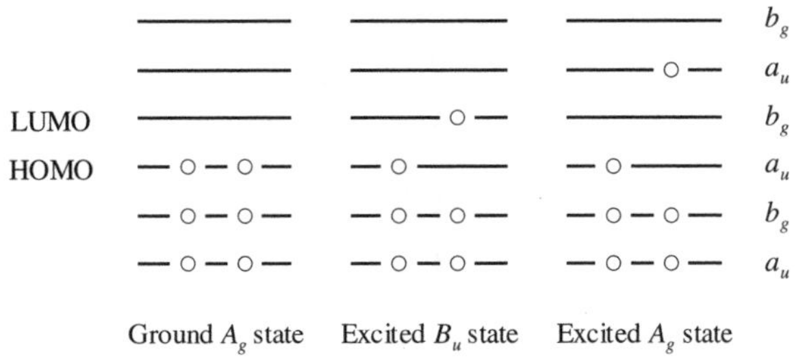

Ground A_g state   Excited B_u state   Excited A_g state

Figure 11.17 Symmetry of configuration of $C_{2n}H_{2n+2}$ (in the case of $n = 3$). The molecule has $2n$ energy levels of π electrons, and, in the ground state, $2n$ energy levels are occupied by $2n$ π electrons from the bottom (the left). The middle and right configurations are examples of excited states of the molecule

$$a \times a = b \times b = a, \quad g \times g = u \times u = g$$
$$a \times b = a \times b = b, \quad g \times u = u \times g = u$$

(11.12)

From this simple calculation, we find that a system with C_{2h} symmetry has only B_u and A_g states (here, we use capital letters for the symmetry of configurations). In the ground state, $2n$ π electrons fill only half of the $4n$ states starting from the bottom, as shown in the left side of Figure 11.17. When one of the π electrons from the HOMO is excited into the LUMO, the symmetry of the configuration becomes B_u, as shown in the middle of Figure 11.17. Furthermore, when the excited π electron is further excited into the next lowest unoccupied level, the symmetry becomes A_g again. Therefore, the system has an electronic energy structure as $1A_g$ (ground state), $1B_u$, $2A_g$, $2B_u$, ..., nA_g, and nB_u appear from the lowest to the highest levels. This is always true whenever a one-electron theory, such as the Hartree–Fock approximation, is valid for the system. However, if electron–electron interaction is not negligible, the electronic configuration on the right-hand side can become more stable state than that in the middle in Figure 11.17, and then the $2A_g$ state appears as the lowest excited state.

Experimentally, the two-photon absorption (TPA) technique is more suitable to investigate A_g states than is conventional linear spectroscopy, such as one-photon absorption and PL measurements, because transitions from the ground ($1A_g$) state to higher excited A_g states are dipole-forbidden. In contrast, in the two-photon absorption process the simultaneous absorption of two photons can directly excite the A_g state (see Figure 11.18), and the B_u state remains silent. Thus, it is possible to determine the energy of the A_g state without any ambiguity. In this process, the two-photon absorption cross-section increases linearly with the laser intensity, whereas the one-photon absorption cross-section remains constant. Therefore, a TPA coefficient is obtained from the intensity dependence of the absorption coefficient. Alternatively, for samples with strong PL, a two-photon excitation (TPE) measurement is performed. In this process, the PL intensity is proportional to the square of the excitation intensity. From this intensity dependence, a TPE coefficient is obtained. Technically, TPE measurement is much easier than TPA measurement, and in many studies of π-conjugated polymers TPE measurement has been performed. Figure 11.19 (a) shows the absorption and TPE spectra of thin films of a PPV derivative [11.11]. In this figure, the A_g state at around 3.2 eV lies between the $1B_u$ (I) and $2B_u$ (II) bands in the absorption spec-

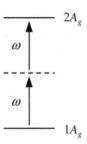

Figure 11.18   A two-photon absorption process. In a one-photon process, transition between states with the same symmetry is forbidden and transition between states with opposite symmetries is allowed. In the two-photon process, this selection rule reverses: Transition from the ground ($1A_g$) state to a higher A_g state is allowed, but in this process it is required that the energy interval between the two A_g states be equal to the sum of photon energies of the two incident photons

trum. On the other hand, Figure 11.19 (b) shows absorption and TPA spectra of a single crystal of polydiacetylene [11.12]. Polydiacetylene and its derivatives are nonluminescent π-conjugated polymers; their fluorescence yield is estimated to be less than 10^{-5} [11.13]. In this case [Figure 11.19 (b)], the A_g state is of lower energy than the lowest $1B_u$ state. Such an excited-state structure is also observed in other nonluminescent π-conjugated polymers, e.g., polyacetylenes [11.14]. Thus, the absence of PL in these polymers is attributed to A_g states being below the lowest B_u state.

The electro-absorption (EA) measurement has also been used as an experimental method to investigate A_g states in π-conjugated polymers. In this measurement, an AC electric field with frequency f is applied to the film, and any change in the absorption spectrum at frequency $2f$ is detected. In this process, the applied electric field couples A_g and B_u states and induces a separation between the ground and excited energy levels, which results in very small absorption spectral change (the absorption coefficient changes by about 0.1% at most). As an example, we show in Figure 11.20 the EA spectrum of thin films of poly(p-phenylene-ethynylene) (PPE) [11.15]. In the EA spectrum of many π-conjugated polymers, a signal due to the red-shift of the lowest absorption band appears (at around 2.5 eV in the case of Figure 11.20). This red-shift is called a Stark shift and results from the strong repulsion between the $1B_u$ state and higher A_g states due to the applied electric field. Although the applied electric field also induces repulsion between the $1A_g$ and $1B_u$ states, this contribution is weaker than the others from higher A_g states because the energy interval between the $1A_g$ and $1B_u$ states is relatively large. Therefore, to explain the Stark shift, a higher A_g state is necessary, at least.

Theoretically, the EA spectrum is described by the third-order nonlinear susceptibility, which is calculated using the third-order perturbation theory and is proportional to the product of four transition dipole moments [11.16]. In the two-level model, the dipole transition moment between the two states has to be used four times for the calculation [see Figure 11.21 (a)] and only repulsion between $1A_g$ and $1B_u$ states is taken into consideration. Consequently, the wrong conclusion, i.e., blue-shift of the absorption band, always results from use of the two-level model. On the other hand, in the three-level model two kinds of transition dipole moments can be used [see Figure 11.22 (b)], and a term using both tran-

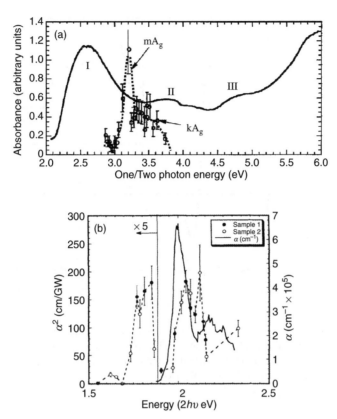

Figure 11.19 (a) The solid and dotted lines indicate absorption and TPE spectra of PPV, respectively. In this figure, 'I', 'II', and 'III' are major absorption bands, i.e. B_u states, whereas the dotted line shows that two A_g states lie between the I and II bands. (b) The solid line is the absorption spectrum of a single crystal of polydiacetylene (left scale). The empty and filled circles are TPA coefficients obtained from two different samples, and the dashed line is a guide to the eye. Below 1.9 eV, the vertical scale is expanded to make clearer the two resonances below the lowest B_u state at 2.0 eV [Reprinted Fig. 1a with permission from S.V. Frolov et al., *Phys. Rev. Lett.*, **85**, 2196. Copyright (2000) by the American Physical Society. Reprinted Fig. 2 with permission from B. Lawrence et al., *Phys. Rev. Lett.*, **73**, 597. Copyright (1994) by the American Physical Society]

sition dipole moments twice each contribute to the third-order nonlinear susceptibility. In fact, this term becomes dominant in the case where the energy difference between $1A_g$ and $1B_u$ states is much larger than that between $1B_u$ and mA_g states, and the observed Stark shift can be explained. When inhomogeneous broadening and vibronic replicas are taken into consideration in the simulation using the asymmetric Gaussian function and Franck–Condon factor, respectively, an experimental EA spectrum of π-conjugated polymers can be well reproduced (see the dashed lines in Figure 11.20) [11.15]. From fitting the simulated spectrum to experimental one, it is possible to determine the energy levels essential to describe the nonlinear properties and dipole moments between these levels.

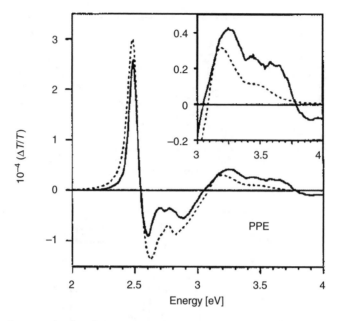

Figure 11.20 An example of an EA spectrum of π-conjugated polymers. The solid and dashed lines are the experimental and theoretical ones, respectively. The inset shows in more detail the EA feature at high photon energy [Reprinted Fig. 5 with permission from M. Liess et al., *Phys. Rev. B*, **56**, 15712. Copyright (1997) by the American Physical Society]

11.8 INTERCHAIN INTERACTION

The interchain interaction should be included as an important interaction to help us understand the optoelectronic properties of π-conjugated polymers. However, it has not been yet established how to treat this interaction in systems having significant inhomogeneous broadening, such as π-conjugated polymers. Here, we briefly review a theory developed for single crystals of small molecules to get a basic picture from the intermolecular interaction that will be helpful in further understanding the optoelectronic properties of π-conjugated polymers.

Each organic molecule has many electronic energy levels and associated vibrational modes that are determined from the chemical structure. Most of these original characters are preserved even during intermolecular interactions. Therefore, we can describe the effect of intermolecular interactions using the perturbation theory. At first, we consider a simple system consisting of two identical molecules (see Figure 11.22). When the molecules are close enough, each energy level of the isolated molecules splits into two levels, which is called Davydov splitting [11.17]. The energy of the interacting system can be written as:

$$E_{\text{system}} = E_{\text{molecule}} + \Delta D \pm \Delta E \tag{11.13}$$

where E_{molecule} is the energy of the isolated molecule, ΔE is the splitting energy, and ΔD indicates an energy shift due to other effects, such as van der Waals forces. Generally, ΔD cannot

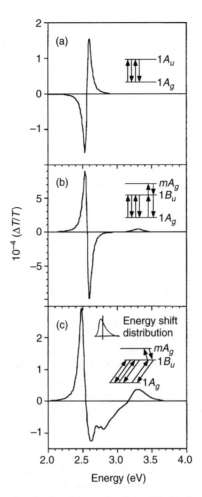

Figure 11.21 EA spectra simulated using (a) two-level model, (b) three-level model, and (c) three-level model taking into consideration asymmetric inhomogeneous broadening and vibronic structure. In the insets, energy levels and the ways to choose four transition dipole moments to simulate the EA spectra are depicted. In (c), an example of asymmetric inhomogeneous broadening is also illustrated [Reprinted Fig. 7 with permission from M. Liess et al., *Phys. Rev. B*, **56**, 15712. Copyright (1997) by the American Physical Society]

be estimated quantitatively and is treated as a negative constant. If the intermolecular interaction is described by the dipole–dipole interaction, we obtain

$$\Delta E = \frac{\mathbf{M}_1 \cdot \mathbf{M}_2}{r^3} - 3\frac{(\mathbf{M}_1 \cdot \mathbf{r})(\mathbf{M}_2 \cdot \mathbf{r})}{r^5} \tag{11.14}$$

where $\mathbf{M}_1$ and $\mathbf{M}_2$ are transition dipole moments of the molecules and $|\mathbf{r}| = r$ is the distance from one molecule to the other. From this equation, we find that ΔE depends not only on the distance between two molecules but also on the relative angle between them. Three arrangements of a pair of molecules and their energy diagrams are illustrated in Figures

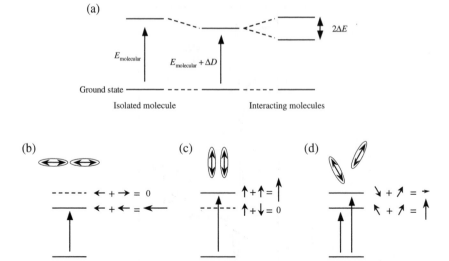

Figure 11.22   (a) Intermolecular interaction and Davydov splitting. On the left are shown the energy levels of an isolated molecule, and on the right are those of the two interacting molecules. $2\Delta E$ is the Davydov splitting energy due to the intermolecular interaction and ΔD is the energy shift due to other effects. (b)–(d) Molecular arrangements and dipole moments. The ellipses and the inside arrows indicate the interactive molecules and their transition dipole moments. Below the ellipses, their energy levels are depicted. In (b) and (c), one of the split branches does not have any transition dipole moment because the two transition dipole moments of individual molecules cancel each other, and only one band is observed in their absorption spectra. On the other hand, in (d), both branches have some transition dipole moments, and two absorption bands are observed

11.22 (b)–(d). In Figure 11.22 (b), the molecules are in a line, in (c) they are aligned parallel to each other, and in (d) they are in a nonparallel arrangement. In part (b), the lower branch corresponds to the parallel arrangement of two dipole moments, and the higher branch corresponds to antiparallel arrangement. Since the transition dipole moments are vectors, they cancel each other in the antiparallel arrangement, and then the parallel arrangement has twice the oscillator strength of the isolated molecule. Consequently, in part (b), only the lower branch is observed in the absorption spectrum. In part (c), the antiparallel arrangement appears in the lower branch and the parallel arrangement appears in the higher branch. In this case, the lower branch has no transition dipole moment, which results in a significantly low fluorescence yield of the system; most of the excited states preferably decay through nonradiative channels, such as by phonon emission. Real materials are more complicated than the examples shown in Figures 11.22 (b) and (c) because a nonparallel arrangement as shown in part (d) is more stable in energy for many organic molecules. In this case, an energy level splits into two branches that have some transition dipole moments, and both branches are observed in the absorption spectrum. In any case, transition dipole moments of a system are calculated by the linear combination of transition dipole moments of individual molecules.

The concept shown in Figure 11.22 can be applied to single crystals of organic molecules. If a single crystal is perfect and large, we do not have to consider all of the con-

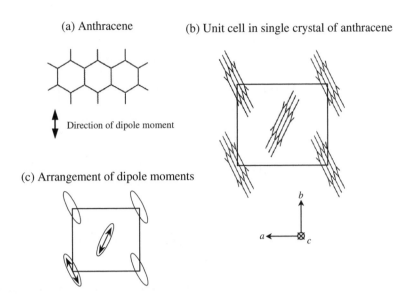

Figure 11.23   (a) The molecular structure of anthracene and transition dipole moment of the lowest excited state. (b) The crystal structure of anthracene. a, b, and c represent the crystallographic axes. (c) The crystal structure depicted to emphasize two representative molecules and their transition dipole moments. This arrangement is almost identical to that shown in Figure 11.22 (d). Thus, the crystal has two transition dipole moments parallel to the a and b axes because of the molecular interaction

stituent molecules. According to the Frenkel exciton theory, all unit cells have the same transition dipole moment in exciton levels with $k = 0$, which is the essential condition for optical responses. Therefore, it is possible to study the macroscopic response of the crystal by considering the intermolecular interaction within a unit cell. Figure 11.23 shows the chemical structure of anthracene and its transition dipole moment in the lowest excited state. An anthracene crystal has two molecules in each unit cell, as shown in Figure 11.23 (b). If we choose two molecules, one at the center and the other at the lower-left corner, as representatives of all molecules in the crystal, we find that their arrangement is identical to that shown in Figure 11.22 (d). Thus, a single crystal of anthracene has two split branches having transition dipole moments along the a and b axes. (More precisely, the former transition dipole moment is on the a–c plane.) These branches are identified from the optical polarization measurements. Figure 11.24 shows the polarization absorption spectra of a crystal of anthracene and its absorption spectrum in solution [11.18]. The absorption spectra of the crystal are red-shifted from that in solution, and ΔD is estimated to be around $1000\,\mathrm{cm}^{-1}$. The Davydov splitting energy in the crystal can be estimated to be around $200\,\mathrm{cm}^{-1}$ from comparison between the polarization absorption spectra.

Although it is impossible to obtain single crystals of many π-conjugated polymers, some of their oligomers have been reported to form single crystals. Single crystals of an oligothiophene that consists of 6 thiophene units have been considerably investigated and have shown a Davydov splitting of about $2600\,\mathrm{cm}^{-1}$ [11.19]. Therefore, there is every possibility that optical properties of π-conjugated polymers are also influenced by the interchain interaction.

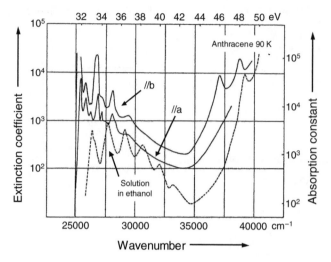

Figure 11.24   The solid lines are polarization absorption spectra of crystalline anthracene parallel to the *a* and *b* axis, respectively. The dashed line is the absorption spectrum of anthracene in ethanol. The absorption spectra of the crystal are red-shifted with respect to that in solution, and the polarization absorption spectra parallel to the *a* and *b* axes are different from each other in their intensity and spectral shape because of the interchain interaction. [Reproduced from H.C. Wolf, *Z. Naturforsch., Teil A*, **13**, 414 (1958), with permission of Verlag Z. Naturforsch]

11.9 CONCLUSIONS

In π-conjugated polymers, carbon atoms are linked with each other to form a one-dimensional chain, where π electrons are delocalized. The π electrons govern their optical properties in the visible range. To treat the π electrons theoretically, electron–phonon and electron–electron interactions should be included in the Hamiltonian. However, it is still difficult to resolve such Hamiltonians because of the significant structural disorder in actual π-conjugated polymers.

From a comparison between absorption and photoconductivity-yield spectra, the exciton-binding energy of π-conjugated polymers are usually estimated to be around 0.1 eV. Ordered films of π-conjugated polymers show clear vibronic structure in their absorption spectrum, which results from coupling between the electronic states and a phonon mode due to electron–phonon interaction.

In π-conjugated polymers, PL originates after several relaxation processes, such as formation of self-trapped excitons and exciton migration within segments with various lengths of π conjugation. PL also shows clear vibronic structure, which indicates that electron–phonon interaction is essential in the PL process, as well. A photoexcitation in these polymers generates not only emissive excited states but also long-lived nonemissive states, such as polarons and triplet excitons. In a system that has strong electron–electron interaction, the dipole-forbidden (A_g) state could be of a lower energy than the lowest dipole-allowed ($1B_u$) state, and their fluorescence yields are quite low. This is the reason for the absence of photoluminescence in polyacetylene and polydiacetylene.

Experimentally, two-photon absorption, two-photon excitation, and electro-absorption measurements are powerful tools to investigate A_g states in π-conjugated polymers.

Intermoleular interaction splits energy levels whose transition dipole moments are linear combinations of transition dipole moments of the constituent molecules. However, this energy splitting (Davydov splitting) and polarization dependence of the absorption spectrum have not yet been recognized in π-conjugated polymers because of their significant structural disorder.

REFERENCES

[11.1] W. Drenth and E.H. Wiebenga, *Acta Crystallogr.*, **8**, 755 (1955).

[11.2] A.J. Heeger, S. Kivelson, J.R. Schrieffer, and W.-P. Su, *Rev. Mod. Phys.*, **60**, 781 (1988).

[11.3] *Conjugated Conducting Polymers*, edited by H.G. Kiess (Springer-Verlag, Berlin, 1992).

[11.4] K. Pakbaz, C.H. Lee, A.J. Heeger, T.W. Hagler, and D. McBranch, *Synth. Met.*, **64**, 295 (1994). A. Köhler et al., *Nature*, **392**, 903 (1998).

[11.5] B. Henderson and G.F. Imbusch, *Optical Spectroscopy of Inorganic Solids* (Clarendon Press, Oxford, 1989).

[11.6] T. Kobayashi, J. Hamazaki, M. Arakawa, H. Kunugita, K. Ema, K. Ochiai, M. Rikukawa, and K. Sanui, *Phys. Rev. B*, **62**, 8580 (2000).

[11.7] S. Heun, R.F. Mahrt, A. Greiner, U. Lemmer, H. Bässler, D.A. Halliday, D.D.C. Bradley, P.L. Burn, and A.B. Holmes, *J. Phys.: Condens. Matter*, **5**, 247 (1993).

[11.8] O.J. Korovyanko, R. Österbacka, X.M. Jiang, Z.V. Vardeny, and R.A.J. Janssen, *Phys. Rev. B*, **64**, 235112 (2001).

[11.9] R. Österbacka, C.P. An, X.M. Jiang, and Z.V. Vardeny, *Science*, **287**, 839 (2000).

[11.10] S. Abe, *Conjugated Polymers: Molecular Exciton versus Semiconductor Band Model*, edited by N.S. Sariciftci (World Scientific, Singapore, 1997), Ch. 5.

[11.11] S.V. Frolov, Z. Bao, M. Wohlgenannt, and Z.V. Vardeny, *Phys. Rev. Lett.*, **85**, 2196 (2000).

[11.12] B. Lawrence, W.E. Torruellas, M. Cha, M.L. Sundheimer, G.I. Stegeman, J. Meth, S. Etemad, and G. Baker, *Phys. Rev. Lett.*, **73**, 597 (1994).

[11.13] Z.G. Soos, S. Etemad, D.S. Galvao, and S. Ramasesha, *Chem. Phys. Lett.*, **194**, 341 (1992).

[11.14] G. Orlandi, F. Zerbetto, and M.Z. Zgierski, *Chem. Rev.*, **91**, 867 (1991).

[11.15] M. Liess, S. Jeglinski, Z.V. Vardeny, M. Ozaki, K. Yoshino, Y. Ding, and T. Barton, *Phys. Rev. B*, **56**, 15712 (1997).

[11.16] R.W. Boyd, *Nonlinear Optics* (Academic Press, San Diego, 1992).

[11.17] A.S. Davydov, *Theory of Molecular Excitons* (McGraw-Hill, New York, 1971).

[11.18] H.C. Wolf, *Z. Naturforsch.*, *Teil A*, **13**, 414 (1958).

[11.19] M. Muccini, E. Lunedei, C. Taliani, D. Beljonne, J. Cornil, and J.L. Brédas, *J. Chem. Phys.*, **109**, 10513 (1998).

12 Organic Semiconductors and Applications

F. Zhu

Opto- & Electronic Cluster
Institute of Materials Research and Engineering
3 Research Link, Singapore 117602

12.1 INTRODUCTION

Silicon-based transistors and integrated circuits are of central importance in the current microelectronics industry, which serves as an engine to drive progress in today's electronics technology. However, there is a great need for significant new advances in the rapidly expanding field of electronics. New materials and innovative technologies are predicted to lead to developments beyond anything we can imagine today. The demand for more user-friendly electronics is propelling efforts to produce head-worn and hand-held devices that are flexible, lighter, more cost-effective, and more environmentally benign than those currently available. Electronic systems that use organic semiconductor materials offer an enabling technology base. This technology has significant advantages over the current silicon-based technology because it allows for an astonishing amount of electronic complexity to be integrated onto lightweight, flexible substrates for the production of a wide

Optical Properties of Condensed Matter and Applications   Edited by J. Singh
© 2006 John Wiley & Sons, Ltd

range of entertainment, wireless, wearable-computing, and network-edge devices. For example, organic light-emitting devices (OLEDs) that use a stack of organic semiconductor layers have the potential to replace liquid crystal displays (LCDs) as the dominant flat-panel display device. This is because OLEDs have high visibility by self-luminescence, do not require backlighting, and can be fabricated into lightweight, thin, and flexible displays. A combination of the organic electronics and existing microelectronics technologies also opens a new world of potential for electronics. The possible uses include a wide variety of industrial, medical, military and other consumer-oriented applications.

12.1.1 OLED architecture and operation principle

A typical OLED is constructed by placing a stack of organic electroluminescent and/or phosphorescent materials between a cathode layer that can inject electrons, and an anode layer that can inject 'holes'. Polymeric electroluminescent materials have been used in OLEDs, and are referred to as polymer light-emitting devices. A conventional OLED has a bottom-emitting structure, which includes a metal or metal alloy cathode, and a transparent anode on a transparent substrate, enabling light to be emitted from the bottom of the structure [Figure 12.1 (a)]. An OLED may also have a top-emitting structure, which is formed on either an opaque substrate or a transparent substrate. A top-emitting OLED (TOLED) has a relatively transparent top cathode so that light can emit from the side of the top electrode [Figure 12.1 (b)].

Figure 12.1 shows a cross-sectional view of (a) typical bottom-emitting OLED and (b) a top-emitting OLED. In a multi-layered OLED, the organic medium consists of a hole-

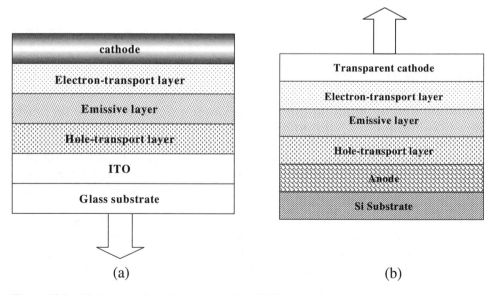

(a)   (b)

Figure 12.1   (a) A conventional bottom-emission OLED is made on a transparent substrate, e.g., glass or clear plastic substrate, and (b) a TOLED requires a semitransparent top cathode. TOLEDs can be made on both transparent and opaque substrates

transporting layer (HTL), a light-emissive layer (EL), and an electron-transporting layer (ETL). Indium tin oxide (ITO) is often used as the transparent anode due to its high optical transparency and electric conductivity. The cathode in the OLEDs is made of low-work-function metals or their alloys, e.g., MgAg, Ca, LiAl, etc.

A schematic energy-level diagram for an OLED under bias is shown in Figure 12.2. When a voltage of proper polarity is applied between the cathode and anode, 'holes' injected from the anode and electrons injected from the cathode combine radiatively to release energy as light, thereby producing electroluminescence. The light usually escapes through the transparent substrate. Different organic semiconductor functional layers in an OLED can be optimized separately for carrier transport and luminescence.

A large number of conducting small molecular materials and conjugated polymers have been used as the charge-transporting or emissive layer in OLEDs. The molecular structures of some commonly used organic semiconductor materials are illustrated in Figure 12.3. A small-molecule-based OLED device has a typical configuration of glass/ITO/HTL/organic emissive layer/ETL/Mg:Ag mixture/Ag. The multilayer thin-film device is fabricated using the thermal evaporation method in a vacuum system. The thermal evaporations usually start at a base pressure of 10^{-6} mbar or lower. The fabrication of polymeric OLEDs involves solution processes and the devices can be made using spin-coating or inkjet printing methods. After the deposition of organic semiconductor layers, a thin electron injector is evaporated following a 150–300 nm thick Al or Ag layer to inhibit the oxidation of the cathode contact.

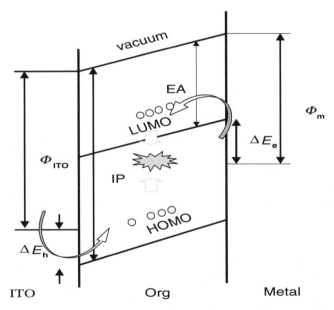

Figure 12.2 Schematic energy-level diagram of an OLED, showing the highest occupied molecular orbital (HOMO) and the lowest unoccupied molecular orbital (LUMO) of the HTL and ETL, which are also referred to as affinitive energy level (EA) and ionization potential (IP). Φ_{ITO} and Φ_m represent the work function of the ITO anode and metallic cathode. ΔE_h and ΔE_e are the barrier height at ITO/organic and organic/cathode interfaces. Hole–electronic current balance in an OLED is set by the size of the barriers at the two electrodes

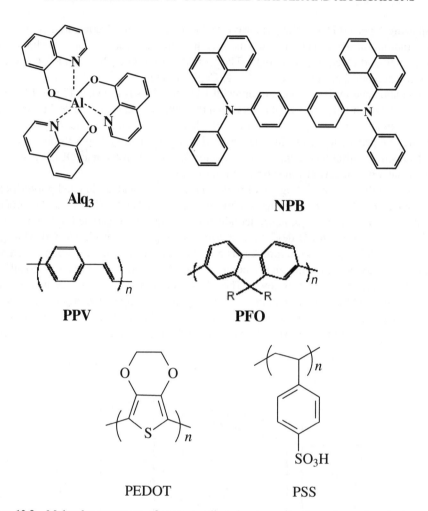

Figure 12.3 Molecular structures of some small molecular and polymeric semiconductors commonly used in OLEDs. Tris(8-hydroxyquinoline)aluminum (Alq_3) is used as an electron-transporting and emissive layer. N,N'-Di-α-napthyl-N,N'-di:phenylbiphenyl-4,4'-diamine (NPB) is used as a hole-transport layer. Poly(p-phenylenevinylene) (PPV) and polyfluorene (PFO) are typical fluorescent polymers. Poly(styrenesulfonate) (PSS)-doped poly(3,4-ethylenedioxythiophene) (PEDOT) is often used as HTL in polymeric OLEDs

The device's operation principle and the fabrication process of an organic-semiconductor-based LED are different from those known for a conventional LED made with an inorganic semiconductor. It is very difficult to form a stable organic semiconductor p–n junction as the organic materials are unable to be doped reproducibly to form p-type and n-type semiconductors. The interfaces of organic semiconductor p–n junctions easily deteriorate or can even be destroyed by chemical reaction and/or interdiffusion. For this reason, the OLEDs are usually designed having a p–i–n configuration, where the emissive layer is nominally intrinsic although in practice it becomes automatically doped. The second major difference between and an OLED and an inorganic LED is in the nature of the charge-carrier transport, recombination, and luminescence processes. In an inorganic semiconductor, charge transport is delocalized and described in terms of Bloch states within the single-electron-band approximation. However, charge transport in amorphous organic semiconductors is characterized by the localization of electronic states to individual molecules and occurs via a thermally activated hopping process [12.1].

12.1.2 Technical challenges and process integration

Organic semiconductors are finding increasing use in plastic electronics including flat-panel displays, organic transistors, photodetectors, and photovoltaic cells, etc. However, processing these materials with the desired uniform thin-film patterns or multi-layer structure is challenging. A spin-coating process is probably the simplest method to produce thin films of a few hundred angstroms but these films are amorphous with low carrier mobility and may have pinholes. Films produced using the Langmuir–Blodgett method are typically highly ordered, but this method is best suited for thin-film formation. Inkjet printing can also be used to produce patterned polymeric thin films, but such films are amorphous. Although the printable transparent conducting and the functional organic semiconductor thin-film technologies are still well in the research stage and have not been used in the immediate products yet, the related technologies thus developed are promising for next-generation productions. Each currently available thin-film deposition method has advantages and disadvantages, depending on the requirements. In addition, for many applications it would be desirable to produce patterned films, but photolithographic techniques cannot be used to pattern these polymeric films. Thus the challenge is to obtain highly ordered, patterned, polymeric thin films. New techniques need to be developed for depositing patterned, high-quality, pinhole-free, ultra-thin organic films for electronics.

12.2 ANODE MODIFICATION FOR ENHANCED OLED PERFORMANCE

Transparent conducting oxide (TCO) thin films have widespread applications due to their unique properties of high electrical conductivity and optical transparency in the visible spectrum range. The distinctive characteristics of TCO films have been applied in anti-static coatings, heat mirrors, solar cells [12.2, 12.3], flat-panel displays [12.4], sensors [12.5], and OLEDs [12.6–12.8]. A number of materials such as ITO, tin oxide, zinc oxide, and cadmium stannate are used as TCOs in many optoelectronic devices. It is well known that the optical, electrical, structural, and morphological properties of TCO films have direct implications for determining and improving device performance. The properties of TCO films are usually optimized accordingly to meet the requirements in various applications involving TCOs. The light-scattering effect due to the usage of textured TCO substrates shows an enhanced absorbance in thin-film amorphous silicon solar cells [12.9, 12.10]. The ITO contact used in LCDs comprises a relatively rough surface in order to promote the good adhesion of the subsequently coated polymeric layer on its surface. However, a rough ITO surface is detrimental for OLED applications. The high electric fields created by the rough anode can cause shorts in thin functional organic layers.

12.2.1 Low-temperature high-performance ITO

ITO is one of the widely used materials for a TCO. Thin films of ITO can be prepared by various techniques, including thermal evaporation deposition [12.11, 12.12], direct current (dc) and radio frequency (rf) magnetron sputtering [12.13, 12.14], electron-beam evaporation [12.15], spray pyrolysis [12.16], chemical vapor deposition [12.17], dip-coating techniques [12.18, 12.19], and the recently developed pulsed laser deposition method [12.20, 12.21]. Among these techniques, magnetron sputtering is one of the more versatile techniques for the ITO film preparation. This technique has the advantage of fabricating uniform ITO films reproducibly. Both reactive and nonreactive forms of dc/rf magnetron sputtering can be used for film preparation.

ITO films prepared by the dc/rf magnetron sputtering method often require heating of the substrate at an elevated temperature during the film deposition or an additional post-annealing treatment at temperature of over 200 °C. High-temperature processes for ITO preparation is unsuitable in some applications. For instance, the organic color-filter-coated substrates for flat-panel displays and flexible OLEDs made with polyester, poly(ethylene terephthalate) (PET), and other plastic foils are not compatible with a high-temperature plasma process. Therefore the development of high-quality ITO films with smooth surfaces, low resistivity, and high transmission over the whole visible spectrum range at low processing temperatures for flat-panel displays and flexible OLEDs is quite a challenge indeed. A number of techniques have been used to prepare ITO films at low processing temperatures. Ma et al. have deposited ITO films on polyester thin films over the substrate temperature of 80–240 °C by reactive thermal evaporation [12.22, 12.23]. Laux et al. have prepared ITO films on glass substrates at room temperature by plasma ion-assisted evaporation [12.24]. Wu et al. [12.25] have used pulsed laser ablation to fabricate ITO films on glass substrates at room temperature. ITO films prepared by the radio frequency (rf) and direct current (dc) magnetron sputtering methods on polycarbonate [12.26] and glass substrates [12.27] at low processing temperatures are also reported. ITO films fabricated by the rf and dc magnetron sputtering methods usually require a low oxygen partial pressure in the sputtering gas mixture when both alloy and oxidized targets are used [12.26, 12.27].

In the following discussion, a comprehensive study on the morphological, electrical, and optical properties of ITO films fabricated in our lab by rf magnetron sputtering using hydrogen–argon mixtures at a low processing temperature will be described. The addition of hydrogen to the sputtering gas mixture affects the overall optical and electric properties of ITO films considerably. Atomic-force microscopy (AFM), X-ray photoelectron spectroscopy (XPS), secondary-ion mass spectroscopy (SIMS), four-point probe technique, Hall effect, and optical measurements were used to characterize the morphological, electrical, and optical properties of the ITO films thus made. The mechanism of ITO film quality improvement due to addition of hydrogen to the sputtering-gas mixture is discussed.

(a) Experimental methods

Thin films of ITO were prepared by rf magnetron sputtering on microscopic glass slides using an oxidized ITO target with In_2O_3 and SnO_2 in a weight ratio of 9 : 1. The background pressure in the sputter chamber was lower than 1.0×10^{-7} Torr. The deposition rate of the films prepared by rf magnetron sputtering can be controlled by the sputtering power and the substrate temperature [12.14]. In this study, a fixed power density of about $1.2\,W/cm^2$ for ITO film preparation was used. The deposition process was carried out in a hydrogen–argon gas mixture at low temperature, i.e., the substrate was not heated during or after the film deposition. The total pressure of sputtering gas was kept constant at 3.0×10^{-3} Torr during the film preparation. The hydrogen partial pressure was varied over the range 1×10^{-5}–2.0×10^{-5} Torr. The thickness of ITO films deposited on glass substrates for morphological, electrical, optical, and spectroscopic characterizations was maintained at the same value of about 250 nm so that the measured properties of the ITO films are comparable.

The thickness of the ITO films was measured by a Tencor alpha-step 500 profilometer. The sheet resistance of the films was determined using a four-point probe method. The charged-carrier concentration and mobility of the films were characterized by Hall effect measurements using the van der Pauw technique. Wavelength-dependent absorption and transmission of ITO films were measured by a Perkin-Elmer spectrophotometer over the wavelength range of 0.3 to 2.0 μm. The surface morphology of the ITO films was investigated in a DI Dimension 3000 atomic-force microscope. The SIMS depth profiles were acquired using a CAMECA ims 6f ion microprobe. Cs^+ primary ions with energy of 15 keV and negatively charged secondary ions were used in the SIMS analyses. XPS measurements were performed using a VG ESCALAB 220-*i* electron spectrometer. The Mg-$K\alpha$ line at 1253.6 eV was chosen as the X-ray source in the XPS measurements. The position of all XPS peaks was calibrated using C 1 s with binding energy $E_b = 284.6$ eV.

(b) Morphological properties

The rf magnetron sputtering method to deposit ITO films on glass substrates over the hydrogen partial-pressure range 0–2.0×10^{-5} Torr was used. The substrate holder was not heated and the substrate temperature during the film preparation was observed to be less than 50 °C. The influence of the hydrogen partial pressure on the surface morphological properties of the ITO films was investigated by AFM. Figure 12.4 shows typical AFM image of ITO films prepared with (a) argon, (b) hydrogen at a partial pressure of 7.0×10^{-6} Torr, and (c) hydrogen at 2.0×10^{-5} Torr. Figure 12.4 shows the similar grainy surface of the ITO films prepared at different hydrogen pressures. The average granule size of ITO films prepared at the hydrogen pressures of 7.0×10^{-6} Torr [Figure 12.4 (b)] and 2.0×10^{-5} Torr [Figure 12.4 (c)] had smaller dimensions than those of ITO film sputtered with argon as shown in Figure 12.4 (a). The decrease of the granule size corresponds to an increase in film smoothness. As the thickness of ITO films prepared at different hydrogen partial pressures was maintained at the same value of about 250 nm, the possible morphological difference due to thickness variations is negligible. The results, shown in Figure 12.4, actually reflect the influence of the hydrogen partial pressure on the morphological property of ITO films.

Root mean square (rms) roughness of ITO films was also estimated from AFM images measured over an area of 300 nm × 300 nm as illustrated in Figure 12.4. The corresponding rms values of ITO films prepared with argon (a), and hydrogen at a partial pressure of (b) 7.0×10^{-6} Torr and (c) 2.0×10^{-5} Torr are 1.44 nm, 1.13 nm, and 0.92 nm respectively. It is clear that the ITO film prepared with argon gas had a higher rms roughness value than that of ITO films prepared with hydrogen–argon mixtures using the same deposition parameters. This effect of hydrogen partial pressure on film morphology could be used to produce smooth ITO film with suitable optoelectrical properties in applications that are not

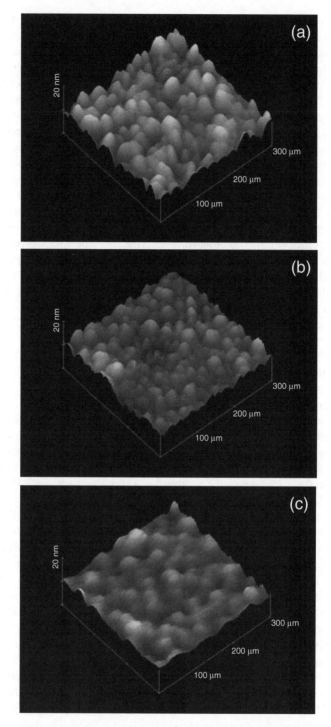

Figure 12.4   AFM images obtained over an area of $300\,\mu m \times 300\,\mu m$ for ITO films grown on glass substrates prepared at a low processing temperature with (a) argon gas, and hydrogen under a partial pressure of (b) 7.0×10^{-6} Torr and (c) 2.0×10^{-5} Torr

sustainable to the high-temperature process. In the application of flexible panel displays, for example, ITO is often required to be coated on the transparent plastic substrates at low processing temperatures to avoid the deformation of the plastic substrates. A smooth ITO anode is also desired in a flexible OLED with a multi-layered thin-film configuration. An ITO anode with a smooth surface can minimize electrical shorts in the thin functional organic layers in OLED that are very often in the range of 100–200 nm.

ITO films formed by rf magnetron sputtering at low temperatures are usually amorphous. There is no lattice matching for ITO growth on glass substrates. Under this circumstance, three-dimensional (3D) nucleation is the dominant film-formation mechanism. The film grown on glass substrate is most likely formed by the coalescence of islands from nucleation sites [12.28]. Sun et al. [12.29] have investigated the initial growth mode of ITO on glass over the substrate temperature range 20–400 °C. They suggested that ITO films with an amorphous structure formed at the substrate temperature below 150 °C are due to a 3D nucleated growth mode similar to the Volmer–Weber mechanism. The layer-by-layer growth mode only took place at substrate temperatures over 200 °C. In this study, the ITO films were fabricated at a low processing temperature of about 50 °C. The 3D-growth mode from nucleation sites is therefore expected although the nucleation density may be varied due to different hydrogen partial pressures used. The initial growth mode may affect the ultimate properties of thin films. However, the surface morphological property of ITO films thus prepared is also related to the energetic ion particles and reactive sputter species in the plasma atmosphere created by the rf magnetron sputtering. Under such a plasma environment the growing film is subjected to various forms of bombardment involving ions, neutral atoms, molecules, and electrons [12.30].

The morphological changes that occurred in ITO films prepared with hydrogen–argon mixtures, as shown in Figure 12.4, were probably due to the presence of the additional reactive hydrogen species in the sputtering atmosphere. As hydrogen was introduced to the sputtering gas mixture during the film preparation, the growing flux during the magnetron sputtering created a significant amount of energetic hydrogen species with energies over the range 10–250 eV [12.31]. These reactive hydrogen species could remove weakly bound oxygen from the depositing film [12.13]. The bombardment of the sputtering particles and hydrogen species on the depositing film could cause the reduction of indium atoms in the ITO film [12.32]. The presence of energetic hydrogen species could also react with the growing clusters containing intermediates such as In_xO_y, adsorbed O, reduced indium atoms, and sub-oxides like In_2O. The weakly adsorbed oxygen and possible reduced interstitial metal atoms in ITO film may be removed or re-sputtered by the reactive hydrogen species in the plasma process. Therefore the nucleation growth kinetics and surface reaction rates on ITO islands formed via the nucleation sites are altered by reactive hydrogen species. The effect of reactive hydrogen species on the depositing ITO film can reduce the size of clusters in comparison with that of the ITO film prepared with only argon gas under the same conditions. As a consequence, the addition of the hydrogen in the sputtering mixture shows a reduced effect on oxide, and ITO films deposited with hydrogen–argon mixtures had a smoother morphology.

(c) Electrical properties

Figure 12.5 shows the dependence of electrical resistivity of ITO films on the hydrogen partial pressure in the gas mixture. It can be seen that the resistivity of the ITO films changed

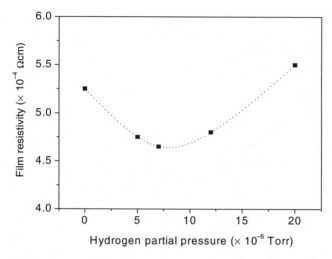

Figure 12.5   The resistivity of ITO films as a function of hydrogen partial pressure

considerably over the hydrogen partial pressure range 0–2.0×10^{-5} Torr used in the film's preparation. The resistivity of the ITO films decreased initially with the hydrogen partial pressure and reached its minimum value of $4.66 \times 10^{-4} \Omega$cm at an optimal hydrogen pressure of about 7.0×10^{-6} Torr. Further increase of the hydrogen partial pressure above the optimal value was shown to increase the film resistivity. The existence of a minimum resistivity was also observed in ITO films prepared at an elevated temperature of 300 °C [12.13]. Figure 12.5 shows that the relative minimum resistivity of ITO film prepared at the optimal condition was about 11% lower than that of the film deposited with argon gas under the same conditions. This shows that the usage of the hydrogen–argon mixture had a direct effect in improving the electrical properties of ITO films fabricated by rf magnetron sputtering at a low processing temperature.

The charge-carrier mobility and concentration in ITO films were measured by Hall Effect using the van der Pauw technique. Measured carrier mobility, μ, and concentration, N, in ITO films as functions of hydrogen partial pressure are plotted in Figure 12.6. It can be seen that both μ and N are very sensitive to hydrogen partial pressures used in the film preparation. Results in Figure 12.6 show that ITO films prepared at the optimal hydrogen partial pressure of 7.0×10^{-6} Torr, which produced ITO films with the lowest resistivity shown in Figure 12.5, had the maximum carrier concentration and minimum mobility values. As the electrical conductivity is proportional to the product of μ and N, this implies that the low resistivity of ITO films prepared at the optimal hydrogen partial pressure was due to the higher carrier concentration in the conduction mechanism. Figure 12.6 shows that ITO films fabricated at hydrogen partial pressure of 7.0×10^{-6} Torr usually have a high carrier concentration of 4.59×10^{20} cm^{-3}.

In ITO films, both tin dopants and ionized oxygen vacancy donors provide the charge-carriers for conduction. The number of oxygen vacancies that provide a maximum of two electrons per oxygen vacancy plays a dominant role in determining the charge-carrier density in an ITO film with high oxygen deficiency. It is also affected by the deposition conditions such as sputtering power, substrate temperature, and Sn/In composition in target

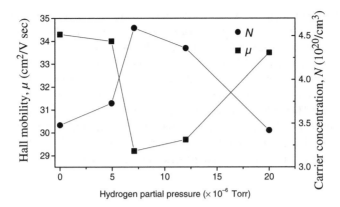

Figure 12.6   Carrier mobility and concentration of ITO films as functions of hydrogen partial pressure

and sputtering species in the plasma during the film preparation. Banerjee et al. [12.33] investigated the effect of oxygen partial pressure prepared by electron-beam evaporation from a hot pressed powder of In_2O_3–SnO_2 mixture (9:1 w/w). They found that the increased film conductivity was due to an enhancement in the Hall mobility, but carrier concentration decreased with the oxygen partial pressure. A similar correlation between oxygen partial pressure and carrier concentration in ITO films prepared by pulsed laser deposition was also observed by Kim et al. [12.34]. Experimental results reveal that the improved electrical properties of ITO films made at the optimal oxygen partial pressure were due to increased carrier mobility in the film. The decrease in carrier concentration was attributed to the dissipation of oxygen vacancies when oxygen was used in the gas mixture during the preparation. When the ITO films prepared at the presence of reactive hydrogen species, however, the carrier mobility of the ITO film had a relative low value. This implies that the mechanism of improvement in the conductivity of ITO films is dependent on the deposition process. The enhancement in the conductivity of ITO films prepared under a presence of hydrogen can be attributed to a high carrier concentration in comparison with those ITO films made without hydrogen in the gas mixture. The above analyses are consistent with the previous results obtained from ITO films prepared at high substrate temperatures [12.13].

The above analyses based on the electrical measurements and morphological results suggest that the film quality improvement was due to the presence of reactive energetic hydrogen species in the sputtering plasma when a hydrogen–argon mixture was used. The relative low film resistivity was attributed to the higher oxygen deficiency and possible good contacts between different domains, as the film was denser when it was prepared with hydrogen. Thin films of ITO with high charge-carrier concentration may have some advantages for OLED applications. ITO is an *n*-type wide-bandgap semiconductor. The Fermi level, E_f, of ITO films is located at about 0.03 eV below the conduction-band minimum [12.35]. It has an upward surface band bending with regard to its E_f [12.36] due to a surface Fermi-level-pinning mechanism. Therefore, the effective barrier for hole injection at the interface of ITO HTL should include both the upward surface band bending of ITO and band offset between E_f of ITO and ionization potential of HTL. It has been reported that the surface band bending of ITO decreases with the increase in the carrier concentration in ITO films [12.37]. Lesser surface band bending lowers the effective energy barrier for carrier injection when it is used as anode in OLED [12.38]. It was demonstrated that the electroluminescence (EL) efficiency of OLEDs made with ITO having a higher carrier concentration was always higher than that of those made with ITO having a lower carrier concentration [12.39]. The increase in EL efficiency reflects enhanced hole injection in the device. It can be considered that ITO anode with a high carrier concentration has a smaller surface band bending, which lowers the effective energy barrier for hole injection in OLED. Therefore ITO films prepared under optimal hydrogen partial pressure of 7.0×10^{-6} Torr by the rf magnetron sputtering method at a low processing temperature are preferable in practical applications.

(d) Optical properties

In parallel with the morphological and electrical analyses, the transmission spectra of the ITO films deposited at different hydrogen partial pressures were examined over the wavelength range 0.3–2.0 µm. Figure 12.7 shows the wavelength-dependent transmittance, $T(\lambda)$, of the ITO films prepared with argon gas (solid curve), hydrogen at a partial pressure of 7.0 $\times 10^{-6}$ Torr (dashed curve), and hydrogen at 2.0×10^{-5} Torr (dotted curve). Except for obvious deviations in the IR region, $T(\lambda)$ of the films prepared at different hydrogen partial pressures also shows a slight difference over the short-wavelength range. Figure 12.7 reveals that the short-wavelength cutoff in the $T(\lambda)$ of ITO films prepared at the optimal hydrogen partial pressure of 7.0×10^{-6} Torr shifts towards shorter wavelengths in comparison with that in the $T(\lambda)$ of ITO films prepared with argon gas. The shift of short-wavelength cutoff in $T(\lambda)$ is related directly to the variation of band gap in the ITO films.

In order to better understand the shift of the short-wavelength cutoff in $T(\lambda)$, the wavelength-dependent absorbance, $A(\lambda)$, of ITO films was measured to estimate their optical

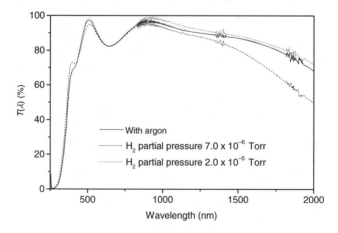

Figure 12.7 Wavelength-dependent transmittance, $T(\lambda)$, of ITO films prepared at a low processing temperature with argon gas (solid curve), and hydrogen at a partial pressure of 7.0×10^{-6} Torr (dashed curve) and 2.0×10^{-5} Torr (dotted curve)

bandgaps. Using the absorption coefficient, α, derived from the measured $A(\lambda)$ of ITO films, the optical bandgap E_g can be estimated. ITO is an ionic-bound degenerate oxide semiconductor. Usually the following relation is used to derive E_g for heavily doped oxide semiconductors [12.3, 12.13]:

$$\alpha^2 \approx (h\nu - E_g) \qquad (12.1)$$

where $h\nu$ is the photon energy. Figure 12.8 shows the photon-energy dependence of α^2 for ITO films prepared at different hydrogen partial pressures. Extrapolation of the linear region of the plot to α^2 at zero gives the value of E_g. It can be seen from Figure 12.8 that E_g values for ITO films prepared with argon, the hydrogen pressures of 7.0×10^{-6} Torr and 2.0×10^{-5} Torr are 3.75 eV, 3.88 eV, and 3.73 eV, respectively. Figure 12.8 shows that ITO films prepared with argon and a hydrogen partial pressure of 2.0×10^{-6} Torr have similar optical bandgaps as both films also have similar carrier concentrations as shown in Figure 12.6. A higher E_g value of 3.88 eV is obtained for the film prepared at the optimal hydrogen pressure of 7.0×10^{-6} Torr.

The widening of the bandgap can be attributed to the increase in the carrier concentrations in ITO film prepared at the optimal hydrogen pressure. ITO is a degenerate semiconductor. The conduction band is partially filled with electrons. The Fermi level, E_f, of ITO is very close to the conduction-band minimum. The Fermi level shifts upward and overlaps or even locates within the conduction band when the charge-carrier concentration increases. As E_f moves to higher energy in the conduction band, the electronic states near the conduction-band minimum are fully occupied. Thus the energy level of the lowest empty states in the conduction band moves to the higher energy positions as well. Therefore, electrons excited from the valence band to those available electronic states in the conduction band require a higher energy. This implies that the effective increase in the energy gap of ITO films prepared at the optimal hydrogen partial of 7.0×10^{-6} Torr was due to an increase in

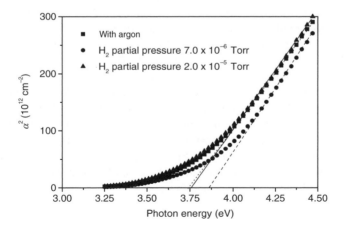

Figure 12.8 Square of absorption coefficient, α^2, plotted as a function of the photon energy for ITO films prepared at different conditions. Extrapolation of the straight region of the plot to $\alpha^2 = 0$ gives bandgap $E_g = 3.88$ eV for ITO films prepared at the optimal hydrogen partial pressure of 7.0×10^{-6} Torr

the carrier concentration. The bandgap broadening due to an increased carrier concentration in ITO film is also known as the Moss–Burstein effect, which means that the lowest states in the conduction band are filled by an excess of charge carriers [12.3]. In this case the increase in the charge carriers was due to an increase in the oxygen vacancies in the ITO films. This analysis is in a good agreement with the results obtained from electrical measurements shown in Figure 12.6.

The transmission spectrum of ITO films prepared at the hydrogen partial pressure of 7.0 $\times$ 10^{-6} Torr [Figure 12.7] shows a considerable decrease in the near-IR region in comparison with that measured for ITO films prepared with argon. In this region, the free-carrier absorption becomes important for the transmittance and reflectance of ITO films. The optical behavior of ITO films in the IR region can be explained by the Drude theory for free charge carriers [12.3, 12.40]. The appreciable reduction in transmission over the IR range for ITO films prepared at optimal hydrogen partial pressure was due to an increased carrier concentration. Since the thickness of ITO films was maintained at the same value, the difference in $T(\lambda)$ observed in Figure 12.5 is an indication of the variation in the refractive index of the ITO films. Bender et al. [12.41] have calculated the refractive index of ITO films prepared by a dc magnetron sputtering method at different oxygen partial pressures. They found that the refractive index of ITO films increases with increasing oxygen flow used in the film preparation. A similar result was also obtained by Wu et al. [12.25] in showing that the refractive index of ITO films decreases with increasing carrier concentration in ITO films. Generally, the index of refraction of an indium tin oxide film can be represented by Equation (12.2) [12.42]:

$$n^2 = \varepsilon_{opt} - \frac{4\pi Ne^2}{m^* \omega_0^2} \tag{12.2}$$

where n is the index of refraction, ε_{opt} is the high-frequency permittivity, m^* is the effective mass of the electron, and ω_0 is the frequency of the electromagnetic oscillations. ITO films prepared with hydrogen–argon mixtures in this work had high carrier concentrations in their conduction mechanism. According to Equation (12.2), a reduction in the refractive index of ITO films would be expected due to an increase in the carrier concentration, N, in films prepared in the presence of hydrogen. As such, the refractive index of an ITO film prepared at the optimal hydrogen partial pressure of 7.0 $\times$ 10^{-6} Torr would have the lowest n value as it has the maximum carrier concentration as shown in Figure 12.6. This implies that a film becomes denser as its refractive index decreases. Based on the information obtained from the above morphological and electrical studies on ITO films, it can be considered that the average density of the film prepared at the optimal hydrogen partial pressure of 7.0 $\times$ 10^{-6} Torr is higher than that of the film sputtered with argon gas. Then, based on the information obtained from the AFM measurements, it may be inferred that a denser ITO film will have fewer internal voids in the bulk and fewer irregularities on its surface when it is fabricated under optimal conditions.

(e) Compositional analysis

The chemical binding energies of In 3d$_{5/2}$ and Sn 3d$_{5/2}$ for different ITO films were examined using XPS measurements. The energy positions of In 3d$_{5/2}$ and Sn 3d$_{5/2}$ peaks measured for the films deposited at different hydrogen partial pressures were all constant at 445.2 eV and 487.2 eV, respectively. There were no evident shoulders observed at the high binding-energy side of the In $3d_{5/2}$ peaks, which may relate to the formation of In—OH-like bonds in the ITO films prepared in the presence of the hydrogen [12.43]. The same binding-energy positions and almost identical symmetric XPS peak shapes of In $3d_{5/2}$ and Sn $3d_{5/2}$ observed from different ITO films suggest that chemical states of indium and tin atoms in the films remained in an ITO form. The atomic concentration of ITO films prepared at different hydrogen partial pressures was also estimated. The result shows that the stoichiometry of ITO films prepared at different hydrogen pressures was very similar. The variation of the hydrogen partial pressure used in this work did not seem to affect the chemical structure of ITO films significantly. The bulk composition of ITO films was also examined by SIMS measurements. Figure 12.9 shows a typical depth profile of ITO films prepared at the optimal hydrogen partial pressure of 7.0×10^{-6} Torr.

The x-axis in Figure 12.9 shows the sputter depth of films that was converted using sputtering time at a sputtering rate of ~0.36 nm/s. The depth profile steps occurring at the ITO surface and boundary between ITO and glass substrate were due to the influence of the interfacial effects. It is obvious that the profiles of ITO elements O, In, and Sn had the stable counts through the whole depth profile region measured by SIMS. The steady distribution of ITO elements shown in Figure 12.9 confirms that ITO films thus prepared were very uniform.

It is well known that the tin dopants and ionized oxygen vacancy donors govern the charge-carrier density in an ITO film. In order to understand the mechanism of the carrier-concentration variations in different ITO films, SIMS is also used to measure the relative concentration of oxygen and tin in films prepared at different hydrogen partial pressures. In order to compare the relative oxygen and tin contents in different films, the intensities of oxygen and tin ions measured by SIMS are normalized to the corresponding intensities of indium acquired in the same measurements. Figure 12.10 plots the normalized depth profiles of oxygen in ITO films prepared with (a) argon gas, (b) the hydrogen partial pressure

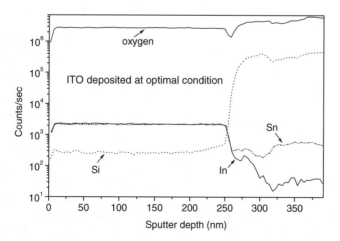

Figure 12.9 SIMS depth profile of typical ITO film prepared at an optimal hydrogen partial pressure of 7.0×10^{-6} Torr

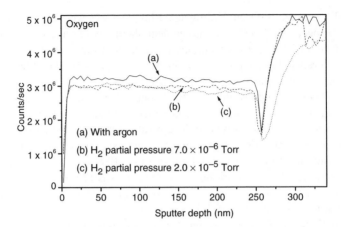

Figure 12.10 Comparison of SIMS depth profiles of normalized relative oxygen concentration in ITO films prepared at a low processing temperature with pure argon gas (a), and hydrogen at a partial pressure of (b) 7.0×10^{-6} Torr and (c) 2.0×10^{-5} Torr

of 7.0×10^{-6} Torr, and (c) hydrogen at 2.0×10^{-5} Torr. It can be seen that there is a slight difference in normalized SIMS oxygen counts measured for the films fabricated at different hydrogen partial pressures. Figure 12.10 shows clearly that the ITO film deposited with only argon gas had relatively higher oxygen content than those measured from the films prepared with hydrogen–argon mixtures. This indicates that the addition of hydrogen to the sputtering gas mixture has the effect of removing oxygen from depositing ITO films and leads to an increase in the number of oxygen vacancies. This result is similar to the effect of hydrogen on ITO films prepared at a higher processing temperature of 350 °C as reported in a previous study [12.13].

SIMS was also used to examine the relative concentration of tin element for ITO films prepared at different hydrogen partial pressures. The corresponding normalized tin concentration in films prepared with argon, hydrogen partial pressures of 7.0×10^{-6} Torr and 2.0×10^{-5} Torr are plotted as curves (a), (b), and (c) in Figure 12.11, respectively. Figure 12.11 reveals that ITO films fabricated with hydrogen–argon mixture had a slightly higher relative tin content than that obtained from an ITO film sputtered with argon gas. In particular, the ITO film prepared at the optimal hydrogen partial pressures of 7.0×10^{-6} Torr, curve (b) in Figure 12.11, had the highest tin content. A tiny deviation of normalized tin concentration in different ITO films may not change the chemical structure in ITO films considerably, but it impacts on the efficient doping level in the material.

During the film's formation tin atoms will substitute for indium atoms in the lattice, leaving the tin atoms with one electron more than the requirement for bonding. This electron will be free in the lattice and act as a charge carrier. Apart from an increase in the number of oxygen vacancies and hence an increase in the number of charge carriers, the substitution of indium by tin atoms also improves the film's conductivity. Based on measured carrier concentration results, as given in Figure 12.6, it can be inferred that an enhanced doping level was achieved in an ITO film prepared at the optimal hydrogen pressure of 7.0×10^{-6} Torr as shown in curve (b) in Figure 12.11.

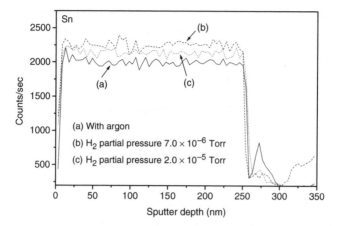

Figure 12.11 Comparison of SIMS depth profiles of normalized relative tin concentration in ITO films prepared at a low processing temperature with pure argon gas (a), hydrogen at a partial pressure of 7.0×10^{-6} Torr (b), and hydrogen at a partial pressure of 2.0×10^{-5} Torr (c)

ITO films fabricated by rf or dc magnetron sputtering method usually require an annealed substrate at temperatures in the range 200–300 °C when both alloy and oxidized targets are used. A high processing temperature enhances the crystallinity in the ITO film and hence increases the carrier mobility and film conductivity. Low-processing temperature ITO is a prerequisite for fabricating flexible OLEDs or top-emitting OLEDs that preclude the use of a high-temperature process. An ITO anode with a smooth surface can minimize electrical shorts in the thin functional organic layers in OLEDs that are very often in the range of 100–200 nm. Our results demonstrated that the process thus developed offers an enabling approach to fabricate ITO films with smooth morphology and low sheet resistance on flexible plastic substrate at a low processing temperature. As such, a process that produces smooth ITO films with high optical transparency and high electric conductivity at a low processing temperature will be of practical and technical interest.

12.2.2 Anode modification

Much progress has been made in OLEDs since the discovery of light emission from electroluminescent polymers. It is appreciated that the physical, chemical, and electrical properties at both anode and cathode interfaces in OLEDs play important roles, though they are still not well understood, in determining the device-operating characteristics and the stability of these devices. Different substrate-treatment techniques, which may involve ultrasonic cleaning of the ITO anode in organic solutions, the exposure of the pre-cleaned ITO to either ultraviolet (UV) irradiation or oxygen-plasma treatment, are used for device fabrication [12.44–12.46]. It has been reported that oxidative treatments, such as oxygen plasma or UV–ozone, can effectively increase the work function of ITO [12.6, 12.47] and also form a negative surface dipole facing towards the ITO. This may then lead to enhanced hole injection and increased device reliability [12.8, 12.48, 12.49]. UV photoelectron spectroscopy and a Kelvin probe are often employed to investigate the changes in work function or surface dipole of ITO due to different surface treatments [12.50, 12.51]. It shows that an increase in ITO work function is closely related to the increase in its surface oxygen content due to oxidative treatment [12.6, 12.52, 12.53]. The effect of ITO surface treatments on improvement of OLEDs, although widely investigated, is both important and still not fully understood. A better understanding of the mechanism of oxygen plasma treatment on an ITO anode in OLED performance is of practical interest.

In addition to the appropriate ITO anode cleaning and treatment, it is also shown that charge-injection properties in OLEDs can be modified in different ways. At the cathode side, the use of low-work-function metals [12.7], and the introduction of electron-transporting materials with high electron affinity [12.54] or a few hundred angstroms thickness of conducting-polymer layers have been investigated [12.55–12.67]. At the same time, the charge injection can also be improved at the anode side by adding an HTL [12.58] by various treatments of the ITO [12.6, 12.8, 12.59, 12.60]. An ITO anode with self-assembled monolayers can bring significant enhancements in luminous efficiency [12.61–12.64]. Modifying the electrode/organic interface using various insulating layers can also substantially enhance the electroluminescence (EL) performance of the devices [12.65–12.69].

The following discussion will focus on an understanding of the anode modification for enhanced carrier-injection properties. *In situ* four-point probe methods in conjunction with XPS and time-of-flight secondary-ion mass spectrometry (TOF-SIMS) measurements were used to explore the relationship between bilayer ITO/insulating-interlayer anode and improvement of OLED performance. It is found that oxidative treatment induces a nanometer-thick oxygen-rich layer on the ITO surface. The thickness of this ultra-thin oxygen-rich layer, deduced from a dual-layer model, is consistent with the XPS and TOF-SIMS measurements. The enhancement in OLED performance correlates directly with such an interlayer of low conductivity. It serves as an efficient hole-injecting anode.

ITO-coated glass with a thickness of 120 nm and a sheet resistance, R_s, of about $20\,\Omega$/square is used in the device's fabrication. The ITO used for OLEDs underwent wet-cleaning processes including ultrasonication in the organic solvents, followed by oxygen plasma or UV–ozone treatment. In this work, wet-cleaned specimens that did not undergo any further oxidative treatment are marked as 'non-treated' ITO. Those wet-cleaned specimens further treated by oxygen plasma at different conditions prior to OLED fabrication are marked as 'treated' ITO. The experiments were carried out in a multi-chamber vacuum system equipped with an ITO sputter chamber, an oxygen plasma treatment chamber, and an evaporation chamber. The system is also connected to a glove box purged by high-purity nitrogen gas to keep oxygen and moisture levels below 1 ppm. The system allows the substrate to be transferred among different chambers without breaking the vacuum. The samples can also be moved from the evaporation chamber to the glove box without exposure to air. The four-point probe is placed inside the glove box for an *in situ* measurement of any changes in ITO sheet resistance, R_s, due to the oxygen plasma treatment. A 15-nm-thin ITO film was also deposited on glass for study. This is to boost the effect of oxygen-plasma treatment on the variations in R_s for a more accurate measurement. A phenyl-substituted poly(*p*-phenylenevinylene) (Ph-PPV) [12.70] was used as an emissive polymer layer. The single-layer testing device has a configuration of ITO (100 nm)/Ph-PPV (80 nm)/Ca (5 nm)/Ag (300 nm).

The change in R_s of ITO films due to different oxygen-plasma treatments was monitored by the *in situ* four-point probe measurement. All plasma-treated ITO films were found to have higher sheet resistance than untreated ones. The increase in sheet resistance observed

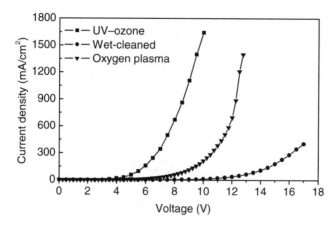

Figure 12.12 Current density vs voltage characteristics of identical devices made on wet-cleaned, oxygen-plasma and UV–ozone-treated ITO substrates

in a 15-nm-thick ITO was more apparent than that in a 120-nm-thick ITO treated under the same conditions. The process for oxygen-plasma treatment was optimized based on the EL performance of OLEDs, and was chosen accordingly by varying the oxygen flow rates at a constant plasma power of 100 W and a fixed exposure time of 10 min.

12.2.3 Electroluminescence performance of OLEDs

Figure 12.12 shows the current density–voltage (I–V) characteristics of a set of identical devices built on ITO treated with wet cleaning, oxygen plasma, and UV–ozone. It shows the effect of different treatments on the devices' performance. For ITO treated with oxygen plasma and UV–ozone, the devices generally have a similar maximum luminance of ~50 000 cd/m². For the wet-cleaned device, it shows a luminance of ~20 000 cd/m². In terms of the current density, the devices treated with oxygen plasma or UV–ozone had lower values compared with the wet-cleaned ITO. In this case, the oxygen plasma treatment was not optimized.

The current–voltage characteristics measured for a set of identical OLEDs made on ITO treated by oxygen plasma at different oxygen flow rates of 0, 40, 60, and 100 standard cubic centimetres per minute (sccm) are illustrated in Figure 12.13. There were obvious differences in the EL performance of OLEDs with regard to the anode treatments. For instance, at a given constant current density of 20 mA/cm², the luminance and efficiency of the identical devices made with oxygen plasma treatments fall within the range of 600–1,000 cd/m² and 5.0–11.0 cd/A, respectively. These values were 560 cd/m² and 5.0 cd/A, respectively, for the same device fabricated on an untreated ITO anode. The enhancement in hole injection in the device made on a treated ITO anode is clearly demonstrated. I–V Characteristics, shown in Figure 12.13, indicate that there is an optimal oxygen-plasma treatment process. The best EL performance was found in the OLED made on an ITO anode treated with an oxygen flow rate of 60 sccm used in this study.

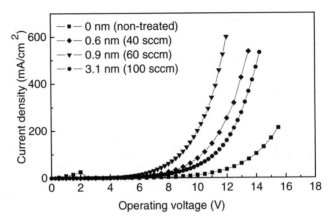

Figure 12.13 Current density vs operating voltage characteristic of identical devices made on ITO anodes treated under different oxygen plasma conditions

It is considered that the enhancement in EL performance of OLEDs due to oxygen plasma treatment is attributed to the improvement in the ITO surface properties. The improvement in the device performance is explained by a low barrier height at the ITO/polymer interface and a better ITO/polymer adhesion. This includes an increased work function, an improved surface morphology [12.6, 12.71], and a less affected bulk ITO property [12.8]. In order to explore the correlation between the variations in R_s and the possible compositional changes on the treated ITO surfaces, the surface contents of untreated and treated ITO films were examined using XPS and TOF-SIMS. For surface compositional analyses, both untreated and treated ITO samples were covered with a 5-nm-thick lithium fluoride (LiF) capping layer before the samples were taken out for XPS and TOF-SIMS measurements. This protective LiF layer was to prevent any possible contamination on the ITO surfaces in air and was removed by argon ion sputtering in the XPS and TOF-SIMS measurements. Therefore, the changes in R_s observed by *in situ* four-point probe measurements and variations in the surface contents obtained by *ex situ* spectroscopic analyses on ITO surfaces due to oxygen plasma are being compared.

In the XPS measurements, the In $3d_{5/2}$, Sn $3d_{5/2}$ and O 1s peaks of untreated and treated ITO films were examined. The binding energies of these peaks are found to be 445.2 eV, 487.2 eV, and 530.3 eV, respectively, for all ITO surfaces. This implies that no major chemical changes occur on an ITO surface during the oxygen plasma treatment. However, there was a considerable increase in the ratio of O/(In + Sn) for an ITO as the oxygen flow rate was increased. For example, the ratio of O/(In + Sn) obtained from an ITO surface treated by oxygen plasma with a flow rate of 60 sccm is almost 10% higher than the value obtained from an untreated ITO surface. It implies that the change in R_s corresponds closely with the increase of the oxygen content on an ITO surface. The increase of the oxygen concentration on an ITO surface is found to correlate strongly with the increase in its work function [12.44, 12.45].

The comparison of TOF-SIMS depth profiles of normalized relative oxygen concentration from the surfaces of untreated ITO and an ITO-treated film with a flow rate of 60 sccm is shown in Figure 12.14. It has been reported that oxygen plasma-treated ITO has a

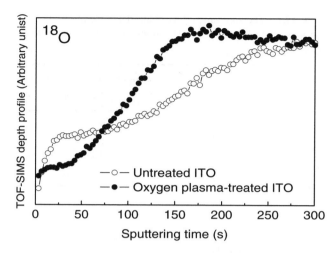

Figure 12.14 Comparison of TOF-SIMS depth profiles of relative oxygen concentration on the untreated and oxygen plasma-treated ITO surfaces

smoother surface than does untreated ITO [12.6]. In this work, the depth profile for ^{18}O ion was used for analyses because the intensity of ^{16}O ion counts was too high that it saturated machine. As shown in Figure 12.14, from the ^{18}O ion the relative oxygen concentration on plasma-treated ITO surface is obviously higher. Based on the sputter rate used in the depth-profile measurements, it appears that oxygen-plasma treatment can induce an oxygen-rich layer of a few nanometers thickness in the near ITO surface region. However, the precise thickness of this region is difficult to determine directly by XPS or TOF-SIMS due to the influence of the interfacial effects occurring during the argon ion sputtering.

ITO is a ternary ionic-bound degenerate semiconducting oxide. The conductivity of oxygen-deficient ITO is governed by the tin dopants and ionized oxygen vacancy donors. In an ideal situation, free electrons can result either from the oxygen vacancies acting as doubly charged donors, providing two electrons each, or from the electrically active tin ionized donor on an indium site [12.11, 12.72]. The additional oxidation on an ITO surface by oxygen plasma may cause the dissipation of oxygen vacancies. Therefore such an oxygen-plasma treatment results in a decrease in electrically active ionized donors in a region near the ITO surface, leading to an overall increase in R_s as manifested by the *in situ* four-point probe measurement. By comparing the highly conducting bulk ITO, it can be considered that the treated ITO anode can be portrayed using a dual-layer model.

In order to gain an insight into the relation between variation in sheet resistance and an optimal oxygen-plasma process for enhanced hole injection, a set of 15-nm-thick ITO films was coated on glass substrates using the same deposition conditions in the sputter chamber. Each of the thin ITO films was transferred to a connected chamber for oxygen-plasma treatment and was then passed to the glove box for an *in situ* sheet resistance measurement. The changes in the sheet resistance, ΔR, between the treated and untreated ITO were 19, 30, and 89 Ω/square at an oxygen flow rate of 40, 60, and 100 sccm, respectively. The thickness of oxygen plasma-induced low-conductivity layer, x, is estimated to be 0.6, 0.9, and 3.1 nm at the oxygen flow rate of 40, 60, and 100 sccm, respectively.

Table 12.1 The oxygen plasma-induced low-conductivity layer thickness and ΔR for the ITO films treated under different conditions

Oxygen flow rate (sccm)	ΔR (Ω/sq)	Oxygen plasma-induced low conductivity layer thickness, x (nm)
0		0
40	19	0.6
60	30	0.9
100	85	3.1

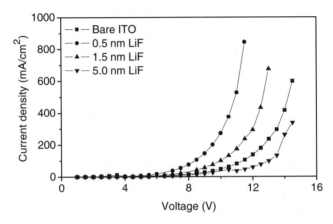

Figure 12.15 Current density vs bias voltage characteristics of devices made with different LiF inter-layer thickness

Table 12.1 summarizes the changes in ΔR and the corresponding estimated thicknesses of oxygen plasma-induced low-conductivity layer, x, obtained for ITO films treated under different conditions. The measured ΔR correlates directly with the thickness of this low-conductivity layer, and both of them increase with the oxygen flow rate. The oxygen-plasma treatment that was optimized for the OLED performance, in this case 60 sccm, induces a nanometer-thick resistive layer on an ITO surface. The above analysis based on the XPS, TOF-SIMS, and the electrical results suggests that the presence of an oxygen-plasma-induced nanometer-thick resistive layer on an ITO surface can also be accounted for the enhanced OLED performance.

Identical OLEDs made on ITO, modified with LiF layers of thicknesses in the range of 0–5.0 nm, were fabricated. Figure 12.15 shows the current density vs bias voltage characteristics of devices made with different LiF layer thicknesses. Comparing with the devices made without this layer, the results demonstrate that the former has a higher EL brightness when operated at the same current density. At a given constant current density of 20 mA/cm², the luminance and efficiency for devices with 1.5 nm LiF-coated ITO were 1600 cd/m² and 7 cd/A, respectively. The corresponding values were 1170 cd/m² and 5.7 cd/A, respectively, for the same devices made with only an ITO anode. The results demonstrate that the presence of an ultra-thin LiF layer between the ITO and polymer favors the efficient operation of LEDs. These improvements are attributed to an improved ITO/polymer interface quality and a more balanced carrier injection that improves the device's efficiency.

One possible mechanism for the above enhancement in the device's performance can be obtained from tunneling theory. LiF is an excellent insulator with a large bandgap of about 12 eV. However, the presence of an ultra-thin insulating interlayer between ITO and polymer enhances hole injections. This indicates that the potential barrier for hole injection that was present in the device with the 1.5-nm-thick LiF interlayer was thus decreased via tunneling. Recently Zhao et al. [12.73] reported that insertion of an insulating LiF interlayer between the ITO anode and the HTL induces an energy-level realignment at the anode/HTL interface. There exists a triangle barrier at the ITO/HTL interface, when a forward bias is applied on an OLED with a presence of a LiF interlayer, a voltage drop across the LiF lowers the ITO work function and hence also reduces the triangle barrier at ITO/HTL interface. Although LiF also introduces an additional barrier, the overall effective interfacial barrier height can be reduced at an optimal interfacial modification condition, leading to an enhanced carrier injection. The possible chemical reactions at the interface should also be taken into account for a more consistent explanation. It reveals that the enhanced hole injection with bilayer ITO/LiF anode altered the internal electric-field distribution in the device. The enhanced injection for holes due to the tunneling effect could induce the change in the potential difference across both electrodes for carrier injection, leading to an improvement in the balance of the hole and electron injections.

As the thickness of the insulating interlayer increases, the probability of carrier tunneling decreases, leading to a weaker carrier-injection process. To achieve a given luminance, therefore, the applied voltage needs to be increased with increasing interlayer thickness. As this becomes thicker, both current density and EL brightness are decreased because of reduced tunneling, but more balanced hole and electron injections were achieved so that EL efficiency can reach its maximum value. For instance, an increased EL efficiency of 7.9 cd/A for a device with 1.5-nm-thick LiF indicated a more balanced injection of both types of carriers, which was less optimal in the case of the device with bare ITO as shown in Figure 12.15. The devices presented in this study clearly portray this behavior.

An ITO anode modified with an insulating interlayer for enhanced carrier injection in OLEDs has been demonstrated [12.69, 12.73, 12.74]. Also, attempts have been made to explore the use of ultra-thin inorganic insulating layers to modify ITO in OLEDs. Ho et al. [12.63] have reported that a 1–2-nm-thick insulating self-assembled monolayer on ITO significantly alters the injection behavior and enhances EL efficiency, which is consistent with our result presented here. These improvements are attributed to an improved ITO/polymer interface quality and a more balanced carrier injection that improves device efficiency. Apart from the current understanding of the formation of a negative dipole facing towards the ITO due to oxygen plasma or UV–ozone treatment, it seems that oxidative treatment modifies an ITO surface effectively by reducing the oxygen deficiency to produce a low-conductivity region. As such, in this case, oxygen-plasma-treated ITO behaves somewhat similarly to the specimens where there is an ultra-thin insulating interlayer serving as an efficient hole-injection anode in OLEDs. The improvement in the EL performance of OLEDs correlates directly to the thickness of low conductive interlayer between anode and the polymer layer. The results show that the best EL performance comes from the device with a nanometer-thick low-conductivity layer induced by an optimal oxygen-plasma treatment. This is consistent with the use of an ultra-thin parylene or LiF layer modified bilayer ITO/interlayer anode for enhanced performance of OLEDs [12.69, 12.75].

12.3 FLEXIBLE OLED DISPLAYS

Current OLED technologies employ rigid substrates, such as glass, which limits the 'mould-ability' of the device, restricting the design and spacing where OLEDs can be used. The demand for more user-friendly displays is propelling efforts to produce head-worn and hand-held devices that are flexible, lighter, more cost-effective, and more environmentally benign than those at present available. Flexible thin-film displays enable the production of a wide range of entertainment-related, wireless, wearable-computing, and network-enabled devices. The display of the future requires that it should be thin in physical dimensions, have both small and large formats, be flexible, and have full color capability at a low cost. These demands are sorely lacking in today's display products and technologies such as the plasma display and LCD. OLEDs [12.76–12.79] have recently attracted attention as display devices that can replace LCDs because OLEDs can produce high visibility by self-lumi-nescence. The OLED stands out as a promising technology that can deliver the above chal-lenging requirements.

The next generation of flexible displays are going to be commercially competitive due to their low power consumption, high contrast, light weight, and flexibility. The use of thin flexible substrates in OLEDs will significantly reduce the weight of flat-panel displays and provide the ability to bend or roll a display into any desired shape. Up to the present time, much effort has been focused on fabricating OLEDs on various flexible substrates [12.80–12.84]. The plastic substrates usually do not have negligible oxygen and moisture permeability. The barrier properties of these substrates are not sufficient to protect the elec-troluminescent polymeric or organic layers in OLEDs from penetration of chemically reac-tive oxygen and water molecules into the active layers of devices. Therefore, plastic substrates with an effective barrier against oxygen and moisture penetration have to be obtained before this simple vision of a flexible display can become a reality [12.85]. Polymer-reinforced ultra-thin glass sheet is one of the alternative substrates for flexible OLEDs. In this section, we will discuss the results of OLEDs fabricated on flexible ultra-thin glass sheet with polymer-reinforcement coating and flexible plastic substrates.

12.3.1 Flexible OLEDs on ultra-thin glass substrate

Polymer-reinforced ultra-thin glass sheet is one of the alternative substrates for flexible OLEDs. It has been found that the flexibility and handling ability of ultra-thin glass sheet can be improved significantly when it is reinforced. The reinforcement polymer layer has the same shrinkage direction as that of an ITO layer deposited on the opposite side of the substrate. ITO-coated ultra-thin glass with reinforcement-polymer layer has high optical transmittance and is suitable for device fabrication.

The work presented in Section 2.1 in this chapter indicates that ITO film developed at a low processing temperature is suitable for OLED applications. The improvement in OLED performance correlates directly with the ITO properties. Plastic foils and ultra-thin glass sheets with reinforced polymer layers are not compatible with a high-temperature plasma process. Therefore, the successful development of high-quality ITO film with smooth surface, high optical transparency, and electrical conductivity at a low temperature provides the possibility for developing flexible displays.

We have used flexible ultra-thin glass sheets with a 250-nm-thick smooth ITO coating and sheet resistance of about $20\,\Omega$/square to fabricate phenyl alkoxyphenyl PPV copolymer-based OLEDs. In order to compare the EL performance of the OLEDs thus fabricated, we also fabricated an OLED with an identical structure using a commercial ITO-coated rigid glass substrate and used it as a reference. Both OLEDs were fabricated using the same conditions and had an active emitting area of $4\,\text{mm}^2$ on glass substrates in a size of $5\,\text{cm} \times 5\,\text{cm}$. Figure 12.16 shows the current and luminance of both types of OLEDs as functions of operating voltage, i.e. $I\text{–}V$ and $L\text{–}V$ plots. The solid curves in Figures 12.16 (a) and (b) represent the device characteristics measured from the reference OLED and the corresponding dashed curves represent those of the OLED made with ITO-coated flexible thin glass sheet. As can be seen from Figure 12.16, there are some differences in the $I\text{–}V$ curves of the two OLEDs at operating voltages more than 8 V. This is probably attributable to the less uniform organic films on ITO-coated ultrathin glass sheets. There was a warp in ITO-coated thin glass sheets

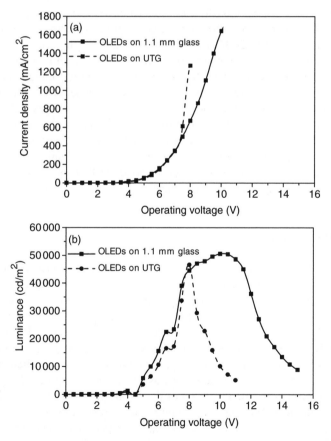

Figure 12.16 (a) Current–voltage and (b) luminance–voltage characteristics of OLEDs measured from OLEDs fabricated with $50\,\mu\text{m}$ flexible ultra-thin glass (UTG) sheets and ITO-coated rigid glass substrate

due to a mismatch in the coefficients of thermal expansion between ITO film and ultra-thin glass sheet. The polymer films spun on warped substrates over an area of 5 cm × 5 cm might be less uniform in comparison with those coated on the rigid substrate under the same conditions. This problem will be overcome by attaching the flexible glass sheet to a rigid substrate. In general, the forward current density measured from both OLEDs, shown in Figure 12.16 (a), is similar in the operating voltage range of 2.5–7.5 V.

A maximum luminance of 4.8×10^4 cd/m^2 and an efficiency of 5.8 cd/A, measured for an OLED fabricated on reinforced flexible glass sheet, at an operating voltage of 7.5 V was obtained. The electroluminescent performance of OLEDs made with 50 μm-thick flexible borosilicate glass sheets is comparable to that of identical devices made with the commercial ITO coated on rigid glass substrates. It is envisaged that further improvements in the devices' performance are possible by optimizing the fabrication processes and encapsulation techniques.

12.3.2 Flexible top-emitting OLEDs on plastic foils

In the past decade, the display industries have experienced an extremely high growth rate. In the recent market report by Standford Research, the projected compounded annual growth rate of OLEDs from 2003 to 2009 is 56%, which means that it will grow from US$ 300 millions in 2003 to US$ 3.1 billions by 2009. OLEDs have recently attracted attention as display devices that can replace LCDs because OLEDs can produce high visibility by self-luminescence. OLEDs do not require backlighting, which is necessary for LCDs, and can be fabricated into lightweight, thin, and flexible display panels. A typical OLED is constructed by placing a stack of organic electroluminescent and/or phosphorescent materials between a cathode layer that can inject electrons and an anode layer that can inject holes. When a voltage of proper polarity is applied between the cathode and anode, holes injected from the anode and electrons injected from the cathode combine radiatively and emit energy as light, thereby producing electroluminescence.

(A) Top-emitting OLEDs

A conventional OLED has a bottom-emitting structure, which includes a reflective metal or metal alloy cathode, and a transparent anode on a transparent substrate, enabling light to be emitted from the bottom of the structure. An OLED may also have a top-emitting structure, which is formed on either an opaque substrate or a transparent substrate. Unlike the conventional OLED structure, top-emitting OLEDs (TOLEDs) can be made on both transparent and opaque substrates. One important application of the top device structure is to achieve monolithic integration of a TOLED on a polycrystalline or amorphous silicon thin-film transistor (TFTs) used in active matrix displays, as illustrated in Figure 12.17. The TOLED structures therefore increase the flexibility of device integration and engineering.

Efficient and durable TOLEDs are required in high-resolution display applications. TOLEDs can be incorporated into an active-matrix display integrated with amorphous or poly-silicon TFTs, which have the important advantage of having on-chip data and scan drivers, allowing for ultra-high pixel resolution (<10–20 μm), e.g, TOLEDs can be made on an opaque substrate to which the electrode adheres, such as the drain contact of *n*-channel

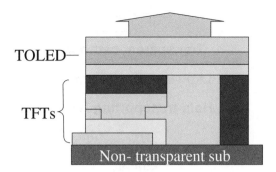

TOLED—

TFTs

Non- transparent sub

Figure 12.17 A cross-sectional view of active-matrix displays with a TOLED on TFTs, a new display architecture with a large aperture

a-Si TFTs or poly-Si TFTs used in active-matrix displays. It is necessary to provide a transparent or semitransparent electrode on top of an active organic layer in TOLEDs. To achieve an efficient TOLED display, injection of holes and electrons and their transport in the emissive layer should be balanced to combine them radiatively in an efficient manner. Semitransparent top electrodes may consist of a stack of thin films with an efficient carrier injector to prevent damage to the OLED layer structure, particularly the underlying organic emissive layer.

In comparison with conventional bottom-emitting OLEDs, less progress has been made on TOLEDs due to the limited success in fabricating a suitable transparent or semitransparent electrode on active organic layers. TOLEDs will become more and more important because the top-emitting architecture is definitely required for high-resolution active-matrix OLED displays. Research into developing an efficient transparent cathode has recently attracted intensive attention worldwide. The resulting device designs are finding use in a wide range of plastic electronic applications such as the fabrication of low-cost thin-film organic photosensors, field-effect transistors, organic/polymeric solar cells, and flexible light-emitting devices. If ITO is sputtered directly onto organic layers, then the transparent electrode may cause deterioration on cathode contact for electron injection. It is indeed quite a challenge to obtain an efficient TOLED with a desired hole–electron current balance. A fundamental understanding of the morphological and electronic properties at the electrode/organic interface is a prerequisite for achieving an efficient TOLED.

Figure 12.18 shows a cross-sectional view of the corresponding TOLED. With this structure, a device can emit light from both the semitransparent top cathode and the transparent substrate. The TOLED design includes a semitransparent cathode, a stack of organic electroluminescent layers, and an anode. The organic emissive and hole-transport layers are deposited on an anode using a spin-coating method. The semitransparent cathode consists of an electron injector and an optical-index-matching layer. The optical-index-matching layer may consist of a stack of organic and inorganic thin films, which can be readily prepared by thermal evaporation without incurring radiation damage to the OLED layer structure, particularly the underlying active organic emissive layer. The performance of TOLEDs made with multilayer structures is also compared with that of the devices fabricated using a conventional structure. The feasibility of employing an organic/inorganic multi-layer structure to improve TOLED performance has been investigated and demonstrated [12.86, 12.87].

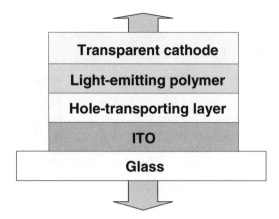

Figure 12.18   A cross-sectional view of a typical TOLED. When a TOLED is fabricated on a transparent substrate, e.g., a glass or a clear plastic foil, the device can emit the light from both the cathode and the anode, forming a see-through OLED

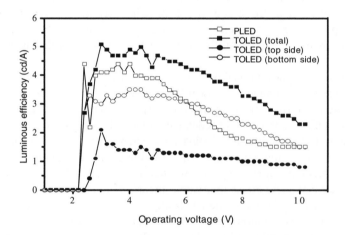

Figure 12.19 Luminous efficiency–voltage characteristics of a bottom emission OLED, ITO/PEDOT/PPV/Ca/Ag, and a TOLED, ITO/PEDOT/PPV/semitransparent cathode

It has been demonstrated that anode modification plays a critical role in determining the electroluminescence efficiency and stability of OLEDs. The ITO surface modification is also applied for TOLED fabrication. In fabricating the TOLED shown in Figure 12.18, spin-coated poly(3,u-ethylenedioxythiophene) (PEDOT) and (Ph-PPV) were used as hole-transporting and emissive layers, respectively. Before spin-coating of the polymers, the anode specimen was pre-treated in oxygen plasma. The semitransparent cathode was then deposited on the active polymeric stack to form a TOLED. The semitransparent cathode has a multilayer architecture consisting of organic and inorganic layers.

The luminous efficiency–voltage characteristics of a conventional bottom-emitting device and the novel top-emitting device are shown in Figure 12.19. The performance of a TOLED, ITO/PEDOT/PPV/semitransparent cathode, can be optimized by choosing an appropriate layer structure for the upper semitransparent cathode. The semitransparent cathode consists of a stack of organic and inorganic thin films of LiF/Ca/Ag/Alq$_3$. At an operating voltage

Figure 12.20   A TOLED, with a transparent anode and a semitransparent cathode on transparent substrate, a see-through TOLED

of 4 V, the conventional bottom-emitting OLED with the identical organic stack, ITO/PEDOT/PPV, had a luminescence efficiency of 4.4 cd/A. At the same operating voltage, a luminescence efficiency of 3.4 cd/A was measured from the bottom side of the TOLED and 1.3 cd/A was obtained from the top side of the TOLED. The total EL efficiency of the TOLED from both top and bottom emission, i.e. 1.3 cd/A + 3.4 cd/A, = 4.7 cd/A, is comparable to that measured for a conventional bottom-emitting OLED. The results indicate that the semitransparent cathode with a multilayer structure of LiF/Ca/Ag/Alq$_3$ is suitable and very promising for the TOLED. It has been demonstrated [12.86, 12.87] that the fabrication process for TOLEDs thus developed is very reliable, reproducible, and can be used for making OLED display units and transparent passive-matrix displays. As a TOLED has a relatively transparent upper electrode, light can be emitted from the side of the top electrode. A see-through or dual-sided OLED display can also be fabricated when an OLED has a transparent anode and a transparent cathode on a transparent substrate. Figure 12.20 is a photograph of a transparent TOLED.

(B) Flexible top-emitting OLEDs on plastic foils

In the previous section, we have discussed the fabrication of flexible OLEDs using polymer-reinforced ultra-thin glass sheet due to its high optical transparency and superior barrier properties. Although the flexibility and handling ability of ultra-thin glass sheet can be improved significantly when it is reinforced [12.88, 12.89], it still has limited application for displays that can be flexed in use. In this section, we will discuss the results related to flexible OLEDs using a TOLED on an Al-laminated plastic substrate. It is known that most metals possess lower gas permeability than plastics by 6–8 orders of magnitude [12.90]. Therefore a metal layer several micrometers thick can serve as a highly effective barrier to minimize the permeation of oxygen and moisture. Hence, the combination of plastic–metal materials is extremely promising for flexible OLED applications. The flexible substrate consists of a plastic layer laminated onto or coated with a metal layer, and could be one of the possible solutions for flexible OLEDs. For example, aluminum-laminated poly(ethylene terephthalate) (Al-PET) foil has very good mechanical flexibility and superior barrier properties. This substrate has the potential to meet permeability standards in excess of the most demanding display requirements of $\sim 10^{-6}\,g/m^2\,day^6$. The robustness of this substrate is also very high. A flexible OLED using Al-PET as substrate may provide a cost-effective approach for mass production, such as roll-to-roll processing, which is a widely used industrial process.

Low-temperature deposition of ITO is a prerequisite for fabricating flexible OLEDs or TOLEDs that preclude the use of a high-temperature process. An ITO anode with a smooth surface can minimize electrical shorts in OLEDs, which have a stack of functional organic layers in the thickness range 100–200 nm. Our results have demonstrated that the low-temperature deposition technique offers an enabling approach to fabricate ITO films with smooth morphology and low sheet resistance on flexible plastic substrate at a low processing temperature [12.91]. Thin films of high performance low-temperature ITO were deposited on a 0.1-mm-thick PET substrate by radio frequency magnetron sputtering using an oxidized target with In_2O_3 and SnO_2 in a weight ratio of 9:1. The substrate was not heated during or after the film deposition. The actual substrate temperature, which might rise due to the plasma process during the film deposition, was less than 60 °C. Sputtering power was kept constant at 100 W. The base pressure in the sputtering system was approximately $2.0 \times 10^{-4}\,Pa$. During the film deposition, the argon–hydrogen gas mixture was employed. The argon partial pressure was set at $2.85 \times 10^{-1}\,Pa$ and the hydrogen partial pressure was varied from 1.0–$4.0 \times 10^{-3}\,Pa$ to modulate and optimize the properties of ITO films. The use of a hydrogen–argon gas mixture allowed a broader process window for the preparation of ITO films having high optical transparency and high conductivity [12.13, 12.89, 12.91]; e.g., a 130-nm-thick ITO film with a sheet resistance of $\sim 25\,\Omega/sq$ and an optical transparency of 80% in the visible light range can be fabricated at a substrate temperature of 60 °C.

We measured the transmittance spectra of a bare PET substrate as a function of wavelength over the range 300–800 nm, which clearly showed that the average transmittance of ITO-coated PET foil is above 80% over the visible light spectrum, indicating that the optical transparency of the low-temperature ITO is suitable for OLED application. The optical transparency of ITO film deposited on such a substrate is comparable that deposited on the polymer-reinforced ultra-thin and normal rigid glass substrates [12.88, 12.91]. An acrylic layer was spin-coated on a PET surface to form the smoother surface for the subsequent film deposition process. It was also found that the presence of an acrylic interlayer increases the adhesion between the film and the substrate. A bare PET surface has rms roughness of ~ 6.0 nm, whereas PET with an acrylic planarization layer has a much lower rms roughness of ~ 0.35 nm. Our results show that the ITO deposited on an acrylic-layer-coated PET foil has a rms roughness of ~ 0.37 nm, which is suitable for OLED fabrication.

Prior to the spinning coating of polymeric materials, the ITO specimens were treated by oxygen plasma. In our work, spin-coated PEDOT and (Ph-PPV) were used as hole-transporting (HTL) and emissive layers, respectively. The samples with polymeric layers were then transferred to an electrode chamber for semitransparent cathode depositions with a multilayer architecture consisting of organic and inorganic layers [12.86]. The semitransparent cathode was then deposited on the active polymeric stack to form a TOLED. The deposition of organic and cathode materials was controlled at a constant rate of about 1 Å/s, and the thickness of the organic and metal layers was estimated and controlled by the deposition time. Figure 12.21(a) shows a cross-sectional view of a TOLED on an Al-PET and a

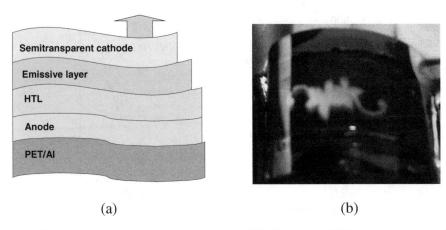

(a)   (b)

Figure 12.21   (a) A cross-sectional view of a typical TOLED on an Al-PET and (b) a photograph of a flexible TOLED, demonstrating a glowing tattoo image, with a configuration of Ag/CFX/Ph-PPV/semitransparent cathode on an Al-PET substrate

control device with a configuration of ITO/HTL/emissive layer/semitransparent cathode on a glass. In a control device, light can be emitted from both the upper semitransparent cathode and the bottom transparent substrate, as illustrated in Figure 12.18. After the device had been fabricated, the samples were transferred to a connected glove box with oxygen and moisture levels lower then 1.0 ppm for current density–voltage (J–V) and luminance–current density (L–J) characterizations.

The architecture of fabricating a TOLED is extended to the design and fabrication of a flexible OLED, an integration of TOLED on Al-PET substrate, as shown in Figure 12.21(a). Figure 12.21(b) is a photograph of a flexible OLED on Al-PET showing a glowing tattoo image. It is found that the performance of OLEDs made on Al-PET does not deteriorate after repeated bending. Although only Al-PET foil was tested for flexible OLEDs in this work, the technique also applies to other plastics. For example, the flexible substrate can be a plastic layer laminated onto or coated with a metal layer or a metal film sandwiched between two plastic foils. When a TOLED is formed on a metal surface of a flexible substrate, the metal surface can also serve as part of the anode for the TOLED and a barrier to minimize the permeation of oxygen and moisture. This substrate has the potential to meet permeability standards in excess of the most demanding display and organic electronics requirements.

The current density–voltage (J–V), luminance–voltage (L–V), and luminous efficiency–voltage characteristics of a TOLED fabricated on both Al-PET and glass substrates are plotted in Figures 12.22 (a), (b) and (c), respectively. For flexible OLEDs made with Ag/ITO and Ag/CF$_X$, the device with a single-layer Ph-PPV on a CF$_X$-modified Ag anode displayed a slightly higher operating voltage to achieve the same current density. This may be attributed to the thicker Ph-PPV layer (110 nm) used. The results (shown in Figure 12.22) also indicate that devices made with both Ag/ITO and Ag/CF$_X$ exhibit similar hole-injection characteristics. However, in comparison to an OLED made with a bilayer anode of Ag/ITO, the identical device with an Ag/CF$_X$ anode exhibited higher luminance at the same current density. This demonstrates that if an efficient flexible ITO-free OLED is fabricated using

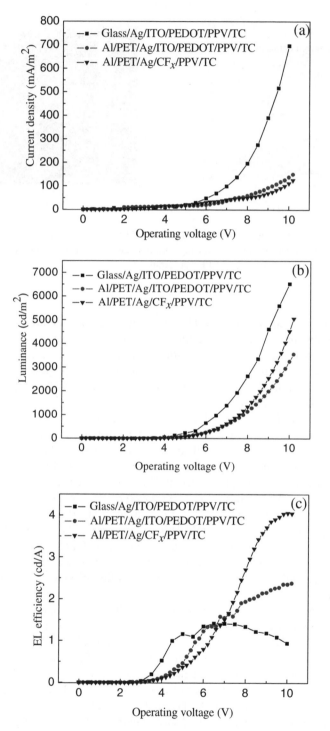

Figure 12.22 Current density–voltage (J–V) (a), luminance–voltage (L–V) (b), and EL efficiency–voltage (c) characteristics measured for TOLEDs on glass and Al-PET. TOLEDs made on glass and flexible substrates had an identical semitransparent top cathode

Ag/CF$_X$ on Al-PET foil, then a luminous efficiency of 4.05 cd/A at an operating voltage of 10 V can be obtained.

Most current OLEDs are based on rigid substrates, such as glass, which limits the 'mould-ability' of the device, restricting the design and spacing where OLEDs can be used. The use of Al-PET in OLEDs will significantly reduce the weight of flat-panel displays and endow the ability to bend a display into any desired shape. Imagine display panels that can be wrapped around the circumference of a pillar as foldable and 'rollable' TVs. Flexible OLEDs using Al-PET, demonstrated and discussed in this chapter, will also make it possible to fabricate displays by continuous roll processing, thus providing the basis for very low-cost mass production.

12.4 CONCLUSIONS

The development of organic semiconductors and related devices is still in its growing stage, particularly the design and optimization of both their structures and their performance. Organic semiconductors as the active components in the devices have many advantages, e.g., thin, lightweight, large area, cost effectiveness, chemical tenability, and mechanical flexibility. Recently, multilayer heterojunction high-performance organic optoelectronic devices such as multiphoton emission devices, organic photovoltaic cells, and organic photodetectors with external quantum efficiencies up to 75% across the visible spectrum and bandwidths approaching 450 MHz have been reported. The optical integrated organic bifunctional matrix arrays, bistable optical switches, have also been demonstrated.

The advantages of organic semiconductors such as low cost due to possible solution-processing technologies render organic semiconductor technology attractive for specialized or cost-sensitive applications. In particular, signage, flexible information display boards, smart cards, organic logic circuits, and other applications based on organic electronics have become widespread, and the need for integrated, high-performance organic optoelectronic devices performing specialized functions on the organic 'chip' will increase accordingly. There is increasing activity in this area, and the prospect for organic semiconductor integrated circuits provides a realistic goal for eventual applications.

REFERENCES

[12.1] M. Pope and C.E. Swenberg, *Electronic Processes in Organic Crystals* (Oxford University Press, New York, 1982).

[12.2] K.L. Chopra, S. Major, and D.K. Pandya, *Thin Solid Films*, **102**, 1 (1983).

[12.3] I. Hamburg and C.G. Granvist, *J. Appl. Phys.*, **60**, R123 (1986).

[12.4] B.H. Lee, I.G. Kim, S.W. Cho, and S.H. Lee, *Thin Solid Films*, **302**, 25 (1997).

[12.5] B.J. Luff, J.S. Wilkinson, and G. Perrone, *Appl. Optics*, **36**, 7066 (1997).

[12.6] J.S. Kim, M. Granström, R.H. Friend, N. Johansson, W.R. Salaneck, R. Daik, W.J. Feast, and F. Cacialli, *J. Appl. Phys.*, **84**, 6859 (1998).

[12.7] I.D. Parker, *J. Appl. Phys.*, **75**, 1656 (1994).

[12.8] C.C. Wu, C.I. Wu, J.C. Sturm, and A. Kahn, *Appl. Phys. Lett.*, **70**, 1348 (1997).

[12.9] B. Schröder, *Mater. Sci. Eng. A*, **139**, 319 (1991).

[12.10] F.R. Zhu, T. Fuyuki, H. Matsunami, and J. Singh, *Sol. Energy Mater. Sol. Cells*, **39**, 1 (1995).

[12.11] F.R. Zhu, C.H.A. Huan, K.R. Zhang, and A.T.S. Wee, *Thin Solid Films*, **359**, 244 (1999).

[12.12] A. Salehi, *Thin Solid Films*, **324**, 214 (1998).

[12.13] K.R. Zhang, F.R. Zhu, C.H.A. Huan, and A.T.S. Wee, *J. Appl. Phys.*, **86**, 974 (1999).

[12.14] K.R. Zhang, F.R. Zhu, C.H.A. Huan, A.T.S. Wee, and T. Osipowicz, *Surf. Interface Anal.*, **28**, 271 (1999).

[12.15] J.K. Sheu, Y.K. Su, G.C. Chi, M.J. Jou, and C.M. Chang, *Appl. Phys. Lett.*, **72**, 3317 (1999).

[12.16] S. Major and K.L. Chopra, *Sol. Energy Mater.*, **17**, 319 (1988).

[12.17] J. Hu and R.G. Gordon, *J. Appl. Phys.*, **72**, 5381 (1992).

[12.18] Y. Takahashi, S. Okada, R.B.H. Tahar, K. Nakano, T. Ban, and Y. Ohya, *J. Non-Cryst. Solids*, **218**, 129 (1997).

[12.19] K. Nishio, T. Sei, and T. Tsuchiya, *J. Mater. Sci.*, **31**, 1761 (1996).

[12.20] H.S. Kwok, X.W. Sun, and D.H. Kim, *Thin Solid Films*, **335**, 299 (1998).

[12.21] H. Kim, A. Piqué, J.S. Horwitz, H. Mattoussi, H. Murata, Z.H. Kafafi, and D.B. Chrisey, *Appl. Phys. Lett.*, **74**, 3444 (1999).

[12.22] J. Ma, S.Y. Li, J.Q. Zhao, and H.L. Ma, *Thin Solid Films*, **307**, 200 (1997).

[12.23] J. Ma, D.H. Zhang, S.Y. Li, J.Q. Zhao, and H.L. Ma, *Jpn. J. Appl. Phys.*, **37**, 5614 (1998).

[12.24] S. Laux, N. Kaiser, A. Zöller, R. Götzelmann, H. Lauth, and H. Bernitzki, *Thin Solid Films*, **335**, 1 (1998).

[12.25] Y. Wu, C.H.M. Marée, R.F. Haglund, Jr., J.D. Hamilton, M.A. Morales Paliza, M.B. Huang, L.C. Feldman, and R.A. Weller, *J. Appl. Phys.*, **86**, 991 (1999).

[12.26] W.F. Wu and B.S. Chiou, *Thin Solid Films*, **298**, 221 (1997).

[12.27] L. Davis, *Thin Solid Films*, **236**, 1 (1993).

[12.28] L. Mao, R.E. Benoit, and J. Proscia, in *Mechanism of Thin Film Evaluation*, edited by Materials Research Society Symposium Proceeding No. 317 S.M. Yalisove, C.V. Thompson, and D.J. Eaglesham (Boston, U.S.A., 1994), p.181.

[12.29] X.W. Sun, H.C. Huang, and H.S. Kwok, *Appl. Phys. Lett.*, **68**, 2663 (1996).

[12.30] J.E. Greene, *Solid State Technol.*, **30**, 115 (1987).

[12.31] M. Katiyar, Y.H. Yang, and J.R. Ableson, *J. Appl. Phys.*, **77**, 6247 (1995).

[12.32] J.H. Lan and J. Kanicki, *Thin Solid Films*, **304**, 127 (1997).

[12.33] R. Banerjee, D. Das, S. Ray, A.K. Batabyal, and A.K. Barua, *Sol. Energy Mater.*, **13**, 11 (1986).

[12.34] H. Kim, C.M. Gilmore, A. Piqué, J.S. Horwitz, H. Mattoussi, H. Murata, Z.H. Kafafi, and D.B. Chrisey, *J. Appl. Phys.*, **86**, 6451 (1999).

[12.35] J.C.C. Fan and J.B. Goodenough, *J. Appl. Phys.*, **48**, 3524 (1977).

[12.36] A.C. Arias, J.R. de Lima, and I.A. Hummelgen, *Adv. Mater.*, **10**, 392 (1998).

[12.37] J.E.A.M. ven den Meerakker, E.A. Meulenkamp, and M. Scholten, *J. Appl. Phys.*, **74**, 3282 (1993).

[12.38] X. Zhou, J. He, L.S. Liao, M. Lu, Z.H. Xiong, X.M. Ding, X.Y. Hou, F.G. Tao, C.E. Zhou, and S.T. Lee, *Appl. Phys. Lett.*, **74**, 609 (1999).

[12.39] F.R. Zhu, K. Zhang, E. Guenther, and S.J. Chua, *Thin Solid Films*, **363**, 314 (2000).

[12.40] L.J. Meng and M.P. dos Santos, *Thin Solid Films*, **322**, 56 (1998).

[12.41] M. Bender, W. Seelig, C. Daube, H. Frankenberger, B. Ocker, and J. Stollenwerk, *Thin Solid Films*, **326**, 72 (1998).

[12.42] R.T. Chen and D. Robinson, *Appl. Phys. Lett.*, **60**, 1541 (1992).

[12.43] Y. Shigesato, Y. Hayashi, A. Masui, and T. Haranou, *Jpn. J. Appl. Phys.*, **30**, 814 (1991).

[12.44] S.A. Van Slyke, C.H. Chen, and C.W. Tang, *Appl. Phys. Lett.*, **69**, 2160 (1996).

[12.45] J.S. Kim, R.H. Friend, and F. Cacialli, *Appl. Phys. Lett.*, **74**, 3084 (1999).

[12.46] Y. Kurosaka, N. Tada, Y. Ohmori, and K. Yoshino, *Jpn. J. Appl. Phys.*, **37**, L872 (1998).

[12.47] K. Sugiyama, H. Ishii, Y. Ouchi, and K. Seki, *J. Appl. Phys.*, **87**, 295 (2000).

[12.48] D.J. Milliron, I.G. Hill, C. Shen, A. Kahn, and J. Schwartz, *J. Appl. Phys.*, **87**, 572 (2000).

[12.49] F. Steuber, J. Staudigel, M. Stössel, J. Simmerer, and A. Winnacker, *Appl. Phys. Lett.*, **74**, 3558 (1999).

[12.50] Y. Park, V. Choong, Y. Gao, B.R. Hsieh, and C.W. Tang, *Appl. Phys. Lett.*, **68**, 2699 (1997).

[12.51] I.H. Campbell, J.D. Kress, R.L. Martin, D.L. Smith, N.N. Barashkov, and J.P. Ferraris, *Appl. Phys. Lett.*, **71**, 3528 (1997).

[12.52] M.G. Mason, L.S. Hung, C.W. Tang, S.T. Lee, K.W. Wong, and M. Wang, *J. Appl. Phys.*, **86**, 1688 (1999).

[12.53] J.A. Chaney and P.E. Pehrsson, *Appl. Surf. Sci.*, **180**, 214 (2001).

[12.54] H.M. Lee, K.H. Choi, D.H. Hwang, L.M. Do, T. Zyung, J.W. Lee, and J.K. Park, *Appl. Phys. Lett.*, **72**, 2382 (1998).

[12.55] Y. Yang and A.J. Heeger, *Appl. Phys. Lett.*, **64**, 1245 (1994).

[12.56] S. Karg, J.C. Scott, J.R. Salem, and M. Angelopoulos, *Synth. Met.*, **80**, 111 (1996).

[12.57] J.C. Carter, I. Grizzi, S.K. Heeks, D.J. Lacey, S.G. Latham, P.G. May, O.R. Delospanos, K. Pichler, C.R. Towns, and H.F. Wittmann, *Appl. Phys. Lett.*, **71**, 34 (1997).

[12.58] N.C. Greenham, S.C. Moratti, D.D.C. Bradley, R.H. Friend, and A.B. Holmes, *Nature*, **365**, 628 (1993).

[12.59] F. Nüesch, L.J. Rothberg, E.W. Forsythe, Q.T. Le, and Y.L. Gao, *Appl. Phys. Lett.*, **74**, 880 (1999).

[12.60] J.S. Kim, F. Cacialli, A. Cola, G. Gigli, and R. Cingolani, *Appl. Phys. Lett.*, **75**, 17 (1999).

[12.61] I.H. Campbell, J.D. Kress, R.L. Martin, and D.L. Smith, *Appl. Phys. Lett.*, **71**, 3528 (1997).

[12.62] S.F.J. Appleyard and M.R. Willis, *Opt. Mater. (Amsterdam)*, **9**, 120 (1998).

[12.63] P.K.H. Ho, M. Granström, R.H. Friend, and N.C. Greenham, *Adv. Mater.*, **10**, 769 (1998).

[12.64] J.E. Malinsky, G.E. Jabbour, S.E. Shaheen, J.D. Anderson, A.G. Richter, T.J. Marks, N.R. Armstrong, B. Kippelen, P. Dutta, and N. Peyghambarian, *Adv. Mater.*, **11**, 227 (1999).

[12.65] G.E. Jabbour, B. Kippelen, N.R. Armstrong, and N. Peyghambarian, *Appl. Phys. Lett.*, **73**, 1185 (1998).

[12.66] G.E. Jabbour, Y. Kawabe, S.E. Shaheen, J.F. Wang, M.M. Morrell, B. Kippelen, and N. Peyghambarian, *Appl. Phys. Lett.*, **71**, 1762 (1997).

[12.67] L.S. Hung, C.W. Tang, and M.G. Mason, *Appl. Phys. Lett.*, **70**, 152 (1997).

[12.68] H. Tang, F. Li, and J. Shinar, *Appl. Phys. Lett.*, **71**, 2560 (1997).

[12.69] F.R. Zhu, B.L. Low, K. Zhang, and S.J. Chua, *Appl. Phys. Lett.*, **79**, 1205 (2001).

[12.70] H. Becker, H. Spreitzer, W. Kreuder, E. Kluge, H. Schenk, I. Parker, and Y. Cao, *Adv. Mater.*, **12**, 42 (2000).

[12.71] J.S. Kim, F. Cacialli, M. Granström, R.H. Friend, N. Johansson, W.R. Salaneck, R. Daik, and W.J. Feast, *Synth. Met.*, **101**, 111 (1999).

[12.72] R.B.H. Tahar, T. Ban, Y. Ohya, and Y. Takahashi, *J. Appl. Phys.*, **83**, 2631 (1998).

[12.73] J.M. Zhao, S.T. Zhang, X.J. Wang, Y.Q. Zhan, X.Z. Wang, G.Y. Zhong, Z.J. Wang, X.M. Ding, W. Huang, and X.Y. Hou, *Appl. Phys. Lett.*, **84**, 2913 (2004).

[12.74] X.M. Ding, L.M. Hung, L.F. Cheng, Z.B. Deng, X.Y. Hou, C.S. Lee, and S.T. Lee, *Appl. Phys. Lett.*, **76**, 2704 (2000).

[12.75] S.J. Chua, L. Ke, R.S. Kumar, and K.R. Zhang, *Appl. Phys. Lett.*, **81**, 1119 (2002).

[12.76] L.S. Hung and C.H. Chen, *Mater. Sci. Eng., R*, **39**, 143 (2002).

[12.77] R.H. Friend, R.W. Gymer, A.B. Holmes, J.H. Burroughes, R.N. Marks, C. Taliani, D.D.C. Bradley, D.A. Dos Santos, J.L. Brédas, M. Lögdlund, and W.R. Salaneck, *Nature*, **397**, 121 (1999).

[12.78] C.W. Tang and S.A. Van Slyke, *Appl. Phys. Lett.*, **51**, 913 (1987).

[12.79] S.R. Forrest, *Org. Electron.*, **4**, 45 (2003).

[12.80] C. Fou, O. Onitsuka, M. Ferreira, M.F. Rubner, and B.R. Hsieh, *J. Appl. Phys.*, **79**, 7501 (1996).

[12.81] N. Krasnov, *Appl. Phys. Lett.*, **80**, 3853 (2002).

[12.82] G. Gustafsson, G.M. Treacy, Y. Cao, F. Klavetter, N. Colaneri and A.J. Heeger, *Synth. Met.*, **57**, 4123 (1993).

[12.83] G. Gu, P.E. Burrows, S. Venkatesh, and S.R. Forrest, *Opt. Lett.*, **22**, 172 (1997).

[12.84] R. Paetzold, K. Heuster, D. Henseler, S. Roeger, G. Weittmann, and A. Winnacker, *Appl. Phys. Lett.*, **82**, 3342 (2003).

[12.85] A.B. Chwang, M.R. Rothman, S.Y. Mao, R.H. Hewitt, M.S. Weaver, J.A. Silvermail, K. Rajan, M. Hack, J.J. Brown, X. Chu, L. Moro, T. Krajewski, and N. Rutherford, *Appl. Phys. Lett.*, **83**, 413 (2003).

[12.86] Y.Q. Li, J.X. Tang, Z.Y. Xie, L.S. Hung, and S.S. Lau, *Chem. Phys. Lett.*, **386**, 128 (2004).

[12.87] F.R. Zhu, X.T. Hao, K.S. Ong, Y.Q. Li, and L.W. Tan, *Proc. IEEE*, **93**, 1440 (2005).

[12.88] K.S. Ong, J.Q. Hu, R. Shrestha, F.R. Zhu, and S.J. Chua, *Thin Solid Films*, **477**, 32 (2005).

[12.89] A. Plichta, A. Weber, and A. Habeck, *Mater. Res. Soc. Symp. Proc.*, **769**, H9.1 (2003).

[12.90] R.R. Tummala and E.J. Rymaszewski, *Microelectronics Packaging Handbook* (Van Nostrand Reinhold, New York, 1989), Ch. 10.

[12.91] K.R. Zhang, F.R. Zhu, C.H.A. Huan, and A.T.S. Wee, *Thin Solid Films*, **376**, 255 (2000).

13 Optical Properties of Thin Films

V.-V. Truong[*] and S. Tanemura[†]

[*]Concordia University, 1455 de Maisonneuve Blvd W, GM 910-3, Montreal (Quebec)
Canada H3G-1M8
Tel: 1-514-848-2424 ext. 5873
Fax: 1-514-848-7399
e-mail: tvovan@alcor.concordia.ca
†Nagoya Institute of Technology, Magoya, Gokisho-cho, Showa-ku, Nagoya 466-8555,
Japan
e-mail: tanemura.sakae@nitech.ac.jp

Optical Properties of Condensed Matter and Applications   Edited by J. Singh

13.1 INTRODUCTION

One of the classical treatments of the optical properties of thin films found in the literature remains in the book by O.S. Heavens first published by Butterworths Scientific Publications in 1955 and subsequently reviewed and republished by Dover in 1965 and in 1991 [13.1]. Besides devoting many sections of the book to the experimental and practical aspects of thin films, the author has also presented Maxwell's equations to calculate the intensity of light transmitted and reflected by single or multi-layered films. Other comprehensive treatments of optical properties of thin films can also be found in references such as Born and Wolf [13.2], Azzam and Bashara [13.3], Palik [13.4], Klein and Furtak [13.5], Ward [13.6], and Bass [13.7]. Although photometry was frequently used to characterize the optical properties of thin films, ellipsometry, or the study of the state of polarization of the reflected or transmitted light and their relation with material properties, has become a standard and powerful characterization method as more computing power and techniques are readily available. Ellipsometric parameters are very sensitive to changes in material properties and it is well recognized that ellipsometry is capable of detecting a few angstroms' change in material layers and can be used to study the surface structures in the subnanometer range.

In the present chapter, expressions for reflection and transmission coefficients are first reported for an isotropic thin film on a substrate. Comments are then made on the matrix method for layered structures and on anisotropic films before presenting an original approach that combines reflection and transmission photoellipsometry for studying a multi-layer structure. Finally, the chapter is completed with a few selected applications of thin films.

13.2 OPTICS OF THIN FILMS

13.2.1 An isotropic film on a substrate

A situation often encountered is the case of an isotropic film on an infinite isotropic substrate. The film is assumed to have parallel-plane boundaries and a thickness D_1. The ambient (medium 0) is characterized by a real index of refraction N_0, whereas the film (medium 1) and the substrate (medium 2) have complex indices of refraction, N_1 and N_2, respectively. Let us consider a plane wave incident at an angle θ_0 from medium 0 that results in a refracted wave at angle θ_1 in medium 1 and a transmitted wave at angle θ_2 in medium 2 (Figure 13.1).

By taking into account multiple reflections at the film–ambient and film–substrate interfaces, the reflection and transmission amplitudes can be calculated. Let β be the phase change that occurs when the light wave traverses the film once from one boundary to the other:

$$\beta = 2\pi \left(\frac{D_1}{\lambda} \right) N_1 \cos \theta_1 \tag{13.1}$$

or

$$\beta = 2\pi \left(\frac{D_1}{\lambda} \right) (N_1^2 - N_0^2 \sin^2 \theta_0)^{\frac{1}{2}} \tag{13.2}$$

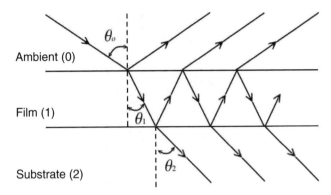

Figure 13.1 Schematic presentation of reflection, refraction, and transmission of an incident wave on an isotropic film deposited on an isotropic substrate

where λ is the wavelength of the incident light, and Equation (13.2) results from Equation (13.1) by applying Snell's law, $N_0 \sin \theta_0 = N_1 \sin \theta_1$.

Summation of all partially reflected and transmitted waves within the film leads to an infinite geometric series and the total reflected amplitude R is given by:

$$R = r_{01} + \frac{t_{01} t_{10} r_{12} e^{-j2\beta}}{1 - r_{10} r_{12} e^{-j2\beta}} \tag{13.3}$$

or

$$R = \frac{r_{01} + r_{12} e^{-j2\beta}}{1 + r_{01} r_{12} e^{-j2\beta}} \tag{13.4}$$

when r_{10} is replaced by $-r_{10}$ and $(t_{01} t_{10})$ by $(1 - r_{10}^2)$.

In a similar way, the total transmitted amplitude can also be obtained as:

$$T = \frac{t_{01} t_{12} e^{-j\beta}}{1 + r_{01} r_{12} e^{-j2\beta}} \tag{13.5}$$

where r_{01} and r_{12} are, respectively, the Fresnel reflection coefficients at the ambient–film (0–1) and film–substrate (1–2) interfaces, and t_{01} and t_{12} are the Fresnel transmission coefficients defined in a similar manner. Equations (13.3)–(13.5) are valid when the incident wave is linearly polarized either parallel (p) or perpendicular (s) to the plane of incidence and a subscript p or s can be added to the formulae in order to account for the polarization of the incident light. Thus, the reflectance amplitude resulting from a p-polarized light will be denoted by R_p and the reflectance amplitude resulting from s-polarized light by R_s. Similarly, the transmission amplitudes taking into account the p and s polarizations will be T_p and T_s, respectively. The proper expressions for the Fresnel coefficients then have to be used and they are as follows [13.3].

$$r_{01p} = \frac{N_1 \cos\theta_0 - N_0 \cos\theta_1}{N_1 \cos\theta_0 + N_0 \cos\theta_1} \tag{13.6}$$

$$r_{12p} = \frac{N_2 \cos\theta_1 - N_1 \cos\theta_2}{N_2 \cos\theta_1 + N_1 \cos\theta_2} \tag{13.7}$$

$$r_{01s} = \frac{N_0 \cos\theta_0 - N_1 \cos\theta_1}{N_0 \cos\theta_0 + N_1 \cos\theta_1} \tag{13.8}$$

$$r_{12s} = \frac{N_1 \cos\theta_1 - N_2 \cos\theta_2}{N_1 \cos\theta_1 + N_2 \cos\theta_2} \tag{13.9}$$

$$t_{01p} = \frac{2N_0 \cos\theta_0}{N_1 \cos\theta_0 + N_0 \cos\theta_1} \tag{13.10}$$

$$t_{12p} = \frac{2N_1 \cos\theta_1}{N_2 \cos\theta_1 + N_1 \cos\theta_2} \tag{13.11}$$

$$t_{01s} = \frac{2N_0 \cos\theta_0}{N_0 \cos\theta_0 + N_1 \cos\theta_1} \tag{13.12}$$

$$t_{12s} = \frac{2N_1 \cos\theta_1}{N_1 \cos\theta_1 + N_2 \cos\theta_2} \tag{13.13}$$

It is noted that angular quantities and the indices of refraction of the media involved are related through Snell's law:

$$N_0 \sin\theta_0 = N_1 \sin\theta_1 = N_2 \sin\theta_2 \tag{13.14}$$

The above expressions are valid when the substrate is semi-infinite, i.e., reflections at the lower surface of the substrate are not taken into account. In the case of a finite-thickness substrate, the system can be considered to be an ambient medium (medium 0) followed by two films (medium 1 and medium 2, respectively) and a third infinite medium is added (medium 3). By assuming coherent combination of multiply reflected and transmitted waves within the two films, summation of those waves will lead to final expressions for the total reflected and transmitted amplitudes.

13.2.2 Matrix methods for multi-layered structures

Obviously, the summation of multiply reflected and transmitted waves within multiple layers is a very cumbersome process and alternative methods can be sought in calculating the optical properties of multi-layered structures. Fortunately, the generally known matrix methods offers a very elegant solution [13.8, 13.9] and a detailed description of such a method can be found in ref. [13.3]. It makes use of the fact that the equations for the propagation of light are linear and the continuity of the tangential fields across an interface can be expressed as a 2×2 linear-matrix transformation.

Let us consider a multi-layered system consisting of a stack of $1, 2, 3, \ldots, j, \ldots m$ homogeneous and isotropic layers sandwiched between two semi-infinite ambient (0) and substrate $(m + 1)$ media. It can be shown [13.3] that the overall reflection and transmission properties of such a structure are represented by a scattering matrix $\mathbf{S}$ that can be expressed as the product of each of the interface and layer matrices $\mathbf{I}$ and $\mathbf{L}$, which describe the effects of the individual interfaces and layers constituting the whole system. The scattering matrix is thus given by:

$$\mathbf{S} = \mathbf{I}_{01}\mathbf{L}_1\mathbf{I}_{12}\mathbf{L}_2 \ldots \mathbf{I}_{(j-1)j}\mathbf{L}_j \ldots \mathbf{L}_m\mathbf{I}_{m(m+1)} \tag{13.15}$$

where $\mathbf{I}_{(j-1)j}$ refers to the interface matrix for the interface between medium $(j - 1)$ and medium (j), and $\mathbf{L}_j$ is the layer matrix for layer (j). The matrix for a plane interface has the form:

$$\mathbf{I} = \begin{bmatrix} 1/t & r/t \\ r/t & 1/t \end{bmatrix} = \left(\frac{1}{t}\right)\begin{bmatrix} 1 & r \\ r & 1 \end{bmatrix} \tag{13.16}$$

where r and t are, respectively, the amplitudes of reflection and transmission coefficients. The matrix for a layer of thickness D on the other hand is given by:

$$\mathbf{L} = \begin{bmatrix} e^{j\beta} & 0 \\ 0 & e^{-j\beta} \end{bmatrix} \tag{13.17}$$

where the phase shift (layer phase thickness) β is expressed as a function of the layer thickness D, the layer refraction index N, the wavelength λ, and the local angle of incidence θ as:

$$\beta = \frac{2\pi D N}{\lambda}\cos\theta \tag{13.18}$$

Let us consider, as an example, the explicit case of two films (1 and 2) between a semi-infinite ambient (0) and a substrate (3) media. According to Equation (13.15), the scattering matrix for this case is given by:

$$\mathbf{S} = \mathbf{I}_{01}\mathbf{L}_1\mathbf{I}_{12}\mathbf{L}_2\mathbf{I}_{23} \tag{13.19}$$

Considering the expressions for the interface and layer matrices expressed by Equations (13.16) and (13.17), the scattering matrix is now:

$$\mathbf{S} = \left(\frac{1}{t_{01}t_{12}t_{23}}\right) \times \begin{bmatrix} 1 & r_{01} \\ r_{01} & 1 \end{bmatrix} \begin{bmatrix} e^{j\beta_1} & 0 \\ 0 & e^{-j\beta_1} \end{bmatrix} \begin{bmatrix} 1 & r_{12} \\ r_{12} & 1 \end{bmatrix}$$
$$\times \begin{bmatrix} e^{j\beta_2} & 0 \\ 0 & e^{-j\beta_2} \end{bmatrix} \begin{bmatrix} 1 & r_{23} \\ r_{23} & 1 \end{bmatrix} \tag{13.20}$$

or

$$\mathbf{S} = \left(\frac{e^{j(\beta_1+\beta_2)}}{t_{01}t_{12}t_{23}}\right) \begin{bmatrix} \begin{aligned} &[(1+r_{01}r_{12}e^{-j2\beta_1}) \\ &+(r_{12}+r_{01}e^{-2j\beta_1}) \\ &\times r_{23}e^{-j2\beta_2}] \end{aligned} & \begin{aligned} &[(1+r_{01}r_{12}e^{-j2j\beta_1})r_{23} \\ &+(r_{12}+r_{01}e^{-2j\beta_1}) \\ &\times e^{-j2\beta_2}] \end{aligned} \\ \begin{aligned} &[(r_{01}+r_{12}e^{-j2\beta_1}) \\ &+(r_{01}r_{12}+e^{-2j\beta_1}) \\ &\times r_{23}e^{-j2\beta_2}] \end{aligned} & \begin{aligned} &[(r_{01}+r_{12}e^{-j2\beta_1})r_{23} \\ &+(r_{01}r_{12}+e^{-2j\beta_1}) \\ &\times e^{-j2\beta_2}] \end{aligned} \end{bmatrix} \tag{13.21}$$

As the reflection and transmission coefficients of the multi-layered structure are respectively given by $R = (S_{21}/S_{11})$ and $T = (1/S_{11})$ [13.3], expressions for those coefficients can be derived from Equation (13.21). R and T are thus obtained as:

$$R = \frac{\left(r_{01}+r_{12}e^{-j2\beta_1}\right)+\left(r_{01}r_{12}+e^{-j2\beta_1}\right)r_{23}e^{-j2\beta_2}}{\left(1+r_{01}r_{12}e^{-j2\beta_1}\right)+\left(r_{12}+r_{01}e^{-j2\beta_1}\right)r_{23}e^{-j2\beta_2}} \tag{13.22}$$

and

$$T = \frac{t_{01}t_{12}t_{23}e^{-j(\beta_1+\beta_2)}}{\left(1+r_{01}r_{12}e^{-j2\beta_1}\right)+\left(r_{12}+r_{01}e^{-j2\beta_1}\right)r_{23}e^{-j2\beta_2}} \tag{13.23}$$

The above formulae apply to both polarizations p and s when the proper subscript is added to R, T, r, and t.

Results in Equations (13.22) and (13.23) can be generalized to a larger number of layers although the expressions derived can be quite lengthy. Matrix calculations for a multi-layer structure can, however, be handled conveniently by a computer.

13.2.3 Anisotropic films

Many films that are of interest are anisotropic and a proper treatment of their properties must be made. Examples of such films are Langmuir–Blodgett layers, aggregated or island-like structure films, columnar structure films, semiconductor superlattices, and complex oxide films. The optical function for an anisotropic film can be described by a dielectric tensor with three principal components, which are referenced to the crystal coordinate system. Discussions on anisotropic films can be found in various references [13.3, 13.10–13.14].

In the case of aggregated films, we have a uniaxially anisotropic film that has its optical axis perpendicular to its boundaries with the ambient and the substrate. Gold aggregated films, for example, have been studied recently and their optical properties in the direction perpendicular to the film plane were found to be significantly different from the optical properties in the direction parallel to it [13.15].

Optical thin films composed of a columnar structure also show a large refractive anisotropy. A biaxial model for these films has been proposed in which three main refractive indices correspond to a direct reference with one axis in the direction of the columns and the two other axes in the perpendicular plane [13.16]. With this model and a 4×4 transfer matrix theory, the reflectance as well as the transmittance of a multi-layer stack of anisotropic films can be computed.

13.3 REFLECTION–TRANSMISSION PHOTOELLIPSOMETRY FOR OPTICAL-CONSTANTS DETERMINATION

As mentioned previously, ellipsometry has become a standard tool for characterizing thin films and many excellent reviews on ellipsometry exist in the literature [13.17–13.20]. In the following, we choose to present a method that combines reflection and transmission photoellipsometry as proposed by Bader et al. [13.21]. This is a most convenient and powerful way to study practical samples consisting of a combination of thin and thick films. The fact that the transmission-mode photoellipsometry is considered together with the reflection photoellipsometry eliminates the need to use a completely absorbing substrate or to grind the backside of semitransparent substrates for only-reflection ellipsometric measurements. In this method, the role of the substrate is examined, and contributions from multiple reflections in the substrate are taken into account. The addition of transmission to reflection measurements, on the other hand, can also improve accuracy in determining the optical constants of the thin films studied.

For simplicity, all thin films considered here are assumed to be perfect, isotropic, and uniform, i.e., they have completely smooth and parallel surfaces. It is also assumed that the ambient surrounding the samples has no absorption. In this context, a thick film is defined as a layer for which no interference effects can be observed for a given wavelength resolution. As is well known, this situation is obtained when an average of the reflectance and transmittance on the phase shift related to the thick film is taken. Generally, the reason behind the averaging is the existence of a wavelength resolution, but it can also be related to film imperfections or experimental procedures. A substrate can thus be considered as a thick film. For easy reference, the notations adopted in this section are the ones used in ref. [13.21].

13.3.1 Photoellipsometry of a thick or a thin film

Let N_i, D_i, θ_i, and β_i be the refractive index (complex), the thickness (real), the angle of propagation (complex), and the phase shift related to film number i, respectively. The subscript 0 refers to ambient in. Proper care must be taken in the following formulae when one calculates $\cos(\theta_i)$:

$$\cos\theta_i = +\left(1 - \frac{N_0^2 \sin^2 \theta_0}{N_i^2}\right)^{1/2} \quad \text{if Im } [N_i^2 - N_0^2 \sin^2 \theta_0^{1/2}] \geq 0$$

$$= -\left(1 - \frac{N_0^2 \sin^2 \theta_0}{N_i^2}\right)^{1/2} \quad \text{if Im } [N_i^2 - N_0^2 \sin^2 \theta_0^{1/2}] < 0 \qquad (13.24a)$$

$$\beta = 2\pi\left(\frac{D_i}{\lambda}\right)N_i \cos\theta_i \qquad (13.24b)$$

where λ is the light wavelength. Note that $Im(\beta_i) \geq 0$.

We now consider an ellipsometer that consists of a rotating polarizer of angle θ_p, one film sample, and a rotating analyser of angle θ_a (for a practical description of ellipsometry and ellipsometers, readers may refer to ref. [13.18]). For the case of $\theta_p = \theta_a$, one can write the reflected and transmitted signals for the detector of such ellipsometer by using the Jones or Stokes–Mueller formalism respectively as:

$$I^R = k[R_p \cos^4 \theta_a + R_s \sin^4 \theta_a$$
$$+2\,\text{Re}\,(R_{ps})\sin^2 \theta_a \cos^2 \theta_a] \qquad (13.25a)$$

$$I^T = k[T_p \cos^4 \theta_a + T_s \sin^4 \theta_a$$
$$+2\,\text{Re}\,(T_{ps})\sin^2 \theta_a \cos^2 \theta_a] \qquad (13.25b)$$

where k is a normalizing constant that depends on the intensity of the incident light and the detector. The subscripts p and s have the usual meaning of p and s polarizations, respectively, and angle θ_a is referenced to the incidence plane. Here $R_{p(s)}$ (real), $T_{p(s)}$ (real), R_{ps} (complex), and T_{ps} (complex) refer to reflection and transmission experiments and are defined on the following pages.

As in photometry, we can normalize this signal by dividing it by the signal measured by the detector without the sample, and then we obtain:

$$I^R = R_p \cos^4 \theta_a + R_s \sin^4 \theta_a$$
$$+ 2\,\text{Re}\,(R_{ps})\sin^2 \theta_a \cos^2 \theta_a \qquad (13.26a)$$

$$I^T = T_p \cos^4 \theta_a + T_s \sin^4 \theta_a$$
$$+ 2\,\text{Re}\,(T_{ps})\sin^2 \theta_a \cos^2 \theta_a \qquad (13.26b)$$

It is well known that the introduction of a compensator (retarder) in the optical path can replace the real part in Equations (13.25) by an imaginary part. Thus complete photoellipsometry permits the determination of the following quantities that can be expressed in terms of the total reflection and transmission amplitudes $r_{p(s)}$ and $t_{p(s)}$:

$$R_{p(s)} = |r_{p(s)}|^2 \tag{13.27a}$$

$$T_{p(s)} = |t_{p(s)}|^2 \tag{13.27b}$$

$$R_{ps} = \left|r_p r_s^*\right| \tag{13.27c}$$

$$T_{ps} = |t_p t_s^*| \tag{13.27d}$$

for a thin film, and

$$R_{p(s)} = \left\langle |r_{p(s)}|^2 \right\rangle \tag{13.28a}$$

$$T_{p(s)} = \left\langle |t_{p(s)}|^2 \right\rangle \tag{13.28b}$$

$$R_{ps} = \left(r_p r_s^*\right) \tag{13.28c}$$

$$T_{ps} = (t_p t_s^*) \tag{13.28d}$$

for a thick film.

Quantities $r_{p(s)}$ and $t_{p(s)}$ are the total reflection and transmission amplitudes, respectively, as given in Equation (13.29) for both polarizations. It must be noted that the ambient in and ambient out are supposed to be identical. Otherwise the usual correction factor must be applied to all transmission terms. The fences denote the well known averaging, which deletes the interferences between the multiple beams. The total reflection and transmission amplitudes are given for both polarizations by:

$$r = r_{0,1}^+ + \frac{t_{0,1} + t_{0,1} - r_{1,2} + \exp(2i\beta_1)}{1 - r_{0,1} - r_{1,2} + \exp(2i\beta_1)} \tag{13.29a}$$

$$t = \frac{t_{0,1} + t_{1,2} + \exp(i\beta_1)}{1 - r_{0,1} - r_{1,2} + \exp(2i\beta_1)} \tag{13.29b}$$

In Equations (13.29a, b), the bottom indices refer to the interface and we have omitted the polarization index. The plus or minus superscript refers to the direction of the incident light on the interface (forward and backward directions). In this simple case, $r_{ij}^{+(-)}$ and $t_{ij}^{+(-)}$ are the well known Fresnel amplitudes. It is then easy to show, by using the series expansion, that the averages are obtained for both polarizations in Equations (13.28a) and (13.28b) as:

$$R_{p(s)} = \langle |r_{p(s)}|^2 \rangle = |r_{0,1,p(s)}^{+}|^2$$

$$+ \frac{|t_{0,1,p(s)}^{+}|^2 |t_{0,1,p(s)}^{-}|^2 |r_{1,2,p(s)}^{+}|^2 \exp(-4 \text{ Im } \beta_1)}{1 - |r_{0,1,p(s)}^{-}|^2 |r_{1,2,p(s)}^{+}|^2 \exp(-4 \text{ Im } \beta_1)} \tag{13.30a}$$

$$T_{p(s)} = \langle |t_{p(s)}|^2 \rangle$$

$$= \frac{|t_{0,1,p(s)}^{+}|^2 |t_{1,2,p(s)}^{+}|^2 \exp(-2 \text{ Im } \beta_1)}{1 - |r_{0,1,p(s)}^{-}|^2 |r_{1,2,p(s)}^{+}|^2 \exp(-4 \text{ Im } \beta_1)} \tag{13.30b}$$

For the case of Equations (13.28c) and (13.28d), the averages are also easily obtained as:

$$R_{ps} = \langle (r_p r_s^*) \rangle = (r_{0,1,p} + r_{0,1,s}^{-*})$$

$$+ \frac{(t_{0,1,p} + t_{0,1,s}^{+*})(t_{0,1,p} - t_{0,1,s}^{-*})(r_{1,2,p} + r_{1,2,s}^{+*})\exp(-4 \text{ Im } \beta_1)}{1 - (r_{0,1,p} - r_{0,1,s}^{-*})(r_{1,2,p} + r_{1,2,s}^{+*})\exp(-4 \text{ Im } \beta_1)} \tag{13.30c}$$

$$T_{ps} = \langle (t_p t_s^*) \rangle$$

$$= \frac{(t_{0,1,p} + t_{0,1,s}^{+*})(t_{1,2,p} + t_{1,2,s}^{+*})\exp(-2 \text{ Im } \beta_1)}{1 - (r_{0,1,p} - r_{0,1,s}^{-*})(r_{1,2,p} + r_{1,2,s}^{+*})\exp(4 \text{ Im } \beta_1)} \tag{13.30d}$$

It is to be noted that the mathematical structure of Equations (13.29) and (13.30) are similar in the sense that an iterative process can be used to compute the reflection and transmission through a stack of thin films, and the same iterative procedure can also be used for a stack of thick films (with appropriate changes). Moreover, this procedure computes not only $< |r|^2 >$ and $< |t|^2 >$ but also $< r_p r_s^* >$ and $< t_p t_s^* >$ for a stack of thick films. This will be shown in following section.

13.3.2 Photoellipsometry for a stack of thick and thin films

Here we consider a layered structure consisting of thick and thin films, as shown in Figure 13.2, with ambient in denoted by 0 and ambient out denoted by N. The main idea introduced here is to define pseudointerfaces between thick and thin films.

The pseudointerface between thick film (i) and thick film ($i + 1$) is defined by the reflection and transmission amplitudes through the layered structure of thin films between the two thick films, and thus for both polarizations and incident directions on the layered thin-film structure we get: $r_{i,i+1,p(s)}^{+(-)}$, $t_{i,i+1,p(s)}^{+(-)}$, where the subscript ($i, i + 1$) denotes the interface and the subscript $p(s)$ denotes the polarization. The usual thin-film theory is used to obtain these amplitudes. The computation of quantities $R_{p(s)}$, $T_{p(s)}$, R_{ps}, and T_{ps} uses the same iterative procedure. Let us define $R_{p(s)}(i, N)$ as the reflectance for incident light coming in the positive direction on a pseudointerface ($i, i + 1$), reflected on stack ($i + 1$) to ($N - 1$) of

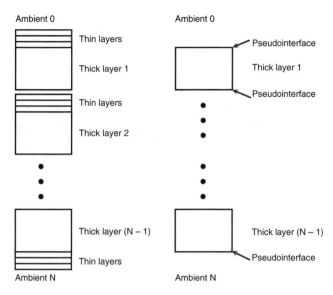

Ambient 0 — Thin layers — Thick layer 1 — Thin layers — Thick layer 2 — Thick layer (N − 1) — Thin layers — Ambient N

Ambient 0 — Pseudointerface — Thick layer 1 — Pseudointerface — Thick layer (N − 1) — Pseudointerface — Ambient N

Figure 13.2 Schematic diagram of a multi-layer structure that consists of thin and thick layers. Thin and thick layers (or substrates) can be replaced by a pseudointerface as defined in the text [Reprinted with permission from G. Bader et al., *Appl. Optics*, **34**, 1684. Copyright (1995) the Optical Society of America]

thick films, and let $T_{p(s)}(i, N)$ be the square modulus of the transmission amplitude through the same stack for both polarizations. Quantities $R_{ps}(i, N)$ and $T_{ps}(i, N)$ are defined in the same manner. We thus obtain the iterative equations for these quantities as:

$$R(i-1,N) = \left| r_{i-1,i}^{+} \right|^2$$
$$+ \frac{\left| t_{i-1,i}^{+} \right|^2 \left| t_{i-1,i}^{-} \right|^2 R(i,N) \exp(-4 \operatorname{Im} \beta_i)}{1 - \left| r_{i-1,i}^{-} \right|^2 R(i,N) \exp(4 \operatorname{Im} \beta_i)} \qquad (13.31a)$$

$$T(i-1,N) = \frac{\left| t_{i-1,i}^{+} \right|^2 T(i,N) \exp(-2 \operatorname{Im} \beta_i)}{1 - \left| r_{i-1,i}^{-} \right|^2 R(i,N) \exp(4 \operatorname{Im} \beta_i)} \qquad (13.31b)$$

for both polarizations and the following expressions for R_{ps} and T_{ps}:

$$R_{ps}(i-1,N) = \left(r_{i-1,i,p}^{+} + r_{i-1,i,s}^{+*} \right)$$
$$+ \frac{\left(t_{i-1,i,p}^{+} + t_{i-1,i,s}^{+*} \right)\left(t_{i-1,i,p}^{-} - t_{i-1,i,s}^{-*} \right) R_{ps}(i,N) \exp(-4 \operatorname{Im} \beta_t)}{1 - \left(r_{i-1,i,p}^{-} - r_{i-1,i,s}^{-*} \right) R_{ps}(i,N) \exp(-4 \operatorname{Im} \beta_t)} \qquad (13.31c)$$

$$T_{ps}(i-1,N) = \frac{(t_{i-1,i,p} + t_{i-1,i,p}^{+*})T_{ps}(i,N)\exp(-2\,\text{Im}\,\beta_i)}{1 - (r_{i-1,i,p} - r_{i-1,i,s}^{-*})R_{ps}(N)\exp(-4\,\text{Im}\,\beta_i)} \tag{13.31d}$$

It must be pointed out that Equations (13.30) are a direct consequence of Equations (13.30), which are easily obtained by considering the series developments of r and t. In the series multiplication, the averaging process eliminates all but the diagonal terms. This corresponds to the elimination of the interference terms between different beams. One can construct Equations (13.31) in the same way by simply adding a layer iteratively, and then using these iterative equations to evaluate:

$$\begin{aligned} R_{p(s)} &= R_{p(s)}(0,N) & T_{p(s)} &= T_{p(s)}(0,N) \\ R_{ps} &= R_{ps}(0,N) & T_{ps} &= T_{ps}(0,N) \end{aligned} \tag{13.32}$$

by using the starting values

$$\begin{aligned} R_{p(s)}(N-1,N) &= |r_{N-1,N,p(s)}^{+}|^2 \\ T_{p(s)}(N-1,N) &= |t_{N-1,N,p(s)}^{+}|^2 \\ R_{ps}(N-1,N) &= (r_{N-1,N,p}^{+}\, r_{N-1,N,s}^{+*}) \\ T_{ps}(N-1,N) &= (r_{N-1,N,p}^{+}\, r_{N-1,N,s}^{+*}) \end{aligned} \tag{13.33}$$

As already stated, if the ambient in and ambient out are different, the usual factor $(N_N\cos\theta_N / N_0\cos\theta_0)$ must be applied to the final results for $T_{p(s)}$ and T_{ps}.

It must be stressed that the same iterative procedure as described in Equations (13.31a) and (13.31b) can be used to compute the total reflection and transmission amplitudes in both polarizations for a stack of thin films, noting that we now have Fresnel interfaces and providing that $|x|^2$ is replaced by x, and $-2\text{Im}(\beta)$ by $i\beta$.

13.3.3 Remarks on the reflection-transmission photoellipsometry method

The above method for reflection and transmission photoellipsometry has been used successfully in the study of thin-film layers on semitransparent substrates [13.21–13.23]. Instead of defining the effective ellipsometric parameters when dealing with transparent thick layers, it is simpler and physically more appropriate to derive such quantities as R_s, R_p, R_{ps}, T_s, T_p, and T_{ps}, which are smoothly varying, have no singularities, and from which effective ellipsometric parameters could be derived [13.24]. The approach proposed permits the use of any number of semitransparent thick films (or substrates) in a multi-layer system, eliminating the need for an opaque or nonback-reflecting substrate in conventional reflection ellipsometry. The experimental method uses reflection and transmission over a wide range of incident angles. Such a range is critical in detecting any anisotropy in samples being studied. In the case of isotropic samples, a wide range of incident angles usually permits a fast and unambiguous determination of optical constants and film thicknesses.

13.4 APPLICATIONS OF THIN FILMS TO ENERGY MANAGEMENT AND RENEWABLE ENERGY TECHNOLOGIES

Applications of thin films based on their specific optical properties abound. One area of particular great interest is the use of thin films in energy management and renewable-energy technologies. Various applications include selective solar-absorbing surfaces (surfaces absorbing energy in the solar spectrum and having a low emittance in the blackbody range defined by their temperature) [13.25], heat mirrors (films having a high transmittance in the visible region and a high reflectance in the infrared) [13.26], photovoltaic films (films that convert light into electricity) [13.27], and smart window coatings (films with reversible changing optical properties for energy management) [13.28]. Within the context of this chapter and the authors' research interests, our review will concentrate on thin films used for smart window applications and some new developments in a sky radiator film and an optical functional film.

Coatings of materials that can change their optical properties upon the application of a stimulus (light, electrical voltage, or heat) are called chromogenic coatings and they can be used to control the flow of light and heat into and out of a window system in buildings, vehicles, and aircraft. The energy efficiency of such chromogenic windows with a control capacity based on occupancy, temperature, and solar radiation has been investigated, showing a great potential for energy saving and increased comfort [13.29].

13.4.1 Electrochromic thin films

The most studied thin films for smart window applications are the electrochromic films, whose optical properties change drastically upon the application of an electric field, from the transparent to a colored state. Inorganic materials exhibiting this optical behavior are metal oxides such as WO_3, MoO_3, Nb_2O_5, V_2O_5, NiO_x, and IrO_x [13.30]. It is generally accepted that the electrochromic effect is caused by a double injection of electrons and positive ions in the case of cathodic materials (WO_3, MoO_3, Nb_2O_5, V_2O_5), and by a double ejection of positive ions and electrons in the case of anodic materials (NiO_x, IrO_x).

For WO_3, the reversible coloring and bleaching process is governed by:

$$xM^+ + xe^- + WO_{3-y} \text{ (transparent)} \leftrightarrow M_xWO_{3-y} \text{ (colored)}$$

where x is normally less than 0.5 and y less than 0.03 [13.31]. When the WO_3 films are amorphous, the colored state corresponds to an absorbing state whereas it is a reflective state if the films are originally in the crystalline state. Depending on the nature of the films used, it is thus possible to modulate either the absorption or the reflectance of the films with a proper double-injection mechanism. A thin-film electrochromic device may have the following configuration:

$$\text{Glass substrate/TC/CE/IC/WO}_3\text{/TC}$$

where TC stands for 'Transparent Conductor', CE for 'Counter-Electrode', and IC for 'Ion Conductor'. The transparent conductors serve as electrodes for applying the needed electric field for performing injection of electrons and ions (originally stored in the CE layer) to color the electrochromic WO_3 film. By reversing the electric field, the colored WO_3 will switch back to its clear state, thus allowing the control of the optical transmission of the device. An all-solid-state device has been proposed by us, using amorphous WO_3, indium tin oxide (ITO) films as electrodes, V_2O_5 as counter-electrode for lithium-ion storage, and $LiBO_2$ as an ion conductor [13.32]. Such a device exhibited integrated solar and visible transmittances of 58% and 65%, respectively, in the bleached state, and corresponding values of 9% and 13% in the colored state when a 3 V voltage was applied. Many other types of smart window systems using thin films have been studied by different groups. A recent survey of electrochromic coatings and devices can be found in ref. [13.33].

13.4.2 Pure and metal-doped VO_2 thermochromic thin films

Vanadium dioxide (VO_2) thin films are well known for thermochromic phenomena found in transition metal oxides [13.34], exhibiting a phase transition upon heating at approximately 67 °C [13.35]. A substantial change in conductivity is observed as the film changes from semiconductor to metallic behavior accompanied by a change in optical properties. Using specific chemical substitutions for the vanadium cation, this transition temperature can be lowered. The substituted cations should behave as electron donors for vanadium V^{4+}. The doping effect is the origin of the electrical, magnetic, and optical modifications [13.34]. Although several studies were previously published, most of them were essentially concerned with near-infrared (NIR) and infrared (IR) transmittance potentialities. Applications include smart windows for thermal regulation of buildings (transmittance contrast) [13.36,13.37], adjustable IR for laser applications (reflectance contrast) [13.38–13.40], optical storage media [13.41], uncooled micro-bolometer [13.42], and optical switching [13.43].

The transmittance and reflectance of VO_2 and metal-doped $V_{1-x}M_xO_2$ (M: W or Mo) thin films prepared by reactive radio frequency (RF) magnetron sputtering in our laboratory [13.44–13.48] are summarized below. The fabrication apparatus is shown schematically in Figure 3.

Water-cooled metal targets (V and M: 99.9%) were used and an RF power of 200 W was applied to the V target and a fraction of it to the M target. For the reduction of the sputtering rate for metal M, the grounded cover made of the same metal element was also provided to reduce the actual target area appropriately. With this apparatus, the x values were controlled between 0 and 0.26 for the W element, and between 0 and 0.04 for the Mo element. The target was biased as needed and the substrate temperature was controlled between room temperature and 900 °C by a SiC back heater. The oxygen flow rate, defined as the ratio between the reactive O_2 gas flow rate and the total gas flow rate, and the substrate temperature were identified to be quite important parameters to obtain a good VO_2 thin film. The optimal oxygen rate and the substrate temperature were confirmed to be 2.7% and 400–500 °C, respectively. If the oxygen flow rate was deviated from the optimal value by 0.1–0.2%, mixed phases of VO_2, V_2O_7, V_6O_{13}, and V_2O_5 were grown. The oxygen-rate

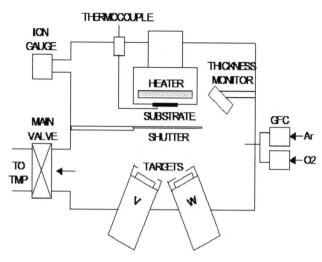

Figure 13.3 Schematic representation of a dual-target radio frequency (RF) magnetron sputtering apparatus used in preparing thermochromic VO_2 thin films doped with W. Metal mode reactive sputtering is assured after removing the surface oxides by the presputtering under the shutter application

value was valid even for lower substrate temperatures such as 250–350 °C. The deposition rate of pure VO_2 thin film was a function of oxygen flow rate. It decreased from 19 nm/min to 7 nm/min as the oxygen flow rate was varied from 0 to 2%, then drastically dropped to about 5 nm/min with an oxygen flow rate of 3.5% and reached the steady rate. The observed distribution of the horizontal diameter of the oblate VO_2 polycrystallites grown vertically in the columnar structure of the films with a thickness of about 60 nm was the mixture of 1000–1500 nm and 200–400 nm for samples fabricated at the substrate temperature of 500 °C, and 100–300 nm for samples prepared at 400 °C. The root mean square of the surface roughness of these films was smaller than 10 nm.

Figures 13.4a and 13.4b, respectively, show the temperature-dependent hemispherical spectral transmittance and/or hemispherical spectral reflectance of a polycrystalline VO_2 thin film 65 nm thick on a pyrex glass substrate. The transmittance was obtained for normal incidence and the reflectance was taken at an incident angle of 12°. Although the change of the transmittance and reflectance due to the metal-to-semiconductor transition in the visible and/or solar spectral ranges was not significant, the one observed in both NIR and IR ranges was large enough to produce an optical switching behavior. The experimental transmittance and reflectance curves were well reproduced by calculations using complex refractive indices obtained by Tazawa et al. [13.49] for both pure VO_2 and $V_{1-x}W_xO_2$ thin films on a glass substrate and the Equations given in Section 2.1 of this chapter. In Table 13.1, the numerical optical-switching behavior in the luminous and solar spectral ranges for VO_2 and $V_{0.986}W_{0.014}O_2$ thin films are summarized.

The transition temperature (τ_c) and its width ($\Delta\tau_c$) evaluated from the temperature-dependent reflectance and transmittance at a wavelength of 2000 nm under a temperature-increasing rate of +2 °C/min and for a temperature-decreasing rate of −2 °C/min, which are not shown here, are 67 °C and 12 °C, respectively. The hysteresis-loop width (Δ_{hys}) observed on the designated spectral curves is 31 °C. When τ_c, $\Delta\tau_c$, and Δ_{hys} are compared with the corresponding values obtained for epitaxially grown VO_2 thin films on a sapphire (110)

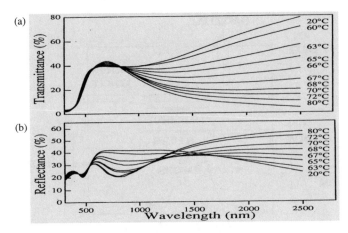

Figure 13.4 (a) spectral transmittance and (b) spectral reflectance, both as a function of an applied external temperature between 20 °C and 80 °C in NIR and IR regions

Table 13.1 Difference in solar transmittance (T_{sol}), luminous transmittance (T_{lum}), solar reflectance (R_{sol}), and luminous reflectance (R_{lum}) for VO_2 and $V_{0.986}W_{0.014}O_2$ between the metal and semiconductor phases. The values at the sample temperatures of 20 and 100 °C are represented

	Thickness (nm)	T_{sol} (%) 20 °C/100 °C	R_{sol} (%) 20 °C/100 °C	T_{lum} (%) 20 °C/100 °C	R_{lum} (%) 20 °C/100 °C
VO_2	65	36.9/30.1	35.5/26.8	37.6/35.5	32.0/28.5
$V_{0.986}W_{0.014}O_2$	80	32.7/22.7	29.5/21.9	32.7/29.5	15.1/13.3

substrate [13.47–13.48], while τ_c is found to be in agreement, the obtained $\Delta\tau_c$ and Δ_{hys} values are larger by 12 and 23 times, respectively. Consequently, the sharp phase-transition is confirmed for epitaxial single-crystalline films as expected.

The decreasing of the transition temperature τ_c that depended on the doping levels elemental of W and Mo in atomic% for polycrystalline VO_2 films is reproduced in Figure 13.5. The decreasing rates for W and Mo are identified as −23 °C/at.% and −11 °C/at.% respectively.

13.4.3 Temperature-stabilized $V_{1-x}W_xO_2$ sky radiator films

The composition of the atmosphere (H_2O, CO_2, etc.) causes a specific spectral range of emittance for the atmosphere in the spectral range 8,000–13,000 nm, termed the atmospheric window [13.50]. The atmospheric window is characterized by a low radiance and a high transmittance of the atmosphere in the mid-IR region where the peak of blackbody radiation is matched to ambient temperature. The low radiance is the origin of the nocturnal radiative cooling observed under clear sky radiation. If the humidity and the cloudiness are

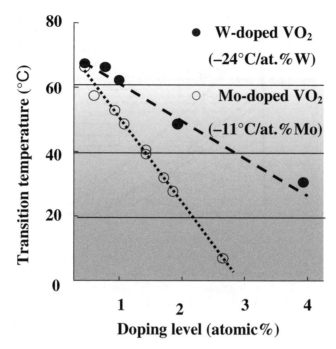

Figure 13.5 Transition temperature as a function of doping level in atomic %. W-doped samples have larger decreases in transition temperature as compared with Mo-doped samples

low, a blackbody may reach a temperature between 15 and 25 °C below ambient temperature as a result of radiative cooling [13.50].

Spectral selective radiating materials (SSRMs) can enhance the radiative cooling effect because of their high emittance in the atmospheric window and their high reflectance outside this window. Consequently, the SSRM can be used to obtain a low temperature in a sky radiator for passive cooling in buildings, with no energy consumption [13.51]. SiO films [13.50–13.52], Si-based films such as SiO_xN_y [13.52], or multi-layered films [13.53–13.56] on highly reflective substrates are regarded as the most feasible SSRMs. These Si-based films with an adequate thickness on metallic substrates usually show an emission only in the atmospheric window, so that an efficient low surface temperature (T_s) from the ambient temperature (T_a) is realized, particularly at night under clear sky conditions. In Figure 13.6, our results from computational simulations of a composite film of SiO and Si 1000 nm thick on an Al substrate [13.54, 13.55] are exemplified. In the simulation, the spectral reflectance and emittance of the multi-layered films were essentially required and calculated by following the procedures given in Section 2.2 of this chapter. The radiative cooling power depends on the compositional ratio (x) between SiO and Si, and can be increased by about 20% as compared with pure SiO films (when $x = 0.4$–0.6). This leads to the cross-point value of the power curve on the ($T_a - T_s$) axis moving to the night by 1–2 °C.

Thermochromic materials can control the energy throughout the IR and NIR parts of the solar radiation as well as the environmental radiation in response to the environmental temperature as shown in the previous section. As the metallic phase of $V_{1-x}W_xO_2$ film has a high reflectance in the IR region, a film made with this material might be a good candidate for

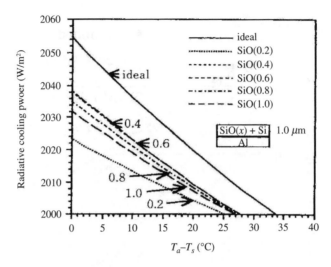

Figure 13.6   Radiative cooling power of SSRM with composite films of SiO and Si on Al substrate as a function of $T_a - T_s$. The film configuration and the SiO ratio are indicated. The ideal case is when the spectral emittance of the SSRM is unity at wavelengths in the atmospheric window, and null for other wavelengths

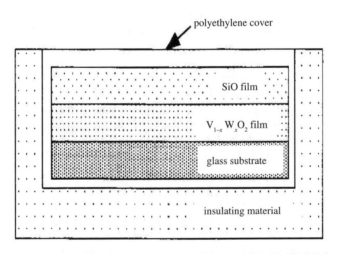

Figure 13.7   A model of a sky radiator with SSRM consisting of SiO/$V_{1-x}W_xO_2$/glass

the substrate and/or the second layer of SSRMs for particular applications beyond τ_c. Furthermore, the expected changes of its optical properties at τ_c near or below the ambient temperature T_a have the potential to add another function to the SSRM as described below.

We recently proposed a unique and interesting SSRM that consisted of an SiO film 1000 nm thick as the top layer and a $V_{1-x}W_xO_2$ film as the second layer on a black substrate for a sky radiator as shown in Figure 13.7 [13.57–13.59]. The newly designed SSRM is capable of attaining a stable surface temperature with an appropriate adjustment of the transition temperature from a metallic to a semiconductor state through the value of x in $V_{1-x}W_xO_2$.

The temperature-stabilized mechanism of this newly designed SSRM film is explained by simulation studies, as follows.

The simulation to obtain the radiative cooling power was performed using the published refractive indices for SiO [13.60] and $V_{1-x}W_xO_2$ [13.49]. Assuming $T_a = 27\,°C$, Figure 13.8 shows the calculated radiative cooling power the SSRM temperature. The solid and dotted lines stand for the low-temperature semiconductor and high-temperature metallic phases of $V_{1-x}W_xO_2$, respectively. As the optical properties of $V_{1-x}W_xO_2$ film change at the transition temperature τ_c, accordingly the radiative cooling power also changes at τ_c. Thus, the radiative cooling power at temperatures lower than τ_c is represented by a solid line and that at higher temperatures by a dotted line.

Figure 13.9 shows the change of the radiative cooling power accompanied by a transition occurring at the SSRM surface temperature $T_s = 19.5\,°C$ ($=\tau_c$) as indicated by a thin vertical dotted line. The radiative cooling power is positive (cooling potential) at temperatures higher than T_s ($=\tau_c$), and negative (heating potential) at lower temperatures. Therefore, the surface temperature of the SSRM is kept almost constant at τ_c.

Strictly speaking, it is relatively difficult to adjust in an accurate way the transition temperature (τ_c) for the $V_{1-x}W_xO_2$ film to a designated value in the usual experimental cases. This interpretation would be reasonable because of the possibility of adjusting the doping level x to the allowed range, and the small dependency of the optical constants of $V_{1-x}W_xO_2$ film on the x-value [13.57]. Such a fact was confirmed by our experiments [13.61]. One of the promising applications of the new SSRM would be in the design of containers for medicines in remote areas, where a refrigerator may not be available. If we can adjust $x > 2.5\%$, it will give $\tau_c < 7\,°C$.

13.4.4 Optical functional TiO_2 thin film for environmentally friendly technologies

Since the discovery of the photo-induced decomposition of water on TiO_2 electrodes by Fujishima and Honda [13.62] in 1971, TiO_2 has become a promising material as a photo-catalyst [13.63] whose function is to decompose and oxidize various organic and/or inorganic chemicals in waste and chemical emissions with its strong oxidation activity based on the created superoxide anion radical ($O_2^{\cdot-}$) and $OH^{\cdot}$ radical. Its high chemical stability and inexpensive cost are other benefits that favor its use as a photocatalyst as well as for photocatalytic microbiocidal effects in killing viruses, bacteria, fungi, algae, and cancer cells [13.64–13.66]. This material has the additional great potential for application to dye-sensitized photovoltaic cells [13.67, 13.68], energy-efficient windows and/or other optical coatings [13.69–13.71], and capacitors in large-scale integration (LSI) [13.72]. The wide-bandgap semiconductor TiO_2 exists in three different crystalline polymorphs: rutile, anatase, and brookite. Among them, rutile (R-titania) and anatase (A-titania) are the most common and widely used phases in applications.

The optical properties, such as complex refractive indices $N = (n - ik)$, complex dielectric constants $\varepsilon = (\varepsilon_r - i\varepsilon_i)$ for a certain range of wavelengths between the UV and the NIR, and optical bandgap values E_g, are very important criteria for the selection of films for different applications. We have recently reported the simultaneous growth of epitaxial thin films of both A-titania on $SrTiO_3$ (STO) substrates and R-titania on sapphire substrates, and the growth of polycrystalline A- and/or R-titania thin films on silicon and/or glass by helicon

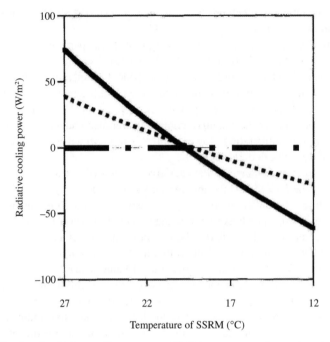

Figure 13.8   Simulated radiative cooling power vs SSRM temperature for semiconductor (solid line) and/or metallic (dotted line) phases of $V_{1-x}W_xO_2$ films in the modeled sky radiator. The horizontal chain line represents power zero

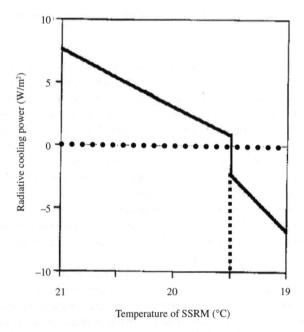

Figure 13.9   Simulated radiative cooling power vs SSRM temperature, taking account the metal–semiconductor transition for $V_{1-x}W_xO_2$ films in the modeled sky radiator

RF magnetron sputtering [13.73–13.75]. We also reported the optical properties of poly-crystalline and/or epitaxial A- and/or R-titania thin films [13.76], the optical bandgaps E_g as extrapolated by the Tauc plot using the obtained extinction coefficient [13.76], and the interpretation of the observed wider bandgap values [13.77]. The complex refractive indices from 0.75 eV to 5 eV (1653–248 nm wavelength range) were obtained by spectroscopic ellipsometry (SE) (Horiba, Jobin-Yvon, UVISE) employing the F&B formula referred to in ref. [13.78], as corrected by Jellison et al. [13.79], for the optical function. In the present SE, only the reflection mode is used under a basic mathematical structure slightly modified from that described in Sections 13.3.1 and 13.3.2 of this chapter. Important results are reproduced in Table 13.2.

The typical refractive index n at a wavelength 500 nm, the maximum value of this index in the wavelength region 400–800 nm, and the maximum value in the wavelength region 248–1653 nm are given in Table 13.2. The values higher than the cited ones in recent articles for thin films [13.80–13.82] and bulk materials [13.82–13.84] reveal the good crystal quality of the films produced.

The real and imaginary parts of the complex dielectric function of rutile are shown in Figures 13.10a and 13.10b, respectively, and they are compared with Jellison et al.'s recent results for bulk materials in both o and e polarization states [13.85]. It is to be noted that Jellison et al. have used an optical function similar to ours in SE to eliminate the dependency on the different optical functions used. Below the band edge (<3.5 eV), all curves are found to be similar. Likewise, above the band edge (>3.5 eV), ε_r of epitaxial films exhibits a sharp peak at 3.85 eV, which agrees with that at 3.9 eV found by Jellison et al. for bulk materials in both o and e states. However, the polycrystalline film shows one broad peak at 3.75 eV and a shoulder at 4.3 eV, which are not present in other curves, and these are due to the applied two-term model in the F&B formula. This feature is clearly observed in ε_i where sharp peaks are found at 4.1 eV (Epi-rutile), 4.3 eV (Jellison, bulk rutile, e state), and 4.1 eV (Jellison, bulk rutile, o state), and a shoulder and/or a broad peak at 4.0 and 4.5 eV, respectively, for the polycrystalline film. The shape of the ε_i curve directly relates to the combined density-of-state (DOS) for the interband transition. Structures observed at $E > 3.5$ eV in epitaxial films represent well the DOS, which has a sharp absorptive transition at around 4.0 eV as theoretically predicted by Shelling et al. [13.86].

A similar comparison for the anatase case is made in Figure 13.11a for ε_r and in Figure 13.11b for ε_i. Here we compare our results directly with those of anatase thin films studied by Jellison et al. [13.85]. Two critical points in ε_r are observed as a sharp peak at 3.75 eV and a rather broad one at 4.4 eV for an epitaxial film, and a broad peak at 3.8 eV and a shoulder at 4.42 eV for a polycrystalline film. Those peaks are again due to the applied two-term model in the F&B formula. The values for the epitaxial film agree well with those found by Jellison et al. denoted as the Jell-anatase film. For ε_i, only a single broad peak at 4.0 eV and a knick feature at 4.8 eV are observed with our epitaxial film, while a shoulder at 4.0 eV and a broad peak at 4.65 eV are observed with a polycrystalline film. Jellison et al.'s thin film results show a behavior similar to that of our epitaxial films. The position and the strength of the peaks of ε_i for our thin films in the energy range $E > 3.0$ eV agree satisfactorily with those given by theoretical calculations by Asahi et al. [13.87], and they correspond to absorptive transitions from valence bands to d_{xy} crystal-field split levels of the t_{2g} orbital of Ti 3d atomic level in the conduction band.

The optical bandgap E_g of the polycrystalline and epitaxial TiO_2 films is determined using the extinction coefficient k from the Tauc expression [13.88], which is expressed as $[\alpha \hbar \omega]^{1/2}$

Table 13.2   Refractive index n of rutile and anatase TiO$_2$ at the designated wavelength

	Rutile		Anatase	
	Polycrystalline	Epitaxial	Polycrystalline	Epitaxial
n at 500 nm	2.85	2.84	2.66	2.64
Maximum n at 400–800 nm	3.18 (400)	3.17 (400)	2.90 (400)	2.89 (400)
Maximum n at 250–1600 nm	3.95 (330)	4.14 (320)	3.61 (320)	3.79 (330)

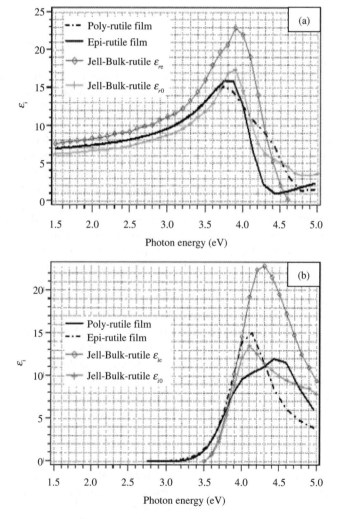

Figure 13.10   Real part (a) and imaginary part (b) of the complex dielectric function of rutile TiO$_2$ as compared with Jellison et al.'s bulk results (Jell–Bulk rutile) for both o (suffux o in ε_r and ε_i) and e polarization states (suffix e in ε_r and ε_i)

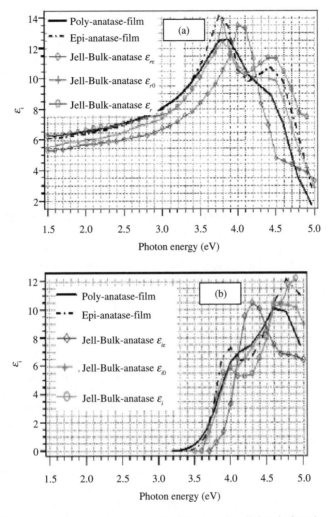

Figure 13.11 Real part (a) and imaginary part (b) of the complex dielectric function of anatase TiO_2 as compared with Jellison et al.'s bulk results (Jell–Bulk anatase) for both o and e polarization states and their thin-film results (Jell anatase film)

$= B^{1/2} (\hbar\omega - E_g)$ where α is the absorption coefficient ($= 2\kappa/\lambda$) and $\hbar\omega$ is the absorbed photon energy under the *ad hoc* assumption that the indirect allowed transition [13.89] is predominant as the optical-transition mode. The parameters used are: E, photon energy ($h\nu$); B, constant; $h\nu$, photon energy; ($4\pi\kappa/\lambda$), absorption coefficient at wavelength λ; and κ, extinction coefficient. The bandgap value E_g is given as the intercept of the photon energy axis with the asymptotic linear line of the right hand of the above equation assumed for a certain range of E. The results are summarized in Table 13.3. It is worthwhile to note that the optical bandgap obtained by a Tauc plot always differs from that parameterized in the F&B formula. All values are larger than those for bulk materials [13.83–13.84] by about 10%.

Table 13.3 Bandgap of anatase and rutile TiO_2 films extrapolated by Tauc's plot in comparison to the bulk material

Structure	Polycrystalline rutile film	Polycrystalline anatase film	Epitaxial rutile film	Epitaxial anatase film	Bulk rutile	Bulk anatase
Band gap E_g (eV)	3.34	3.39	3.37	3.51	3.03	3.20

It is well known that the optical properties of intrinsic semiconductors can be changed significantly by the application of an external strain to the nondegenerate state [13.90]. In order to understand the increase of the bandgap value observed in the currently strained TiO_2 epitaxial films, an *ab initio* simulation based on density functional theory (DFT) to obtain the DOS and the energy of the electronic orbital that depends on lattice strains was performed by our group [13.77]. Consequently, the theoretical correlation between lattice constants of the thin films and the optical bandgap can predict satisfactorily the increase of E_g with respect to bulk values both for epitaxial and polycrystalline anatase and/or rutile films as listed in Table 13.3.

13.5 CONCLUSIONS

Basic notions on the optical properties of thin films are given in this chapter, introducing the usual tools for treating single layers as well as multiple-layer structures. A section is devoted to presenting a practical method for reflection-transmission photoellipsometry that permits the analysis of samples consisting of thin films combined with semitransparent thick layers or substrates in the form of multi-layer structures. This method has the advantage of being nondestructive and can be used, for example, in studying actual samples of energy-efficient coatings for windows, monitoring useful phenomena such as aging effects. We then illustrate the use of optical thin films in a few applications in energy management and renewable-energy technologies. These applications have the potential to be of much interest in the present context when concerns for a greener environment are raised by the ratification by many countries of the Kyoto Protocol on global warming.

REFERENCES

[13.1] O.S. Heavens, *Optical Properties of Thin Solid Films* (Butterworths Scientific Publications, London, 1955; Dover, Toronto, 1991).

[13.2] M. Born and E. Wolf, *Principles of Optics* (Pergamon, Oxford, 1970).

[13.3] R.M.A. Azzam and N.M. Bashara, *Ellipsometry and Polarized Light* (North Holland Publishers, Amsterdam, 1977, and 1984).

[13.4] *Handbook of Optical Constants of Solids*, edited by E.D. Palik (Academic Press, Orlando, 1985).

[13.5] M.V. Klein and T.E. Furtak, *Optics* (John Wiley & Sons, Inc., New York, 1986).

[13.6] L. Ward, *The Optical Constants of Bulk Materials and Films* (Adams Hilger, Bristol and Philadelphia, 1988).

[13.7] *Handbook of Optics*, edited by M. Bass (McGraw-Hill, New York, 1995), Vol 1.

[13.8] F. Abeles, *Ann. Phys.*, **5**, 595 (1950).

[13.9] P.C.S. Hayfield and G.W.T. White, in *Ellipsometry in the Measurement of Surfaces and Thin Films*, edited by E. Passaglia, R.R. Stromberg, and J. Kruger (National *Bureau of* Standards Miscellaneous Publication 256, U.S. Govt. Printing Office, Washington DC, 1964).

[13.10] M. Schubert, B. Rheinlander, B. Johs, and J.A. Woollam, *Proc. SPIE-Int. Soc. Opt. Eng.*, **3094**, 255 (1997).

[13.11] M.I. Alonso and M. Garriga, *Thin Solid Films*, **455–456**, 124 (2004).

[13.12] U. Zhokhavets, R. Goldhahn, G. Gobsch, M. Al-Ibrahim, H.-K. Roth, S. Sensfuss, E. Klemm, and D.A.M. Egbe, *Thin Solid Films*, **444**, 215 (2003).

[13.13] M. Losurdo, *Thin Solid Films*, **455–456**, 301 (2004).

[13.14] N.J. Podraza, Chi Chen, Ilsin An, G.M. Ferreira, P.I. Rovira, R. Messier, and R.W. Collins, *Thin Solid Films*, **455–456**, 571 (2004).

[13.15] Vo-Van Truong, R. Belley, G. Bader, and Al. Hache, *Thin Solid Films*, **212–213**, 140 (2003).

[13.16] I.J. Hodgkinson, F. Horowitz, H.A. Macleod, M. Sikkens, and J.J. Wharton, *J. Opt. Soc. Am. A*, **2**, 1693 (1985).

[13.17] R.M.A. Azzam, *Selected Papers on Ellipsometry, SPIE Milestone Series*, edited by (SPIE, Bellingham, WA, 1991), Vol. MS27.

[13.18] H.G. Tompkins and W.A. McGahan, *Spectroscopic Ellipsometry and Reflectometry, A User's Guide* (John Wiley & Sons, Inc., New York, 1999).

[13.19] J.B. Theeten and D.E. Aspnes, *Annu. Rev. Mater. Sci.*, **11**, 97 (1981).

[13.20] J. Rivory, in *Thin Films for Optical Systems, Optical Engineering Series No. 49*, edited by F. Flory (Marcel Dekker, New York, 1995).

[13.21] G. Bader, P.V. Ashrit, F.E. Girouard, and Vo-Van Truong, *Appl. Opt.*, **34**, 1684 (1995).

[13.22] P.V. Ashrit, G. Bader, S. Badilescu, F.E. Girouard, L.Q. Nguyen, and Vo-Van Truong, *J. Appl. Phys.*, **74**, 602 (1993).

[13.23] G. Bader, P.V. Ashrit, and Vo-Van Truong, *Proc. SPIE-Int. Soc. Opt. Eng.*, **2531**, 70 (1995).

[13.24] J.-Th. Zettler and L. Schrottke, *Phys. Status Solidi B*, **163**, K69 (1991).

[13.25] A. Othonos, M. Nestoros, D. Palmerio, C. Christofides, R.S. Bes, and J.P. Traverse, *Sol. Energy Mater. Sol. Cells*, **51**, 171 (1998); R. Joerger, R. Gampp, A. Heinzel, W. Graf, M. Kohl, P. Gantenbein, and P. Oelhafen, *Sol. Energy Mater. Sol. Cells*, **54**, 351 (1998).

[13.26] M. Tazawa, M. Okada, K. Yoshimura, and S. Ikezawa, *Sol. Energy Mater. Sol. Cells*, **84**, 159 (2004); T. Naganuma and Y. Kagawa, *Acta Mater.*, **52**, 5645 (2004).

[13.27] A. Jager-Waldau, *Sol. Energy*, **77**, 667 (2004); J. Nelson, *Curr. Opin. Sol. Stat. Mater. Sci.*, **6**, 87 (2002).

[13.28] *Large Area Chromogenics: Materials and Devices for Transmittance Control*, edited by C.M. Lampert and C.G. Granqvist (SPIE Institute Series, Bellingham, 1990), Vol. IS4.

[13.29] See, for example, J. Karlsson, *Control System and Energy Saving Potential for Switchable Windows* (Proceedings of the 7th International IBPSA Conference, Rio de Janeiro Brazil, 2001); A. Azens and C.G. Granqvist, *J. Solid State Electrochem.*, **7**, 64 (2003).

[13.30] C.G. Granqvist, *Handbook of Inorganic Electrochromic Materials* (Elsevier, Amsterdam, 1995).

[13.31] J.S.E.M. Svensson and C.G. Granqvist, *Appl. Phys. Lett.*, **45**, 828 (1984).

[13.32] P.V. Ashrit, K. Benaissa, G. Bader, F.E. Girouard, and Vo-Van Truong, *Proc. SPIE-Int. Soc. Opt. Eng.*, **1728**, 232 (1992).

[13.33] C.G. Granqvist, E. Avendano, and A. Azens, *Thin Solid Films*, **442**, 201 (2003).

[13.34] (a) D. Addler, *Insulating and Metallic State in Transition Metal Oxides*, in *Solid State Physics* (Academic Press, New York, 1968) Vol. 21, pp. 1–132; (b) D. Addler *Rev. Mod. Phys.*, **40**, 714 (1968).

[13.35] J.B. Goodenough, *J. Solid State Chem.*, **3**, 490 (1971).

[13.36] C.G. Granqvist, *Thin Solid Films*, **193/194**, 730 (1990).

[13.37] G. Garry, O. Durand, and A. Lordereau, *Thin Solid Films*, **453**, 427 (2004).

[13.38] O.P. Konovalova, A.I. Sidrof, and I.I. Shgnanov, *J. Opt. Technol.*, **62**, 41 (1995).

[13.39] A.I. Sidrov and E.N. Sosnov, *Proc. SPIE-Int. Soc. Opt. Eng.*, **3611**, 323 (1999).

[13.40] O.P. Mikheeva and A.I. Sidnov, *J. Opt. Technol.*, **68**, 278 (2001).

[13.41] V.L. GalPrin, I.A. Khakhaev, F.A. Chudovskii, and E.B. Shadrin, *Proc. SPIE-Int. Soc. Opt. Eng.*, **2969**, 270 (1996).

[13.42] C. Chen, X. Yi, X. Zhao, and B. Xiong, *Sens. Actuators. A: Phys.*, **90**, 212 (2001).

[13.43] M.F. Becker, A.B. Buckman, R.M. Walser, T. Iepine, P. Georges, and A. Burn, *Appl. Phys. Lett.*, **65**, 1507 (1994).

[13.44] P. Jin and S. Tanemura, *Jpn. J. Appl. Phys.*, **33**, 1478 (1994).

[13.45] P. Jin and S. Tanemura, *Jpn. J. Appl. Phys.*, **34**, 2459 (1995).

[13.46] P. Jin and S. Tanemura, *Thin Solid Films*, **281/282**, 239 (1996).

[13.47] P. Jin and S. Tanemura, *J. Vac. Sci. Technol., A*, **15**, 113 (1997).

[13.48] P. Jin and S. Tanemura, *J. Vac. Sci. Technol., A*, **17**, 1817 (1999).

[13.49] M. Tazawa, P. Jin, and S. Tanemura, *Appl. Opt.*, **37**, 1858 (1998).

[13.50] C.G. Granqvist and T.S. Erikson, *Materials for Radiative Cooling to Low Temperature*, in *Material Science for Solar Energy Conversion Systems*, edited by C.G. Granqvist (Pergamon Press, Oxford, 1991), pp. 168–203.

[13.51] M. Martin, *Radiative Cooling*, in *Passive Cooling*, edited by J. Cook (MIT Press, Cambridge, MA, 1989), pp. 138–196.

[13.52] C.G. Granqvist and A. Hjortsberg, *J. Appl. Phys.*, **52**, 4205 (1981).

[13.53] T.S. Eriksson, S.J. Jiang, and C.G. Granqvist, *Sol. Energy Mater.*, **12**, 319 (1985).

[13.54] M. Tazawa, K. Yoshimura, T. Miki, and S. Tanemura, *A Computational Design of SiO Based Selectively Radiative Film*, Proceedings of the ISES Congress, Budapest 2, (1993), pp. 333.

[13.55] M. Tazawa, P. Jin, Y. Tai, T. Miki, K. Yoshimura, and S. Tanemura, *Proc. SPIE-Int. Soc. Opt. Eng.*, **2255**, 149 (1994).

[13.56] M. Tazawa, T. Miki, P. Jin, and S. Tanemura, *Trans. Mater. Res. Soc. Jpn.*, **18A**, 525 (1994).

[13.57] M. Tazawa, P. Jin, and S. Tanemura, *Proc. SPIE-Int. Soc. Opt. Eng.*, **2531**, 326 (1995).

[13.58] M. Tazawa, P. Jin, and S. Tanemura, *Thin Solid Films*, **281/282**, 232 (1996).

[13.59] M. Tazawa, P. Jin, K. Yoshimura, T. Miki, S. Tanemura, *Sol. Energy*, **64**, 3 (1998).

[13.60] H.R. Philipp, *Silicon Monoxide*, in *Handbook of Optical Constants of Solids*, edited by E.D. Palik (Academic Press, New York, 1985), p. 765.

[13.61] M. Tazawa, P. Jin, T. Miki, K. Yoshimura, K. Igarashi, and S. Tanemura, *Thin Solid Films*, **375**, 100 (2000).

[13.62] A. Fujishima and K. Honda, *Bull. Chem. Soc. Jpn.*, **44**, 1148 (1971).

[13.63] A. Fujishima and K. Honda, *Nature (London)*, **238**, 37 (1972).

[13.64] T. Matunaga, R. Tomoda, T. Nakajima, and H. Wake, *FEMS Microbiol. Lett.*, **29**, 211 (1985).

[13.65] Z. Huang, P.C. Maness, D.M. Blake, E.J. Wolfrum, S.L. Smolinski, and W.A. Jacoby, *J. Photochem. Photobiol. Chem.*, **130**, 163 (2000).

[13.66] See, for example, L. Miao, S. Tanemura, Y. Kondo, M. Iwata, S. Toh, and K. Kaneko, *Appl. Surf. Sci.*, **238**, 125 (2004).

[13.67] B.O. Regan and M. Gratzel, *Nature (London)*, **353**, 737 (1991).

[13.68] A. Hagfelt and M. Gratzel, *Chem. Rev.*, **95**, 49 (1995).

[13.69] C.G. Granqvist, *Energy Efficient Windows: Present and Forthcoming Technology*, in *Materials Science for Solar Energy Conversion Systems*, edited by C.G. Granqvist (Pergamon Press, Oxford, 1991), p. 106.

[13.70] P. Jin, L. Miao, S. Tanemura, G. Xu, M. Tazawa, and K. Yoshimura, *Appl. Surf. Sci.*, **212–213**, 775 (2003).

[13.71] G.S. Brady, *Materials Handbook*, 10th Edition (McGraw-Hill, NewYork, 1971).

[13.72] Y.H. Lee, *Vacuum*, **51**, 503 (1998).

[13.73] L. Miao, P. Jin, K. Kaneko, and S. Tanemura, *Sputter Deposition of Polycrystalline and Epi-taxial TiO$_2$ Films with Anatase and Rutile Structures*, Proceedings of the 8th IUMRS International Conference on Electronic Materials: Advanced Nanomaterials and Nanodevices, edited by H.J. Gao, H. Fuchs, and D.M. Chen, (Institute of Physics, Bristol and Philadelphia, 2002), p. 943.

[13.74] L. Miao, S. Tanemura, P. Jin, K. Kaneko, A. Terai, and N. Nabatova-Gabain, *J. Crystal Growth*, **254**, 100 (2003).

[13.75] L. Miao, P. Jin, K. Kaneko, A. Terai, N. Nabatova-Gabain, and S. Tanemura, *Appl. Surf. Sci.*, **212–213**, 255 (2003).

[13.76] S. Tanemura, L. Miao, P. Jin, K. Kaneko, A. Terai, and N. Nabatova-Gabain, *Appl. Surf. Sci.*, **212–213**, 654 (2003).

[13.77] W. Wunderlich, L. Miao, S. Tanemura, M. Tanemura, P. Jin, K. Kaneko, A. Terai, N. Nabatova-Gabin, and R. Belkada, *Int. J. Nanosci.*, **3**, 439 (2004).

[13.78] A.R. Forouhi and I. Bloomer, *Phys. Rev. B*, **38**, 1865 (1988).

[13.79] G.E. Jellison, Jr. and F.A. Modie, *Appl. Phys. Lett.*, **69**, 371 (1996).

[13.80] C.C. Ting and S.Y. Chen, *J. Appl. Phys.*, **88**, 4628 (2000).

[13.81] M.H. Suhail, G. Mohan Rao, and S. Mohan, *J. Appl. Phys.*, **71**, 1421 (1992).

[13.82] T.M.R. Viseu, B. Almeida, M. Stchakovsky, B. Drevillon, M.I.C. Ferreira, and J.B. Sousa, *Thin Solid Films*, **401**, 216 (2001).

[13.83] D.C. Cronemeyer, *Phys. Rev.*, **87**, 876 (1952).

[13.84] H. Tang, F. Levy, H. Berger, and P.E. Schmid, *Phys. Rev. B*, **52**, 7771 (1995).

[13.85] G.E. Jellison, Jr, L.A. Boatner, J.D. Budai, B.S. Jeong, and D.P. Norton, *J. Appl. Phys.*, **93**, 9537 (2003).

[13.86] P.K. Shelling, N. Yu, and J.W. Halley, *Phys. Rev. B*, **58**, 1279 (1998).

[13.87] R. Asahi, Y. Taga, W. Mannstadt, and A.J. Freeman, *Phys. Rev. B*, **61**, 7459 (2000).

[13.88] A.R. Frouhi and I. Bloomer, *Calculation of Optical Constants, n and k, in the Interband Region*, in *Optical Properties of Solids II*, edited by E.D. Palik (Academic Press, New York, 1991), p. 163.

[13.89] N. Daude, C. Gout, and C. Jouanin, *Phys. Rev. B*, **15**, 3229 (1977).

[13.90] See, for example, T. Hamaguchi, *Change of Energy Band Structure by Crystal Strain*, in *Semiconductor Physics* (Asakura Publishers, Tokyo, 2001), p. 101.

14 Negative Index of Refraction: Optics and Metamaterials

J.E. Kielbasa, D.L. Carroll, and R.T. Williams

Department of Physics Center for Nanotechnology and Molecular Materials, Wake Forest University, Winston-Salem, NC 27109 USA

14.1 INTRODUCTION

Some of the astonishing optical possibilities of a material with negative refractive index were discussed by Veselago 30 years before any such medium existed [14.1]. Negative n does not occur naturally. Partly because of the lack of materials for experimentation, the field remained fairly dormant with only about 3 publications between 1968 and May 2000. In 2000, Smith et al. [14.2] constructed a metamaterial of macroscopic imbedded elements producing a negative index in the 6 cm (5 GHz) microwave range. Furthermore, Pendry [14.3] discussed in October 2000 the possibility of fabricating a 'perfect lens' based on negative refractive index, by which super-resolution beyond the transverse spatial frequency limit of propagating rays might be achieved. The experimental breakthrough achieved by the metamaterials approach, the super-resolution possibility, and the chance for nanoscale

Optical Properties of Condensed Matter and Applications Edited by J. Singh
© 2006 John Wiley & Sons, Ltd

engineering of metamaterials to open up their infrared (IR) and visible spectra ignited a burst of activity which is now ongoing with more than 570 citations of Pendry's paper through 2005. In this chapter, starting from the electromagnetic response of materials, the optical properties of negative-refractive-index materials are discussed. Consequent to the sign reversal in Snell's law, a flat slab acts as a lens. Not only does it focus propagating waves, it also amplifies evanescent waves carrying transverse spatial frequencies beyond the diffraction limit. Some of the current and prospective methods for building such a material are presented with a focus on achieving negative refracting index in the visible range.

14.1.1 Electric and magnetic response

The refractive index is defined as the ratio of light velocities in vacuum and in a polariz-able medium given in the electromagnetic wave equation in terms of ε, μ, and ε_o, μ_o:

$$n^2 = \frac{c^2}{v^2} = \frac{\varepsilon\mu}{\varepsilon_0\mu_0}$$ (14.1)

in SI units. We will henceforth use the convention of expressing relative permittivity and permeability by $\varepsilon = \varepsilon/\varepsilon_o$ and $\mu = \mu/\mu_o$. Then Equation (14.1) gives:

$$n = \pm\sqrt{\varepsilon\mu}$$ (14.2)

One is accustomed to taking the positive square root when dealing with conventional optical materials. Veselago [14.1] was the first to analyse when it is appropriate to take the nega-tive square root in Equation (14.2).

The electric permittivity ε and magnetic permeability μ are, in general, complex func-tions of frequency (see, e.g., Chapter 1), and hence so is n. The imaginary part of complex $n(\omega)$ is proportional to the optical loss for propagating waves at frequency ω. To the degree that a material is transparent in some spectral range, the corresponding $n(\omega)$ is real. If ε and μ are also real in that range, real n implies ε and μ are either both positive or both nega-tive. If only one is negative, n would be imaginary and the fields would be exponentially damped rather than propagating. An example of the latter is the model of a nonmagnetic plasma with negligible collision losses,

$$n^2 = \varepsilon\mu = 1 - \frac{\omega_p^2}{\omega^2}$$ (14.3)

for frequencies below the plasma frequency ω_p.

Since we are about to examine consequences of positive or negative real n, we focus our attention at present on real ε and real μ, such as are encountered (approximately) in trans-parent, nonmagnetic optical materials. In nonmagnetic materials, $\mu \approx 1$ throughout the spec-trum, and $\varepsilon > 1$ on the low-frequency side of the nearest resonance yet outside the absorption band, i.e., the useful transparency range. Both ε and μ are real and positive in such con-ventional optical materials.

Maxwell's (curl) equations with harmonic plane-wave solutions proportional to $e^{i(\bar{k}\cdot\bar{r}-\omega t)}$ dictate the following consequences for **k** (using our convention for relative ε and μ):

$$\bar{\nabla}\times\mathbf{E} = -\frac{\partial\mathbf{B}}{\partial t} \Rightarrow \mathbf{k}\times\mathbf{E} = \omega\mu\mu_0\mathbf{H} \tag{14.4}$$

and

$$\bar{\nabla}\times\mathbf{H} = -\frac{\partial\mathbf{D}}{\partial t} \Rightarrow \mathbf{k}\times\mathbf{H} = -\omega\varepsilon\varepsilon_0\mathbf{E} \tag{14.5}$$

With ε and μ positive, **k**, **E**, and **H** thus form a right-handed set:

$$\mathbf{k} = \frac{n\omega}{c}\frac{\mathbf{E}\times\mathbf{H}}{|\mathbf{E}||\mathbf{H}|} = \frac{n\omega}{c}\hat{\mathbf{S}} \tag{14.6}$$

where $\hat{\mathbf{S}}$ is a unit vector in the direction of the Poynting vector. The choice of the positive square root in Equation (14.2) is seen to be consistent with **k** being in the direction of the Poynting vector:

$$\mathbf{S} = \mathbf{E}\times\mathbf{H} \tag{14.7}$$

which is itself independent of the sign of ε or μ.

Veselago pointed out that if ε and μ are both negative, Equations (14.4) and (14.5) then dictate that **k**, **E**, and **H** form a left-handed set, or equivalently −**k**, **E**, and **H** are a right-handed set [14.1]. The sign reversal of **k** relative to **S** amounts to a negative n in Equation (14.6), i.e., taking the negative square root in Equation (14.2). The antiparallelism of **k** and **S** is the central aspect of negative refractive index for many applications. To summarize, simultaneous negative real values of $\varepsilon(\omega)$ and $\mu(\omega)$ specify through Maxwell's equations for harmonic plane wave solutions that **k** is antiparallel to the direction of energy transport, which is equivalent to choosing the negative square root in Equation (14.2).

But how does one obtain simultaneous negative real values of ε and μ at a given ω? One may use the fact that ε and μ dip to (possibly[1]) negative values on the high-frequency sides of electric and magnetic resonances, respectively. This is a general consequence of the fact that a driven undamped oscillator responds 180° out of phase with the driving force above resonance, and a damped oscillator tends toward a 180° response above resonance. Figure 14.1 illustrates the desired goal of matching electric and magnetic resonance frequencies to achieve a spectral range on the high-frequency side where both ε and μ are negative and approximately real. Unfortunately a matching of resonances sufficiently strong to produce negative response has not been found in naturally occurring materials. The main reason is that typical atomic and molecular electric dipole resonances occur at terahertz and higher frequencies (taking dipole-active longitudinal optical (LO) phonons as one of the lowest frequency responses in solids); while magnetic dipole resonances are typically tens of giga-hertz and lower. Certainly for visible or near-IR use, the magnetic resonances of bound electrons are far too low in frequency. The main requirement reduces to creating magnetically

---

[1]Depending on the amount of damping and accumulation of the residual real tails of higher resonances.

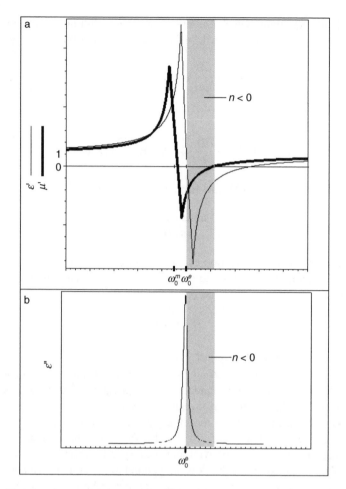

Figure 14.1   The electric permittivity $\varepsilon = \varepsilon' + i\varepsilon''$ and magnetic permeability $\mu = \mu' + i\mu''$ plotted as a function of the frequency ω. ω_0^e and ω_0^m denote the resonance frequencies of the electric permittivity and magnetic permeability, respectively. (a) The real part of each term goes out of phase with the driving force above the resonance, resulting in a negative response. The goal of fabricating metamaterials is to overlap the electric and magnetic resonances so that ε and μ are simultaneously negative in a range of ω (shaded), resulting in negative n. (b) The imaginary part of ε (shown here) has a peak at ω_0^e, causing strong absorption. The usable spectral range of negative-index transmissive materials would be the part of the shaded region where ε'' and μ'' are small

driven oscillators with resonances at higher frequencies. This can be addressed by fabricating an assemblage of synthetic oscillators, each of which is much smaller than a wavelength of the target spectrum, but not necessarily atomic or molecular in size. If the spatially averaged response of the assemblage can be represented by a function μ or ε, as appropriate, one calls it a metamaterial. This is the route by which the negative refractive index was first demonstrated (in the microwave range) [14.2], and it is also a main route by which the continuing work in this field seeks to push negative refractive response into the IR/visible region. These topics will be discussed in Section 14.4.

First we pause to discuss some of the main points of imaging with negative-refractive-index materials (NRIMs),[2] including the role of nonpropagating high-spatial-frequency components giving rise to super-resolution [14.3]. This will include the extension by Pendry to systems that, although not having negative n, can convey and amplify the super-resolution spatial frequencies and do exist naturally for experimentation, i.e., thin metal films.

14.1.2 Veselago's slab lens and Pendry's perfect lens

Veselago showed that Snell's law in NRIM implies refraction to the opposite side of the normal from that occurring in positive-index media, i.e., $\theta_t^- = -\theta_t^+$. This is illustrated in Figure 14.2. Although this result follows simply from reversing the sign of n_2 in $n_1 \sin \theta_i = n_2 \sin \theta_t$, it is worthwhile to look at the derivation of Snell's law for confirmation of when θ_t should be measured to one side of the normal or the other. The phases of the incident, reflected, and transmitted waves must be equal at the interface, which means $\mathbf{k}_i \cdot \mathbf{x} = \mathbf{k}_r \cdot \mathbf{x} = \mathbf{k}_t \cdot \mathbf{x}$. For positive n_1 and n_2, this leads to Snell's law in the usual way and identifies the direction in which positive θ_t should be measured relative to the normal. When n_1 is positive but n_2 is negative, the consequent reversal of $\mathbf{k}_t$ with respect to the energy-propagation direction $\hat{\mathbf{S}}$ in Equation (14.6) now requires, via $\mathbf{k}_i \cdot \mathbf{x} = \mathbf{k}_t \cdot \mathbf{x}$, that $\theta_t^- = -\theta_t^+$, where θ_t is the angle of transmission measured from the normal to the interface between $n(+)$ and $n(-)$. This may be simplest to see when one considers that $\mathbf{k} \cdot \mathbf{x}$ is the component along the positive $\mathbf{x}$ axis in the interface plane.

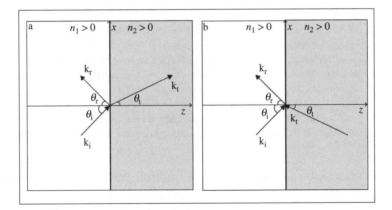

Figure 14.2 Illustration of waves incident from a medium with $n_1 > 0$ on a dielectric being refracted according to Snell's law. (a) The familiar refraction occurs for positive n_2. (b) If n_2 is negative, the wave-vector is reversed and continuity of phase $(\mathbf{k} \cdot \mathbf{r} - \omega t)$ across the interface requires that the electromagnetic field be refracted in the opposite direction

---

[2]Other names found in the literature are: left-handed media (LHM), negative-refraction material (NRM), negative index of refraction (NIR) material (NIRM), double negative (DNG) material (DNM), backward wave (BW) medium, and negative phase-velocity medium. LHM was used in Veselago's initial paper and is perhaps the most fundamentally descriptive as discussed around Equations (14.4)–(14.6). It has enjoyed fairly wide use, but risks confusion with issues of chirality and optical activity. NRIM seems unambiguous and contains all aspects of the perfect lens (see Section 14.1.2).

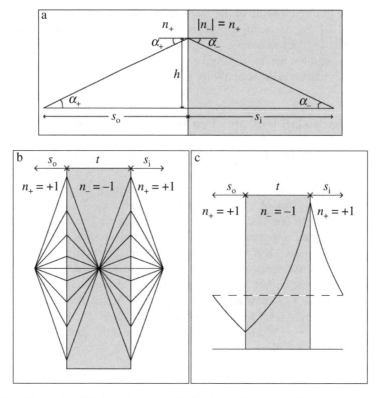

Figure 14.3 The basis of flat slab imaging. (a) At an interface separating n_+ and n_-, all rays will bend toward an axis connecting object and image points. If $|n_+| = |n_-|$ the angle of refraction α_- is equal to the angle of incidence α_+. As the triangles formed share a common side h and a common angle $\alpha_+ = \alpha_-$ the image distance s_i and object distance s_o must be equal. (b) This is true for all rays from the same object point. A second interface, as in the flat slab, will likewise form an image of the intermediate focus outside the lens as shown. (c) The evanescent waves are amplified in NRIMs. In the case of the perfect lens, the phases of the propagating waves and the amplitudes of the evanescent waves acquire values at the image point that exactly correspond to their values at the object point

Reasoning from Snell's law with negative n, Veselago showed that a flat-sided slab of NRIM will focus propagating waves from an object point to a corresponding image point [14.1]. The principle of the slab lens is illustrated in Figure 14.3. Any ray incident on an NRIM slab is refracted back toward the optical axis. If $|n_+| = |n_-|$ (such as $|n_+| = |n_-| = 1$, as specified by Veselago) Snell's law dictates $\alpha_+ = \alpha_-$ in Figure 14.3a, so the two triangles are similar with a common side h. This is true for all rays, regardless of their intersection height h with the interface. By construction all rays from an image point will intersect an optical axis[3] at the same distance $s_i = s_o$. This constitutes point-to-point imaging. The image is real and upright. The uprightness of the real image, the lack of a unique optical axis, and

---

[3]Imaging will occur likewise for any point in the object plane (as long as the thickness of the lens is greater than the object distance, see Section 14.2.2). In that sense there is no unique optical axis and there are no off-axis aberrations.

convenient lack of monochromatic aberration are among the distinctive features of the slab lens.

The NRIM slab lens, comprising two interfaces (Figure 14.3b), always includes formation of an intermediate image inside the slab by the first interface. This is re-imaged by the exit interface at the final image plane in the surrounding medium. The thickness of the slab must be equal to $s_i + s_o$, as evident from Figure 14.3b. If the equality of $|n_+|$ and $|n_-|$ is relaxed, aberrations will occur. An NRIM flat slab cannot focus collimated radiation, and in that sense does not have its focal length defined in the traditional way. Details of the slab lens and its aberrations for nonideal parameters will be discussed in Sections 14.2.2 and 14.2.3.

The predictions of slab lens properties given in 1968 were fascinating, but still dealt only with propagating waves and thus did not yet promise image resolution better than the diffraction limit. Thirty-two years after Veselago's paper, Pendry published a paper entitled *Negative Refraction Makes a Perfect Lens* [14.3], igniting a flurry of new inquiry. The evanescent waves in the near-field of an object contain the finest detail, i.e., transverse spatial wave-vectors larger than $n\omega/c$. Whereas such high spatial frequencies are nonpropagating and decay exponentially in a positive index, Pendry showed that they should grow exponentially in a slab of NRIM. Upon exiting into an ambient positive-index medium, they begin to decay exponentially again, but within the near-field space on the exit side they reconstruct a very sharp edge-defined image with spatial definition in excess of the diffraction limit for propagating rays. In principle, there is no limit to the spatial definition until the wavelength becomes smaller than the building blocks of the medium, hence the word 'perfect' in the title of Pendry's paper. To reconstruct a faithful image the low spatial frequencies represented by propagating rays (see Section 14.2.1) are also needed. Pendry showed that an idealized flat-slab NRIM brings together propagating and evanescent waves with the proper phase and amplitude, respectively, to form a perfectly resolved image if the magnitude of the NRIM's refractive index matches that of its surrounding medium, which will generally be free space (i.e., $n_- = -1$). Figures 14.3b and 14.3c represent an illustration of the propagating and evanescent waves in this situation, respectively.

Pointing the way to a more immediately realizable experimental test of super-resolution with just the high-spatial-frequency evanescent waves, Pendry analysed the behavior of the nonpropagating high-spatial-frequency components in a metal film approximated as a collisionless plasma, i.e., with imaginary n below the plasma frequency. As discussed in connection with Equation (14.3), such a metal has a negative ε and positive μ, within a reasonable approximation. Note again that this is distinct from the negative index, and does not support propagating waves, as is well known. What Pendry showed was the surprising result that when p-polarized radiation is transmitted through a slab of negative-ε material (NεM) the evanescent waves corresponding to spatial frequencies $k_x > n\omega/c$ are amplified exponentially vs thickness of slab of the imaginary index medium, rather than decaying as they would in a real-index medium or in a semi-infinite slab of NεM. Details of metal-film imaging with NεMs will be discussed in Section 14.3, including results of several experimental tests, which confirm the principle behind super-resolution.

We should note that the 'perfect lens' does not operate in the far-field. The high spatial frequencies giving super-resolution come from the evanescent waves, which begin decaying immediately upon exiting the NRIM, with decay lengths which are spatial-frequency dependent, but typically tens of nanometers for visible light. Thus the super-resolution is in a near-field region, although it is relayed some distance and amplified by the imposition of the NRIM layer between object and image planes. Super-resolution with NRIM slab lenses retains much in common with super-resolution in near-field microscopy.

14.2 OPTICS OF PROPAGATING WAVES WITH NEGATIVE INDEX

14.2.1 Foundation in Fourier optics

The electromagnetic field representing an object scene can be decomposed in a Fourier spectrum of transverse spatial-frequency components. Considering a single transverse dimension x and an optical axis along z, the spatial frequency k_x is carried by a plane wave propagating with wave-vector of magnitude $k = n\omega/c$ directed at an angle $\theta = \sin^{-1}(k_x/k)$. In other words, the information on more sharply modulated features is carried by plane waves directed at larger angles to the optical axis. A lens which reconstructs an image in the image plane also projects a Fourier transform that displays focused spots for each plane-wave spatial frequency in the focal plane of the lens.

According to the Abbe theory of imaging, the spatial resolution in the image is determined by the ability of the system to transmit the higher spatial frequencies. The highest transverse spatial frequency that can be carried by propagating waves of wavelength λ is given by

$$k_{x\max} = k = \frac{2\pi}{\lambda} \tag{14.8}$$

A faithful image at the diffraction limit will be formed if the optical transfer function (OTF) is unity up to $k_{x\max}$. To summarize, this diffraction limit arises because no *propagating* wave can carry transverse spatial modulation greater than $k_{x\,max} = 2\pi/\lambda$. It is the propagating waves that can be used to construct far-field images, and in that instance the above limit is genuine. However, in the near-field, not all electric-field components need be propagating waves.

The electric field of an electromagnetic plane wave has the form:

$$\mathbf{E}(\mathbf{r},t) = \mathbf{E}_0 e^{i(\mathbf{k}\cdot\mathbf{r} - \omega t)} \tag{14.9}$$

The wave vector, $\mathbf{k}$, is defined as:

$$\mathbf{k} = \frac{n\omega}{c}\hat{\mathbf{n}} \tag{14.10}$$

Assuming the optical axis parallel to the z-axis in a two-dimensional optical system, as above, the transverse spatial frequency (or transverse wave vector) is $\mathbf{k}_x$ and the longitudinal propagation vector is $\mathbf{k}_z$, i.e., $k^2 = k_x^2 + k_z^2$. The electric field in this case is:

$$\mathbf{E}(\mathbf{r},t) = \mathbf{E}_0 e^{i(k_z z - \omega t)} \tag{14.11}$$

with k_z given by:

$$k_z = \pm\sqrt{k^2 - k_x^2} = \pm\sqrt{\left(\frac{n\omega}{c}\right)^2 - k_x^2} \tag{14.12}$$

In accordance with Equation (14.11), the positive root indicates phase propagation in the $+z$ direction, and the negative root indicates phase propagation in the $-z$ direction. If k_x is greater than $n\omega/c$, k_z becomes imaginary and can be written as:

$$k_z = \pm i \sqrt{k_x^2 - \left(\frac{n\omega}{c}\right)^2} \tag{14.13}$$

In this case there is no propagation along z, but the positive imaginary value gives a field decaying exponentially toward positive z, and the negative imaginary value gives one that is growing exponentially toward positive z. In order to avoid infinite energies the positive sign must be chosen in any semi-infinite medium.

We see that there are two kinds of wave solutions depending on the magnitude of the transverse spatial frequency (transverse component of $\mathbf{k}$) relative to the magnitude of $\mathbf{k}$ $(= n\omega/c)$:

$$k_x \leq \frac{n\omega}{c} \Rightarrow \text{propagating waves} \tag{14.14a}$$

$$k_x > \frac{n\omega}{c} \Rightarrow \text{evanescent waves} \tag{14.14b}$$

The evanescent waves are inhomogeneous plane waves propagating transversely to the optical axis and nonpropagating, exponentially damped, parallel to the optical axis (Figure 14.4). The propagating waves are plane waves with phase propagation in the direction of $\mathbf{k}$, which is determined by the amplitudes of $\mathbf{k}_x$ and $\mathbf{k}_z$. One must be within a fraction of a wavelength from a scene for an appreciable signal to exist. Traditional imaging systems cannot use these high spatial frequencies, which is the source of the diffraction limit. This has been remedied somewhat by near-field scanning optical microscopy (NSOM), which collects light nanometers from a surface and uses the evanescent field of the object to achieve an order of magnitude improvement over optical microscopes ($\sim$10 nm).

The positive sign must be chosen in Equation (14.13) inside a semi-infinite medium to avoid infinite energies. Over a finite distance this restriction is lifted. Specifically, the NRIM flat slab may have a growing 'evanescent' field inside the slab because it can only reach finite amplitudes over finite distances. Amplification of the evanescent field, resulting in recovery of the high spatial frequencies, is the basis for the sub-diffraction-resolution super-lens as will be discussed in Section 14.3.1.

Next we will discuss how an NRIM will focus propagating waves to see how to build a lens that will appropriately correct the phase of propagating fields and amplitude of evanescent fields.

14.2.2 Fermat's principle in a slab lens

Fermat's Principle states that a ray will travel the path for which the optical path length (OPL) is extremal with respect to small excursions. This of course is the condition for the wavefronts normal to the rays in a bundle to interfere constructively at the destination point. As shown in Figure 14.5, if rays originating at the point O converge at point P, and an equal number of phase fronts exist between the points for every ray, an image is formed. This is the same as saying each ray travels an equal OPL, given by:

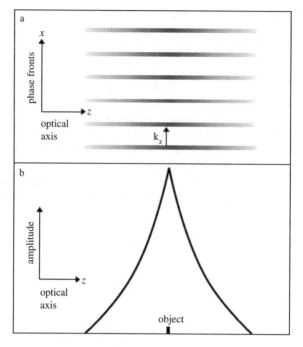

Figure 14.4 A transverse evanescent field component in the near-field of the object is represented in (a) by inhomogeneous plane-wave phase fronts propagating along the x-axis, transverse to the optical axis (z) along which real propagating waves transport energy. The evanescent-wave amplitudes are exponentially damped along z, as shown by the amplitude plot in (b) and the shading in (a)

$$OPL = \int_{initial}^{final} n dl = n_+ l_+ + n_- l_- \tag{14.15}$$

where l_+ and l_- are the distances traveled by a ray in positive- and negative-index media, respectively. Because of the symmetry evident in Figure 14.3b the accumulation of phase along the positive index portions is exactly canceled by the subtraction of phase along the negative portions. The result is that the total OPL is zero for every ray trajectory from object to image. This is an especially strong consequence of Fermat's principle for imaging by a slab lens without aberrations.

The conclusion just discussed of zero total OPL means that:

$$l_+ n_+ = l_- |n_-| \tag{14.16}$$

which is true for all rays. We can divide both sides of the equation by h, the height that the ray intersects the boundary (see Figure 14.6), to get:

$$\frac{l_+ n_+}{h} = \frac{l_- |n_-|}{h} \tag{14.17}$$

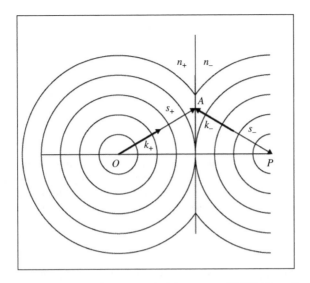

Figure 14.5 Phase fronts from a point source incident on an NRIM. Ray diagrams, which obey Snell's law, are determined by lines perpendicular to the fronts of constant phase. Positive phase accumulated along the path from O to A is offset by negative phase accumulated along the path from A to P (k is reversed in the negative n). The total phase change from O to P is zero along any path taken. P is the image of O. Wavefronts coming from the entire infinite interface comprise hemispheres in the image space

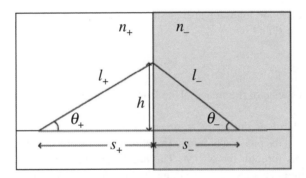

Figure 14.6 Geometry of a ray incident on an NRIM

With the help of Figure 14.6, this gives:

$$\frac{n_+}{\sin\theta_+} = \frac{|n_-|}{\sin\theta_-} \tag{14.18}$$

Therefore the flat slab has a condition for imaging without aberration. If we combine this with Snell's law we obtain:

$$n_+^2 = n_-^2 \tag{14.19}$$

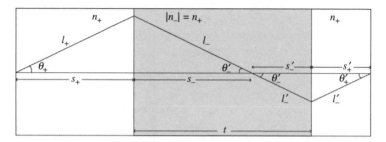

Figure 14.7 Geometry of the flat slab lens. If the perfect-lens condition is satisfied the NRIM acts as a negative-phase velocity space, and all phase accumulated during traversal of the object and image spaces (n_+) are canceled by negative phase accumulated in traversal of the slab, for every possible ray trajectory. From this geometry, the slab thickness t equals the sum of the object and image distances, $t = s_+ + s'_+$

which means that the magnitudes of the indices of refraction for the positive and negative materials must be equal to get an image without aberrations. In this case the impedances of the media are matched, which means no reflection will occur at the interfaces.

Figure 14.7 and Equations (14.16) and (14.18) show that if the magnitudes of the indices of refraction are equal, $l_+ = l_-$ and $\theta_+ = \theta_-$, and therefore:

$$s_+ = s_- \qquad (14.20)$$

This holds for the second interface also as shown in Figure 14.7, so we get a condition for the thickness of the slab:

$$t = s_+ + s'_+ \qquad (14.21)$$

where s'_+ is the distance of the image plane from the NRIM's second interface (see Figure 14.7).

Within in the slab lens the phase argument decreases for propagation in the positive z direction. As a result the phase acquired in traveling from the object in free space is unwound as the wave travels through the NRIM. The net accumulated phase becomes zero at the internal image point. This process occurs again (in reverse order) from the internal image to the external image. This is a particularly strong instance of Fermat's Principle where the accumulated phase on all ray paths contributing to the image is not just extremal, but zero.

14.2.3 Ray tracing with negative index and aberrations

We discussed above how Snell's law applies to NRIMs, so ray-tracing programs can be used to analyse the propagating rays in a slab lens system. We have conducted a number of ray-tracing simulations for slab lenses where simply negative n was supplied as a parameter. The results are consistent with the analytical treatments discussed here and permit graphical representations of the aberrations.

If the indices of the two media do not have n's of equal magnitude, rays at different heights will not converge in the slab, as seen in Figure 14.8a. We can determine the dis-

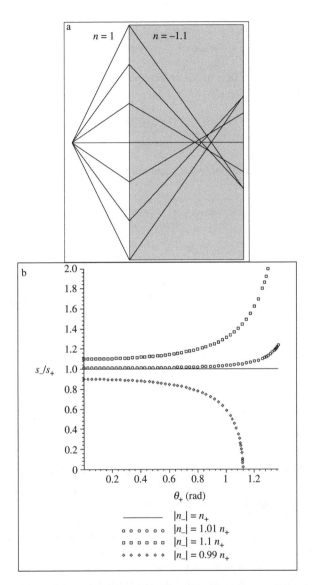

Figure 14.8 (a) Ray diagram with $|n_-| \neq |n_+|$. (b) The ratio of the image distance inside the slab s_- to object distance s_+ vs the angle of incidence (longitudinal aberration). It is unity for all spatial frequencies in the perfect lens ($|n_+| = |n_-|$). The aberrations grow significantly with increasing angle when there is a 10% difference between $|n_+|$ and $|n_-|$

tance from the interface that different rays will intersect in the NRIM. Using the following relationships from Figure 14.6:

$$\tan\theta_+ = \frac{h}{s_+}, \quad \tan\theta_- = \frac{h}{s_-} \tag{14.22}$$

and then equating the expressions of h obtained from these relations [Equation (14.22)] one gets:

$$\frac{s_- |\sin\theta_-|}{\cos\theta_-} = \frac{s_+ |\sin\theta_+|}{\cos\theta_+} \tag{14.23}$$

Our goal is to get s_- as a function of the indices of refraction, s_+ and θ_+. Using Snell's law in Equation (14.23) we get:

$$s_- = s_+ \frac{\cos\theta_-}{\cos\theta_+} \frac{n_-}{n_+} \tag{14.24}$$

Expressing $\cos(\theta_-)$ as a function of $\sin(\theta_-)$ and using Snell's law again in Equation (14.24) we get:

$$s_- = s_+ \frac{\sqrt{\left(\frac{n_-}{n_+}\right)^2 - \sin^2\theta_+}}{\cos\theta_+} \tag{14.25}$$

Figure 14.8b shows a plot of s_-/s_+ as a function of θ_+ for various values of n_- and n_+. Note that for $n_- < n_+$, there is an incident angle at which s_- is zero. This corresponds to the condition of total internal reflection, with a critical angle of incidence, θ_c given by:

$$\theta_+ = \theta_c = \sin^{-1}\left|\frac{n_-}{n_+}\right| \tag{14.26}$$

A corresponding analysis would apply when tracing from the intermediate image considered here to a final image formed by the second interface.

14.3 SUPER-RESOLUTION WITH THE SLAB LENS

14.3.1 Amplification of the evanescent waves

In the discussion of Fourier optics (Section 14.2.1), we noted that the electric fields of high spatial frequencies given by Equation (14.14b) decay exponentially with distance from an object and do not propagate power, and therefore are termed evanescent fields. The exponential decay with distance severely limits the contributions of these near-field components to an image produced at some distance from the object, unless some means of amplifying the evanescent field and/or relaying it over some distance can be arranged. For insight into the basis for amplification and relay of evanescent fields carrying high transverse spatial frequencies, consider first a negative-permittivity material (NɛM), not necessarily NRIM, such as a metal below its plasma frequency [Equation (14.3) with positive μ]. It is shown in the theory of surface plasmon polaritons (SPP) [14.4] at an interface between media of permittivity ε_1 and ε_2 that the continuity of **E** and **H** at the interface and Ampere's law demand that the SPP wave-vector components perpendicular to the interface must satisfy:

$$\frac{k_{zi}}{\varepsilon_i} + \frac{k_{zt}}{\varepsilon_2} = 0 \tag{14.27}$$

where k_{zi} and k_{zt} are the magnitudes of the incident and transmitted longitudinal propagation vectors, respectively. This requires that there must be a change in the sign of permittivity across the interface if a surface plasmon resonance is to exist at the interface. Furthermore, applying Faraday's law leads to the dispersion relation for the SPP transverse wave-vector component along the interface as:

$$k_x = \frac{\omega}{c} \sqrt{\frac{\varepsilon_1 \varepsilon_2}{\varepsilon_1 + \varepsilon_2}} \qquad (14.28)$$

This requires that for a real transverse propagating wave-vector along the interface when ε_2 < 0, the magnitude of ε_2 must exceed that of ε_1. Equality of magnitudes (but opposite signs) of ε_1 and ε_2 in Equation (14.28) results in a divergent wave-vector for the SPP (in the absence of damping, i.e., assuming real ε). This just corresponds to infinite amplitude of an undamped driven resonance. The SPP momentum diverges in such a case. k_{zi} and k_{zt} become imaginary so that the SPP field is exponentially damped perpendicular to the interface. The illustration of phase fronts and amplitudes of the SPP resonance (i.e., an inhomogeneous plane-wave evanescent field) is shown in Figure 14.9.

Consider first the near-field of an isolated object point, for which the evanescent-field amplitude for a value of $k_x > n\omega/c$ is illustrated as a function of z in Figure 14.10a. Now

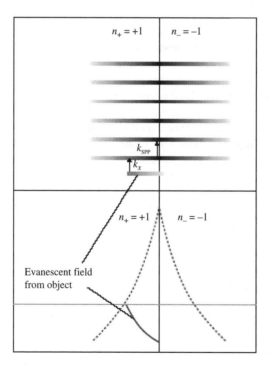

Figure 14.9 Surface plasmon polariton (SPP) stimulated by evanescent field of an object. The top figure shows that SPPs are inhomogeneous plane waves propagating along an interface as shown by the surfaces of constant phase. Exponential damping in the direction perpendicular to the interface is shown by the shading of the phase fronts. The bottom figure plots the amplitudes of the evanescent fields of the object alone (solid line) and of the SPP (dotted line)

place the $\varepsilon_+/\varepsilon_-$ interface into the near-field of the object, so that there is an oscillating electric field distinguishable above the noise on the interface. Assuming that these values satisfy the resonance conditions [Equations (14.27) and (14.28)] for the surface plasmon polariton, the oscillating field will drive the plasmon resonance. Depending on the Q (or 1/losses) of the resonance, the amplitude of the driven resonance can become much larger than the driving field after some number of cycles. The steady-state field amplitude distribution corresponding to the object and a metal plane supporting a high-Q plasmon resonance is illustrated schematically in Figure 14.10b. The point we want to illustrate is that the amplification has already been accomplished simply by introducing a single interface ($\varepsilon_+/\varepsilon_-$), which could be a metal surface or a negative-index surface. In the case of a metal surface, this is in fact an illustration of a host of surface-enhanced field phenomena such as surface-enhanced Raman scattering. In the present case the oscillating near-field of the object excites an SPP (a strong coherent resonator) and the superposition field of both grows to a steady-state value limited only by properties of the driven resonator. The object point has acquired an increased *oscillator strength* by coherent coupling to a larger antennae structure. The evanescent field of the combined object and interface decays again on going farther into the metal, and eventually reaches the value of the field at the object point. We could call this an *image* of the high spatial frequency k_x from the object point produced by the interface at some extended (tens of nanometers) distance from the object, except that it is not of much use because it exists within a metal.

To make it useful, we need to bring up a second interface to complete the *slab* (or in the metal case, *film*) lens. If the second interface is the exit back into positive ε_1 (the common object and image medium), SPP resonances of the same frequency (not considering interaction yet) will exist at the exit and entrance interfaces. Let the film thickness be such that the evanescent field of the first-surface-driven resonance is still above the noise level when it intersects the second interface. Then again we have a small driving field exciting a resonant response which can grow very large over a number of cycles to reach a steady state. This picture suggests a field distribution as illustrated in Figure 14.10c. This is not fully correct, as it ignores important details of mutual coupling of the oscillators. The picture does illustrate in a simple and mechanistic way how it is basically the availability of a resonant structure within the oscillating near-field of the object which allows the superposition field of the object and resonators to grow to a larger (amplified) amplitude than was present in the object field alone. This is how the oscillating near-field can be relayed some distance (still only a fraction of a wavelength) from plasmon resonance to plasmon resonance.

Figure 14.10b illustrates the qualitative field distribution at a single metal interface. The resonator field overlaps the object point and forms a superposition at that point, which changes the amplitude of the object field. If one could probe the field strength between the object point and the single interface, one would find the field growing (amplifying) on the vacuum side as the interface is approached. Figure 14.10c has illustrative value, but is incorrect in detail as it neglects details of coupled states of the two resonators at the entrance and exit surfaces. This aspect has been treated by multiple scattering from plane interfaces (i.e., reflection and transmission of a slab) in the famous paper by Pendry [14.3], and alternatively by matching boundary conditions on **E** and **H** at the interfaces in the work by Ramakrishna and Pendry [14.5]. The result is not intuitively obvious, but for the case $\varepsilon_2 = -\varepsilon_1 = -1$ assumed by Pendry, it leads to cancellation of the resonant field overlapping the object field in front of the slab interface, and a region of exponential growth between the

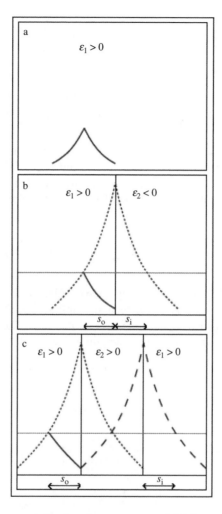

Figure 14.10   (a) The exponentially decaying evanescent field from an object. (b) An interface between positive and negative permittivity and supporting resonant SPPs is introduced within the object's evanescent field. The amplitude of the in-phase resonance grows to large values in the absence of significant loss. The electromagnetic field from an SPP decays exponentially away from the surface on both sides, as indicated by the dashed lines. For simplicity we have chosen this distance to be equal to the object distance, which is not necessarily true. (c) A second interface that supports resonant states can be driven by SPPs on the first interface, relaying the high spatial frequency image outside the NεM. The picture in part (c) is incomplete, however, in that it neglects mutual coupling of the oscillators

slab interfaces. Pendry's multiple-scattering approach is summarized in the following paragraphs.

Pendry [14.3] presents an argument for evanescent wave growth in an NRIM slab which allows calculation of the amplification. In both the positive-refractive-index material (PRIM) (n_+ = +1) and NRIM (n_- = −1) he chooses the decaying solution for evanescent fields. In this case the propagation vectors are:

$$k_z^+ = i\sqrt{k_x^2 - \left(\frac{n_+\omega}{c}\right)^2} = i\sqrt{k_x^2 - \left(\frac{\omega}{c}\right)^2} \qquad (14.29)$$

and

$$k_z^- = i\sqrt{k_x^2 - \left(\frac{n_-\omega}{c}\right)^2} = i\sqrt{k_x^2 - \varepsilon_-\mu_-\left(\frac{\omega}{c}\right)^2} \qquad (14.30)$$

The reason for keeping ε_- and μ_- in Equation (14.30) will become apparent shortly. He then considers the multiple-scattering processes that would occur from both interfaces of the slab as shown in Figure 14.11. In each pass through the lens the field acquires phase $\exp(ik_z^-d)$, where d is the thickness of the NRIM slab. The transmission coefficient T is given by:

$$T = t_{+\to-}t_{-\to+}e^{ik_z^-d} + t_{+\to-}t_{-\to+}r_{-\to+}^2e^{3ik_z^-d} + t_{+\to-}t_{-\to+}r_{-\to+}^4e^{5ik_z^-d} + \ldots = \frac{t_{+\to-}t_{-\to+}e^{ik_z^-d}}{1 - r_{-\to+}^2e^{2ik_z^-d}} \qquad (14.31)$$

where $t_{+\to-}$ and $t_{-\to+}$ are the transmission-amplitude coefficients for a wave at a PRIM/NRIM interface and vice-versa, $r_{-\to+}$ is the reflection amplitude coefficient for a wave at an NRIM/PRIM interface, and d is the thickness of the slab. The Fresnel equations give the relative transmission and reflection amplitudes at each interface for p-polarized light as:

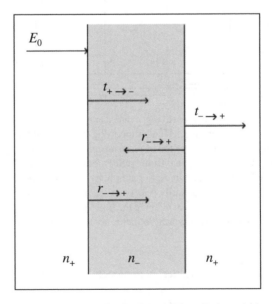

Figure 14.11   Multiple scattering events contributing to the total transmission through an NRIM flat slab. $t_{+\to-}$ and $t_{-\to+}$ are the transmission amplitude coefficients for a wave at a PRIM/NRIM interface and vice-versa. Similarly, $r_{-\to+}$ is the reflection amplitude coefficient for a wave at an NRIM/PRIM interface

$$t_{+\to-} = \frac{2k_z^+}{\dfrac{k_z^-}{\varepsilon_-} + k_z^+}$$

$$t_{-\to+} = \frac{2\dfrac{k_z^-}{\varepsilon_-}}{\dfrac{k_z^-}{\varepsilon_-} + k_z^+} \qquad (14.32)$$

and

$$r_{-\to+} = \frac{\dfrac{k_z^-}{\varepsilon_-} - k_z^+}{\dfrac{k_z^-}{\varepsilon_-} + k_z^+}$$

For the *propagating* waves incident on a perfect lens ($\varepsilon = -1$, $\mu = -1$) the relative reflection amplitude is zero and the relative transmission amplitudes are 1 since $\varepsilon_- = -1$ and $k_z^- = -k_z^+$ (see Section 14.1.1), consistent with impedance being matched. For the *evanescent* waves incident on a perfect lens, Equations (14.29) and (14.30) show that $k_z^- = k_z^+$, since $\varepsilon_- \mu_- = +1$. Therefore all relative transmission and reflection amplitudes in Equation (14.32) are infinite. Thus each term in Equation (14.31) is infinite. As we mentioned in Section 14.3.1, the wave-vector for an SPP is divergent if the magnitudes of the ε are the same on each side of the interface. Pendry's solution to this divergence was to take the limit as ε and μ approach -1. Substituting Equation (14.32) into Equation (14.31) Pendry obtained:

$$\lim_{\varepsilon_- \to -1, \mu_- \to -1} T = \lim_{\varepsilon_- \to -1, \mu_- \to -1} \frac{2k_z^+}{\dfrac{k_z^-}{\varepsilon_-} + k_z^+} \frac{2\dfrac{k_z^-}{\varepsilon_-}}{\dfrac{k_z^-}{\varepsilon_-} + k_z^+} \exp(ik_z^- d) \left[1 - \left(\frac{\dfrac{k_z^-}{\varepsilon_-} - k_z^+}{\dfrac{k_z^-}{\varepsilon_-} + k_z^+}\right)^2 \exp(2ik_z^- d)\right]^{-1}$$

$$= \exp(ik_z^- d) = \exp(ik_z^+ d) \qquad (14.33)$$

This corresponds to an exponentially growing field that is amplified in the medium with $n = -1$ at the same rate as it decays in the medium with $n_+ = +1$. The result is the same for s-polarized evanescent fields via substitution of μ_- for ε_- in Equations (14.32).

Pendry did not directly state if there is an upper limit of slab thickness for which exponential amplification is valid. One supposes that in an ideal case where one can indulge in arbitrarily many optical cycles to let arbitrarily high-Q resonances grow to steady state from arbitrarily small driving fields persisting across to the opposite interface, then there might be no limit in principle until material breakdown occurs due to the high field. But there should be a limit on the slab thickness over which amplification can be sustained as an experimentally realizable matter. We believe that insight on that limit comes from Figure 14.10c and the associated discussion of driven plasmon resonances on the two surfaces. The practical limit is that the driving field should be distinguishable above noise at the position of the resonance it is supposed to excite.

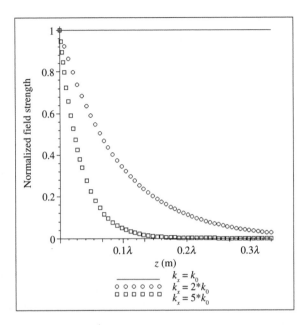

Figure 14.12 Decay of evanescent waves in a medium with $|n| = 1$. The field containing features half the size of the diffraction limit (i.e., $k_x = 2k_0$) decays to e^{-1} at about $z = 0.1\lambda$. The decay is faster for $|n| > 1$ since the rate of decay is proportional to $|n|$

The growth of the evanescent waves in an NRIM flat slab depends on the relative indices of refraction of the NRIM and PRIM. For the scene to be reproduced exactly, the evanescent waves must be equal in amplitude to their value at the scene. If they are larger one will see edge enhancement and if they are smaller the image will be blurred, moving towards the diffraction limit. In the case that the magnitude of the index of refraction matches that of free space, $n_- = -1$, the rate of growth of the evanescent wave will exactly match its rate of decay. Therefore the distance traveled in free space must be equal to that traveled in the NRIM to get the correct strength of the evanescent field. Nature is kind to us in this case. As discussed in Section 14.2.2 the thickness of the NRIM slab must equal $s_o + s_i$ if imaging is to occur. This can be cast even more generally in terms of Fermat's principle (Section 14.2.2) as the statement that the total optical path length from object to image is always zero in an imaging system. The perfect lens correctly adjusts the phase of the propagating rays and amplitude of the evanescent waves to form a perfect image. If we are to use the evanescent waves, the image will have to be produced in the near-field. This is due to the rapid decay of the evanescent wave in free space (Figure 14.12).

Knowing that imaging must be done in the near-field, one may ask if there is significant improvement in resolving power and/or practicality over NSOM to make this worthwhile. The action of NSOM is to get as close to the sample as possible, since the NSOM is picking up evanescent waves that are rapidly decaying from their original amplitude, whereas the goal of an NRIM lens is to reproduce the amplitude at some distance. Researchers have already achieved $\lambda/6$ resolution using evanescent-wave amplification [14.6], and the current limit of NSOM is $\sim\lambda/10$.

14.3.2 Aberrations in the evanescent image

Deviation from the perfect lens condition will produce aberrations in the propagating field as discussed in Section 14.2.3. The evanescent field will also experience aberrations since the amplitude correction to it will not correlate to the phase correction to the propagating field. If the amplification of the evanescent field is not enough the edges will become blurred as not all spatial frequencies contribute. Conversely, if the amplification is too great the edges will be artificially enhanced. This does not hurt resolution, but is a problem if the goal is to exactly reproduce the scene.

In order to examine the effect of deviating from the perfect lens condition, Ramakrishna and Pendry calculated the case where $|n_+| = |n_-|$ at one interface and not at the other, which they called the asymmetric slab [14.5]. In this case the surface plasmon polaritons are more strongly excited on the surface where the magnitudes of the indices of refraction match, as seen in Figure 14.13. They worked in the electrostatic limit and used air as the first medium, lossless silver as the lens, and GaAs as the third medium. The bottom plot of Figure 14.13 shows the symmetric perfect lens (air on both sides). The top plot shows the case where

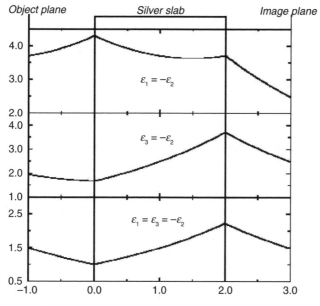

Figure 14.13 (from ref. [14.5]) Evanescent field strength vs distance (arbitrary units) for a silver slab in the electrostatic limit with different arrangements of surrounding media. In the electrostatic limit the electric and magnetic fields are decoupled, so only negative ε is needed for evanescent field growth. *top*: The silver lens' permittivity is matched at the first surface, which is where the plasmon will be strongly excited. Reflection in this case creates strongly altered fields in the object space. *middle*: The permittivity is matched at the second surface and the fields in the object space are not altered as much as in the top diagram. *bottom*: All permittivities are matched, satisfying the perfect-lens condition. There is no reflection and the fields in the object space are unaffected. At the image plane the evanescent field has exactly the same value as at the object plane [Reprinted with permission from S.A. Ramakrishna and J.B. Pendry, *J. Mod. Optics*, **49**, 1747. Copyright (2002) Taylor and Francis Ltd]

impedance was matched at the first interface. This was called the unfavorable condition because the reflected field changed the object field more strongly than the case where impedance is matched at the second interface (middle plot, favorable condition).

In the same paper Ramakrishna and Pendry [14.5] added a small imaginary component to ε to see how the asymmetric lens might realistically be expected to perform. They showed that the favorable case of the lossy asymmetric lens actually beats the performance of the lossy symmetric lens, and was hence called the asymmetric lossy near-perfect lens. They predicted that a resolution of 60–70 nm should be possible in this system using a light of 620 nm wavelength.

Smith et al. [14.7] studied the effect of both loss and deviation of the real part of n_- from -1. They introduced a small lossy component to both ε ($\varepsilon = \varepsilon' + i\varepsilon''$) and μ ($\mu = \mu' + i\mu''$) and then varied μ'. The optical transfer function for various values of μ' exhibited sharp resonances vs k_x/k_0 [14.7]. These are poles of the transfer function arising from surface plasmon polaritons, and are prevented from going to infinity by the loss. These produce a distortion relative to a perfect image, as the corresponding k_x will be overly enhanced, but they aren't undesirable since they are responsible for the resolution enhancement.

They also found the following approximation for the resolution enhancement:

$$R \equiv \frac{\lambda}{\lambda_{\min}} = \frac{k_x}{k_0} = -\frac{1}{2\pi} \ln\left|\frac{\delta\mu}{2}\right| \frac{\lambda}{d} \qquad (14.34)$$

where $\lambda_{\min} = 2\pi/k_x$, d is the thickness of the NRIM slab, and $\delta\mu$ is the deviation of μ' from -1. Notice that the wavelength-to-slab-thickness ratio varies much more quickly than does the term dependent on $\delta\mu$. Thus a small change in slab thickness allows for large changes in μ'.

14.3.3 Experimental results with evanescent waves

The silver film superlens ($\varepsilon < 0$, $\mu = 1$) proposed by Pendry [14.3] has been a useful tool to verify evanescent-wave enhancement. We discussed how the surface plasmon polariton couples with the evanescent field to produce amplification in Section 14.3.1. Pendry simply extended his argument from multiple scattering events to the electrostatic case to reach the conclusion of an NɛM superlens. The key is that in the electrostatic limit all length scales are much smaller than λ. Of particular importance is the fact that if the object size l is much smaller than λ, the transverse spatial frequencies are much greater than the wave-vector k:

$$l \ll \lambda$$
$$\frac{2\pi}{l} \gg \frac{2\pi}{\lambda}$$
$$k_x \gg k$$
$$k_x \gg \varepsilon\mu\frac{\omega}{c}$$

Therefore, in the limit that $k_x \to \infty$, ε_- and μ_- dependence can be neglected in Equation (14.30). The transmission coefficient for p-polarized light, Equation (14.31), depends only on ε_-, and thus the evanescent wave is amplified, as in Equation (14.33). The transmission coefficient for s-polarized light similarly depends only on μ, so one could use a negative-permeability material (NμM) as a superlens. The advantage of NεMs is that negative ε is readily available in simple metals in the visible spectrum, while negative μ is not.

Liu et al. [14.8] directly observed the growth of the evanescent wave in a silver thin film. In their experiment a thin silver film was deposited on the flat side of a glass hemisphere (Figure 14.14a). A far-field collimated, polarized laser beam illuminated the silver film at the air/silver interface. The natural roughness of the silver film scatters the propagating light, creating transverse spatial frequencies k_x (labeled as Δk and indicated by vertical dashed arrows in Figure 14.14b). The scattered waves include components with $k_x > \omega/c$ (i.e., evanescent components). For a given ω, only a particular k_x (or range of transverse spatial frequencies in the damped case) can satisfy the dispersion relation [Equation (14.28)] and thus excite the resonance. As discussed in Section 14.3.1, exciting the SPP resonance corresponds to amplifying the corresponding spatial frequency.[4] All other transverse spatial frequencies decay. The transmittivities of these fields are shown in the silver film (ε_2) portion of Figure 14.14b. Upon exiting the silver film and entering the glass (ε_3), continuity of phase

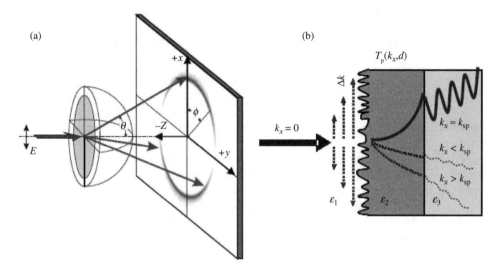

Figure 14.14   (from ref. [14.8], Liu et al. with some label changes) (a) Experimental setup for measurement of evanescent field amplification in a silver film. (b) Incoming collimated polarized radiation is scattered by roughness of the silver film, producing different transverse spatial frequencies (Δk, vertical dashed arrows), some of which are evanescent. Inside the silver film (ε_2) only the evanescent field corresponding to the k_x that satisfies the SPP dispersion relation is amplified (solid curve) while all others decay (dashed curves). Inside the glass (ε_3) some waves that were evanescent in silver are propagating in glass due to the fact that $k_0 = n_{glass}\omega/c$ increases across the boundary while k_x must remain constant. These oscillating fields propagate at an angle $\theta = \sin^{-1}(k_x/k_0)$ with the optical axis [Reproduced from N. Fang et al., *Optics Express*, **11**, 682. Copyright (2003) by permission of the Optical Society of America]

---

[4]Note that for an air ($\varepsilon_1 = 1$)-to-metal interface where $|\varepsilon_2| \gg \varepsilon_1$ (as when the illumination is far below the plasma frequency of the metal) the dispersion relation for the SPP [Equation (14.9)] requires $k_x \approx \omega/c = k_0$ for the spatial frequency to be amplified.

requires k_x is the same across the boundary. But k_0 is *not* constant, as it increases from ω/c to $n_{glass}\omega/c$. Therefore some evanescent waves with transverse spatial frequency $k_x \leq n_{glass}\omega/c$ become propagating when they enter the glass. These waves propagate at an angle $\theta = \sin^{-1}(k_x/k_0)$. This is shown in the glass-hemisphere portion of Figure 14.14b (ε_3), where the direction of the oscillating wave corresponds *only* to propagation direction θ; the direction is not a continuation of the decaying field and does not represent change in amplitude. Only one transverse spatial frequency is amplified, which corresponds to only one angle of propagation θ_{SP} (the surface plasmon coupling angle). This results in a ring of enhanced radiation as seen in Figure 14.14a. Their results for transmissivity at θ_{SP} vs film thickness showed that the amplification of the evanescent field increased with thickness of the film up to 50 nm, after which losses start to dominate. The experiment was done at a wavelength of 524.5 nm, setting the stage for a superlens operating in the visible spectrum. This was realized in 2005 by Fang et al. [14.6]. A silver film of 35 nm thickness was used to image a scene 40 nm away. The silver film was evaporated on an acrylic film that separated the scene and the lens. A photoresist that recorded the image coated the other side of the silver film. They resolved features 60 nm in size with 365 nm illumination, achieving $\lambda/6$ resolution.

14.4 NEGATIVE REFRACTION WITH METAMATERIALS

In our brief review of electromagnetic response of materials in Section 14.1.1, Equation (14.3) was presented to remind us that the collisionless-plasma model of a conductor yields negative ε in a wide range below the plasma frequency. This seems nearly ideal for half of the NRIM challenge, as one is tempted to seek ways of imbedding magnetic resonating structures of appropriate frequency in simple metals. There are several problems with such an approach, however, the main one being that a uniform conducting host material interacts with the magnetic subunits such as split-ring resonators (SRRs) which one wants to introduce. It is a better approach to the NRIM metamaterial if electrically and magnetically polarizable subunits can be imbedded in a normal dielectric. Pendry et al. [14.9] showed, and Ramakrishna [14.10] has further discussed, that an array or mesh of long thin wires displays the average permittivity of a free-electron plasma [Equation (14.3)] with plasma frequency

$$\omega_p = \sqrt{\frac{n_{eff}e^2}{\varepsilon_0 m^*}} \tag{14.35}$$

where n_{eff} is the effective density of conduction electrons and m^* is the effective mass of electrons flowing on the wire surface. The effective mass for this geometry is increased by self-inductance of currents in the wire. For the model of a square array of infinitely long wires of radius r, wire-to-wire spacing a, and density n of conduction electrons in the metal forming the wire, n_{eff} and m^* are given by:

$$n_{eff} = \frac{\pi r^2}{a^2}n$$

$$m^* = \frac{\mu_0 r^2 n e^2}{2}\ln\left(\frac{a}{r}\right) \tag{14.36}$$

Substituting Equations (14.36) into Equation (14.35), the wire metal electron density, n, cancels out and one gets the plasma frequency as:

$$\omega_p = \sqrt{\frac{ne^2}{\varepsilon_0 m^*}} = \sqrt{\frac{2\pi c^2}{a^2 \ln\left(\dfrac{a}{r}\right)}} \qquad (14.37)$$

Thus, the plasma frequency depends only on the wire spacing a and the wire radius r. These are the parameters one can manipulate to set ω_p, below which the wire mesh should exhibit negative ε.

According to Equation (14.37) the strongest dependence is an increase of ω_p with closer wire spacing. Increasing the wire radius, while staying well below a wavelength in size, produces a slower increase in ω_p. As reviewed by Ramakrishna [14.10], these results do not depend strongly on the assumptions made of a square array of aligned wires. A 3-dimensional mesh leads to isotropic response of a qualitatively similar form.

However, the model does depend essentially on the wires being very long relative to the radiation wavelength. If the wires have finite length, charge accumulation (self-capacitance) at the ends leads to an LC resonator response, giving a resonant dispersion curve for $\varepsilon(\omega)$ akin to the illustration in Figure 14.1. One would then look to the negative-ε region above resonance to fabricate an NRIM. This is a more restrictive range than that afforded by the plasma response. Lagarkov et al. built polymer composites filled with conducting fibers randomly arranged in planes and measured resonances of this type [14.11].

The other half of the NRIM metamaterials problem is to fabricate a resonant structure driven by the magnetic field. The basic structure that has been shown successful is the split-ring resonator (SRR), a version of which is illustrated in Figure 14.15a. The ring is simply a conducting ring in which a changing magnetic field can drive a current. The split, or gap, forms a capacitor that is charged by the current. These are the two elements of an LC resonator. The single SRR has a possibly undesirable electric dipole associated with the gap, in addition to the desired magnetic dipole of the ring. The double SRR, which consists of concentric SRRs with gaps opposite each other (Figure 14.15b), solves this problem since

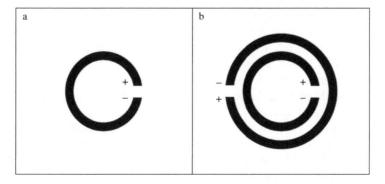

Figure 14.15   Some building blocks of magnetic metamaterials. (a) The single split-ring resonator (SRR) is a LC circuit and as such will have a resonance that depends on its geometry. (b) The double SRR cancels the electric dipoles that form in the gaps, while maintaining the magnetic dipoles

the two electric dipole fields largely cancel each other while the magnetic dipoles do not. As shown by Pendry et al. [14.12] and further elucidated by Ramakrishna [14.10], the resonance frequency of a double SRR with gap width d, ring radius r, and host (background) permittivity ε is:

$$\omega_0^m = \sqrt{\frac{3d}{\mu_0 \varepsilon_0 \varepsilon \pi^2 r^3}}$$ (14.38)

The effective μ of a metamaterial composed of such resonators goes negative above ω_0 and remains negative up to the 'magnetic plasma' frequency of:

$$\omega_p^m = \sqrt{\frac{3d}{(1-f)\mu_0 \varepsilon_0 \varepsilon \pi^2 r^3}}$$ (14.39)

where $f = \pi r^2/a^2$ is the filling factor of such resonators placed with spacing a on a square lattice.

Notice that except for the factor $(1-f)$, ω_0^m and ω_p^m have similar dependence on r, d, and ε, with the strongest dependence on r. A smaller radius of the rings raises the frequency range for negative effective μ. Larger splits d and smaller host permittivity ε also help raise the frequency range, but more slowly.

Smith et al. [14.2] demonstrated negative n in the 6 cm (5 GHz) microwave region with a metamaterial consisting of the electric and magnetic subunits described above. Shelby, Smith, and Schultz [14.13] later verified negative refraction in the 2.9 cm (10.5 GHz) microwave region. The result of negative refraction has been verified experimentally by several other groups as well [14.14], and focusing in a metamaterial slab lens has achieved sub-wavelength resolution at wavelengths of 17 cm (1.76 GHz) [14.15].

In the quest for IR/visible NRIM, microscale and nanoscale fabrication must be used. The maximum frequency response for SRRs in metamaterials is in the mid-IR region (3 μm, 100 THz) [14.16]. Grigorenko et al. [14.17] have observed simultaneously negative ε and μ in the visible region using paired-nanopillar resonators, which are more compatible with 100 nm fabrication techniques (see below). In the case of the mid-IR resonance, single SRRs were used with length scales of 100 nm. According to Equation (14.37), wires with radii of 20 nm and spacing of 0.15 μm will produce an electric-response plasma frequency of 562 THz (534 nm). Using Equation (14.38), ring radii of 35 nm and gap sizes of 20 nm produce a magnetic resonance at 568 THz (528 nm). Obviously it is easier to extend negative ε to the visible spectrum than it is to extend negative μ. It may be possible to make single SRRs of a size used in the above estimate by forming nanotubes on cylindrical templates. A lot of materials research needs to be done for this side of development.

Panina et al. [14.18] performed calculations based on a modified antenna theory on single SRRs made from a metal similar to silver. The magnetic field was set to be perpendicular to the plane of the rings. They considered a composite with single SRRs having a ring radius of 50 nm, wire diameter of 10 nm, and gap size of 35 nm. The resonant wavelength for their case was 2 μm (150 THz, IR). As in Equation (14.38), their equations show that decreasing the ring radius increases resonant frequency. With this in mind, they reduced the ring radius to 30 nm and the gap spacing to 21 nm. In this case the resonant wavelength became 1.3 μm (231 THz, near-IR). Further decrease in the ring radius was not realistic, as it had to be much greater than the radius of the wire which was at its lower limit. At higher frequencies radiation losses prevented significant magnetic response, leading Panina et al. to conclude that the utility of this structure was limited to the IR spectrum. The strength of the resonance depended on the concentration of the SRRs. A concentration of 6% did not produce a negative response, but a concentration of 30% did.

Another route to negative μ at higher frequencies was proposed by Podolskiy et al. [14.19] and was also examined in the above paper by Panina et al. [14.18]. Through simulations, Podolskiy et al. examined metal nanowires, which were found to support SPPs. The conditions Podolskiy imposed were that radius r of the wire is comparable to its skin depth, the wire–wire separation is much greater than r, and its length l is much greater than the wire–wire separation. The length of the wire could be of the order of the wavelength of illuminating light. The SPP resonance for a single nanowire, which provided amplification of the order of 10^3, occurred when the length of the nanowire was an integer multiple of half of the SPP wavelength. A magnetic dipole moment was introduced by placing two wires parallel to each other, forming a paired-nanowire resonator (PNR) as shown in Figure 14.16. In the simulation a changing magnetic field perpendicular to the plane of a pair induces currents in each wire, which flow opposite to each other. The gap between the ends of the wires provides a capacitance, and an LC resonator is formed. A 2-dimensional film of PNRs with magnetic field H perpendicular to the plane of PNRs and electric field E parallel to the wires has an effective ε and μ given by:

$$\varepsilon = 1 + \frac{4p}{\frac{l}{2}rd}\frac{d_E}{E} \tag{14.40}$$

and

$$\mu = 1 + \frac{4p}{\frac{l}{2}rd}\frac{m_H}{H} \tag{14.41}$$

where p is the surface concentration of wires, d is the distance between nanowires in a PNR, and d_e and m_h are the electric and magnetic dipole moments for a PNR, respectively. The variables l, r, and d have equal dependence, and can be adjusted to create overlapping resonances of ε and μ, which may result in negative n. Increasing l, r, and d moves the resonant frequency higher, although l is fixed, as it must be half of the SPP wavelength for the greatest amplification. In a later paper, Podolskiy et al. [14.20] performed more thorough calculations and found that increasing d and r causes greater radiative losses, which weakens and narrows the resonance. An NRIM lens can be formed from these films by stacking them to form a 3-dimensional structure. They speculated that even a random arrangement of PNRs may have negative n. Ref. [14.19] concludes with the suggestion that a random arrangement of unpaired nanowires has a chance of having negative n.

At Wake Forest University we have produced samples containing Ag nanowires (length ~5 μm, diameter ~100 nm) randomly dispersed in a sol–gel dielectric. Ellipsometry and angle-resolved reflectance yield values of complex n/μ $[= (\varepsilon/\mu)^{1/2}]$ that are positive and mostly real. Note that optical data analysed by the Fresnel equations for conventional materials with $\mu = 1$ yield refractive index n directly. But when $\mu \neq 1$, which we must allow in principle if anticipating a magnetic response, the impedance $Z = (\varepsilon/\mu)^{1/2}$ is obtained upon fitting the Fresnel equations. Additional data such as the refraction angle in Snell's law

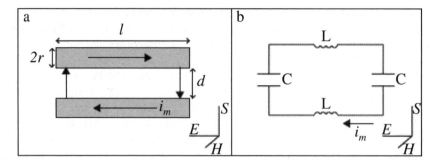

Figure 14.16 (a) The magnetic response of a paired nanowire resonator. The current (i_m) induced by a changing magnetic field perpendicular to the plane of the PNR is shown. The LC resonator is formed by the self-inductance in the individual wires and the capacitance between them. (b) The equivalent LC circuit

would be needed along with the Fresnel equations to separate ε and μ. The samples we have produced so far are too highly scattering for refraction experiments.

Advances are occurring rapidly in the field. As this chapter was being submitted, a ground-breaking paper by Grigorenko et al. [14.17] announced the measurement of simultaneously negative ε and μ in the visible range. The sample was a planar array of lithographically fabricated pairs of Au nanopillars (height, diameter, and spacing of the order of 100 nm). The pillar pair functions as a PNR such as that illustrated in Figure 14.16. The antisymmetric mode of induced currents in a pillar pair into and out of the plane, respectively, represents a current loop with capacitive gaps. Such an antisymmetric current mode is induced by the transverse magnetic (TM) mode of incident light and represents a resonance in μ. The symmetric-current mode (both currents in the same direction) represents an oscillating dipole and contributes a resonance in ε at longer wavelength than the μ resonance. Reflection vs wavelength plats for both polarizations were fitted to determine ε and μ, and results agreed reasonably with simulations solving Maxwell's equations for the specified pillar geometry. The imaginary part of μ was too high to measure negative refraction, but the authors observed that the impedance [$Z = (\varepsilon/\mu)^{1/2}$] of the array at one wavelength matched that of air and the reflectance vanished. The achievement of a negative real part of μ at a visible wavelength (540 nm) is a milestone that should open many possibilities.

14.5 CONCLUSIONS

Following the pioneering work of Veselago [14.1] and Pendry [14.3], it is remarkable how simple and 'perfect' optical imaging becomes if one has a material with negative refractive index, and particularly if one has the ideal material with $n = -1$. A flat slab functions as a lens. There are no monochromatic aberrations – no spherical aberration, no coma, no astigmatism, no field curvature, no distortion – because there is no unique optical axis. All rays propagating from any point of the flat object plane come to a unique focus in the conjugate flat image plane. Fermat's principle has a much stronger statement for the negative-n slab lens than for a positive-n aberration-corrected lens: all optical path lengths between conjugate points are *zero*, not just constant. Furthermore, evanescent waves bearing transverse spatial frequencies beyond the diffraction limit of propagating waves are relayed by such an ideal slab to the same image plane where the propagating rays focus, reconstructing the high spatial frequencies at the same amplitude as in the object plane. Finally, there are no reflections from the surfaces of such a slab lens because the impedances are perfectly matched. Thus, in a striking number of aspects, imaging really becomes simpler and better with negative index.

However, one's enthusiasm about these remarkable improvements and simplifications, even assuming that the ideal materials can be fabricated, needs to be balanced by recognition of some important fundamental restrictions inherent in using negative-n imaging. The range of object and image distances is restricted on two counts. For slab-lens imaging by propagating rays, it was shown that the sum of object and image distances must equal the slab thickness. For example, slab eyeglasses would be prohibitively thick, and a slab telescope is unrealizable. Use of the evanescent waves to achieve super-resolution poses an even more severe limit. The sum of object and image distances cannot exceed a few times the range of the source near-field, i.e., a few hundred nanometers using visible light, since otherwise exponential amplification to unsustainable values of the fields will occur inside the NRIM. Within the limited spectral range satisfying negative real n and acceptably small loss, the spectral requirement for the aberration-free case $|n_+| = |n_-|$ is even more restrictive, i.e., monochromatic. 'Perfect imaging' cannot occur in color.

Progress towards NRIM in the visible region is coming fast. A metamaterial with simultaneously negative ε and μ near 470 nm [14.17] has been fabricated utilizing a variation on the idea of Podolskiy et al. [14.19] and Panina et al. [14.18] that pairs of nanowires can form magnetic resonators. The challenges that remain are extending the system to 3-D and reducing losses. Tackling loss is probably the more difficult of the two. The prospect for realization of usable NRIM in the visible range certainly seems to have crossed from the realm of speculation to the realm of experiment.

REFERENCES

[14.1] V.G. Veselago, *Sov. Phys. Usp.* (*Engl. Transl.*), **10**, 509 (1968).

[14.2] D.R. Smith, W.J. Padilla, D.C. Vier, S.C. Nemat-Nasser, and S. Schultz, *Phys. Rev. Lett.*, **84**, 4184 (2000).

[14.3] J.B. Pendry, *Phys. Rev. Lett.*, **85**, 3966 (2000).

[14.4] H. Raether, *Surface Plasmons on Smooth and Rough Surfaces and on Gratings* (Springer-Verlag, Berlin and New York, 1988).

[14.5] S.A. Ramakrishna and J.B. Pendry, *J. Modern Opt.*, **49**, 1747 (2002).

[14.6] N. Fang, H. Lee, C. Sun, and X. Zhang, *Science*, **308**, 534 (2005).

[14.7] D.R. Smith, D. Schurig, M. Rosenbluth, and S. Schultz, *Appl. Phys. Lett.*, **82**, 1506 (2003).

[14.8] Z.W. Liu, N. Fang, T.J. Yen, and X. Zhang, *Appl. Phys. Lett.*, **83**, 5184 (2003); N. Fang, Z.W. Liu, T.J. Yen, and X. Zhang, *Opt. Express*, **11**, 682 (2003).

[14.9] J.B. Pendry, A.J. Holden, D.J. Robbins, and W.J. Stuart, *J. Phys. Condens. Matter*, **10**, 4785 (1998); J.B. Pendry, A.J. Holden, W.J. Stuart, and I. Youngs, *Phys. Rev. Lett.*, **76**, 4773 (1996).

[14.10] S.A. Ramakrishna, *Rep. Prog. Phys.*, **68**, 449 (2005).

[14.11] A.N. Lagarkov, S.M. Matytsin, K.N. Rozanov, and A.K. Sarychev, *J. Appl. Phys.*, **84**, 3806 (1998).

[14.12] J.B. Pendry, A.J. Holden, D.J. Robbins, and W.J. Stuart, *IEEE Trans. Microwave Theory Tech.*, **47**, 2075 (1999).

[14.13] R.A. Shelby, D.R. Smith, and S. Schultz, *Science*, **292**, 77 (2001).

[14.14] C.G. Parazzoli, R.B. Greegor, K. Li, B.E.C. Koltenbach, and M. Tanielian, *Phys. Rev. Lett.*, **90**, 107401 (2003); A.A. Houck, J.B. Brock, and I.L. Chuang, *Phys. Rev. Lett.*, **90**, 137401 (2003).

[14.15] A.N. Lagarkov and V.N. Kissel, *Phys. Rev. Lett.*, **92**, 077401 (2004).

[14.16] S. Linden, C. Enkrich, M. Wegener, J. Zhou, T. Koschny, and C.M. Soukoulis, *Science*, **306**, 1351 (2004).

[14.17] A.N. Grigorenko, A.K. Geim, H.F. Gleeson, Y. Zhang, A.A. Firsov, I.Y. Khruschev, and J. Petrovic, *Nature*, **438**, 335 (2005).

[14.18] L.V. Panina, A.N. Grigorenko, and D.P. Makhnovskiy, *Phys. Rev. B*, **66**, 155411 (2002).

[14.19] V.A. Podolskiy, A.K. Sarychev, and V.M. Shalaev, *J. Nonlinear Opt. Phys. Mater.*, **11**, 65 (2002).

[14.20] V.A. Podolskiy, A.K. Sarychev, and V.M. Shalaev, *Opt. Express*, **11**, 735 (2003).

15 Excitonic Processes in Quantum Wells

J. Singh and I.-K. Oh

Faculty of Technology, B-41
Charles Darwin University
Darwin, NT 0909, Australia

15.1 INTRODUCTION

During the last couple of decades, there has been astonishing progress in the field of semi-conductor nanostructures due to the development of techniques in the epitaxial crystal-growth of materials such as molecular-beam epitaxy and metal–organic chemical vapor deposition. These techniques have made it possible to manipulate and design semiconduc-tor devices with high precision in an atomic scale and to produce high-quality low-dimensional systems such as quantum wells (QWs), superlattices, quantum wires (QWRs), and quantum dots (QDs) which give rise to several exotic and interesting phenomena [15.1]. These low-dimensional semiconductor systems have important applications in optoelec-tronic devices, e.g., light detectors, LASERs, LEDs, etc. Many physical properties of optoelectronic devices fabricated from such low-dimensional semiconductor systems are determined from information on excitonic processes. One of the most important properties in these semiconductor nanostructures is the exciton–phonon interaction which plays a very significant role in the dynamics of excitons [15.2] associated with the formation [15.3, 15.4],

Optical Properties of Condensed Matter and Applications   Edited by J. Singh
© 2006 John Wiley & Sons, Ltd

relaxation [15.5, 15.6], dephasing [15.7], and localization [15.8] of excitons leading to the study of photoluminescence kinetics [15.9], phonon-induced luminescence [15.10], degenerate four-wave mixing [15.11], etc. In this chapter, we shall discuss the excitonic processes related with charge carrier–phonon interaction in quantum wells.

15.2 EXCITON–PHONON INTERACTION

Excitonic processes such as the formation, relaxation, dephasing, and localization of excitons are influenced by the interaction with phonons. Since an exciton consists of an electron in the conduction band bound with a hole in the valence band through their Coulomb interaction, the exciton–phonon interaction is a sum of the electron–phonon and hole–phonon interactions. In noncentrosymmetric and polar crystals of III–V and II–VI semiconductors, charge carriers as well as excitons interact with acoustic phonons through the deformation potential (DP) and piezoelectric (PE) couplings and with optical phonons though the polar optical (PO) coupling [15.5]. However, it is to be noted that a strong electron–phonon or hole–phonon interaction does not mean a strong exciton–phonon interaction [15.3], which will be explained later.

In general, as the conduction band near the minimum is isotropic, electron–acoustic phonon interaction obtained from the DP has a contribution only from longitudinal acoustic (LA) phonons. However, as the valance band is anisotropic, both holes–LA and holes–transverse acoustic (TA) phonons interactions are nonzero [15.5]. On the other hand, through the PE coupling, interactions of both the charge carriers, electrons in the conduction band and holes in the valence band, are nonzero with both LA and TA phonons. As for the interaction of charge carriers with the optical phonons due to the polar coupling is concerned, the contribution of transverse optical (TO) phonons can usually be ignored because the TO phonons produce negligible electric fields, whereas the longitudinal optical (LO) phonons produce sizeable electric fields in the direction of phonon propagation [15.12]. Accordingly, the exciton–phonon interaction Hamiltonians due to the deformation potential H_I^D, piezoelectric H_I^P, and polar LO phonon H_I^O couplings, can be written, respectively, as [15.5]:

$$H_I^D(\mathbf{r}_\parallel, \mathbf{R}_\parallel, z_e, z_h)_{\lambda, \mathbf{q}} = C_D^\lambda \left[i e^{i \mathbf{q}_\parallel \cdot \mathbf{R}_\parallel} \left(\Xi_c^\lambda(\mathbf{q}) e^{i \alpha_h \mathbf{q}_\parallel \cdot \mathbf{\eta}} e^{i q_z z_e} \right. \right.$$
$$\left. \left. - \Xi_v^\lambda(\mathbf{q}) e^{-i \alpha_e \mathbf{q}_\parallel \cdot \mathbf{\eta}} e^{i q_z z_h} \right) \hat{b}_{\lambda \mathbf{q}} + c.c. \right] \tag{15.1}$$

$$H_I^P(\mathbf{r}_\parallel, \mathbf{R}_\parallel, z_e, z_h)_{\lambda, \mathbf{q}} = C_P^\lambda \left[i e^{i \mathbf{q}_\parallel \cdot \mathbf{R}_\parallel} \left(e^{i \alpha_h \mathbf{q}_\parallel \cdot \mathbf{\eta}} e^{i q_z z_e} - e^{-i \alpha_e \mathbf{q}_\parallel \cdot \mathbf{\eta}} e^{i q_z z_h} \right) \hat{b}_{\lambda \mathbf{q}} + c.c. \right] \tag{15.2}$$

and

$$H_I^O(\mathbf{r}_\parallel, \mathbf{R}_\parallel, z_e, z_h)_{LO, \mathbf{q}} = C_O^{LO} \left[e^{i \mathbf{q}_\parallel \cdot \mathbf{R}_\parallel} \left(e^{i \alpha_h \mathbf{q}_\parallel \cdot \mathbf{\eta}} e^{i q_z z_e} - e^{-i \alpha_e \mathbf{q}_\parallel \cdot \mathbf{\eta}} e^{i q_z z_h} \right) \hat{b}_{LO \mathbf{q}} + c.c. \right] \tag{15.3}$$

Here, c.c. denotes the complex conjugate of the first term, and the center of mass coordinate and the relative coordinate for an exciton in the plane of quantum wells are given, respectively, as:

$$\mathbf{R}_\parallel = \alpha_e \mathbf{r}_{e\parallel} + \alpha_h \mathbf{r}_{h\parallel}, \quad \alpha_e = \frac{m_{e\parallel}^*}{M_\parallel^*}, \quad \alpha_h = \frac{m_{h\parallel}^*}{M_\parallel^*} \tag{15.4}$$

and

$$\mathbf{r}_{\|} = \mathbf{r}_{e\|} - \mathbf{r}_{h\|} \tag{15.5}$$

where $m_{e\|}^*$, $m_{h\|}^*$, and $M_{\|}^*$ are the effective masses of electron, hole, and exciton, respectively, in the plane of QWs. $\hat{b}_{\lambda \mathbf{q}}$ is a λ-mode ($\lambda = \mathrm{LA}, \mathrm{TA}_1, \mathrm{TA}_2$) phonon annihilation operator with wave vector $\mathbf{q} = (\mathbf{q}_{\|}, q_z)$, $\mathbf{q}_{\|}$ in the plane of QWs and q_z perpendicular to the QW walls, z_e and z_h are the z-components of the electron and hole coordinates, respectively. $\Xi_c^{\lambda}(\mathbf{q})$ and $\Xi_v^{\lambda}(\mathbf{q})$ are the effective deformation potentials [15.5] for conduction and valence bands, respectively. C_D^{λ}, C_P^{λ}, and C_O^{LO} are respectively, given by:

$$C_D^{\lambda} = \sqrt{\frac{\hbar q}{2 \rho_m v_{\lambda} V}} \tag{15.6}$$

$$C_P^{\lambda} = i \left(\frac{e h_{\mathbf{q}}^{\lambda}}{\kappa \varepsilon_0} \right) \sqrt{\frac{\hbar}{2 \rho_m v_{\lambda} q V}} \tag{15.7}$$

and

$$C_O^{LO} = i e \sqrt{\frac{\hbar \omega_{LO}}{2 \varepsilon_0 V} \left(\frac{1}{\kappa_{\infty}} - \frac{1}{\kappa_0} \right) \frac{1}{q}} \tag{15.8}$$

where ρ_m denotes the material density, V volume, v_{λ} sound velocity, κ relative dielectric constant, ε_0 electric permittivity in the vacuum, $h_{\mathbf{q}}^{\lambda}$ effective piezoelectric constant for a λ-mode phonon, κ_{∞} and κ_0 high-frequency and static relative dielectric constants, respectively, and ω_{LO} is the LO phonon frequency.

15.3 EXCITON FORMATION IN QUANTUM WELLS ASSISTED BY PHONONS

When the energy of an incident photon on a nanostructure semiconductor is above its bandgap energy, the photon is absorbed and a free electron–hole pair is excited. Such photogenerated electron–hole pairs may relax nonradiatively and form excitons [15.9] by emitting phonons. At excitation energies larger than the bandgap energy, a photoexcited electron–hole pair can form an exciton first with a large total wavevector $\mathbf{K}_{\|}$, corresponding to its center-of-mass motion [15.13, 15.14]. The exciton then relaxes nonradiatively down to $\mathbf{K}_{\|} \sim 0$ state by emitting phonons, and finally it recombines radiatively from the $\mathbf{K}_{\|} \sim 0$ excitonic state by emitting a photon to conserve energy and momentum. Therefore, in exciton luminescence experiments, the information on the formation time of an exciton as a function of exciton wavevector is crucial to the study of luminescence rise time τ_R. At low excitation densities, photoexcited electrons and heavy holes occupy only their first sub-band and so do heavy-hole excitons. In this case, an exciton is considered to be formed from a free quasi-2D electron and hole pair due to phonon emission or absorption in a λ-mode (LA, TA_1, TA_2, and LO) via DP, PE, and PO couplings. Then the interaction Hamiltonian in the second quantized form involving such formation of excitons can be written as (see Refs. [15.3] and [15.4, 15.5]):

$$\hat{H}_{I,\lambda}^J = \frac{1}{\sqrt{A_0}} \sum_{\mathbf{q}_\parallel, q_z, \mathbf{K}_\parallel, \mathbf{k}_\parallel} C_J^\lambda \left[F_{\lambda-}^J(\mathbf{q}_\parallel, q_z, \mathbf{k}_\parallel) \, \hat{B}_{\mathbf{K}_\parallel}^\dagger \hat{b}_{\lambda\mathbf{q}} \hat{d}_{h,\alpha_h(\mathbf{K}_\parallel - \mathbf{q}_\parallel) - \mathbf{k}_\parallel} \hat{a}_{c,\alpha_e(\mathbf{K}_\parallel - \mathbf{q}_\parallel) + \mathbf{k}_\parallel} \right.$$

$$\left. + \left[F_{\lambda+}^J(\mathbf{q}_\parallel, q_z, \mathbf{k}_\parallel) \, \hat{B}_{\mathbf{K}_\parallel}^\dagger \hat{b}_{\lambda\mathbf{q}}^\dagger \hat{d}_{h,\alpha_h(\mathbf{K}_\parallel + \mathbf{q}_\parallel) - \mathbf{k}_\parallel} \hat{a}_{c,\alpha_e(\mathbf{K}_\parallel + \mathbf{q}_\parallel) + \mathbf{k}_\parallel} \right] \right] \tag{15.9}$$

where A_0 is the 2D area of quantum wells, and $\hat{d}_{h,\alpha_h(\mathbf{K}_\parallel - \mathbf{q}_\parallel) - \mathbf{k}_\parallel}$ and $\hat{a}_{c,\alpha_e(\mathbf{K}_\parallel - \mathbf{q}_\parallel) + \mathbf{k}_\parallel}$ are hole and electron annihilation operators in the valence and conduction bands, respectively. $\hat{B}_{\mathbf{K}_\parallel}^\dagger$ is the creation operator of an exciton. $J = D, P,$ and O represent DP, PE, and PO couplings, respectively. The factors $F_{\lambda\mp}^D$, $F_{\lambda\mp}^P$ [15.4], and $F_{LO\mp}^O$ [15.3] are, respectively, obtained as:

$$F_{\lambda\mp}^D(\mathbf{q}_\parallel, q_z, \mathbf{k}_\parallel) = \pm i [\Xi_c^\lambda(\mathbf{q}) F_e(\pm q_z) G_1(\mathbf{k}_\parallel \pm \alpha_h \mathbf{q}_\parallel)$$

$$- \Xi_c^\lambda(\mathbf{q}) F_h(\pm q_z) G_1(\mathbf{k}_\parallel \mp \alpha_e \mathbf{q}_\parallel)] \tag{15.10}$$

$$F_{\lambda\mp}^P(\mathbf{q}_\parallel, q_z, \mathbf{k}_\parallel) = \pm i h_\mathbf{q}^\lambda [F_e(\pm q_z) G_1(\mathbf{k}_\parallel \pm a_h \mathbf{q}_\parallel) - F_h(\pm q_z) G_1(\mathbf{k}_\parallel \mp \alpha_e \mathbf{q}_\parallel)] \tag{15.11}$$

and

$$F_{LO\mp}^O(\mathbf{q}_\parallel, q_z, \mathbf{k}_\parallel) = [F_e(\pm q_z) G_1(\mathbf{k}_\parallel \pm a_h \mathbf{q}_\parallel) - F_h(\pm q_z) G_1(\mathbf{k}_\parallel \mp \alpha_e \mathbf{q}_\parallel)] \tag{15.12}$$

The form factors F_j ($j = e,h$) and G_1 are given as [15.3]:

$$F_j(q_z) = \int dz_j |\phi_j(z_j)|^2 e^{iq_z z_j}, \quad j = e,h \tag{15.13}$$

and

$$G_1(\mathbf{k}_\parallel + \alpha\mathbf{q}_\parallel) = \int d\mathbf{r}_\parallel \phi_x^*(\mathbf{r}_\parallel) e^{i(\mathbf{k}_\parallel + \alpha\mathbf{q}_\parallel)\cdot\mathbf{r}_\parallel} = \sqrt{\frac{8\pi}{\beta^2}} \left[\left(\frac{|\mathbf{k}_\parallel + \alpha\mathbf{q}_\parallel|}{\beta} \right)^2 + 1 \right]^{-3/2} \tag{15.14}$$

where ϕ_j is the electron ($j = e$) and hole ($j = h$) sub-bands and ϕ_x is the 1s exciton wave-function which is chosen to be a trial wave function given by [15.16]:

$$\phi_x(\mathbf{r}_\parallel) = \sqrt{\frac{2\beta^2}{\pi}} e^{-\beta_\eta} \tag{15.15}$$

The first term of Equation (15.9) corresponds to the formation of an exciton due to phonon absorption, and the second term corresponds to that due to phonon emission. It is obvious from Equations (15.11) and (15.12) that the PE and LO phonon contributions to the inter-action Hamiltonian in quantum wells (QWs) are mainly influenced by the form factors F_j and G_1 regardless of the individual strength of electron– and hole–phonon interactions. For instance, when the effective mass of an electron is equal to that of hole and the band offset of the conduction band is equal to that of valence band, i.e., $m_{ez}^* = m_{hz}^*$, $m_{e\parallel}^* = m_{h\parallel}^*$, and $\Delta E_c = \Delta E_v$, the exciton–phonon interaction will become zero. As for the DP coupling [see Equation (15.10)], the signs and magnitudes of each effective DP of both the conduction and the valence band play a very important role too.

Using Fermi's golden rule and the interaction Hamiltonian in Equation (15.9), the rate of formation of an exciton with wave vector $\mathbf{K}_\parallel$ from free electron–hole pairs due to a phonon emission in λ-mode via J ($J = D, P, O$) coupling is obtained as [15.3, 15.4, 15.15]:

$$W_{\lambda+}^J = \frac{1}{A_0} \frac{2\pi}{\hbar} \sum_{\mathbf{q}_\|,q_z,\mathbf{k}_\|} \left| C_J^\lambda F_{\lambda+}^J (\mathbf{q}_\|, q_z, \mathbf{k}_\|) \right|^2 f_{\alpha_h (\mathbf{K}_\|+\mathbf{q}_\|)-\mathbf{k}_\|}^h f_{\alpha_e (\mathbf{K}_\|+\mathbf{q}_\|)+\mathbf{k}_\|}^e$$

$$\times \left(f_{\mathbf{K}_\|}^{ex} + 1 \right)\left(n_\mathbf{q}^\lambda + 1 \right) \delta\left(E_x - E_{e-h} + \hbar\omega_{\mathbf{q},\lambda} \right) \qquad (15.16)$$

where f^h, f^{ex}, f^e, and $n_\mathbf{q}^\lambda$ are the occupation numbers of hole, electron, exciton, and phonon, respectively [15.3, 15.4]. Here exciton energy E_x and electron–hole-pair energy E_{e-h} are, respectively, given by:

$$E_x = \frac{\hbar^2 |\mathbf{K}_\||^2}{2M_\|^*} - E_b \qquad (15.17)$$

and

$$E_{e-h} = \frac{\hbar^2 |\mathbf{K}_\| + \mathbf{q}_\||^2}{2M_\|^*} + \frac{\hbar^2 |\mathbf{k}_\||^2}{2\mu_\|^*} \qquad (15.18)$$

where E_b is the exciton binding energy and $\mu_\|^* = (1/m_{e\|}^* + 1/m_{h\|}^*)^{-1}$ is the reduced mass of electron–hole pair. For simplicity, we assume that there are no excitons in the system in the initial state, i.e., $f_{\mathbf{K}_\|}^{ex} = 0$ The densities of the photogenerated electron and hole can be considered to be the same ($N_e = N_h = N_{e-h}$) in an intrinsic QW. We also assume that electrons and holes are at the same temperature, i.e., $T_e = T_h = T_{e-h}$.

Let us first discuss the rate of formation of an exciton due to emission of an acoustic phonon in [001] GaAs/Al$_{0.3}$Ga$_{0.7}$As QWs of well width $L_z = 0$ Å at a lattice temperature, T, = 4.2 K. For comparing the individual rates due to emission of an LA or TA phonon, we have calculated them from Equation (15.15) for $\lambda = $ LA and TA (TA$_1$ + TA$_2$) separately, and plotted them as a function of $\mathbf{K}_\|$ at $T_{e-h} = 20$ K for charge carrier densities, $N_{e-h} = 1 \times 10^{10}$ cm^{-2} and $N_{e-h} = 5 \times 10^{10}$ cm^{-2} in Figure 15.1 (a) and (b), respectively. It is clear from Figure 15.1 (a) and (b) that the rate of formation due to an LA phonon emission is comparable to that due to a TA phonon emission, though the rate of formation of an exciton via a LA phonon is slightly higher than that via a TA phonon for relatively small $\mathbf{K}_\|$, and the rate due to a TA phonon is slightly higher than that due to an LA phonon for relatively large $\mathbf{K}_\|$. Also, the maximum value of the rate via TA-phonon emission is relatively larger than that via LA-phonon emission. For GaAs QWs, when considering only the deformation-potential coupling, holes in the valance band interact with both TA and LA phonons whereas electrons in the conduction band interact with only LA phonons. Consequently, an exciton, which is a bound state of an electron and hole, interacts with both LA and TA phonons due to deformation potentials. In this case, it is clear from Equation (15.10) that the sign of the deformation potential of charge carriers is very important in determining the strength of exciton–phonon interaction. A very strong electron–phonon interaction or hole–phonon interaction does not automatically mean a very strong exciton–phonon interaction. It also depends on the signs of the two coupling coefficients. For instance, if the effective deformation potentials of electrons and holes are of comparable magnitude for any phonon mode but have opposite signs, then the strength of the corresponding exciton–phonon interaction is enhanced because the coupling coefficients get added, but if they have the same sign then the exciton–phonon coupling will be very weak as the electron–phonon coupling will cancel

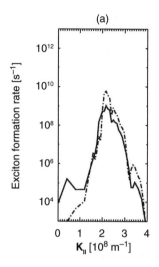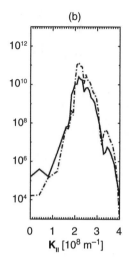

Figure 15.1 The formation rate of an exciton via LA (solid curve) and TA (dash-dotted curve) phonon emissions in GaAs quantum wells plotted as a function of the center-of-mass wavevector $\mathbf{K}_\parallel$ for $L_z = 80$ Å, $T = 4.2$ K, and $T_{e-h} = 20$ K. (a) $N_{e-h} = 1 \times 10^{10}$ cm^{-2} and (b) $N_{e-h} = 5 \times 10^{10}$ cm^{-2} [15.4] [Reprinted from I.-K. Oh and J. Singh, *Superlattices Microstruct.*, **30**, 221. Copyright (2001) Elsevier]

out the hole–phonon coupling. Our results show that τ_f, the time of formation of an exciton defined by $1/\tau_f = \sum_{J,\lambda} W_{\lambda+}^J$ ($J = D$, P, and $\lambda =$ LA, TA$_1$, TA$_2$), at the maximum rate of formation varies from 145 ps to 714 ps for $N_{e-h} = 1 \times 10^{10}$ cm^{-2} and it is 6.2–28 ps for $N_{e-h} = 5 \times 10^{10}$ cm^{-2} in the carrier temperature T_{e-h} ranging from 20 to 80 K. Accordingly, the rate of formation of excitons is very sensitive to both carrier temperatures and densities.

Figures 15.2 (a) and (b) illustrate the dependence of the rate of exciton formation on center-of-mass wave vector $\mathbf{K}_\parallel$ of exciton, charge-carrier temperature T_{e-h}, and density N_{e-h}. We have taken into account both TA and LA phonon due to DP and PE couplings for AC phonon processes. As can be seen from Figures 15.2 (a) and (b), the rate of formation is very sensitive to $\mathbf{K}_\parallel$, T_{e-h}, and n_{e-h}. We find from Figure 15.2 (a) that the rate of formation decreases for $\mathbf{K}_\parallel \lesssim 2.24 \times 10^8$ m^{-1} whereas it increases for $\mathbf{K}_\parallel \gtrsim 2.72 \times 10^8$ m^{-1} with increasing T_{e-h} at the charge carrier density $N_{e-h} = 1 \times 10^{10}$ cm^{-2}.

Figure 15.2 (a) also illustrates that the maximum rate of formation occurs at a nonzero value of $\mathbf{K}_\parallel$ in the range 2.16–2.24 $\times 10^8$ m^{-1} which corresponds to a kinetic energy of center-of-mass motion about 20–22 meV. Figure 15.2 (b) shows the rate of formation due to AC phonon emission as a function of T_{e-h} and N_{e-h} at $\mathbf{K}_\parallel = 2.16 \times 10^8$ m^{-1} and indicates that the rate of formation increases very fast in the low carrier-temperature region T_{e-h}, with increasing photoexcited charge carrier density n_{e-h} in comparison with that in the high temperature region.

At higher excitation densities, Auger processes may also contribute to the formation of excitons. In this case, an excited electron–hole pair may form an exciton by giving the excess of energy to another excited charge carrier but not to phonons. However, the reverse process to dissociate excitons will also be equally probable to take place at a high excitation density,

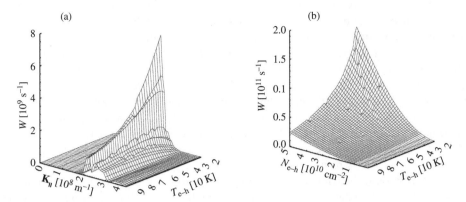

Figure 15.2 The formation rate of an exciton due to acoustic phonon emission in GaAs QWs ($L_z =$ 80 Å) (a) as a function of center-of-mass wave vector $\mathbf{K}_\parallel$ and charge-carrier temperature T_{e-h} at a charge carrier density of $N_{e-h} = 1 \times 10^{10}\,\mathrm{cm}^{-2}$ and as a function of charge-carrier density N_{e-h} and temperature T_{e-h} at $\mathbf{K}_\parallel = 2.16 \times 10^8\,\mathrm{m}^{-1}$ [15.16] [Reprinted from I.-K. Oh and J. Singh, *Int. J. Mod. Phys. B*, **15**, 3660. Copyright (2001) World Scientific]

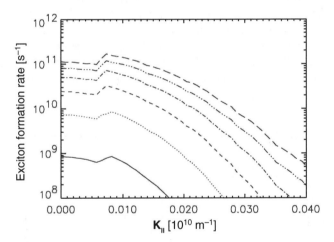

Figure 15.3 The formation rate of an exciton in GaAs quantum wells as a function of the center-of-mass wavevector $\mathbf{K}_\parallel$ for $L_z = 80$ Å, $N_{e-h} = 1 \times 10^{10}\,\mathrm{cm}^{-2}$, and different temperatures, i.e., $T_{e-h} = 30\,\mathrm{K}$ (—), $T_{e-h} = 40\,\mathrm{K}$ (·····), $T_{e-h} = 50\,\mathrm{K}$ (– – –), $T_{e-h} = 60\,\mathrm{K}$ (– · – ·), $T_{e-h} = 70\,\mathrm{K}$ (– ··· – ···), $T_{e-h} = 80\,\mathrm{K}$ (– –) [15.3] [Reprinted with permission from I.-K. Oh et al., *Phys. Rev. B*, **62**, 2045 (2000) with permission of the American Physical Society]

and will deplete the number of excitons thus formed. Therefore, the Auger process of exciton formation may not be able to establish excitonic density as effectively as the phonon-emission processes at low temperatures.

In Figure 15.3 are plotted the calculated values for the formation rate of an exciton as a function of exciton wavevector $\mathbf{K}_\parallel$ for different values of charge carrier temperatures in QWs with well width of $L_z = 80$ Å and charge-carrier density of $N_{e-h} = 1 \times 10^{10}\,\mathrm{cm}^{-2}$. At charge-carrier temperatures $T_{e-h} \lesssim 30\,\mathrm{K}$, the formation of an exciton occurs dominantly at $\mathbf{K}_\parallel = 0$, but for $T_{e-h} \gtrsim 30\,\mathrm{K}$ it occurs at nonzero $\mathbf{K}_{\parallel} \cong 0.0082 \times 10^{10}\,\mathrm{m}^{-1}$. Our results indicate that, first, the formation rate of an exciton decreases with increasing $\mathbf{K}_{\parallel}$, and then it increases to a peak value at about $\mathbf{K}_{\parallel} \cong 0.0082 \times 10^{10}\,\mathrm{m}^{-1}$, after which it decreases continuously as the exciton wavevector $\mathbf{K}_{\parallel}$ increases (see Figure 15.3). Results in Figure 15.3 are in agreement with those of Damen et al. [15.7] that excitons are formed dominantly at $\mathbf{K}_{\parallel} \neq 0$, but our results also suggest that the formation process depends on T_{e-h} and n_{e-h}. For instance, according to Figure 15.3, an exciton can be dominantly formed at $\mathbf{K}_{\parallel} = 0$ at low carrier temperatures $T_{e-h} \lesssim 30\,\mathrm{K}$ for a carrier density of $N_{e-h} = 1 \times 10^{10}\,\mathrm{cm}^{-2}$. However, at higher carrier temperatures the situation can be different. For example, for a fixed carrier density of $N_{e-h} = 1 \times 10^{10}\,\mathrm{cm}^{-2}$, by raising the temperature to $T_{e-h} = 40\,\mathrm{K}$, we find that the formation time $\tau_f = 133\,\mathrm{ps}$ at $\mathbf{K}_{\parallel} = 0$, which is slightly slower than $\tau_f = 118\,\mathrm{ps}$ found at $\mathbf{K}_{\parallel} = 0.0082 \times 10^{10}\,\mathrm{m}^{-1}$. Likewise, if the temperature is raised further to $T_{e-h} = 50\,\mathrm{K}$, it is found again that $\tau_f = 40\,\mathrm{ps}$ at $\mathbf{K}_{\parallel} = 0$, higher than $\tau_f = 30\,\mathrm{ps}$ at $\mathbf{K}_{\parallel} = 0.0074 \times 10^{10}\,\mathrm{m}^{-1}$. This trend seems to continue at even higher temperatures as well (see Figure 15.3).

Figure 15.4 shows the formation rate of an exciton as a function of exciton wavevector $\mathbf{K}_{\parallel}$ at $T_{e-h} = 60\,\mathrm{K}$ for three different well widths, 80, 150, and 250 Å, and two different charge-carrier densities. For all three QWs, the maximum formation rate occurs at nonzero $\mathbf{K}_{\parallel}$. This is obvious from Figure 15.4, which shows that for QW: widths of 80 and 150 Å the maximum formation rate occurs at $\mathbf{K}_{\parallel} = 0.0074 \times 10^{10}\,\mathrm{m}^{-1}$ and $0.0016 \times 10^{10}\,\mathrm{m}^{-1}$, at the carrier density of $N_{e-h} = 1 \times 10^{10}\,\mathrm{cm}^{-2}$, which change to $0.0081 \times 10^{10}\,\mathrm{m}^{-1}$ and $0.0025 \times 10^{10}\,\mathrm{m}^{-1}$, respectively, at the carrier density of $5 \times 10^{10}\,\mathrm{cm}^{-2}$. However, for the well of $L_z = 250\,\text{Å}$, the maximum rate occurs at $\mathbf{K}_{\parallel} = 0.0041 \times 10^{10}\,\mathrm{m}^{-1}$ for both the carrier densities. For QWs of widths $L_z = 80\,\text{Å}$ and $L_z = 250\,\text{Å}$, the formation rates first decrease slightly, and then increase to a maximum, and then decrease again continuously as $\mathbf{K}_{\parallel}$ increases.

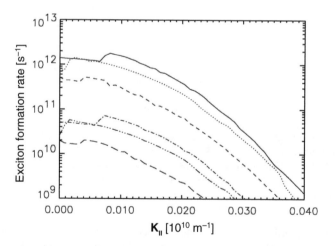

Figure 15.4 The formation rate of an exciton in GaAs QWs as a function of the center-of-mass wavevector $\mathbf{K}_{\parallel}$ at $T_{e-h} = 60\,\mathrm{K}$ for different well widths and charge-carrier densities, $L_z = 80$ Å, $N_{e-h} = 5 \times 10^{10}\,\mathrm{cm}^{-2}$ (—), $L_z = 150$ Å, $N_{e-h} = 5 \times 10^{10}\,\mathrm{cm}^{-2}$ (·····), $L_z = 250$ Å, $N_{e-h} = 5 \times 10^{10}\,\mathrm{cm}^{-2}$ (– – –), $L_z = 80$ Å, $N_{e-h} = 1 \times 10^{10}\,\mathrm{cm}^{-2}$ (–·–·), $L_z = 150$ Å, $N_{e-h} = 1 \times 10^{10}\,\mathrm{cm}^{-2}$ (— ···), and $L_z = 250$ Å, $N_{e-h} = 1 \times 10^{10}\,\mathrm{cm}^{-2}$ (– –) [15.3] [Reprinted with permission from I.-K. Oh et al., *Phys. Rev. B*, **62**, 2045 (2000) with permission of the American Physical Society]

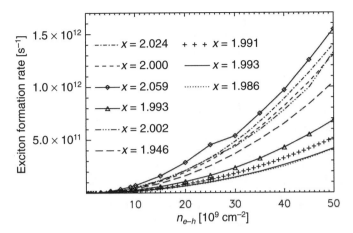

Figure 15.5   The formation rate of an exciton in GaAs QWs as a function of the carrier density N_{e-h} at $T_{e-h} = 60$ K for different well widths and center-of-mass wavevectors, $L_z = 80$ Å $(-\cdot-\cdot)$, $L_z = 150$ Å $(-\triangle-)$, $L_z = 250$ Å $(++++)$ at $\mathbf{K}_{\parallel} = 0$, $L_z = 80$ Å $(---)$, $L_z = 150$ Å $(-\cdots\cdots)$, $L_z = 250$ Å, $(—)$ at $\mathbf{K}_{\parallel} = 0.003 \times 10^{10}$ m^{-1}, and $L_z = 80$ Å $(-\diamond-)$, $L_z = 150$ Å $(--)$, and $L_z = 250$ Å $(\cdots\cdots)$ at $\mathbf{K}_{\parallel} = 0.0075 \times 10^{10}$ m^{-1}. The numbers x in the figure are obtained from the curve fitting $W(N_{e-h}) = b \, N_{e-h}^x$ [15.3] [Reprinted with permission from I.-K. Oh et al., *Phys. Rev. B*, **62**, 2045 (2000) with permission of the American Physical Society]

However, the formation rate for $L_z = 150$ Å QWs first increases up to its maximum and then decreases continuously with increasing $\mathbf{K}_{\parallel}$. The formation time corresponding to the maximum formation rate in the $L_z = 150$ Å QW is obtained as $\tau_f = 17$ ps and 0.7 ps at the carrier density of $N_{e-h} = 1 \times 10^{10}$ cm^{-2} and $N_{e-h} = 5 \times 10^{10}$ cm^{-2}, respectively. Such a trend, that the formation time corresponding to the maximum rate becomes shorter with an increase in the carrier density, is obtained in all the three well widths.

In Figure 15.5, we have plotted the rate of formation of an exciton as a function of charge-carrier density N_{e-h}, by varying it from 1×10^8 to 5×10^{10} cm^{-2} at $T_{e-h} = 60$ K for the three well widths. From the calculated results of the formation rate, we have carried out curve fittings using the relation $W(N_{e-h}) = bN_{e-h}^x$, where b and x are fitting parameters. Our results give $x \approx 2$ as shown in Figure 15.5. In other words, our theory shows a square-law dependence of the formation rate of an exciton on N_{e-h}. This provides a theoretical confirmation of the observed experimental square-law dependence of the photoluminescence on excitation density obtained by Strobel et al. [15.18].

For comparison between acoustic and LO phonon processes, we have plotted in Figures 15.6 (a) and (b) the rate of formation of an exciton calculated as a function of $\mathbf{K}_{\parallel}$ at $T_{e-h} = 20$, 50, and 80 K for two different charge-carrier densities $N_{e-h} = 1 \times 10^{10}$ cm^{-2} and $N_{e-h} = 5 \times 10^{10}$ cm^{-2}, respectively, and at a lattice temperature of 4.2 K. The —, ..., and $-\cdot-\cdot$ curves represent the formation rates due to acoustic phonon emission, and the $-\cdots$, $---$, and $——$ curves are the formation rates due to LO phonon emission at $T_{e-h} = 20$, 50, and 80 K, respectively. Results of Figure 15.6 show that the rate of formation of excitons emitting acoustic phonons first increasing with increaing $\mathbf{K}_{\parallel}$ to a peak value and then decreases with increasing $\mathbf{K}_{\parallel}$. It is also found that the charge-carrier temperature (T_{e-h}) dependence of

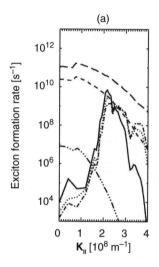
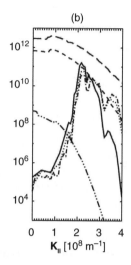

Figure 15.6   The formation rate of an exciton in GaAs QWs is plotted as a function of the center-of-mass wavevector $K_\parallel$ for $L_z = 80$ Å, $T = 4.2$ K, (a) $N_{e-h} = 1 \times 10^{10}$ cm^{-2} and (b) $N_{e-h} = 5 \times 10^{10}$ cm^{-2}. The —, ·····, and – · – · curves correspond to the formation rate due to acoustic phonon emission at T_{e-h} = 20, 50, and 80 K, respectively. The – ···, – – –, and – – curves correspond to the formation rates due to LO phonon emission at T_{e-h} = 20, 50, and 80 K, respectively [Reprinted from I.-K. Oh and J. Singh, *Superlattices Microstruct.*, **30**, 221. Copyright (2001) Elsevier]

the rate of formation changes before and after the peak value. The rate of formation, including its peak position, decreases with increasing T_{e-h}, but at a certain $K_\parallel$ after the peak value, the rate increases with increasing T_{e-h} (see Figure 15.6). This dependence of formation rate of excitons on carrier temperature via acoustic phonons is different from that via LO phonons: in the latter case, the rate of exciton formation increases with increasing T_{e-h} for all $K_\parallel$ [15.3]. As shown in Figure 15.6, the maximum rate of formation due to acoustic phonon emission occurs at relatively larger value of $K \approx 2.24 \times 10$ m^{-1} in comparison with that due to LO phonon emission at $K \approx 0.82 \times 10^8$ m^{-1} for $N_{e-h} = 5 \times 10^{10}$ cm^{-2} and $T_{e-h} = 50$ K. In other words, our results suggest that the rate of formation of hot excitons (with large $K_\parallel$ values) due to acoustic phonon emission is much faster at $T_{e-h} \lesssim 40$ K, but then as T_{e-h} approaches ~50 K it becomes comparable with the rate of formation due to LO phonon. At $T_{e-h} \gtrsim 50$ K, the formation of an exciton via LO phonon emission becomes more efficient in comparison with that via acoustic phonon emission, which agrees reasonably well with Piermarocchi et al.'s result [15.18, 15.19]. Our results also show that the rate of formation due to acoustic phonon emission is relatively more sensitive to the center-of-mass wavevector $K_\parallel$ of excitons compared with that due to LO phonon emission. In particular, we have found that excitons are formed mainly due to LO phonon interaction at relatively small values of $K_\parallel$, but at relatively large $K_\parallel$ the acoustic phonon interaction becomes dominant. For instance, the LO phonon process dominates for $K_\parallel \lesssim 2.0 \times 10^8$ m^{-1} and the acoustic phonon process for $K_\parallel \gtrsim 2.0 \times 10^8$ m^{-1} at $T_{e-h} = 40$ K and $T = 4.2$ K. The maximum rate of formation due to LO phonon emission is at $K_\parallel \sim (0.74–0.82) \times 10^8$ m^{-1} (see Figure 15.6), which corresponds to a much smaller kinetic energy of about 2–3 meV in comparison with

20–22 meV obtained for the acoustic phonon process. This can be explained as follows: when the energy of photoexcited free electron–hole pairs is high enough, excitons formed at relatively small values of $\mathbf{K}_{\parallel}$ are mainly due to the LO phonon process whereas those at relatively large values of $\mathbf{K}_{\parallel}$ are due to the acoustic phonon process because the energy involved in the formation of an exciton due to LO phonon emission is much larger than that due to AC phonon emission. In other words, as an excited electron–hole pair in a given state loses more energy in an LO phonon emission than that in an acoustic phonon emission, the exciton formed due to LO phonon emission has less kinetic energy than that formed due to acoustic phonon emission.

15.4 NONRADIATIVE RELAXATION OF FREE EXCITONS

As discussed in the previous section, the formation of excitons from free electron–hole pairs has nonzero $\mathbf{K}_{\parallel}$ exciton states. These excitons are 'hot' so they will relax, first, nonradiatively by emitting phonons, and then recombine radiatively by emitting photons. Therefore, the process of relaxation plays a significant role in the properties of photoexcited semiconductor systems. There are two important processes in the relaxation in QWs, intraband and interband. Here we first consider intraband transitions of relaxation and then interband transitions.

15.4.1 Intraband processes

The interaction Hamiltonian associated with the intraband transition of a 1 s exciton by interaction with a λ-mode phonon via J ($J = D, P, O$) couplings can be written in the second quantized form as [15.5, 15.20]:

$$\hat{H}^{J}_{ex-ph,\lambda}(\mathbf{q}_{\parallel},q_z) = \sum_{\mathbf{K}_{\parallel}} C^{\lambda}_{J} \left[G^{J}_{\lambda-}(\mathbf{q}_{\parallel},q_z) \hat{B}^{\dagger}_{\mathbf{K}_{\parallel}+\mathbf{q}_{\parallel}} \hat{B}_{\mathbf{K}_{\parallel}} \hat{b}_{\lambda\mathbf{q}} + G^{J}_{\lambda+}(\mathbf{q}_{\parallel},q_z) \hat{B}^{\dagger}_{\mathbf{K}_{\parallel}-\mathbf{q}_{\parallel}} \hat{B}_{\mathbf{K}_{\parallel}} \hat{b}^{\dagger}_{\lambda\mathbf{q}} \right] \quad (15.19)$$

where the subscripts + and − correspond to phonon emission and absorption processes, respectively. It should also be noted that $G^{J}_{\lambda+}(\mathbf{q}_{\parallel},q_z) = G^{J*}_{\lambda-}(\mathbf{q}_{\parallel},q_z)$.[1] $G^{D}_{\lambda-}(\mathbf{q}_{\parallel},q_z)$, $G^{P}_{\lambda-}(\mathbf{q}_{\parallel},q_z)$, and $G^{O}_{LO-}(\mathbf{q}_{\parallel},q_z)$ are obtained as [15.5]:

$$G^{D}_{\lambda-}(\mathbf{q}_{\parallel},q_z) = i\left[\Xi^{\lambda}_{c}(\mathbf{q}) F_{e}(q_z) G(\alpha_{h}\mathbf{q}_{\parallel}) - \Xi^{\lambda}_{v}(\mathbf{q}) F_{h}(q_z) G(-\alpha_{e}\mathbf{q}_{\parallel})\right] \quad (15.20)$$

$$G^{P}_{\lambda-}(\mathbf{q}_{\parallel},q_z) = i\left[F_{e}(q_z) G(\alpha_{h}\mathbf{q}_{\parallel}) - F_{h}(q_z) G(-\alpha_{e}\mathbf{q}_{\parallel})\right] \quad (15.21)$$

and

$$G^{O}_{LO-}(\mathbf{q}_{\parallel},q_z) = \left[F_{e}(q_z) G(\alpha_{h}\mathbf{q}_{\parallel}) - F_{h}(q_z) G(-\alpha_{e}\mathbf{q}_{\parallel})\right] \quad (15.22)$$

---

[1]In references [15.5] and [15.20], we have used $F^{J}_{\lambda\mp}$ for $G^{J}_{\lambda\mp}$.

Here the form $G(\alpha \mathbf{q}_\parallel)$ is obtained as [15.5]:

$$G(\alpha \mathbf{q}_\parallel) = \int d\mathbf{r}_\parallel e^{i\alpha \mathbf{q}_\parallel \cdot \mathbf{r}_\parallel} |\phi_x (\mathbf{r}_\parallel)|^2 \qquad (15.23)$$

Using the variational wavefunction of Equation (15.15), we get an analytical expression for the form factor G as [15.5]:

$$G(\alpha \mathbf{q}_\parallel) = \left[\left(\frac{\alpha q_\parallel}{2\beta} \right)^2 + 1 \right]^{-3/2} \qquad (15.24)$$

Then from Equation (15.19) and Fermi's golden rule, the relaxation rate of an exciton with wavevector $\mathbf{K}_\parallel$ via J ($J = D, P, O$) couplings due to emission (absorption) of a phonon is given by [15.5]:

$$W_\pm^{J,\lambda} (\mathbf{K}_\parallel) = \sum_{\mathbf{q}_\parallel, q_z} W_\lambda^J (\mathbf{q}_\parallel, q_z) \left(f_{\mathbf{K}_\parallel \mp \mathbf{q}_\parallel}^{ex} + 1 \right) f_{\mathbf{K}_\parallel}^{ex}$$

$$\times \left(n_q^\lambda + \frac{1}{2} \pm \frac{1}{2} \right) \delta (E(\mathbf{K}_\parallel \mp \mathbf{q}_\parallel) - E(\mathbf{K}_\parallel) \pm \hbar \omega_{\mathbf{q},\lambda}) \qquad (15.25)$$

where the upper (lower) sign cooresponds to the emission (absorption) of a phonon and $W_\lambda^J(\mathbf{q}_\parallel, q_z)$ is given by [15.5]:

$$W_\lambda^J (\mathbf{q}_\parallel, q_z) = \frac{2\pi}{\hbar} |C_J^\lambda G_{\lambda+}^J (\mathbf{q}_\parallel, q_z)|^2 = \frac{2\pi}{\hbar} |C_J^\lambda G_{\lambda-}^J (\mathbf{q}_\parallel, q_z)|^2 \qquad (15.26)$$

Here we study the total phonon emission from all excitonic relaxation processes as functions of the phonon wavevector $\mathbf{q}$, exciton density, and exciton temperature. Taking into account the reverse excitonic process involving phonon emission, the net phonon-emission rate of a λ-mode phonon with wavevector $\mathbf{q}$ can be written as:

$$\frac{\partial n_\mathbf{q}^\lambda}{\partial t} = \sum_{J, \mathbf{K}_\parallel} W_\lambda^J (\mathbf{q}_\parallel, q_z) \{ [(f_{\mathbf{K}_\parallel - \mathbf{q}_\parallel}^{ex} + 1) f_{\mathbf{K}_\parallel}^{ex} (n_q^\lambda + 1) - (f_{\mathbf{K}_\parallel}^{ex} + 1) f_{\mathbf{K}_\parallel - \mathbf{q}_\parallel}^{ex} n_q^\lambda]$$

$$\times \delta(E(\mathbf{K}_\parallel - \mathbf{q}_\parallel) - E(\mathbf{K}_\parallel) + \hbar \omega_{\mathbf{q},\lambda}) - [(f_{\mathbf{K}_\parallel + \mathbf{q}_\parallel}^{ex} + 1) f_{\mathbf{K}_\parallel}^{ex} n_q^\lambda - (f_{\mathbf{K}_\parallel}^{ex} + 1) f_{\mathbf{K}_\parallel + \mathbf{q}_\parallel}^{ex} (n_q^\lambda + 1)]$$

$$\times \delta(E(\mathbf{K}_\parallel + \mathbf{q}_\parallel) - E(\mathbf{K}_\parallel) - \hbar \omega_{\mathbf{q},\lambda}) \} \qquad (15.27)$$

The phonon-emission rate obtained in Equation (15.27) is the same as that derived by Vass [15.21] for $f_{\mathbf{K}_\parallel}^{ex} \ll 1$ and using the same coupling constants. Furthermore, if we ignore the excitonic transitions $\mathbf{K}_\parallel \rightarrow \mathbf{K}_\parallel - \mathbf{q}_\parallel$ and $\mathbf{K}_\parallel \leftarrow \mathbf{K}_\parallel - \mathbf{q}_\parallel$ and use the same coupling constants then the rate in Equation (15.27) also agrees with that obtained by Takagahara [15.22] provided that $f_{\mathbf{K}_\parallel}^{ex} \ll 1$. Thus the present result can be regarded to be more general and applicable to calculating the rates of all three types of phonon emissions.

In Figure 15.7 (a) we show the rate of an LA phonon emission due to the deformation-potential (DP) interaction as a function of the direction of phonon emission in spherical polar coordinates (angles θ and ϕ). In this case, the rate is not very sensitive to the ϕ angle, but it is quite sensitive to the θ angle. The rate has a maximum value of about $2.1 \times 10^9 \, \text{s}^{-1}$ at

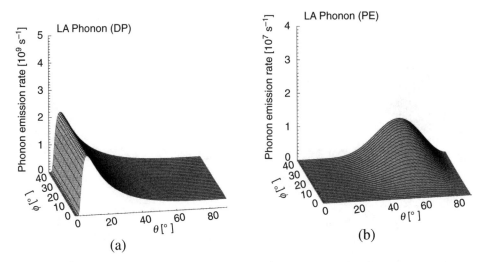

Figure 15.7 LA phonon-emission rates due to (a) deformation potential (DP) and (b) piezoelectric (PE) couplings, as a function of phonon-emission angles for GaAs/Al$_{0.3}$Ga$_{0.7}$As QWs with well width $L_z = 80$Å, phonon energy $E_{ph} = 0.2$ meV, exciton density $N_{ex} = 5 \times 10^{10}$ cm^{-2}, exciton temperature $T_{ex} = 20$ K, and lattice temperature $T = 4.2$ K [15.5] [Reprinted from I.-K. Oh and J. Singh, *J. Lumin.*, **85**, 233. Copyright (2000) Elsevier]

$\theta \sim 7°$, but it is nearly independent of ϕ. In Figure 15.7 (b) we show the rate of LA phonon emission using the PE interaction. Unlike Figure 15.7 (a) with deformation potential, the rate of scattering depends on both θ and ϕ angles. The maximum rate of phonon emission is about 1.1×10^7 s^{-1} at $\theta \sim 61°$ and $\phi \sim 45°$.

Likewise we have also plotted the rate of emission of TA phonons in Figure 15.8 (a) with DP and in Figure 15.8 (b) with PE coupling. Although the maximum rate of scattering with the emission of TA phonons due to DP is higher ($\approx 5.4 \times 10^9$ s^{-1} at $\theta \sim 4°$), the directional dependence is similar to that of LA phonons shown in Figure 15.7 (a). However, TA phonon emission [Figure 15.8 (b)] due to PE coupling is quite different from LA phonon emission [Figure 15.7 (b)]. In Figure 15.8 (b), one can see three peaks of emission of TA phonons. The first maximum at $\theta \sim 37°$ and $\phi \sim 45°$, the third maximum at $\theta = 90°$ and $\phi = 45°$ correspond to TA$_1$ phonons, and the second maximum at $\theta \sim 37°$ and $\phi \sim 0°$ is from TA$_2$ phonons. From our calculations, we find that LA and TA phonons are equally effective in excitonic processes of GaAs QWs. Furthermore, the emission rate of TA phonons due to deformation potential is slightly higher than that of LA phonons. This result is quite different from those obtained earlier [15.21–15.23], because the anisotropy effect has not been taken into account previously. As a result only the contribution of exciton–LA phonon interaction has been taken into account in earlier works [15.21–15.23].

In Figure 15.9 we have plotted the rate of LO phonon emission as a function of angle θ and the phonon wavevector q, which gives a peak rate of about 1.3×10^9 s^{-1} at $q = 3 \times 10^8$ m^{-1} and $\theta \sim 90°$, which is the plane of the QW. As the exciton–LO phonon interaction is isotropic in the x–y plane, the rate of scattering is independent of angle ϕ. The results shown in Figures 15.7–15.9 illustrate the point that exciton scattering is dominated by the

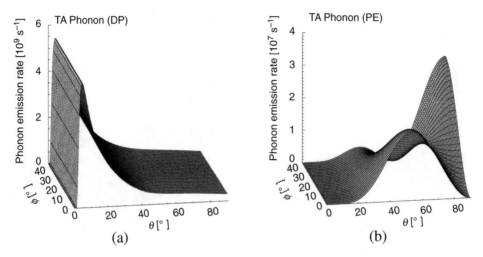

Figure 15.8 TA phonon emission rates due to (a) deformation potential (DP) and piezoelectric (PE) couplings, as a function of phonon-emission angles for GaAs/Al$_{0.3}$Ga$_{0.7}$As QWs with well width $L_z = 80$ Å, phonon energy $E_{ph} = 0.2$ meV, exciton density $N_{ex} = 5 \times 10^{10}$ cm^{-2}, exciton temperature $T_{ex} = 20$ K, and lattice temperature $T = 4.2$ K [15.5] [Reprinted from I.-K. Oh and J. Singh, *J. Lumin.*, **85**, 233. Copyright (2000) Elsevier]

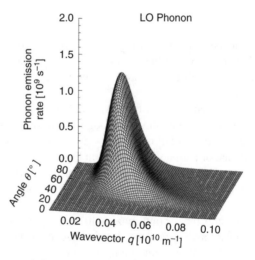

Figure 15.9 LO phonon-emission rate, as a function of θ and q for GaAs/Al$_{0.3}$Ga$_{0.8}$As QWs with $L_z = 80$ Å, $N_{ex} = 8 \times 10^{10}$ cm^{-2}, $T_{ex} = 80$ K, and $T = 4.2$ K [15.5] [Reprinted from I.-K. Oh and J. Singh, *J. Lumin.*, **85**, 233. Copyright (2000) Elsevier]

emission of acoustic phonons due to deformation potential at low θ angles, whereas LO phonon processes become dominant at high θ angles. The exciton scattering is dominated by acoustic phonons at small phonon wavevector $q \ll 3 \times 10^8$ m^{-1}, but LO phonons dominate near $q = 3 \times 10^8$ m^{-1}. It is to be remembered that the phonon wavevector is obtained from the momentum conservation equal to the difference of exciton wavevectors in the

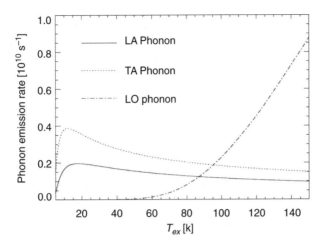

Figure 15.10   Phonon-emission rate, as a function of T_{ex} for GaAs/Al$_{0.3}$Ga$_{0.7}$As QWs with L_z = 80 Å, N_{ex} = 5 × 10^{10} cm^{-2}, and T = 4.2 K [15.5] [Reprinted from I.-K. Oh and J. Singh, *J. Lumin.*, **85**, 233. Copyright (2000) Elsevier]

initial and final states ($|\mathbf{K}_{\|} - \mathbf{K}_{\|}'| = q$). Therefore, for LO phonon emission we have used a fixed energy (e.g., $\hbar\omega_{LO}$ = 36.2 meV for GaAs) and hence the rate has a peak value at a specific q value obtained from the value of momentum conservation. However, in the case of acoustic phonons, phonon energy depends linearly on the phonon wavevector, and therefore we get the phonon emission rate as a decreasing function of q.

The calculated results of the dependence of phonon-emission rate on the exciton temperature suggest that the rate of LO phonon emission increases as exciton temperature increases from 5 K to 200 K. However, the rate of acoustic phonon emission increases first rapidly for exciton temperature $5 < T_{ex} < 17$ K at exciton densities in the range of 5 × 10^9 to 8 × 10^{11} cm^{-2}, and then it decreases for $17 \approx 19 < T_{ex} < 200$ K. It is also to be noted that our results suggest that the LO phonon-emission rate is dominant at higher exciton temperatures, whereas the acoustic phonon-emission rate is dominant at lower temperatures. We have plotted rates of phonon emission in Figure 15.10 as a function of T_{ex} at an exciton density N_{e-h} = 5 × 10^{10} cm^{-2}. Accordingly, TA phonon processes are relatively more efficient at temperatures $T_{ex} \leq 87$ K and $T_{ex} \leq 95$ K, respectively, in comparison with LO phonon processes. However, as temperature increases, LO phonon processes dominate both LA and TA phonon processes. As the exciton temperature is directly proportional to the exciton kinetic energy at a given exciton density, this suggests that if the the exciton scattering takes place at higher exciton wavevector, the LO phonon emission will be dominant. At lower exciton wavevectors, acoustic phonon emission will be dominant. This agrees very well with the experimental results reported for GaAs QWs [15.24], where a faster exciton formation time is observed due to the involvement of LO phonons at higher exciton wavevectors.

It is important to discuss the dependence of the phonon-emission rate on the exciton density N_{ex}. Our results suggest that at a given temperature of exciton and lattice, the rate of phonon emission first increases with N_{ex} and then shows a kind of saturation at higher exciton densities. For example, for $T_{ex} = 20$ K and lattice temperature 4.2 K, the rate of emission of LA phonons with energy 0.2 meV becomes saturated at an exciton density $N_{ex} \sim 2.5 \times 10^{12}$ cm^{-2}. Likewise the rate of LO phonon emission becomes saturated at $N_{ex} \sim 2.5 \times 10^{12}$ cm^{-2} for $T_{ex} = 50$ K, and lattice temperature 4.2 K.

15.4.2 Interband processes

In semiconductor QWs, depending on the well width and structure of the confinement potential, there exist several electron and hole sub-bands through which the exciton formation and relaxation may occur. For example, in GaAs/Al$_x$ Ga$_{1-x}$As QWs ($x = 0.3$), there is only one electron sub-band and only one heavy-hole sub-band for well width $L_z \lesssim 41$Å.[2] In this case, the relaxation by inter-sub-band transition is not possible. In QWs of width $L_z \approx 42$–47 Å, there are two electron sub-bands ($n_e = 1, 2$) and one heavy-hole sub-band ($n_h = 1$). Then there are two electron sub-bands ($n_e = 1, 2$) and two heavy-hole sub-bands ($n_h = 1, 2$) in QWs of well width $L_z \approx 48$–82 Å. QWs of even larger widths have increasingly more sub-bands. The number of sub-bands increases with increasing well width but the energy separation between adjacent sub-bands decreases. This is an important point to consider in conserving the energy and momentum in inter-sub-band transitions. In the study of dynamics of excitons in intrinsic QWs, both processes, exciton formation and relaxation, are important. When charge carriers in QWs are created by optical excitations, the transitions occurring at $\Delta n = n_e - n_h = 0$ are called dipole allowed, and for $\Delta n \neq 0$ dipole forbidden [15.25]. Among the transitions with $\Delta n \neq 0$, those with $\Delta n =$ even are parity-allowed and with $\Delta n =$ odd are parity-forbidden [15.26]. Thus, $\Delta n =$ even ($\neq 0$) transitions are called parity-allowed forbidden transitions and $\Delta n =$ odd are called parity-forbidden forbidden transitions. The parity-allowed forbidden transitions with $\Delta n =$ even have been observed in parabolic QWs [15.27] and the parity-forbidden forbidden transitions with $\Delta n =$ odd have been observed in square-well QWs at very high excitation intensity [15.26]. This illustrates quite clearly that an exciton can be excited through several possible combinations of electron and hole sub-bands. Each excitonic state corresponding to an optically allowed transition will have its own discrete Rydberg series. Only the exciton state formed from the lowest electron ($n_e = 1$) and hole ($n_h = 1$) sub-band is well separated from the continuum states of free electron–hole pairs (see Figure 15.11). In other words, an exciton formed from higher electron or/and hole sub-bands, e.g., $n_e = 2$ and $n_h = 1$ or $n_e = 2$ and $n_h = 2$ sub-bands, is embedded in the continuum states of free electron–hole pairs formed from lower electron and hole sub-bands [15.28] [see Figure 15.11 (b)]. It is known that a resonant Coulomb interaction between the discrete exciton states and the continuum states of free electron–hole pairs results in Fano interference [15.29], which gives rise to an asymmetric absorption line shape called the Fano profile. As a result, the study of the dynamics of excitons or free electron–hole pairs in QWs involving inter-sub-band transitions due to LO phonon interaction becomes quite complicated.

The relaxation of excitons by emitting photons (photoluminescence) also follows the same selection rules as in excitation. However, the excitonic relaxation by emitting phonons

---

[2]For the calculations of electron and hole sub-band structures, the band discontinuities $\Delta E_c = 1.1x$ eV and $\Delta E_v = 0.17x$ eV ($x = 0.3$) for conduction and valence bands, respectively, have been used as given in ref. [15.25].

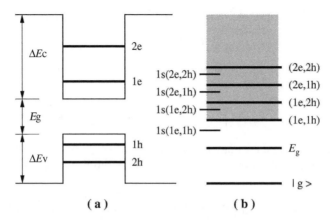

Figure 15.11 Schematic illustration of (a) the structure of electron and heavy-hole sub-bands and (b) energy levels of the 1s exciton states and continuum states of free electron–hole pairs (shaded) formed from possible combinations of electron and hole sub-bands for a GaAs QW $L_z = 48$–82 Å ref. [15.6] [Reprinted with permission from I.-K. Oh, J. Singh, and A.S. Vengurlekar, *J. Appl. Phys.*, **91**, 5796. Copyright (2002) American Institute of Physics]

is different. The relaxation of a higher-order exciton 1s (n_e e, n_h h) to any lower-order free electron–hole pair (n'_e e, n'_h h) by emitting phonons can be a multi-channel process. Here 1s (n_e e, n_h h) denotes the 1s excitonic state created by exciting an electron in the n_e-th and a hole in the n_h-th sub-bands, and (n'_e e, n'_h h) denotes a free electron–hole pair with an electron in the n'_e-th and a hole in the n'_h-th sub-bands. For example, when an 1s (2e, 2h) exciton formed in $n_e = n_h = 2$ sub-bands relaxes to a (1e, 1h) free electron–hole pair in $n_e = n_h = 1$ sub-bands, it can do so in one step as ($n_e = n_h = 2 \rightarrow n_e = n_h = 1$) or in two steps with two possibilities ($n_e = n_h = 2 \rightarrow n_e = 1$, $n_h = 2 \rightarrow n_e = n_h = 1$) or ($n_e = n_h = 2 \rightarrow n_e = 2$, $n_h = 1 \rightarrow n_e = n_h = 1$). However, as explained later, the orthogonality of sub-band states does not allow a direct excitonic relaxation from 1s (2e, 2h) to (1e, 1h) by emitting an LO phonon. Here we consider only LO phonon processes but the same selection rules apply to acoustic (AC) phonons as well. In the two-step process, energy conservation may exclude one of the steps. This happens when the energy separation between the adjacent sub-bands is less than the LO phonon energy.

The interaction Hamiltonian for the relaxation of an exciton into a free electron–hole pair associated with interband transitions due to LO phonon coupling can be written in the second quantized form as [15.6]:

$$\hat{H}_I^{ex \rightarrow eh} = \frac{1}{\sqrt{A_0}} \sum_{\mathbf{K}\,\mathbf{k}\,\mathbf{q}\,q_z} C_0^{LO} \Big[F_{n_f m_{f-}}^{n_i m_i}(\mathbf{q}\,,q_z,\mathbf{k})\hat{d}^\dagger_{m_f,\alpha_{h(\mathbf{K}\,-\mathbf{q}\,)-\mathbf{k}}}\,\hat{a}^\dagger_{n_f,\alpha_{e(\mathbf{K}\,-\mathbf{q}\,)+\mathbf{k}}}\,\hat{b}_\mathbf{q}\hat{B}_\mathbf{K}^{n_i m_i}$$
$$+F_{n_f m_{f+}}^{n_i m_i}(\mathbf{q}\,,q_z,\mathbf{k})\hat{d}^\dagger_{m_f,\alpha_{h(\mathbf{K}\,+\mathbf{q}\,)-\mathbf{k}}}\,\hat{a}^\dagger_{n_f,\alpha_{e(\mathbf{K}\,+\mathbf{q}\,)+\mathbf{k}}}\,\hat{b}^\dagger_\mathbf{q}\hat{B}_\mathbf{K}^{n_i m_i} \Big] \qquad (15.28)$$

where $\hat{B}_{\mathbf{K}_\parallel}^{nm}$ is the annihilation operator of an exciton formed from the n-th electron and m-th hole sub-bands with center-of-mass momentum wavevector $\mathbf{K}_\parallel$ in the plane of QW, and $\hat{a}^\dagger_{n,\alpha_e\mathbf{K}_\parallel+\mathbf{k}_\parallel}$ and $\hat{d}^\dagger_{m,\alpha_h\mathbf{K}_\parallel-\mathbf{k}_\parallel}$ denote electron and hole creation operators in the n-th electron and m-th hole sub-bands, respectively. $F_{n_j m_{f_z}}^{n_i m_i}$ is given by [15.6]:

$$F_{n_f m_f \mp}^{n_i m_i}(\mathbf{q}_\|, q_z, \mathbf{k}_\|) = \left[F_e^{n_i n_f}(\pm q_z) G_1^{n_i m_i}(\mathbf{k}_\| \pm \alpha_h \mathbf{q}_\|) \delta_{m_i, m_f} \right.$$
$$\left. - F_h^{m_i m_f}(\pm q_z) G_1^{n_i m_i}(\mathbf{k}_\| \pm \alpha_e \mathbf{q}_\|) \delta_{n_i, n_f} \right] \tag{15.29}$$

Here $F_j^{nm}(q_z)$ ($j = e, h$) are the form factors associated with the inter-sub-band transitions and they are obtained as [15.6]:

$$F_j^{nm}(q_z) = \int dz_j \phi_m^{j*}(z_j) \phi_n^j(z_j) e^{iq_z z_j}, \quad j = e, h \tag{15.30}$$

and G_1^{nm} is obtained using the variational wavefunction[3] $\phi_x^{nm} = (\mathbf{r}_\|) = \beta_{nm} \sqrt{2/\pi} \, e^{-\beta_{nm} r_\|}$ of the 1 s exciton formed from the n-th electron and m-th hole sub-bands as [15.6]:

$$G_1^{nm}(\mathbf{K}_\| + \alpha \mathbf{q}_\|) = \int d\mathbf{r}_\| \phi_x^{nm}(\mathbf{r}_\|) e^{i(\mathbf{k}_\| + \alpha \mathbf{q}_\|) \cdot \mathbf{r}_\|} = \sqrt{\frac{8\pi}{\beta_{nm}^2}} \left[\left(\frac{|\mathbf{K}_\| + \alpha \mathbf{q}_\||}{\beta_{nm}} \right)^2 + 1 \right]^{-3/2} \tag{15.31}$$

The first term of Equation (15.28) corresponds to the inter-sub-band transition of an exciton due to an LO phonon absorption and the second term corresponds to that due to an LO phonon emission. However, in a relaxation process at low crystal temperatures as considered here, only the second term of emission of an LO phonon needs to be considered. It is to be noted here that, according to Equation (15.29), the inter-sub-band relaxation Hamiltonian is nonzero only if one of the charge carriers (electron or hole) remains in its sub-band during the transition and the other goes through an inter-sub-band transition. In Equation (15.30), the first term is zero if $m_i \neq m_f$ and the second term is zero if $n_i \neq n_f$. In other words, a direct inter-sub-band relaxation from an 1 s (2e, 2h) ($n_i = m_i = 2$) exciton to a (1e, 1h) ($n_f = m_f = 1$) free electron–hole pair is not allowed. This is due to the orthogonality of sub-band states. Then the total rate of relaxation of an exciton with wavevector $\mathbf{K}_\|$ is obtained by [15.6]:

$$W_{n_f m_f +}^{n_i m_i}(\mathbf{K}_\|) = \frac{1}{A_0} \frac{2\pi}{\hbar} \sum_{\mathbf{q}_\|, q_z, \mathbf{k}_\|} \left| C_0^{LO} F_{n_f m_f +}^{n_i m_i}(\mathbf{q}_\|, q_z, \mathbf{k}_\|) \right|^2 \left[1 - f_{m_i}^h(\alpha_h(\mathbf{K}_\| + \mathbf{q}_\|) - \mathbf{k}_\| x) \right]$$
$$\times \left[1 - f_{n_i}^e(\alpha_e(\mathbf{K}_\| + \mathbf{q}_\|) + \mathbf{k}_\|) \right] f_{n_i m_i}^{ex}(\mathbf{K}_\|) [n(\mathbf{q}) + 1] \delta \left(E_{e-h}^{n_f m_f} - E_{ex}^{n_i m_i} + \hbar \omega_{LO} \right) \tag{15.32}$$

where f_n^e, f_m^h, f_{nm}^{ex}, and $n(\mathbf{q})$ are the occupation numbers of electrons in the n-th sub-band, holes in the m-th sub-band, excitons formed from them, and phonons, respectively, and:

$$E_{e-h}^{n_f m_f} = \frac{\hbar^2 |\mathbf{K}_\| + \mathbf{q}_\||^2}{2M_\|^*} + \frac{\hbar^2 |\mathbf{K}_\||^2}{2\mu_\|^*} + E_{n_f}^e + E_{m_f}^h \tag{15.33}$$

and

$$E_{ex}^{n_i m_i} = \frac{\hbar^2 |\mathbf{K}_\||^2}{2M_\|^*} - E_b^{n_i m_i} + E_{n_i}^e + E_{m_f}^h \tag{15.34}$$

---

[3] In reference [15.6], we have used β for β_{nm}.

where $E_b^{n_i m_i}$ is the exciton binding energy, and E_n^j ($j = e, h$) the energy of n-th sub-band. The total rate $W_{n_f m_{f+}}^{n_i m_i}$ ($\mathbf{K}_\parallel$) thus obtained gives the inverse of the time of relaxation, $1/\tau_0$ ($\mathbf{K}_\parallel$) = $W_{n_f m_{f+}}^{n_i m_i}$ ($\mathbf{K}_\parallel$), of an exciton with wavevector $\mathbf{K}_\parallel$ [15.6].

Using the conduction- and valence-bands discontinuities [see Figure 15.11 (a)] as $\Delta E_c = 1.1x\,\mathrm{eV}$ and $\Delta E_v = 0.17x\,\mathrm{eV}$ [15.25], respectively, we have calculated the structure of electron and valence-bands discontinuities [see Figure 15.11 (a)] as $\Delta E_c = 1.1x\,\mathrm{eV}$ and $\Delta E_v = 0.17x\,\mathrm{eV}$ [15.25], respectively, we have calculated the structure of electron and heavy-hole sub-bands, binding energy, and wave functions of a 1 s exciton formed from the possible combinations of electron and hole sub-bands. There are two electron sub-bands and one hole sub-band at a well width in the range $L_z = 42$–$47\,\text{Å}$, two electron and two hole sub-bands in the range $L_z = 48$–$82\,\text{Å}$, and four electron and three hole sub-bands at $L_z = 130\,\text{Å}$. Therefore, for $L_z = 42$–$47\,\text{Å}$ QWs, there is only one possible inter-sub-band relaxation of an exciton [1 s (2e, 1h)] formed initially with an excited electron in $n_e = 2$ and hole in $n_h = 1$ sub-bands to a free electron–hole pair [(1e, 1h)] with the electron moving down to $n_e = 1$ sub-band by emitting an LO phonon. For $L_z = 48$–$82\,\text{Å}$ QWs, there are in principle four possible 1 s excitonic states and thus four channels of relaxation of an exciton formed initially in a higher sub-band (see Figure 15.11). If the exciton is formed initially in the exciton state [1 s (2e, 2h)] with an electron in the $n_e = 2$ and a hole in the $n_h = 2$ sub-bands, then there are in principle two possible channels of relaxations: (1) The exciton relaxes to a free electron–hole pair [(2e, 1h)], where the electron remains in $n_e = 2$ sub-band but the hole relaxes from the $n_h = 2$ to the $n_h = 1$ sub-band and (2) the exciton relaxes to a free electron–hole pair [(1e, 2h)], where the hole remains in the $n_h = 2$ sub-band but the electron relaxes from the $n_e = 2$ to the $n_e = 1$ sub-band. The third possible channel is when the exciton is initially created in the state [1 s (2e, 1h)] by exciting an electron in the $n_e = 2$ and a hole in the $n_h = 1$ sub-bands, it can then relax to a free electron–hole pair [(1e, 1h)] such that the hole remains in the $n_h = 1$ sub-band but the electron relaxes from the $n_e = 2$ to the $n_e = 1$ sub-band. The fourth possible channel of relaxation is when an exciton excited initially in the state [1 s (1e, 2h)] with an electron in the $n_e = 1$ and a hole in the $n_h = 2$ sub-bands relaxes to a free electron–hole pair [(1e, 1h)], where the electron remains in the $n_e = 1$ sub-band but the hole relaxes from the $n_h = 2$ to the $n_h = 1$ sub-band. However, since the energy separation between the first and second hole sub-bands for such QWs is in the range of 24–34 meV, the inter-sub-band relaxation of excitons by LO phonon emission involving hole sub-bands is not possible. Thus out of the four channels, only the second and third channels are possible in practice and satisfy the condition of energy conservation for the relaxation of excitons by LO phonon emission.

For comparison with recent experimental results [15.30], we have also considered QWs of well width $L_z = 130\,\text{Å}$ where, as stated above, there are four electron and three hole sub-bands. In this case, we have only considered the cares where excitons are excited to sub-bands $n_e \leq 2$ and $n_h \leq 2$ for comparing our results with the experimental ones. The energy separation between $n_e = 1$ and $n_e = 2$ sub-bands is about 65.1 meV and that between $n_h = 1$ and $n_h = 2$ about 12.7 meV. The binding energies of 1s(1e, 1h), 1s(1e, 2h), 1s(2e, 1h), and 1s(2e, 2h) excitons are 7.0, 6.2, 6.2, and 6.1 meV, respectively. Since the energy separation between $n_h = 1$ and $n_h = 2$ is only 12.7 meV, the inter-sub-band relaxations from 1s(2e, 2h) to (2e, 1h) and from 1s(1e, 2h) to (1e, 1h) by emitting an LO phonon are not possible. The only possible inter-sub-band relaxations by LO phonon emission are from 1s(2e, 2h) to (1e, 2h) and 1s(2e, 1h) to (1e, 1h). Accordingly, as stated above, a direct relaxation from 1s(2e, 2h) to (1e, 1h) by emitting a single LO phonon is not possible: It appears to be a two-phonon two-step process: 1s(2e, 2h) $\xrightarrow{\text{LO phonon}}$ (1e, 2h) $\xrightarrow{\text{AC phonon}}$ (1e, 1h) or 1s(2e, 2h) $\xrightarrow{\text{AC phonon}}$ (2e, 1h) $\xrightarrow{\text{LO phonon}}$ (1e, 1h). In Figure 15.12, we have plotted the rate of

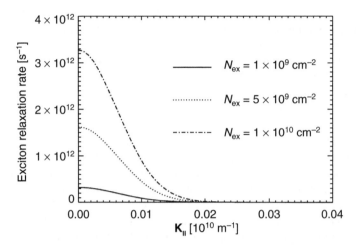

Figure 15.12   The relaxation rate of a 1 s (2e, 2h) exciton to a (1e, 2h) free electron–hole pair by LO phonon emission in GaAs QWs as a function of the center-of-mass wavevector $\mathbf{K}_{\parallel}$ for $L_z = 130$ Å, $T_{ex} = 20$ K, $T = 8$ K, and three different exciton densities, $N_{ex} = 1 \times 10^9\,cm^{-2}$ (—), $N_{ex} = 5 \times 10^9\,cm^{-2}$ (·····), and $N_{ex} = 1 \times 10^{10}\,cm^{-2}(-\cdot-\cdot)$ [15.6] [Reprinted with permission from I.-K. Oh, J. Singh, and A.S. Vengurlekar, *J. Appl. Phys.*, **91**, 5796. Copyright (2002) American Institute of Physics]

relaxation of 1s(2e, 2h) excitons into (1e, 2h) free electron–hole pairs by emitting an LO phonon as a function of the center-of-mass wavevector $\mathbf{K}_{\parallel}$ for three different exciton densities and at a crystal temperature of $T = 8$ K and exciton temperature $T_{ex} = 20$ K in QWs of width 130 Å. Similar results are obtained for the relaxation of a 1s(2e, 1h) exciton state to a (1e, 1h) electron–hole pair state by LO phonon emission. As illustrated in Figure 15.12, the rate decreases with increasing center-of-mass wavevector $\mathbf{K}_{\parallel}$ and decreasing exciton density N_{ex}. The maximum rate of relaxation is found to occur at $\mathbf{K}_{\parallel} = 0$. From the calculation, we have thus obtained values for the relaxation time of 305 fs at an exciton density of $N_{ex} = 1 \times 10^{10}\,cm^{-2}$, 616 fs at $N_{ex} = 5 \times 10^9\,cm^{-2}$, and 3.11 ps at $N_{ex} = 1 \times 10^9\,cm^{-2}$. The corresponding times for relaxation from 1s(2e, 1h) to (1e, 1h) are obtained as 304 fs, 614 fs, and 3.10 ps. Although these relaxation times for exciton densities in the range of $N_{ex} = 5$–10 $\times 10^9\,cm^{-2}$ are in a good agreement with those obtained by Pal and Vengurlekar [15.30], these are not times of relaxation from 1s(2e, 2h) to (1e, 1h). Our calculations also reveal that the relaxation time is very sensitive to the density of excitons, which has not been studied experimentally. However, as the chemical potential depends on the exciton density, the relaxation time depends quite sensitively on the exciton density [15.6].

In Figure 15.13, we have plotted the dependence of the relaxation rate for the same process as in Figure 15.12 [1s(2e, 2h) to (1e, 2h)] on the exciton temperature in the range of $T_{ex} = 10$–50 K at $T = 8$ K, $\mathbf{K}_{\parallel} = 0$, and at three different exciton densities, $N_{ex} = 1 \times 10^9\,cm^{-2}$, $5 \times 10^9\,cm^{-2}$ and $1 \times 10^{10}\,cm^{-2}$ for $L_z = 130$ Å QWs. The results show that the rate at $\mathbf{K}_{\parallel} = 0$ decreases as the temperature of excitons T_{ex} increases at a given exciton density. This is due to the fact that at lower T_{ex} the probability of excitons occupying states at $\mathbf{K}_{\parallel} = 0$ is higher compared with that at higher T_{ex}. As shown in Figure 15.13, the rate of relaxation from a 1s(2e, 2h) exciton to a (1e, 2h) e–h pair at an exciton density of $N_{ex} = 5$

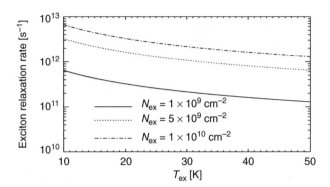

Figure 15.13   The relaxation rate of a 1s(2e, 2h) exciton to a (1e, 2h) free electron–hole pair by LO phonon emission in GaAs QWs as a function of the exciton temperature T_{ex} (K) for $L_z = 130$ Å, $\mathbf{K}_\parallel = 0$, $T = 8$ K, and three different exciton densities, $N_{ex} = 1 \times 10^9$ cm^{-2} (—), $N_{ex} = 5 \times 10^9$ cm^{-2} (·····), $N_{ex} = 1 \times 10^{10}$ cm^{-2} (–·–·) [15.6]. [Reprinted with permission from I.-K. Oh, J. Singh, and A.S. Vengurlekar, *J. Appl. Phys.*, **91**, 5796. Copyright (2002) American Institute of Physics]

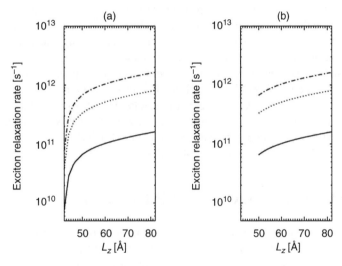

Figure 15.14   The relaxation rate of (a) 1s(2e, 1h) exciton to (1e, 1h) free electron–hole pair and (b) 1s(2e, 2h) exciton to (1e, 2h) free electron–hole pair by LO phonon emission in GaAs QWs as a function of the well width (L_z) of QWs at $T_{ex} = 20$ K, $\mathbf{K}_\parallel = 0$, $T = 4.2$ K, and three different exciton densities, $N_{ex} = 1 \times 10^9$ cm^{-2} (–), $N_{ex} = 5 \times 10^9$ cm^{-2} (·····), $N_{ex} = 1 \times 10^{10}$ cm^{-2} (–·–·) [15.6] [Reprinted with permission from I.-K. Oh, J. Singh, and A.S. Vengurlekar, *J. Appl. Phys.*, **91**, 5796. Copyright (2002) American Institute of Physics]

$\times 10^9$ cm^{-2}, decreases from 3.28×10^{12} s^{-1} to 6.45×10^{11} s^{-1}; as a result, the corresponding relaxation time decreases from 305 fs to 1.55 ps when T_{ex} increases from 10 to 50 K.

In Figure 15.14 (a) and (b), we have plotted the rates of relaxation of 1s(2e, 1h) and 1s(2e, 2h) excitons, respectively, at $T_{ex} = 20$ K and $T = 4.2$ K, $\mathbf{K}_\parallel = 0$ and the above three different exciton densities as a function of L_z from 42 to 82 Å. As there is only one hole sub-band in

QWs of width $L_z = 42$–$47\,\text{Å}$, no relaxation of 1s(2e, 2h) excitons is possible in these QWs, as shown in Figure 15.14 (b). According to Figure 15.14, the rate of relaxation of both 1s(2e, 1h) and 1s(2e, 2h) excitons increases with increasing width of QW. It is found that the relaxation time from 1s(2e, 1h) to (1e, 1h), e.g., at $N_{ex} = 5 \times 10^9\,\text{cm}^{-2}$, decreases from 26.2 ps to 1.23 ps and that from 1s(2e, 2h) to (1e, 2h) decreases from 3.57 ps to 1.24 ps when L_z increases from 48 to 82 Å. Such an increase is also found in the process of inter-sub-band scattering of 2D electrons in QWs because the form factor of the inter-sub-band transitions increases with well width [15.31]. However, this is quite different from the excitonic relaxation through intra-sub-band transitions by emission of an acoustic phonon. In that case, the rate of relaxation decreases with increasing L_z [15.32]. These two distinct features may be used to identify whether the mechanism of relaxation in QWs is through inter-sub-band or intra-sub-band transitions.

In narrow GaAs QWs, the relaxation of an exciton through an inter-sub-band transition by emission of an LO phonon is possible only via the inter-sub-band transitions of electrons. This is because the energy separation between different hole sub-bands is less than the LO phonon energy of $\hbar\omega_{LO} = 36.2\,\text{meV}$. However, for QWs of wider well widths, e.g., 500 Å, the relaxation of excitons through higher inter-sub-band transition of holes involving acoustic phonon emission may be possible. In this case, the energy separation between $n_h = 1$ and $n_h = 2$ hole sub-bands is 1.2 meV and that between $n_h = 1$ and $n_h = 3$ sub-bands is 3.2 meV and then the excitation of electrons from the $n_h = 2$ or $n_h = 3$ sub-band is expected. It is found that the inter-sub-band relaxation of excitons occurs only when one of the charge carriers (electron or hole) remains in the same sub-band during the transition and the other goes through an inter-sub-band transition. A direct transition from 1s(2e, 2h) exciton to (1e, 1h) free electron–hole pair is not allowed through inter-sub-band transitions. Such a selection rule cannot possibly be identified experimentally. The rate of relaxation is found to depend on the excitation density very sensitively, which has not yet been investigated experimentally. We have also found that the rate of inter-sub-band relaxation increases with increasing well width of GaAs QWs from 42–82 Å.

15.5 QUASI-2D FREE-EXCITON LINEWIDTH

In optical spectroscopic experiments such as transmission spectroscopy [15.33], photoluminescence [15.34, 15.35], four-wave mixing [15.36, 15.37], etc., ideal noninteracting excitons show very sharp optical peaks in their line shapes. However, the exciton line in low-dimensional semiconductors becomes broadened due to exciton scattering by acoustic [15.37], longitudinal optical (LO) [15.34] phonons, interface and alloy disorders [15.35], excitons [15.38, 15.39], free charge carriers [15.40], etc. In particular, at low exciton densities, the exciton line shape is determined mainly by exciton–phonon interactions which are responsible for homogeneous line broadening [15.34]. In this section, we present a theory to calculate the free exciton homogeneous linewidth and dephasing r ate at low excitation densities due to exciton–acoustic phonon interaction as a function of the lattice temperature, exciton temperature, well width, and exciton density in QWs.

The total rate of dephasing of a free exciton $W_h^{ph}(\mathbf{K}_\parallel)$ due to acoustic phonon interaction ($J = D,P$) is obtained by calculating the rate of transition from its initial state with $\mathbf{K}_\parallel$ to other exciton states by emitting and absorbing phonons as [15.7]:

$$W_h^{ph}(\mathbf{K}_\parallel) = \sum_{J,\lambda,\mathbf{q}_\parallel,q_z} W_\lambda^J(\mathbf{q}_\parallel,q_z)\big[(f_{\mathbf{K}_\parallel+\mathbf{q}_\parallel}^{ex}+1)f_{\mathbf{K}_\parallel}^{ex}n_\mathbf{q}^\lambda\delta(E(\mathbf{K}_\parallel+\mathbf{q}_\parallel)-E(\mathbf{K}_\parallel)-\hbar\omega_{\mathbf{q},\lambda})$$

$$+\big[(f_{\mathbf{K}_\parallel-\mathbf{q}_\parallel}^{ex}+1)f_{\mathbf{K}_\parallel}^{ex}(n_\mathbf{q}^\lambda+1)(E(\mathbf{K}_\parallel-\mathbf{q}_\parallel)-E(\mathbf{K}_\parallel)+\hbar\omega_{\mathbf{q},\lambda})\big] \tag{15.35}$$

where $E(\mathbf{K}_\parallel) = \hbar K_\parallel^2/2M_\parallel^*$ is the kinetic energy of a free exciton associated with its center of mass, and $\omega_{\mathbf{q},\lambda}$ is the frequency of a λ-mode acoustic phonon with a wavevector $\mathbf{q}$. Having obtained the rate of transition in Equation (15.35), the homogeneous linewidth due to acoustic phonons is given by ref. [15.33]:

$$\Gamma_h^{ph}(\mathbf{K}_\parallel) = \hbar W_h^{ph}(\mathbf{K}_\parallel) \tag{15.36}$$

To interpret the experimental results [15.33, 15.39, 15.41], usually this linewidth is assumed to be linearly dependent on the lattice temperature T as $\Gamma_h^{ph} = \gamma_{AC}T$ [15.40], and γ_{AC} is called the acoustic–phonon scattering parameter [15.38, 15.39]. Assuming that excitons and phonons are in thermal equilibrium before and after transitions, $n_\mathbf{q}^\lambda$ and $f_{\mathbf{K}_\parallel}^{ex}$ can be replaced by the average values obtained from the Bose–Einstein distribution as [15.5]:

$$n_\mathbf{q}^\lambda = \big[e^{\hbar\omega_{\mathbf{q},\lambda}/k_BT}-1\big]^{-1} \tag{15.37}$$

and

$$f_{\mathbf{K}_\parallel}^\lambda = \big[e^{[E(\mathbf{K}_\parallel)-\mu]/k_BT}-1\big]^{-1} \tag{15.38}$$

where μ is the chemical potential of 2D excitons, and T_{ex} is the exciton temperature. As the acoustic phonon energies are small, the exponential in the average phonon occupation in Equation (15.37) can be expanded. Retaining terms only up to the first order, one gets $n_\mathbf{q}^\lambda \propto T$. This implies that the linear relation may be valid only for $\hbar\omega_{\mathbf{q},\lambda} \ll k_BT$. The chemical potential μ is a function of the exciton temperature T_{ex} and 2D exciton density N_{ex} [15.41] as $\mu = k_BT_{ex}\ln[1-e^{-2\pi N_{ex}\hbar^2/(gM_\parallel^*k_BT_{ex})}]$. Therefore, the dephasing rate as derived here is also obtained as a function of the exciton temperature T_{ex} and density N_{ex}, as well as the exciton energy $E(\mathbf{K}_\parallel)$ (or center-of-mass momentum $\hbar K_\parallel$) and lattice temperature T. In this regard the present result of dephasing rate is more general compared with that derived by Takagahara [15.32, 15.40]. It may be noted that the excitation density considered here is relatively low, which enables us to treat excitons through Bose–Einstein distribution.

Here we have calculated the dephasing rate and linewidth for GaAs/Al$_{0.3}$Ga$_{0.7}$As QWs of widths $L_z = 28$ and 80 Å. The QW width of 28 Å is chosen for comparing the present results with experimental ones where the same well width has been used [15.37], and the other width of 80 Å is used because, recently, several papers have appeared on excitonic process in GaAs QWs of this width. In Figures 15.15 and 15.16, we have plotted the dependence of dephasing rate on the center-of-mass momentum at the exciton density $N_{ex} = 5 \times 10^9\,\text{cm}^{-2}$ for QWs of widths $L_z = 28$ and 80 Å, respectively, at three different lattice temperatures (5, 80, and 150 K) and two different exciton temperatures (10 and 40 K). Both Figures 15.15 and 15.16 illustrate that the dephasing rate of free excitons at relatively low exciton temperature is more sensitive to the exciton momentum (or kinetic energy) than that at relatively high exciton temperature. At a given lattice temperature, the dephasing rate of excitons with lower momentum (or kinetic energy) is dominant at relatively low exciton temperatures

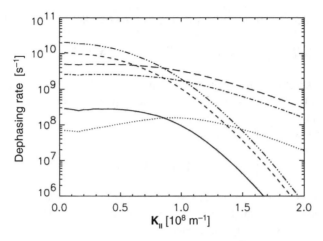

Figure 15.15   The dephasing rate as a function of $\mathbf{K}_\parallel$ for $L_z = 28\,\text{Å}$ QWs with $N_{ex} = 5 \times 10^9\,\text{cm}^{-2}$ at various lattice and exciton temperatures (–), $T = 5$ and $T_{ex} = 10\,\text{K}$; (·····), $T = 5$ and $T_{ex} = 40\,\text{K}$; (– –), $T = 80$ and $T_{ex} = 10\,\text{K}$; (–·–·), $T = 80$ and $T_{ex} = 40\,\text{K}$; (– ···), $T = 150$ and $T_{ex} = 10\,\text{K}$; (– –), $T = 150$ and $T_{ex} = 40\,\text{K}$ [15.7] [Reprinted from I.-K. Oh and J. Singh, *Superlattices Microstruct.*, **30**, 287. Copyright (2001) Elsevier]

whereas that with higher momentum (or kinetic energy) is dominant at relatively high exciton temperatures. As expected, this trend implies that the average exciton kinetic energy is proportional to $k_B T_{ex}$, where k_B is Boltzmann's constant. In particular, in the case at low lattice and high exciton temperatures, e.g., $T = 5$ and $T_{ex} = 40\,\text{K}$, the maximum value of dephasing rate occurs at a high exciton momentum (or kinetic energy) whereas at high lattice and low exciton temperatures it occurs at $\mathbf{K}_\parallel \sim 0$, which means near the exciton-band minimum. Figures 15.15 and 15.16 suggest that the dephasing rate for $L_z = 28\,\text{Å}$ QWs is higher than that for $L_z = 80\,\text{Å}$ QWs for the same n_{ex}, T, and T_{ex}. This agrees with the theoretical prediction for an infinite barrier potential by Borri et al. [15.11]. However, their experimental results for $In_xGa_{1-x}As/GaAs$ QWs are available only for narrow QWs of widths 10–40 Å, in which range the penetration of electronic wavefunctions may play a very significant role. Also the roughness of QW walls due to alloying will become important in such narrow QWs.

The dependence of the dephasing rate on lattice and exciton temperatures obtained from Equation (15.35) for $L_z = 80\,\text{Å}$ and $N_{ex} = 1 \times 10^{10}\,\text{cm}^{-2}$ is shown in Figures 15.17 and 15.18 at two different exciton momenta $\mathbf{K}_\parallel = 0.0004 \times 10^8\,\text{m}^{-1}$ and $\mathbf{K}_\parallel = 1 \times 10^8\,\text{m}^{-1}$, respectively. Here, $\mathbf{K}_\parallel = 0.0004 \times 10^8\,\text{m}^{-1}$ corresponds to the kinetic energy $E(\mathbf{K}_\parallel) = 3.46 \times 10^{-7}\,\text{meV}$ and $\mathbf{K}_\parallel = 1 \times 10^8\,\text{m}^{-1}$ corresponds to $E(\mathbf{K}_\parallel) = 2.16\,\text{meV}$. At a given exciton temperature and $\mathbf{K}_\parallel$, the dephasing rate of an exciton increases linearly as the lattice temperature increases, which agrees very well with the experimental result [15.36]. However, the dependence of dephasing rate on exciton temperature at a given T is different. As can be seen from Figure 15.17 at $\mathbf{K}_\parallel = 0.0004 \times 10^8\,\text{m}^{-1}$, which is near the exciton-band minimum, the dephasing rate decreases with increasing exciton temperature. However, in the high-momentum region, shown in Figure 15.18, the dephasing rate increases first to a peak value and then decreases as the exciton temperature increases. For $L_z = 28\,\text{Å}$ QWs, one also finds a similar trend, although it is not shown here.

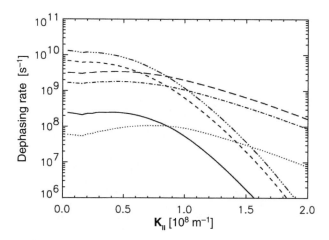

Figure 15.16 The dephasing rate as a function of $\mathbf{K}_\parallel$ for $L_z = 80\,\text{Å}$ QWs with $N_{ex} = 5 \times 10^9\,\text{cm}^{-2}$ at various lattice and exciton temperatures (—), $T = 5$ and $T_{ex} = 10\,\text{K}$; (·····), $T = 5$ and $T_{ex} = 40\,\text{K}$; (– –), $T = 80$ and $T_{ex} = 10\,\text{K}$; (–. –.), $T = 80$ and $T_{ex} = 10\,\text{K}$; (– ···), $T = 150$ and $T_{ex} = 10\,\text{K}$; (– –), $T = 150$ and $T_{ex} = 10\,\text{K}$ [15.7] [Reprinted from I.-K. Oh and J. Singh, *Superlattices Microstruct.*, **30**, 287. Copyright (2001) Elsevier]

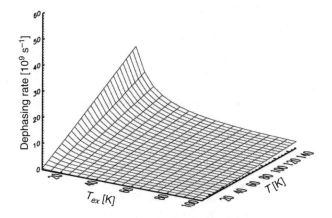

Figure 15.17 The dephasing rate as a function of T and T_{ex} at $\mathbf{K}_\parallel = 0.0004 \times 10^8\,\text{m}^{-1}$ for $L_z = 80\,\text{Å}$ QWs with $N_{ex} = 1 \times 10^{10}\,\text{cm}^{-2}$ [15.7] [Reprinted from I.-K. Oh and J. Singh, *Superlattices Microstruct.*, **30**, 287. Copyright (2001) Elsevier]

Experimentally, the temperature dependence of homogeneous exciton linewidth is determined by fitting the experimental data to the following equation [15.33, 15.42]:

$$\Gamma_h(T) = \Gamma_h(0) + \gamma_{AC}T + \gamma_{LO}\frac{1}{e^{\hbar\omega_{LO}/k_BT} - 1} \tag{15.39}$$

where the first term is the contribution to the linewidth due to impurities, and the second and third terms are due to acoustic and LO phonon scatterings, respectively. For bulk

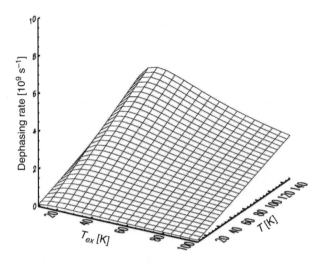

Figure 15.18   The dephasing rate as a function of T and T_{ex} at $K_{\parallel} = 1 \times 10^8\,m^{-1}$ for $L_z = 80\,\text{Å}$ QWs with $N_{ex} = 1 \times 10^8\,cm^{-2}$ [15.7] [Reprinted from I.-K. Oh and J. Singh, *Superlattices Microstruct.*, **30**, 287. Copyright (2001) Elsevier]

crystals of GaAs, the observed linewidth due to acoustic–phonon interaction $\gamma_{AC}T$ from various experiments shows large variations as listed below: $\gamma_{AC} = 3\,\mu eV/K$ for $L_z = 45\,\text{Å}$ [15.43], $1.7\,\mu eV/K$ for $L_z = 150\,\text{Å}$, $3\,\mu eV/K$ for $L_z = 325\,\text{Å}$ [15.44], $2.5\,\mu eV/K$ for $L_z = 135$ Å, $5\,\mu eV/K$ for $L_z = 277\,\text{Å}$ [15.45], $3.8\,\mu eV/K$ for $L_z = 130\,\text{Å}$, $3.8\,\mu eV/K$ for $L_z = 170\,\text{Å}$, 5.2 $\mu eV/K$ for $L_z = 250\,\text{Å}$, and $4.4\,\mu eV/K$ for $L_z = 340\,\text{Å}^4$ QWs, and $14\,\mu eV/K$ [15.46] and 17 $\mu eV/K$ [15.47]. In contrast to the prediction of Borri et al. [15.11] and present theory, these results do not indicate any systematic relation between γ_{AC} and QW width. It may, however, be noted that Equation (15.39) does not take into account any dependence of $\Gamma_h(T)$ on the temperature and density of excitons, which is found to exist quite sensitively from the present theory. This implies that all experimental results are not obtained at the same temperature and density of excitons, and therefore it is not meaningful to expect any systematic relation between $\Gamma_h(T)$ and L_z, as obtained from the present theory.

Recently, Fan et al.[5] have measured $\Gamma \sim (14\text{–}19)\,\mu eV$ at $T = 10\,K$ and $N_{ex} \sim (1\text{–}10) \times 10^9\,cm^{-2}$ in $L_z = 28\,\text{Å}$ GaAs QWs. For the same well width of $L_z = 28\,\text{Å}$ and $T = 10\,K$, our calculation gives $\Gamma = (1.1\text{–}6.4)\,\mu eV$ at $N_{ex} = 5 \times 10^9\,cm^{-2}$ and $(2.4\text{–}16)\,\mu eV$ at $N_{ex} = 10 \times 10^9\,cm^{-2}$ with decreasing T_{ex} from 5 to 1 K. These results agree reasonably well with Fan et al.'s experimental results although our results are slightly lower. Such a discrepancy in theoretical and experimental results is also found in bulk GaAs, where $\gamma_{AC} = 0.64\,\mu eV/K$ [15.48] is obtained from theory, and $14\,\mu eV/K$ [15.46] and $17\,\mu eV/K$ [15.47] are obtained from experiments.

In Figures 15.19 and 15.20, we have plotted the exciton dephasing rate at $T_{ex} = 5\,K$ and $40\,K$, respectively, as a function of exciton density and momentum at a lattice temperature of $10\,K$ for a quantum well of width $28\,\text{Å}$ as used by Fan et al. [15.36]. At a given exciton

---

[4]It should be noted that the linewidths in ref. [15.33] are given in terms of a half-width at half-maximum.
[5]The values of linewidths are estimated from the experimental date given in Figure 2 (b) in ref. [15.36].

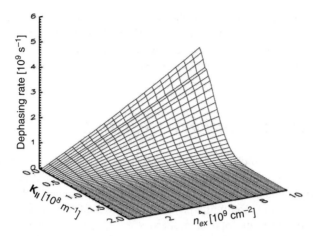

Figure 15.19   The dephasing rate as a function of N_{ex} and $\mathbf{K}_\parallel$ at $T = 10$ and $T_{ex} = 5\,\mathrm{K}$ for $L_z = 28\,\text{Å}$ QWs [15.7] [Reprinted from I.-K. Oh and J. Singh, *Superlattices Microstruct.*, **30**, 287. Copyright (2001) Elsevier]

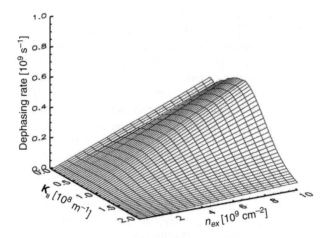

Figure 15.20   The dephasing rate as a function of N_{ex} and $\mathbf{K}_\parallel$ at $T = 10$ and $T_{ex} = 40\,\mathrm{K}$ for $L_z = 28\,\text{Å}$ QWs [15.7] [Reprinted from I.-K. Oh and J. Singh, *Superlattices Microstruct.*, **30**, 287. Copyright (2001) Elsevier]

momentum, the rate of dephasing (homogeneous linewidth) increases linearly with increasing exciton density, which agrees very well with the experimental results of Fan et al. [15.36]. Such a linear relation has also been observed in other QW structures such as $In_xGa_{1-x}As/GaAs$ [15.11], $ZnSe/Zn_{0.94}Mg_{0.06}Se$ [15.38] and $CdTe/Cd_{1-x-y}Mg_xZn_yTe$ [15.37]. The slope of the linear relation, as shown in Figures 15.19 and 15.20, depends on the temperature and momentum of excitons. However, such a dependence cannot be compared with any experimental results, as no experiments have been done in connection with these parameters. As a rough estimate, we have calculated the slope of Γ_h with respect to N_{ex}. Our result at $T = 10\,\mathrm{K}$ and $T_{ex} = 2\,\mathrm{K}$ near exciton band minimum gives a slope of $0.76 \times 10^{-9}\,\mu\mathrm{eV\,cm^2}$, which agrees with the value of $0.75 \times 10^{-9}\,\mu\mathrm{eV\,cm^2}$ estimated from Fan et al.'s result.[6] We have also found that the slope is very sensitive to the exciton temperature. For example, at a lattice temperature of 10 K, we have found that the slope increases from $0.3 \times 10^{-9} \mu eV cm^2$ to $1.9 \times 10^{-9} \mu eV cm^2$ when T_{ex} decreases from 5 K to 1 K. However, more experimental results are required to verify such a dependence on the exciton temperature.

15.6 LOCALIZATION OF FREE EXCITONS

The interfaces between two different semiconductors in QWs can dramatically alter physical properties such as confinements, of charge carriers [15.49] and phonons [15.50]. In addition, there are inevitable structural disorders of the atomic lattice at the interfaces, which give rise to roughness and modify the physical properties of QWs even further. The interface roughness influences the optical properties and plays a very important role in the dynamics of excitons in QWs. The fine structures and splitting in the photoluminescence (PL) signals of excitons are some examples of the effects of interface roughness causing localization in the center-of-mass (COM) motion of an exciton. The aim of this section is to study the localization of free excitons due to interface roughness through the nonradiative processes in QWs as a transition from a free exciton state to a localized exciton state. When a sample of QWs is subjected to photons of energy greater than the bandgap of the well material, several excitonic processes may occur: (1) a free exciton with nonzero COM wavevector $K_\parallel$ can be formed, which then relaxes nonradiatively to another free-exciton state with $K_\parallel \sim 0$ and finally decays radiatively to the ground state, (2) a free exciton is formed and then becomes localized nonradiatively within the interface roughness potential by emitting phonons, and (3) the localized exciton formed due to the interface roughness in (2) is transferred nonradiatively to another localized state. There may be more complicated relaxation and radiative decay processes; however, in this paper, we will focus on process (2), i.e., free excitons initially created by photons of energy higher than the bandgap becoming localized in the localized energy states caused by the roughness potential at the interface by emitting phonons.

Here we assume that free excitons are present in QWs. If the interface of QWs is perfectly smooth, these excitons will relax to exciton states with COM wavevector $K_\parallel \sim 0$ and then recombine radiatively. In this case, the PL signal should be sharp and smooth and its peak should also be well defined although it has a characteristic line-broadening associated with the interaction of excitons with other excitations such as phonons and other excitons. In practice, however, there is always some roughness or disorder at the interfaces of QWs. The potential energy fluctuations due to such interface roughnesses or alloy disorders in semiconductor heterostructures can hinder the smooth motion and cause localization of excitons, giving rise to fine structures and red-shift in PL [15.8]. Here we consider the localization of a free exciton in QWs due to the interface roughness involving acoustic phonons. This is based on the assumption that initially excitons are created with energy higher than the bandgap, as stated earlier, so they are not localized. Thus the initial wave function can be approximated by a plane wave. We assume that the density of free excitons is so low that the exciton–exciton interaction can be ignored.

---

[6]The values of linewidths are estimated from the experimental date given in Figure 2 (b) in ref. [15.36].

For the perpendicular motion of electron and hole in QWs, we consider only the lowest sub-band. We also consider only 1s excitons for the internal motion of an exciton in the QW plane. Using the COM $\mathbf{R}_\|$ and relative $\mathbf{r}_\|$ coordinates for an exciton in the plane of QWs, the field operator $\hat{\psi}^\dagger_{ex}$ describing the creation of an exciton with its COM wavefunction $\psi_{ex}(\zeta, \mathbf{R}_\|)$ can be written as [15.50]:

$$\hat{\psi}^\dagger_{ex} = \sum_\zeta \psi^*_{ex}(\zeta, \mathbf{R})\phi^*_x(\mathbf{r})\phi^*_e(x^e_3)\phi^*_h(x^h_3)\hat{B}^\dagger_\zeta \tag{15.40}$$

where ζ represents the set of quantum numbers for the COM of an exciton. As a convenient notation, we use $\zeta = \mathbf{K}_\|$ for a free-exciton state with the COM wavevector $\mathbf{K}_\|$, and $\zeta = \mathbf{R}_a$ for a localized exciton state at a site $\mathbf{R}_a$. Then the COM wave function $\psi_{ex}(\zeta, \mathbf{R}_\|)$ can be written as [15.50]:

$$\psi_{ex}(\zeta, \mathbf{R}_\|) = \begin{cases} \dfrac{1}{\sqrt{A_0}}e^{i\mathbf{K}_\| \cdot \mathbf{R}_\|} & \text{for} \quad \zeta = \mathbf{K}_\| \\ \psi_{loc}(\mathbf{R}_\| - \mathbf{R}_a) & \text{for} \quad \zeta = \mathbf{R}_a \end{cases} \tag{15.41}$$

In Equation (15.40), ϕ_x, ϕ_e, and ϕ_h are wave functions of the exciton in the relative coordinate, electron, and hole for the motion perpendicular to the well of the QW, respectively, and $\hat{B}^\dagger_\zeta$ is the creation operator of an exciton in state ζ. Likewise, using the annihilation operator $\hat{B}_\zeta$ of an exciton, the field operator $\hat{\psi}_{ex}$ describing the annihilation of an exciton can be written as:

$$\hat{\psi}_{ex} = \sum_\zeta \psi_{ex}(\zeta, \mathbf{R}_\|)\phi_x(\mathbf{r}_\|)\phi_e(x^e_3)\phi_h(x^h_3)\hat{B}_\zeta \tag{15.42}$$

In principle, the free-exciton wave function is influenced by the presence of potential fluctuations due to interface roughness in the QW plane causing an exciton to be localized as it is considered here in the final state. Therefore, in the initial state, we have assumed the free-exciton wavefunction as a plane wave as given in Equation (15.41). We also assume that the energy fluctuations due to interfacial roughness or material compositions are not strong enough to influence the exciton wave function associated with its relative motion and charge carrier sub-band wave functions associated with their motion perpendicular to QWs. That means ϕ_x, ϕ_e, and ϕ_h are assumed to be the same for both free and localized excitons.

According to Citrin [15.51], we consider a model potential V_d due to a well-width fluctuation of lateral size b and thickness $a/2$ where a is the bulk GaAs lattice constant. Then V_d in relative and COM coordinates can be written as:

$$V_d(\mathbf{R}_\|, \mathbf{r}_\|, x^e_3, x^h_3) = -V_e \exp\left(\frac{-|\mathbf{R}_\| + \alpha_\|\mathbf{r}_\||^2}{2b^2}\right)[\Theta(x^e_3 - L_z/2 + a/4) - \Theta(x^e_3 - L_z/2 - a/4)]$$

$$-V_h \exp\left(\frac{-|\mathbf{R}_\| + \alpha_e\mathbf{r}_\||^2}{2b^2}\right)[\Theta(x^h_3 - L_z/2 + a/4) - \Theta(x^h_3 - L_z/2 - a/4)]$$

$$\tag{15.43}$$

with

$$\Theta(x_3) = \begin{cases} 1 & \text{for } x_3 \geq 0 \\ 0 & \text{for } x_3 < 0 \end{cases} \tag{15.44}$$

where V_e and V_h are the band offsets in the conduction and valence bands, respectively. It is to be noted that the model fluctuations considered here are 1 molecular layer (ML) thick from the QW wall, both inside and outside. Then the effective fluctuation potential of an exciton for the COM motion can be expressed as [15.50]:

$$V_d^{eff}(\mathbf{R}_{\parallel}) = \int d\mathbf{r}_{\parallel} \int dx_3^e \int dx_3^h V_d(\mathbf{R}_{\parallel}, \mathbf{r}_{\parallel}, x_3^e, x_3^h) |\phi_x(\mathbf{r}_{\parallel}) \phi_e(x_3^e) \phi_h(x_3^h)|^2 \tag{15.45}$$

The energy eigenvalue E_{loc} and wave function $\psi_{loc}(\mathbf{R}_{\parallel})$ of the localized exciton for its COM motion can be determined from the Schrödinger equation [15.50]:

$$\left(\frac{\hbar^2 \nabla_{\parallel}^2}{2M_{\parallel}^*} + V_d^{eff}(\mathbf{R}_{\parallel}) \right) \psi_{loc}(\mathbf{R}_{\parallel}) = E_{loc} \psi_{loc}(\mathbf{R}_{\parallel}) \tag{15.46}$$

Here we consider only the lowest bound state of a localized exciton due to the interface roughness. As a trial wavefunction for ψ_{loc}, we use the Gaussian type of COM envelope function for the localized state given by:

$$\psi_{loc}(\mathbf{R}_{\parallel} - \mathbf{R}_a) = \frac{1}{\xi\sqrt{\pi}} e^{-|\mathbf{R}_{\parallel} - \mathbf{R}_a|^2/2\xi^2} \tag{15.47}$$

where $\mathbf{R}_a$ is the site center of localization and ξ is the characteristic localization length [15.22] used as a variational parameter [15.52, 15.53]. Then the exciton–phonon interaction Hamiltonian in the second quantized form is [15.50]:

$$\hat{H}_I = \sum_J \int d\mathbf{R}_{\parallel} \int d\mathbf{r}_{\parallel} \int dx_3^e \int dx_3^h \hat{\psi}_{ex}^{\dagger} H_I^J \hat{\psi}_{ex}$$

$$= \sum_{J\lambda\zeta\zeta'\mathbf{q}_{\parallel},q_z} C_J^{\lambda} [F_{\lambda-}^J(\mathbf{q}_{\parallel},q_z) M_-(\mathbf{q}_{\parallel},\zeta',\zeta) \hat{B}_{\zeta'}^{\dagger} \hat{B}_{\zeta} \hat{b}_{\lambda\mathbf{q}}$$

$$+ F_{\lambda+}^J(\mathbf{q}_{\parallel},q_z) M_+(\mathbf{q}_{\parallel},\zeta',\zeta) \hat{B}_{\zeta'}^{\dagger} \hat{B}_{\zeta} \hat{b}_{\lambda\mathbf{q}}^{\dagger}] \tag{15.48}$$

with

$$M_{\pm}(\mathbf{q}_{\parallel},\zeta',\zeta) = \int d\mathbf{R}_{\parallel} \psi_{ex}^*(\zeta',\mathbf{R}_{\parallel}) \psi_{ex}(\zeta,\mathbf{R}_{\parallel}) e^{\mp i\mathbf{q}_{\parallel} \cdot \mathbf{R}_{\parallel}} \tag{15.49}$$

The interaction Hamiltonian $\hat{H}_I$ represents an interaction operator for a transition from one exciton state to another through the interaction with acoustic phonons. Thus, when $\zeta = \mathbf{K}_{\parallel}$ and $\zeta' = \mathbf{K}_{\parallel}'$, i.e., two different free-exciton states, are considered, the Hamiltonian in Equation (15.48) represents a transition between two free excitons involving acoustic phonons as given in Equation (47) of ref. [15.5]. When $\zeta = \mathbf{R}_a$ and $\zeta' = \mathbf{R}_a'$, i.e., two localized-exciton states at two different sites $\mathbf{R}_a$ and $\mathbf{R}_a'$, are considered, the interaction Hamiltonian represents the exciton transfer from site $\mathbf{R}_a$ to $\mathbf{R}_a'$ through the acoustic phonon interaction, which has been studied in detail by Takagahara [15.22]. The case of $\zeta' = \mathbf{R}_a$ and $\zeta = \mathbf{K}_{\parallel}$, one localized-exciton state at site $\mathbf{R}_a$ and the other a free-exciton state with $\mathbf{K}_{\parallel}$, has not been considered before. The interaction Hamiltonian in this case is associated with the localization of a free exciton or delocalization of a localized exciton through the acoustic phonon interaction. Here, we consider the process of transition from a free exciton to a localized exciton by emission of an acoustic phonon. For convenience, we write the localized wave function in Equation (15.47) in $\mathbf{K}_\parallel$-space as:

$$\psi_{loc}(\mathbf{R}_\parallel - \mathbf{R}_a) = \frac{1}{\sqrt{A_0}} \sum_{\mathbf{K}_\parallel'} g(\mathbf{K}_\parallel', \xi, \mathbf{R}_a) e^{i\mathbf{K}_\parallel' \cdot \mathbf{R}_\parallel} \tag{15.50}$$

with

$$g(\mathbf{K}_\parallel', \xi, \mathbf{R}_a) = \left(\frac{1}{2\pi}\right)^2 \int d\mathbf{R}_\parallel \psi_{loc}(\mathbf{R}_\parallel - \mathbf{R}_a) e^{-i\mathbf{K}_\parallel' \cdot \mathbf{R}_\parallel} = \frac{\xi}{2\pi\sqrt{2\pi}} e^{-i\mathbf{K}_\parallel' \cdot \mathbf{R}_a - \xi^2 K_\parallel'^2/2} \tag{15.51}$$

Using Equations (15.41) and (15.49)–(15.51), we get:

$$M_\pm(\mathbf{q}_\parallel, \mathbf{R}_a, \mathbf{K}_\parallel) = g(\mathbf{K}_\parallel \mp \mathbf{q}_\parallel, \xi, \mathbf{R}_a) \tag{15.52}$$

Using the Fermi golden rule and the second term in Equation (15.48), the rate of localization of a free exciton due to emission of an acoustic phonon can be obtained as:

$$W(E_i) = \frac{2\pi}{\hbar} \sum_f |\langle f|\hat{H}_I|i\rangle|^2 \, \delta(E_f - E_i) \tag{15.53}$$

with

$$E_i = \frac{\hbar^2 K_\parallel^2}{2M_\parallel^*} - E_b \tag{15.54}$$

and

$$E_f = E_{loc}(\xi) - E_b + \hbar\omega_{\lambda q} \tag{15.55}$$

where E_b is the binding energy of the exciton for the relative motion, and $\omega_{\lambda q}$ frequency of λ-mode phonon. $E_{loc}(\xi)$ is obtained from Equation (15.46) with the help of Equation (15.47). The initial state $|i\rangle$ and final state $|f\rangle$ in Equation (15.53) are characterized by the occupation numbers of the free exciton, localized exciton, and phonon states involved in the transition as given in ref. [15.54].

We consider the rate of localization of free excitons in [001]-oriented GaAs/Al$_{0.3}$Ga$_{0.7}$As QWs. As we consider the case that the energy of incident photons is greater than the energy gap of QWs, we assume that initially only 1s free excitons are resonantly excited and their density in the QW plane is N_{ex}. The problem considered here is a transition from a free-exciton state to a localized-exciton state caused by the roughness at the QW interfaces. Free excitons thus created experience the potential fluctuations in their COM motion and become localized by emitting acoustic phonons. In this process we ignore the direct creation of localized excitons, which may occur dominantly when the energy of the exciting photons is less than the excitation energy of 1s free excitons. We also ignore the interaction between free excitons. For simplification and without the loss of generality, we consider that the localization occurs on a site at $\mathbf{R}_a = 0$.

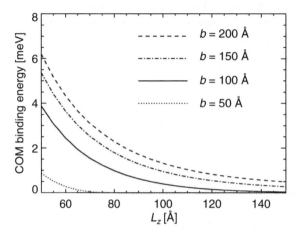

Figure 15.21 The COM binding energy as a function of L_z in the plane QWs due to interface roughness for four different values of b

Starting from the trial wave function in Equation (15.47) and using the variational method and effective potential in Equation (15.45), we have calculated the energy eigenvalue and the corresponding localized eigenfunction from Equation (15.46) using different values of b and assuming that the well-width fluctuation in thickness is $1\,\text{ML}$ ($a/2 = 2.83\,\text{Å}$). This variational calculation is repeated for every well width in the range 50–$150\,\text{Å}$, starting from $50\,\text{Å}$ and increasing in steps of $5\,\text{Å}$. The COM binding energy of a localized exciton, which is thus obtained from the energy eigenvalues, is plotted in Figure 15.21 as a function of the well width for $L_z = 50$–$150\,\text{Å}$ and four different lateral sizes $b = 50, 100, 150,$ and $200\,\text{Å}$. According to the results shown in Figure 15.21, the binding energy in relatively narrow QWs is larger than that in wider QWs at a given lateral size b. This feature can be explained as follows: the penetration of the sub-band wavefunctions ϕ_e and ϕ_h into the barrier becomes large for narrow QWs compared with that for wider QWs. As a result, the effective interaction of charge carriers with potential fluctuations at the interface of narrow QWs becomes larger compared with that for wider QWs. In other words, larger overlap of the sub-band wave functions with the potential fluctuations at the interface enhances the effective potential of an exciton in the COM motion. Our results reveal that the binding-energy difference between different lateral sizes for relatively narrow QWs becomes large enough to observe the splitting of localized excitons even in conventional PL spectroscopy in comparison with that for relatively wide QWs. This feature agrees very well with the experimental results [15.54]. As can be seen clearly from Figure 15.21, the splitting or fine structures of excitons in wide QWs, e.g., $L_z > 100\,\text{Å}$, is so small that it may only be observed through micro-PL spectroscopy. Our results also imply that the localization of the 1s free exciton may occur only due to acoustic phonon emission for QWs with well width $L_z > 50\,\text{Å}$ and $b < 200\,\text{Å}$ because the COM binding energy is much less than the LO phonon energy of $36.2\,\text{meV}$. This is the reason for considering the localization of 1s free excitons only through the acoustic phonon interaction here.

Taking into account both DP and PE couplings, we have plotted in Figure 15.22 the rates of localization of a 1s free, exciton by emission of longitudinal acoustic (LA) and trans-

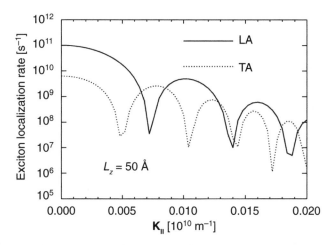

Figure 15.22 Localization rates as a function of $|\mathbf{K}_\parallel|$ due to LA and TA phonon emission, respectively, at $N_{ex} = 5 \times 10^9 \, \text{cm}^{-2}$, $b = 100 \, \text{Å}$, $T_{ex} = 15 \, \text{K}$, and $T = 4.2 \, \text{K}$ for $L_z = 30 \, \text{Å}$ QWs

verse acoustic (TA) phonons, respectively, as a function of the COM wavevector $|\mathbf{K}_\parallel|$ at a crystal temperature $T = 4.2 \, \text{K}$ and exciton temperature $T_{ex} = 15 \, \text{K}$ in QWs of width $L_z = 50$ Å and lateral size $b = 100 \, \text{Å}$. The results show that the rate of localization by emission of acoustic phonons changes by about 10^3 and 10^2 for LA and TA phonon processes, respectively, when $|\mathbf{K}_\parallel|$ increases from 0 to $0.007 \times 10^{10} \, \text{m}^{-1}$. This corresponds to a change in the COM kinetic energies by only 1.06 meV. At low COM kinetic energies (small $|\mathbf{K}_\parallel|$), the rate of localization is dominated by the LA phonon emission. This is due to the combined effect of the anisotropic coupling constants, phonon dispersion relation, form factors, and energy–momentum conservation. For a free exciton near $\mathbf{K}_\parallel = 0$, the acoustic phonons involved in the localization process are dominantly those near $\mathbf{q} = \left(\mathbf{q}_\parallel \sim 0, \ q_z \sim \sqrt{E^2 / (\hbar v_\lambda)^2 - K_\parallel^2} \right)$, $[E = \hbar^2 K_\parallel^2 / (2 M_\parallel^*) - E_{loc}]$. It is found from the phonon-dispersion relation that the velocity of acoustic phonons $v_{LA} > v_{TA}$, which gives smaller q_z for LA phonons than that for TA phonons. Furthermore, near $\mathbf{q}_\parallel = 0$, the localization due to the PE coupling can be ignored because of the PE coupling constants $h_{\mathbf{q}}^\lambda \sim 0$ for both LA and TA phonons (see ref. [15.5]). For the localization process due to the deformation coupling at $\mathbf{q}_\parallel \sim 0$, we can show that $|F_{LA+}^D(\mathbf{q}_\parallel \sim 0, q_z)|^2$ is much larger than $|F_{TA+}^D(\mathbf{q}_\parallel \sim 0, q_z)|^2$.[7] Therefore, for a free exciton near $\mathbf{K}_\parallel = 0$, the rate of the localization due to LA phonons is greater than that due to TA phonons. It is to be noted that, in this case, although the dominant contribution of phonons to the localization is due to those near $\mathbf{q}_\parallel = 0$, the calculation of the rate has to be carried out by integration over all $\mathbf{q} = (\mathbf{q}_\parallel, q_z)$. The structured features appearing in the graphs of rates shown in Figure 15.2 are related to the anisotropy

[7]From ref. [15.5], we can show that $F_{LA+}^D(\mathbf{q}_\parallel \sim 0, q_z) \sim -i[\Xi_c F_e(q)(q_z) - m F_h(q_z)]q_z) \sim$ and $F_{LA+}^D(\mathbf{q}_\parallel \sim 0, q_z) = F_{LA2+}^D(\mathbf{q}_\parallel \sim 0, q_z) \sim -i[n/\sqrt{3})F_h(q_z)]$. Since $|F_j(q_z)|$ ($j = e, h$) decreases as q_z increases and q_z for LA phonons is smaller that for TA phonons as explained in the text, $|F_j(q_z)|$ for LA phonons is greater than that for TA phonons. For GaAs, the deformation potentials are given by $\Xi_c = -8 \, \text{meV}$, $m = 2.86 \, \text{eV}$, and $n = -7.88 \, \text{eV}$. Therefore, $|F_{LA+}^D(\mathbf{q}_\parallel, q_z)|$ is larger than $|F_{TA+}^D(\mathbf{q}_\parallel, q_z)|$.

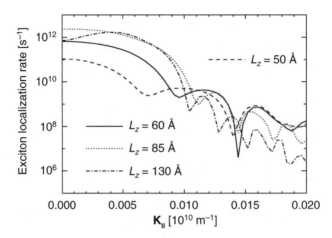

Figure 15.23   Localization rate due to acoustic phonon emission as a function of $|\mathbf{K}_{||}|$ at $N_{ex} = 5 \times 10^9 \, \text{cm}^{-2}$, $b = 100 \, \text{Å}$, $T_{ex} = 15 \, \text{K}$, and $T = 4.2 \, \text{K}$ for four different widths of QWs

in both deformation potential and piezoelectric coupling constants and the energy–momentum conservation. In the process of localization of a 1s free exciton with $\mathbf{K}_{||}$, by emitting acoustic phonons, the contribution comes mainly from phonons with specific wavevectors near $|\mathbf{q}_{||}| = |\mathbf{K}_{||}|$ and $q_z \sim \sqrt{E^2/(\hbar v_\lambda)^2 - K_{||}^2}$, because of the energy–momentum conservation. In other words, the energy–momentum conservation and anisotropy in coupling constants give rise to the structures in the rate of localization as shown in Figure 15.22. The total rate of localization decreases from 1.07×10^{11} to $2.37 \times 10^9 \, \text{s}^{-1}$ and its corresponding time of localization increases from 9.3 ps to 423 ps when the COM kinetic energy increases from 0 to 1.12 meV ($|\mathbf{K}_{||}| = 0.0072 \times 10^{10} \, \text{m}^{-1}$).

In Figure 15.23, we have plotted the rate of localization of a 1s free exciton due to emission of both LA and TA phonons as a function of $|\mathbf{K}_{||}|$ at $N_{ex} = 5 \times 10^9 \, \text{cm}^2$, $T_{ex} = 15 \, \text{K}$, and $T = 4.2 \, \text{K}$ for four different widths $L_z = 50$, 60, 80, and 130 Å of QWs with the same value of $b = 100 \, \text{Å}$. The results show that the dependence of the rates of localization on $|\mathbf{K}_{||}|$ for QWs with $L_z = 60$, 85 and 130 Å is more sensitive than that for QWs with $L_z = 50 \, \text{Å}$. This can be explained as follows. The COM binding energy of a localized exciton decreases with increasing width of QWs as shown in Figure 15.21. This means that the characteristic localization length ξ increases when L_z increases. Therefore, the Fourier transform of the localized wavefunction, $g(\mathbf{K}_{||}, \xi, 0)$ in Equation (15.51), becomes more sensitive to $|\mathbf{K}_{||}|$ in wider QWs (large L_z) compared with that in narrow QWs (smaller L_z). As the rate of localization is proportional to $|g(\mathbf{K}_{||}, \xi, 0)|^2$, the resulting rate also becomes more sensitive to $|\mathbf{K}_{||}|$ in wider QWs than in narrow QWs. The results in Figure 15.23 also indicate that the plotted rates exhibit more fine structure at larger values of $|\mathbf{K}_{||}|$ for all four well widths, as also found in Figure 15.22. As explained above, these structures are also associated with the anisotropy and energy-momentum conservation.

In Figure 15.24 we have shown the dependence of the localization rate of a 1s free exciton on the well width in the range of $L_z = 50$–150 Å for $b = 100 \, \text{Å}$ at $N_{ex} = 5 \times 10^9 \, \text{cm}^{-2}$, $T_{ex} = 15 \, \text{K}$, and $T = 4.2 \, \text{K}$. The (—), (·····), and (−·−·) curves correspond to the rates of localization at $|\mathbf{K}_{||}| = 0$, 0.52×10^8, and $1 \times 10^8 \, \text{m}^{-1}$, respectively and the corresponding COM

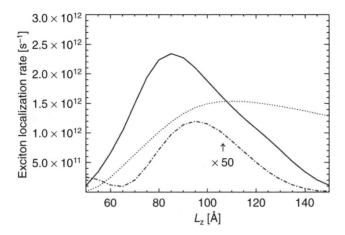

Figure 15.24 Localization rate as a function of L_z for $N_{ex} = 5 \times 10^9\,cm^{-2}$, $b = 100\,\text{Å}$, $T_{ex} = 15\,K$, and $T = 4.2\,K$. The (—) curve is at $|\mathbf{K}_\parallel| = 0$, (·····) curve at $|\mathbf{K}_\parallel| = 0.52 \times 10^8\,m^{-1}$, and (–·–·) curve at $|\mathbf{K}_\parallel| = 1 \times 10^8\,m^{-1}$

kinetic energies are 0, 0.58, and 2.16 meV. The rates of localization show a nonmonotonic behavior with respect to L_z. As explained above, rates depend on $|g|^2$, which in turn depends on ξ, which is a monotonically increasing function of L_z. According to Equation (15.51), g depends on ξ nonmonotonically as $\xi e^{-\xi^2 K_\parallel^2/2}$ and therefore the rates also depend nonmonotonically on L_z as $\xi^2 e^{-\xi^2 K_\parallel^2}$ as shown in Figure 15.24. The results of Figure 15.24 can be summarized as follows: At $|\mathbf{K}_\parallel| = 0$, the rate of localization has a maximum of $2.34 \times 10^{12}\,s^{-1}$ at $L_z = 85\,\text{Å}$ and the corresponding time of the localization is about 427 fs. At $|\mathbf{K}_\parallel| = 0.52 \times 10^8\,m^{-1}$, the maximum rate of $1.54 \times 10^{12}\,s^{-1}$ occurs at $L_z = 110\,\text{Å}$ and the corresponding time of localization is about 651 fs; and at $|\mathbf{K}_\parallel| = 1 \times 10^8\,m^{-1}$, the maximum rate is $2.40 \times 10^{10}\,s^{-1}$ at $L_z = 95\,\text{Å}$ and the corresponding time of localization is 41.6 ps. It is to be noted that, at $L_z = 50\,\text{Å}$, the rates of localization are 1.07×10^{11}, 1.45×10^{10}, and $5.12 \times 10^9\,s^{-1}$ at $|\mathbf{K}_\parallel| = 0$, 0.52×10^8, and $1 \times 10^8\,m^{-1}$, respectively, and their corresponding times of localization are 9.3, 69, and 195 ps.

Figure 15.25 shows the dependence of the rate of localization on $|\mathbf{K}_\parallel|$ in $L_z = 50\,\text{Å}$ QWs for four different lateral sizes of the well-width fluctuations. The values of N_{ex}, T_{ex}, and T are the same as those used in Figure 15.24. The results show that, for a small lateral size of $b = 50\,\text{Å}$, the rate of localization is less sensitive to $|\mathbf{K}_\parallel|$ in the region of relatively small values of $|\mathbf{K}_\parallel|$ but as $|\mathbf{K}_\parallel|$ increases it becomes more sensitive to $|\mathbf{K}_\parallel|$. In this case, the maximum rate of $2.85 \times 10^{12}\,s^{-1}$ occurs at $|\mathbf{K}_\parallel| = 0.002 \times 10^{10}\,m^{-1}$ and the corresponding time of localization is about 350 fs. In the small $|\mathbf{K}_\parallel|$ region, when b changes from 50 to 100 Å, the rate of localization decreases by more than an order of magnitude at $|\mathbf{K}_\parallel| \sim 0$. However, when b changes from 150 to 200 Å, such a drastic change in the rates is not found near $|\mathbf{K}_\parallel| = 0$. For $b = 100\,\text{Å}$, the maximum rate and its corresponding time of localization are $1.07 \times 10^{11}\,s^{-1}$ and 9.30 ps, respectively, at $|\mathbf{K}_\parallel| = 0$. On the other hand, for larger lateral sizes $b = 150$ and 200 Å, the rates of localization are less sensitive to $|\mathbf{K}_\parallel|$ and their maximum rates are $6.74 \times 10^9\,s^{-1}$ at $|\mathbf{K}_\parallel| = 0.0048$ and $7.34 \times 10^9\,s^{-1}$ at $|\mathbf{K}_\parallel| = 0$, respectively. The corresponding times of localization are 148 and 136 ps. Therefore, from Figures 15.23 and 15.25,

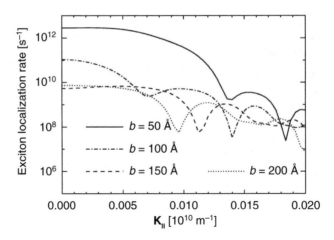

Figure 15.25 Localization rate as a function of $|\mathbf{K}_\parallel|$ for four different lateral sizes $b = 50, 100, 150,$ and $200\,\text{Å}$, at $N_{ex} = 5 \times 10^9\,\text{cm}^{-2}$, $L_z = 50\,\text{Å}$, $T_{ex} = 15\,\text{K}$, and $T = 4.2\,\text{K}$

we can conclude that the rate of localization is high and the time of localization is relatively faster for lateral sizes $b = 100(50)$ Å and well width $L_z = 85(50)$ Å at $|\mathbf{K}_\parallel| \sim 0$.

15.7 CONCLUSIONS

In summary, we have presented in this chapter a comprehensive theory of formation processes of an exciton due to acoustic and LO phonon emission via DP, PE, and PO couplings as a function of the center-of-mass wavevector $\mathbf{K}_\parallel$, charge-carrier temperature T_{e-h}, and charge-carrier density n_{e-h} at a low lattice temperature of $T = 4.2\,\text{K}$. The results in [001]GaAs/Al$_{0.3}$Ga$_{0.7}$As QWs shows that the rate of formation of excitons is very sensitive to $\mathbf{K}_\parallel$, T_{e-h}, and n_{e-h}. We have found from this study that the rate of formation due to LO phonon emission is dominant over that due to acoustic phonon emission at all $\mathbf{K}_\parallel$ for $T_{e-h} \gtrsim 50\,\text{K}$, but for $T_{e-h} \lesssim 50\,\text{K}$ there is a crossover at an exciton wavevector $\mathbf{K}_\parallel^0$ where the formation via acoustic phonon emission becomes dominant for $\mathbf{K}_\parallel > \mathbf{K}_\parallel^0$. It is also found that the rates of exciton formation by LA and TA phonons are comparable.

A theory is presented for calculating the rate of dephasing of free excitons as a function of temperature and density of excitons, lattice temperature, center-of-mass momentum of excitons, and QW width. Our results for GaAs/Al$_{0.3}$Ga$_{0.7}$As QWs show that the dephasing rate and linewidth of free excitons are very sensitive to the temperature and density of excitons. We have also found from our calculations that the rate of dephasing increases linearly with increasing exciton density as well as lattice temperature, which agrees very well with experimental results.

We have also developed a theory to calculate the rate of localization of 1s free excitons as a function of well width, lateral size of well-width fluctuations, center-of-mass kinetic energy, exciton temperature, and exciton density due to acoustic phonon emission in QWs and have applied it to GaAs/Al$_{0.3}$GA$_{0.7}$As QWs. We have found that the time of localization is very sensitive to the COM kinetic energy.

REFERENCES

[15.1] C. Weisbuch, H. Benisty, and R. Houdré, *J. Lumin.*, **85**, 271 (2000).

[15.2] J. Singh, *Excitation Energy Transfer Processes in Condensed Matter* (Plenum, New York, 1994).

[15.3] I.-K. Oh, J. Singh, A. Thilagam, and A.S. Vengurlekar, *Phys. Rev. B*, **62**, 2045 (2000) and references therein.

[15.4] I.-K. Oh and J. Singh, *Superlattices Microstruct.*, **30**, 221 (2001) and references therein.

[15.5] I.-K. Oh and J. Singh, *J. Lumin.*, **85**, 233 (2000) and references therein.

[15.6] I.-K. Oh, J. Singh, and A.S. Vengurlekar, *J. Appl. Phys.*, **91**, 5796 (2002) and reference therein.

[15.7] I.-K. Oh and J. Singh, *Superlattices Microstruct.*, **30**, 287 (2001) and references therein.

[15.8] E. Runge and R. Zimmermann, *Ann. Physik. (Leipzig)*, **7**, 417 (1998) and references therein.

[15.9] L.V. Butov, A. Imamoglu, A.V. Mintsev, K.L. Campman, and A.C. Gossard, *Phys. Rev. B*, **59**, 1625 (1999).

[15.10] A.V. Akimov, E.S. Moskalenko, L.J. Challis, and A.A. Kaplyanskii, *Physica B*, **219/220**, 9 (1996).

[15.11] P. Borri, W. Langbein, J.M. Hvam, and F. Martelli, *Phys. Rev. B*, **59**, 2215 (1999).

[15.12] G.D. Mahan, *Polarons in Heavily Doped Semiconductors*, in *Polarons in Ionic Crystals and Polar Semiconductors*, edited by J.T. Devreese (North-Holland, Amsterdam, 1972).

[15.13] T.C. Damen, J. Shah, D.Y. Oberli, D.S. Chemla, J.E. Cunningham, and J.M. Kuo, *Phys. Rev. B*, **42**, 7434 (1990).

[15.14] M. Gurioli, P. Borri, M. Colocci, M. Gulia, F. Rossi, E. Molinari, P.E. Selbmann, and P. Lugli, *Phys. Rev. B*, **58**, 13403 (1998).

[15.15] I.-K. Oh and J. Singh, *Int. J. Mod. Phys. B*, **15**, 3660 (2001).

[15.16] S.L. Chuang, *Physics of Optoelectronic Devices* (John Wiley & Sons Inc., New York, 1995).

[15.17] R. Strobel, R. Eccleston, J. Kuhl, and K. Köhler, *Phys. Rev. B*, **43**, 12564 (1991).

[15.18] C. Piermarocchi, F. Tassone, V. Savona, A. Quattropani, and P. Schwendimann, *Phys. Rev. B*, **55**, 1333 (1997).

[15.19] C. Piermorocchi, V. Savona, A. Quattropani, P. Schwendimann, and F. Tassone, *Phys. Status Solidi (A)*, **164**, 221 (1997).

[15.20] I.-K. Oh and J. Singh, *J. Lumin.*, **87–89**, 219 (2000).

[15.21] E. Vass, *Z. Phys. B*, **90**, 401 (1993).

[15.22] T. Takagahara, *Phys. Rev. B*, **31**, 6552 (1985).

[15.23] P.K. Basu and P. Ray, *Phys. Rev. B*, **45**, 1907 (1992).

[15.24] R. Kumar, A.S. Vengurlekar, S.S. Prabhu, J. Shah, and L.N. Pfeiffer, *Phys. Rev. B*, **54**, 4891 (1996).

[15.25] S. Adachi, *GaAs and Related Materials: Bulk Semiconducting and Superlattice Properties* (World Scientific, Singapore, 1994).

[15.26] R.C. Miller, D.A. Kleinman, O. Munteanu, and W.T. Tsang, *Appl. Phys. Lett.*, **39**, 1 (1981).

[15.27] R.C. Miller, A.C. Gossard, D.A. Kleinman, and O. Munteanu, *Phys. Rev. B*, **29**, 3740 (1984); R.C. Miller, D.A. Kleinman, and A.C. Gossard, *Phys. Rev. B*, **29**, 7085 (1984).

[15.28] S. Arlt, U. Siegner, F. Morier-Genoud, and U. Keller, *Phys. Rev. B*, **58**, 13073 (1998).

[15.29] S. Glutsch, D.S. Chemla, and F. Bechstedt, *Phys. Rev. B*, **51**, 16885 (1995).

[15.30] B. Pal and A.S. Vengurlekar, *Appl. Phys. Lett.*, **79**, 72 (2001).

[15.31] B.K. Ridley, *Electrons and Phonons in Semiconductor Multilayers* (Cambridge University Press, Cambridge, 1997).

[15.32] T. Takagahara, *J. Lumin.*, **44**, 347 (1989).

[15.33] D. Gammon, S. Rudin, T.L. Reinecke, D.S. Katzer, and C.S. Kyono, *Phys. Rev. B*, **51**, 16785 (1995).

[15.34] A.V. Gopal, R. Kumar, A.S. Vengurlekar, A. Bosacchi, S. Franchi, and L.N. Pfeiffer, *J. Appl. Phys.*, **87**, 1858 (2000).

[15.35] A. Patané, A. Polimeni, M. Capizzi, and F. Matelli, *Phys. Rev. B*, **52**, 2784 (1995).

[15.36] X. Fan, T. Takagahara, J.E. Cunningham, and H. Wang, *Solid State Commun.*, **108**, 857 (1998).

[15.37] D. Brinkmann, J. Kudrna, P. Gilliot, B. Hönerlage, A. Arnoult, J. Cibert, and S. Tatarenko, *Phys. Rev. B*, **60**, 4474 (1999).

[15.38] H.P. Wagner, A. Schätz, R. Maier, W. Langbein, and J.M. Hvam, *Phys. Rev. B*, **57**, 1791 (1998).

[15.39] D.-S. Kim, J. Shah, J.E. Cunningham, T.C. Damen, S. Schmitt-Rink, and W. Schäfer, *Phys. Rev. Lett.*, **68**, 2838 (1992).

[15.40] T. Takagahara, *Phys. Rev. B*, **32**, 7013 (1985).

[15.41] B. Segall, in Proceedings of the IXth International Conference on the Physics of Semiconductors, Moscow, 1968, edited by S.M. Ryvkin (Nauka, Leningrad, 1968), p. 425.

[15.42] A.L. Ivanov, P.B. Littlewood, and H. Haug, *Phys. Rev. B*, **59**, 5032 (1999).

[15.43] T. Ruf, J. Spitzer, V.F. Sapega, V.I. Belitsky, M. Cardona, and K. Ploog, *Phys. Rev. B*, **50**, 1792 (1994).

[15.44] V. Srinivas, J. Hryniewicz, Y.J. Chen, and C.E.C. Wood, *Phys. Rev. B*, **46**, 10193 (1992).

[15.45] L. Schultheis, A. Honold, J. Kuhl, K. Köhler, and C.W. Tu, *Phys. Rev. B*, **34**, 9027 (1986).

[15.46] A. Tredicucci, Y. Chen, F. Bassani, J. Massies, C. Deparis, and G. Neu, *Phys. Rev. B*, **47**, 10348 (1993).

[15.47] L. Schultheis and K. Ploog, *Phys. Rev. B*, **29**, 7058 (1984).

[15.48] S. Rudin, T.L. Reinecke, and B. Segall, *Phys. Rev. B*, **42**, 11218 (1990).

[15.49] F. García-Moliner and V.R. Velasco, *Theory of Single and Multiple Interfaces* (World Scientific, Singapore, 1992).

[15.50] I.-K. Oh and J. Singh, *J. Appl. Phys.*, **95**, 4883 (2004).

[15.51] D.S. Citrin, *Phys. Rev. B*, **47**, 3832 (1993).

[15.52] G. Bastard, C. Delalande, M.H. Meynadier, P.M. Frijlink, and M. Voos, *Phys. Rev. B*, **29**, 7042 (1984).

[15.53] O. Madelung, *Introduction to Solid-State Theory* (Springer, Berlin, 1978).

[15.54] C.P. Luo, M.K. Chin, Z.L. Yuan, and Z.Y. Xu, *Superlattices Microstruct*, **24**, 163 (1998).

16 Optical Properties and Spin Dynamics of Diluted Magnetic Semiconductor Nanostructures

A. Murayama and Y. Oka

Institute of Multidisciplinary Research for Advanced Materials, Tohoku University, Katahira 2-1-1, Aoba-ku, Sendai 980-8577, Japan

16.1 INTRODUCTION

Diluted magnetic semiconductors (DMSs) are compound semiconductors with magnetic ions, such as, Mn, Fe, and Co. They show remarkable spin-dependent properties due to the s, p–d exchange interactions of band-electrons (-holes) with the magnetic ions [16.1, 16.2]. These characteristic properties are very attractive for applications to magneto-optical devices. Much research interest has recently been generated for both III–V and II–VI DMS materials aiming at developing spin-related optical and electronic devices [16.3]. A significant optical property of DMSs is the giant Zeeman splitting in excitonic states, where the energy levels of spin states of the electron and hole in an exciton produce very large splittings in magnetic fields. The spin-splitting energy of the exciton is up to 100 meV. Therefore, the exciton relaxes rapidly into the lowest energy state, which has the specific spin direction, and the excess of energy is given to phonons. As a result, the exciton spins can

Optical Properties of Condensed Matter and Applications   Edited by J. Singh
© 2006 John Wiley & Sons, Ltd

be fully polarized in a DMS, which is a potential advantage in using DMSs for semiconductor devices, where the direction of the electron spin can be controlled by the direction of an external magnetic field. Another important property of DMSs is the giant Faraday rotation, especially for optical applications at room temperature. For example, an optical isolator using a $Cd_{1-x}Mn_xTe$-based crystal has already been developed and used in optical communications [16.4]. In addition to these two properties, several fundamental phenomena relating to the spin dynamics of the exciton are still under investigation. One of the very interesting properties is the formation of excitonic magnetic polarons, where the exciton interacts with many local magnetic moments of the doped magnetic ions through exchange interactions [16.5–16.7]. The exciton spin is significantly affected by the dynamical behavior of the local spins. This cooperative phenomenon of many spins is one of the most interesting properties of spin dynamics in semiconductors.

However, one still needs the basic understanding of quantum-confinement effects, exchange interactions between excitons and magnetic ions, and exciton-spin dynamics in DMS nanostructures. Here, we describe our recent studies on spin-dependent optical processes and the dynamics in several types of nanostructures based on II–VI DMS [16.8, 16.9]. These nanostructures show many attractive spin-related functionalities, such as spin injection, spin separation and switching, and ultrafast spin polarization. Moreover, the coupled nanostructures of the DMS, including self-assembled quantum dots (QDs), provide opportunities for the spin manipulation of carriers based on quantum tunneling. Therefore, the coupled DMS nanostructures are worth investigating for future developments of spintronic devices and quantum computing. Hybrid nanostructures consisting of DMSs and ferromagnetic materials can be used in a new technological development, since the small dimensions of the ferromagnetic materials can generate very strong local magnetic fields, which can align the exciton spins in the DMS nanostructures.

Spin injection and transport of excitons or carriers have been reported, where a DMS is used as the source of the spin-polarized excitons or carriers [16.10–16.13]. Details of the spin-injection process, however, had not been completely explored. Therefore, double-quantum-well (DQW) systems with proper tunneling barriers have been investigated by transient circularly polarized photoluminescence (PL) spectroscopy [16.14–16.17] to understand the dynamical processes of spin injection and relaxation. In a type of nanostructure called the spin-superlattice, modifications of the spin distribution of excitons or carriers are possible by applying magnetic fields [16.18, 16.19]. We have studied the field-induced switching of the spatial spin distribution and the resultant spin injection of excitons in the magnetic DQWs [16.20–16.22]. Transient optical phenomena of electron–hole (e–h) plasma, excitons, and magnetic ions in DMS-QWs involve basic physics on the dynamics of their spin states [16.23–16.25]. To clarify these ultrafast spin dynamics, transient pump-probe spectroscopy has been performed in magnetic fields. These measurements present rich physics of the dynamics of creation and relaxation of e–h plasma, excitons, and subsequent spin alignment of excitons [16.26–16.28].

Nanoscale dot and wire structures of DMS are expected to show efficient magneto-optical properties due to the low-dimensional confinement effects and high quantum yield of exciton luminescence [16.29–16.33]. We have designed and fabricated such DMS nanostructures by means of electron-beam lithography [16.34–16.37]. The exciton dynamics and the giant magneto-optical properties are studied in the DMS nanostructures with lateral dimensions ranging from 20 to 100 nm. Further, in a coupled QD system composed of the DMS and nonmagnetic QDs, the inter-dot injection of spin-polarized excitons is realized.

A new type of coupled system with self-assembled QDs that show strong quantum-confinement effects is also examined.

Hybrid nanostructures composed of DMS and ferromagnetic metals have various properties applicable to spin-related electronics [16.38]. We have fabricated a nanoscale hybrid structure of DMS wires sandwiched with Co wires [16.39–16.41]. The magneto-optical properties are affected by the local magnetic field that the neighboring Co wires generate. This magnetic field induces Zeeman splitting for the electron spins of Mn ions, which is detected by the spin-flip light scattering at the room temperature. The giant Zeeman effect of excitons induced by the Mn-spin alignment is also demonstrated.

In this chapter, in Section 16.2, we describe coupled QWs, and in Section 16.3, nanostructures fabricated by electron-beam lithography are presented. In Section 16.4, we have covered the properties of self-assembled QDs, and finally Section 16.5 presents the hybrid nanostructures with ferromagnetic materials.

16.2 COUPLED QUANTUM WELLS

16.2.1 Spin injection

The spin injection of excitons or carriers has been studied in a coupled DQW system with a magnetic quantum well (MW) and a nonmagnetic well (NW) [16.13]. However, the dynamical processes in the spin injection have not been clarified sufficiently, although knowledge of such detailed dynamics is crucial for the development of ultrafast spin-related devices. For instance, the spin-injection efficiency is determined by the injection rate of spin-polarized carriers from MW to NW relative to the spin-relaxation rate in both wells. The spin-injection efficiency decreases, if the carrier injection to the NW occurs faster than the spin polarization in the MW. The insertion of a proper tunneling barrier between the MW and NW can increase the spin-injection efficiency, since the carrier injection rate into the NW can be controlled by designing the thickness and/or the potential height of the barrier. Therefore, the dynamics of spin injection has been studied in the magnetic DQWs with various thicknesses of the tunneling barrier [16.14–16.17].

The energy diagram of a DQW studied is shown in Figure 16.1, which consists of the NW of $Zn_{0.76}Cd_{0.24}Se$ and MW of $Zn_{0.96}Mn_{0.04}Se/CdSe$. These NW and MW are separated by a barrier of ZnSe. The well widths of NW and MW were fixed to 7 nm and 40 nm, respectively. The barrier width (L_B) was varied from 4 to 8 nm in order to study the variation in the spin-injection dynamics, depending on the carrier tunneling rate. Transient and circularly polarized PL from the DQW was measured at 2 K by a spectrometer and a streak camera. Magnetic fields up to 6 T were applied in the Faraday configuration.

The PL spectrum of the DQWs ($L_B = 8$ nm) is shown in Figure 16.2. The DQW was excited by a linearly polarized light of energy 2.770 eV. The exciton emissions from the NW and MW appear at 2.530 and 2.720 eV at $B = 0$ T. The exciton PL in MW is very weak, although the MW with a width of 40 nm is wide enough to give rise a PL intensity in the MW. It suggests that a tunneling of excitons or carriers occurs from MW to NW in the present DQW. By increasing the magnetic field to 6 T, the PL peak of MW exciton with left-circular (σ^+) polarization at 2.677 eV shows a lower-energy shift (red shift), 43 meV, which is the typical giant Zeeman shift. Here, the σ^+ polarization is defined as the exciton recombination for the down-spin state with lower energy [magnetic quantum number:

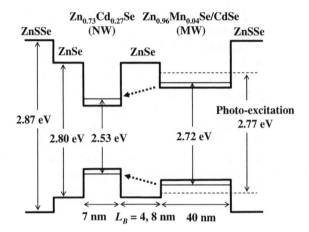

Figure 16.1 Energy diagram of the magnetic DQW with a magnetic quantum well (MW) of $Zn_{0.96}Mn_{0.04}Se/CdSe$ and a nonmagnetic well (NW) of $Zn_{0.73}Cd_{0.27}Se$

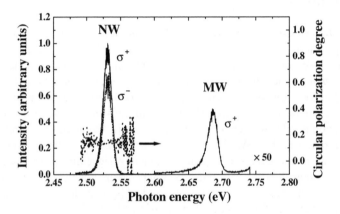

Figure 16.2 Excitonic PL spectra with circular polarizations (σ^+: solid lines and σ^-: a broken line) in the magnetic DQW ($L_B = 8$ nm) at an applied magnetic field of 3 T. Circular polarization degree also is plotted for the exciton PL of NW (a thick dotted line with an arrow). The excitation was made by using a linearly polarized light

$m_j = -1/2$ and $+3/2$ for electron and heavy-hole (hh), respectively]. In the NW, the exciton PL peak energy remains unchanged over the range of the magnetic field. For the optical excitation in magnetic fields, the carriers generated in the MW occupy, mostly, the down-spin states with their spin alignment parallel to the magnetic field direction. This is due to the fast spin relaxation, and the exciton spin is therefore highly polarized. These polarized down-spin excitons or carriers then tunnel into the NW. As a result, the circularly polarized exciton PL arises from the NW, if the spin relaxation is not significant in the NW. As can be seen in Figure 16.2, the circular polarization property appears in the NW with the polarization degree P of 0.20 at the magnetic field $B = 3$ T (shown as a thick dotted line). Here, the circular polarization degree is defined as $P = (I_{\sigma+} - I_{\sigma-}) / (I_{\sigma+} + I_{\sigma-})$, where $I_{\sigma+}$ and $I_{\sigma-}$

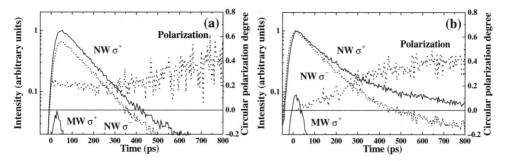

Figure 16.3 Circularly polarized PL intensities (σ^+: solid lines and σ^-: a broken line) and the circular polarization degree (a dotted line) as a function of time in the DQWs with $L_B = 4$ nm (a) and 8 nm (b) in 3 T

are the circularly polarized PL intensities with σ^+ and σ^- components. The observation of P from the NW is experimental evidence of the spin injection through the tunneling process. For studying the spin-injection dynamics, we measured the time-resolved PL of DQW by exciting it with a linearly polarized light pulse of energy 2.774 eV at $B = 3$ T, as shown in Figure 16.3. In the DQW sample with $L_B = 4$ nm shown in Figure 16.3(a), the PL lifetime from NW was 115 ps and the P value is 0.20 just after the excitation. On the other hand, in the sample with $L_B = 8$ nm shown in Figure 16.3(b), the PL lifetime from NW was 138 ps and P increases from 0.03 to 0.40 with a rise time of 400 ps. The increase in P in NW with the time constant of 400 ps can be considered as a result of the exciton tunneling from MW to NW with a long time constant. However, in the present case, the PL lifetime of excitons in MW is much shorter (30 ps), and a model of slow exciton-tunneling cannot be applied to these short-lifetime excitons. Therefore, we have proposed a model of individual tunneling of electrons and holes to interpret the experimental results quantitatively. Using this model, we have calculated the injection and recombination rates of electrons and holes in DQWs with the aid of rate equations. The calculated results of the time developments of MW and NW exciton PL intensities show good agreement with the experimental time evolution of P, where the electron-spin tunneling time is 30 ps and the hh-spin tunneling time is 1 ns. Therefore, when the PL lifetime from MW is short and the P builds up gradually in NW due to the slower hh tunneling, the individual charge-carrier tunneling model reproduces the observed results quantitatively.

16.2.2 Spin separation and switching

Field-induced switching of the spatial spin configuration has been realized in a DQW structure composed of MWs of $Zn_{0.97}Mn_{0.03}Te$ and NWs of ZnTe [16.21, 16.22]. The excitonic PL spectra of the magnetic DQW with the up- (σ^- polarization) and down- (σ^+ polarization) spin states are plotted as a function of the applied magnetic field in Figure 16.4. The energy of the exciton PL from MW shows the typical giant Zeeman shift with increasing magnetic field, while the exciton energy in NW remains constant at below 4 T. As a result, at a magnetic field of 4 T, both exciton energies overlap for the down- (σ^+ polarization) spin state.

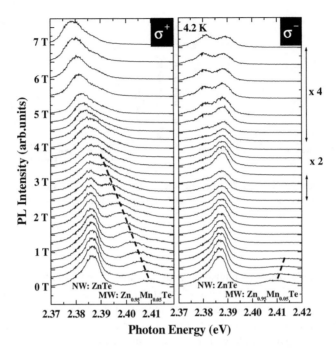

Figure 16.4 Circularly polarized excitonic PL spectra as a function of the applied magnetic field in the magnetic DQW of $Zn_{0.97}Mn_{0.03}Te/ZnTe$ measured at a temperature of 4.2 K. The broken lines indicate the magnetic field dependencies of the PL peak energy of $Zn_{0.97}Mn_{0.03}Te$, as a visual guide

At weak magnetic fields (<4 T), the intense σ^- and σ^+ PL arise from the ZnTe NW. Therefore, both the up- and down-spin (σ^- and σ^+) excitons in MW are injected into NW, since the PL intensity is approximately proportional to the number of exciton. The excitonic PL peak energies are plotted as a function of the applied magnetic field in Figure 16.5. As the magnetic field approaches 4 T, the σ^+ PL peak energy of the MW exciton approaches the PL peak energies of the NW exciton. Then, the σ^+ branch of the MW exciton overlaps at 4 T with the σ^- branch of the NW. However, the anti-crossing behavior is seen between the σ^+ branches of the MW and NW excitons. Solid lines are calculated by taking into account the interactions of the hh-exciton states for the σ^+ branches in the MW and NW. The interaction energy of the hh exciton is deduced to be 1.5 meV, which can be attributed to the exchange interaction between the wave functions of the hh-excitons. At magnetic fields above 4 T, the σ^+-exciton energies in MW and NW become separated again. In this case, the down-spin excitons (σ^+) can be injected from the NW layer to the MW, since the energy of the down-spin branch of the MW is lower than that of the NW.

The time-resolved PL was measured and compared with the calculated result from the rate equations based on the spin-injection model. From this fitting, we have deduced the spin-injection time. Figure 16.6 shows the deduced spin-injection time as a function of magnetic field. For the magnetic field lower than 2 T, the up- and down-spin excitons from MW are injected with the finite injection times into NW. The spin-injection time for the down-spin exciton ($\tau_{inj}^{\downarrow}$) increases from 12 ps to 300 ps with the increase in the magnetic field

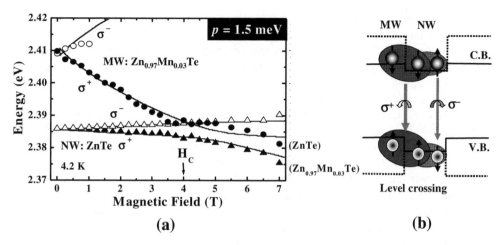

Figure 16.5 (a) Circularly polarized exciton energies as a function of magnetic field in the $Zn_{0.97}Mn_{0.03}Te/ZnTe$ DQW measured at 4.2 K. H_c indicates a level-crossing field. Solid lines are calculated taking into account the anti-crossing behavior due to the *hh*-exciton mixing with the interaction energy $p = 1.5$ meV. (b) A schematic illustration of the level crossing of spin-polarized excitons [Reproduced from S. Shirotori et al., *Phys. Status Solidi C*, **1**, 945 (2004) by permission of Wiley-VCH]

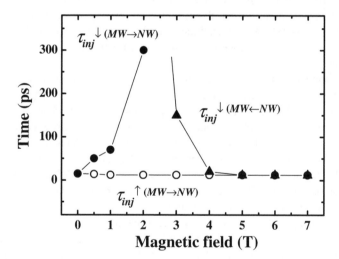

Figure 16.6 Spin-injection time deduced from the rate-equation analysis for the circularly polarized transient PL as a function of the applied magnetic field. For a magnetic field below 2 T, the spin injection of the down-spin exciton (σ^+) occurs from MW to NW, but above 3 T the direction of the spin injection reverses. In this case, the spin injection of the up-spin excitons (σ^-) takes place from MW to NW (open circles) with a constant injection time. The solid lines are drawn only as a visual guide. [Reproduced from S. Shirotori et al., *Phys. Status Solidi C*, **1**, 945 (2004) by permission of Wiley-VCH]

from 0 to 2 T, while the up-spin exciton injection time ($\tau_{inj}^{\uparrow}$) remains constant at 12 ps and does not change with the magnetic field (Figure 16.6). The spin-relaxation times derived for the excitons in MW and NW are 3 ps and 1 ns, respectively, which are functions of the Zeeman splitting energies. At magnetic fields above 3 T, the up- and down-spin excitons are spatially separated in each layer. The down-spin excitons are now more stable in MW than in NW, while the up-spin excitons remain stable in NW. In this range of the magnetic field, the transient PL can also be fitted by the rate equations, where the direction of the down-spin injection becomes opposite to that in the range of magnetic field $B \leq 2$ T (see Figure 16.6). For $B \geq 4$ T (above the level crossing field), the spin-injection time for the down-spin excitons decreases from 300 ps to 12 ps. The smaller energy difference of the down-spin levels between the MW and NW causes the longer injection time due to the reverse-injection process caused by thermalization. Thus, the results of Figure 16.6 show the field-induced switching of the spatial spin configuration and the control of the spin injection direction for the spin-polarized excitons.

16.2.3 Spin dynamics studied by pump-probe spectroscopy

Transient optical processes in the DMS-QWs involve various interesting spin dynamics of electron–hole (e–h) plasma, excitons, and local spins of magnetic ions [16.24, 16.25]. These phenomena are basic physical processes with the potential of spin-related optical applications. To distinguish each spin dynamics that appears in the ultrashort time range, a technique of the transient pump-probe absorption spectroscopy in magnetic fields has been developed [16.26–16.28]. The technique has been applied to study $Cd_{0.95}Mn_{0.05}Te/Cd_{0.80}Mg_{0.20}Te$ multiple quantum wells (MQWs) grown by molecular beam epitaxy on ZnTe (100) substrates.

Figure 16.7 shows the time evolution of the pump-probe absorption spectra in the magnetic MQW at zero magnetic field, with a pump-power density of 1.5 mJ/cm². $\Delta\alpha d$ is a transient differential absorbance with and without pumping. An induced absorption (defined as a positive value of $\Delta\alpha d$), which rises and decays within 3 ps, is observed at 1.735 eV after the pump-pulse excitation. This induced absorption directly reflects the bandgap renormalization (BGR) due to the exchange correlation in the dense e–h plasma excited by the intense pump-pulse. In the meantime, a bleached absorption (negative $\Delta\alpha d$) appears at the exciton-energy position of 1.795 eV. This bleached transient absorption results from the elimination of screening for the Coulomb interactions in excitons due to their high-density effects, after the e–h plasma decays. The renormalized energy bandgap after the optical pump can be extracted from the spectra of transient absorption $\Delta\alpha d$ calculated by taking into account the effects of the Coulomb screening, bandgap renormalization, and carrier heating [16.23]. The increase and decay of this BGR energy without any magnetic field, i.e., 0 T and at a temperature of 1.8 K, can be fitted by a function with two exponential components. The time constants for the increase and decay of this bandgap shift ΔE_g are 0.2 ps and 1.0 ps, respectively. ΔE_g reaches its maximum value at 77 meV in about 0.5 ps after the pumping, which gives the maximum plasma density of 3×10^{12} cm⁻². In DMS materials, the strong s, p–d exchange interactions between carriers and the localized spins of magnetic ions result in the giant Zeeman splitting. The BGR effect was clearly resolved for each spin-split Landau subband when the measurement was carried out in magnetic fields. Following the decay of the dense e–h plasma, the bleached peak of the exciton absorption at 1.795 eV reaches its

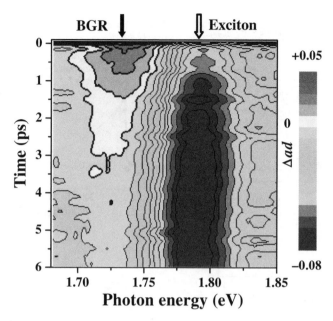

Figure 16.7 Transient differential absorption spectra of a magnetic $Cd_{0.95}Mn_{0.05}Te/Cd_{0.80}Mg_{0.20}Te$ MQW measured by the pump-probe technique with no magnetic field (0 T)

maximum intensity in 4 ps after the pumping. The maximum intensity indicates the maximum population of pumping-created excitons. The excitonic bleached peaks measured at 1.8 K without any magnetic field show an evident red-shift within the exciton lifetime. At a magnetic field of 3 T, the bleached peak splits into two branches of σ^+ and σ^- polarizations, which is caused by the giant Zeeman splitting of excitons.

Next, we discuss the time dependence of the bleached peaks in magnetic fields. Transient differential absorption spectra are shown in Figure 16.8(a) in the longer time range at a magnetic field of 5 T excited by a σ^--circularly polarized pump-pulse of 1.860 eV and with the lower pump-power density of 10 μJ/cm². Two intense bleached peaks were observed with an energy splitting up to 80 meV corresponding to the spin-polarized hh-exciton levels due to the giant Zeeman effect. The field dependence of the energy splitting can be expressed by a Brillouin function. The bleached peak at 1.820 eV corresponding to the up-spin exciton state decays faster than that of the down-spin state. It reflects the spin relaxation via spin-flip processes between those levels accompanied by the energy relaxation. However, a PL peak is observed only from the down-spin hh-exciton state. The disappearance of the up-spin PL spectrum can be attributed to the fast relaxation of the hh spin in the exciton state. The faster decay of the hh spin forms a dark exciton state. Therefore, the electron cannot recombine efficiently at the up-spin exciton branch. After the relaxation of the electron from the up- to down-spin states, the electron and hh can recombine radiatively to emit PL. The differential absorbances with σ^+ and σ^- polarizations are plotted as a function of time in Figure 16.8(b). The observed time dependences are fitted with those calculated by rate equations. From the best-fitted calculated results, we have successfully determined the spin-relaxation time of an electron to be 29 ps at a magnetic field of 5 T. The saturated absorbance

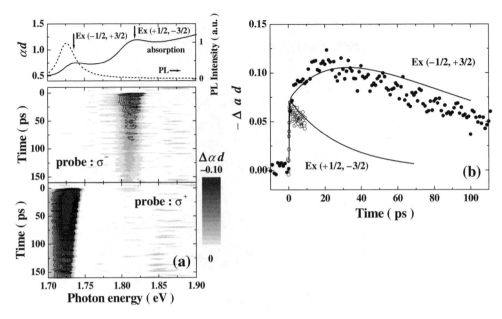

Figure 16.8 (a) Transient differential absorption spectra measured with the probes of σ^+ and σ^- polarizations for the $Cd_{0.95}Mn_{0.05}Te$ MQW in an applied magnetic field of 5 T and excited by a σ^--polarized pump-pulse of 1.860 eV. The linear absorption spectrum and PL spectrum are also shown above the differential absorption spectra. (b) Differential absorbances as a function of time for the higher- (open circles) and lower-energy (closed circles) spin states of excitons in an applied magnetic field of 5 T. Solid lines are best-fitted curves calculated using rate equations

for the down-spin state rises in accordance with the decay of that for the up-spin state. The decay of the saturated absorbance for the up-spin state directly displays the dynamics of the spin-flip process of the electron. The spin-relaxation time of the *hh* is found to be faster than 0.2 ps, which is our instrumental resolution limit. This fast decay of the *hh* spin may be attributed to the participation of multiple LO phonons in the relaxation process in addition to the band mixing of the hole states, because the calculated energy splitting of the *hh* state is found to be 64 meV at a magnetic field of 5 T and the LO-phonon energy is known to be 21 meV for CdTe. On the other hand, the relaxation time of the electron spin is obtained as 17 ps at zero magnetic field, which is shorter than the 46 ps observed in CdTe MQW. Such a short relaxation time can be attributed to the spin relaxation via an s–d exchange mechanism with many Mn spins. At the applied magnetic field of 5 T, the spin-relaxation time of 29 ps is found to be longer than that of 17 ps at 0 T. At such a high magnetic field, the Mn spins are strongly pinned down by the external high magnetic field, which may eliminate the relaxation pathways of the s–d exchange mechanism. As the energy splitting of the electron-spin state is calculated to be 16 meV at the applied magnetic field of 5 T, which is still lower than the LO-phonon energy of 21 meV, the LO phonon cannot participate in the electron-spin relaxation process. Therefore, the electron-spin relaxation caused by the exchange mechanism is experimentally elucidated to be suppressed by magnetic fields.

16.3 NANOSTRUCTURES FABRICATED BY ELECTRON-BEAM LITHOGRAPHY

The fabrication of intentionally designed nanostructures of DMS is important to develop spin electronic devices [16.29, 16.30]. We have fabricated nanostructures of QDs and wires of $Zn_{1-x-y}Cd_xMn_ySe$ [16.34–16.37]. Single QWs of $Zn_{1-x-y}Cd_xMn_ySe$ with thicknesses from 5 to 10 nm were grown on GaAs (100) substrates by molecular beam epitaxy. DMS nanostructures with the lateral dimensions of 20–100 nm were successfully fabricated from the QW by electron-beam lithography followed by chemical wet-etching. Figure 16.9 shows (a) a schematic illustration of the dot structure fabricated from a DMS-QW of $Zn_{1-x-y}Cd_xMn_ySe$ by using electron-beam lithography, (b) a scanning electron microscope (SEM) image of the dot array, (c) a SEM image of the dots with an average diameter of 20 nm, and (d) a SEM image of dots of 80 nm diameter.

An excitonic PL spectrum from a dot sample of diameter 30 nm is shown in Figure 16.10 (solid line) and also shown for comparison is the PL spectrum from the QW before the dot fabrication (dotted curve). The peak of exciton PL in these QDs appears at 2.690 eV, in comparison with that at 2.680 eV in the QW (Figure 16.10). The blue-shift of 10 meV observed in the dot's PL peak can be attributed to the lateral confinement effects for the carriers, while the exciton-binding energy tends to be lowered in the confined structure. The relaxation of the lattice strain in the dot formation, which causes a lower energy shift of the exciton, is

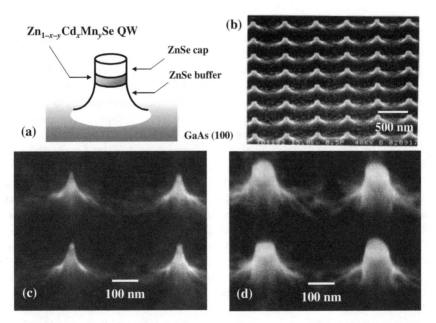

Figure 16.9 (a) A schematic drawing of the dot structure fabricated from a DMS-QW of $Zn_{1-x-y}Cd_xMn_ySe$ by using electron-beam lithography. (b) SEM image of the dot array. (c) SEM image of the dots with an average diameter of 20 nm. (d) SEM image of dots of 80 nm diameter [Reproduced from A. Uetake et al., *Phys. Status Solidi C*, **1**, 941 (2004) by permission of Wiley-VCH]

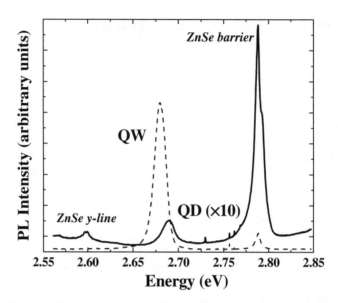

Figure 16.10 Excitonic PL spectra from a DMS-dot of $Zn_{0.80}Cd_{0.15}Mn_{0.05}Se$ with a lateral diameter of 30 nm (a solid line) and from the QW before the dot fabrication (dotted line). PL peaks due to the ZnSe barrier and the defect-related ZnSe y-line are also observed [Reprinted from A. Murayama et al., *J. Lumin.* (ICL2005 Conference Proceedings), in press, with permission from Elsevier]

negligible, since the lattice mismatch between the QW and the buffer- or barrier-layers is relatively small (1.4%). In addition to this, the exciton dynamics is found to be different between the QDs and the QW. The transient PL was measured in a high magnetic field, as shown in Figure 16.11. The exciton PL peak from QDs of diameter 30 nm shows a lower energy shift of 2.5 meV with a time constant of 102 ps, while that from the QW shows a larger energy shift of 6 meV with a shorter time constant of 60 ps. Applications of high magnetic fields remove the transient formation of excitonic magnetic polarons [16.7]. Therefore, the smaller spectral diffusion with a longer time constant for the QD excitons indicates a significant suppression of the localization of excitons, possibly due to the limitation of diffusion length of the excitons in the dots. In the $Zn_{1-x-y}Cd_xMn_ySe$ QW, the potential fluctuations due to the alloying can cause exciton localization and hence spectral diffusion. The exciton-diffusion length is calculated to be 40 nm, which is larger than the dot diameter of 30 nm. Therefore, the exciton diffusion is significantly limited in the dots. The giant Zeeman shift of excitons in the QDs appears with a shift energy of 45 meV at 7 T, which is larger than that observed in the QW before the QD fabrication. By applying a magnetic field, a significant increase in the luminescence intensity is also observed, which originates from the suppression of the Auger energy transfer from excitons to Mn ions [16.42].

A systematic study of the increase in the giant Zeeman effect of excitons was made by using a wire structure of DMS, where the wire width was intended to vary from 30 to 90 nm, as shown in Figure 16.12. Using a Brillouin function, a solid line is fitted by calculation as follows:

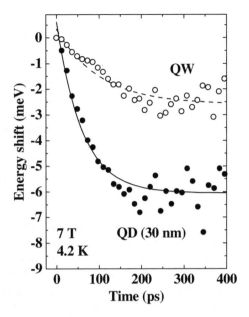

Figure 16.11   Exciton PL-peak energies as a function of time for DMS-QDs with the lateral diameter of 30 nm (closed circles) and DMS-QW (open circles) at an applied magnetic field of 7 T at 4.2 K. Solid and broken lines are calculated by fitting to single-exponential functions

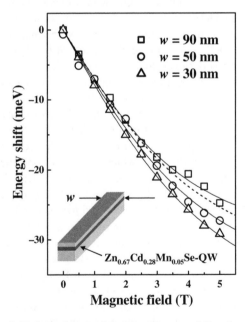

Figure 16.12   Magnetic-field dependences of the giant Zeeman shifts of excitons in the DMS-wires of $Zn_{0.67}Cd_{0.28}Mn_{0.05}Se$ of various wire widths denoted by w. Solid lines are fitted calculations using Brillouin functions. A broken line represents the case of the QW before the wire fabrication [Reprinted from A. Murayama et al., *J. Lumin.* (ICL2005 Conference Proceedings), in press, with permission from Elsevier]

$$\Delta E = \frac{1}{2} x N_0 (\alpha - \beta) S_0 B_{\frac{5}{2}} \left(\frac{5 g \mu B}{2 k_B (T + T_0)} \right) \tag{16.1}$$

where x is Mn concentration, $N_0\alpha$ and $N_0\beta$ are exchange parameters for an electron and heavy hole, respectively [16.43]. g and μ are the g-factor and magnetic moment of Mn spin, respectively. T_0 is an effective temperature expressing antiferromagnetic coupling among Mn spins. As can be seen, the field dependence of the exciton Zeeman shift can be expressed by the Brillouin function. Moreover, the giant Zeeman shift of excitons in high magnetic fields increases with decreasing wire width. The relaxation of lattice strain due to the wire formation in the epitaxial layers is negligible for the increase in the Zeeman shift, since the increase in the Mn–Se bond length reduces the hole-exchange parameter [16.44]. Also, the increase in the confinement effect can decrease the electron-exchange parameter [16.45]. Therefore, the suppression of exciton diffusion is the most reasonable cause, since the exciton-diffusion length of 40 nm is consistent with the experimental wire width of 50 nm, where the Zeeman shift starts to increase. The exciton can migrate and then localize at deeper potential sites at which the Mn concentration is lower in the well structure. Such migration is suppressed in the narrow wire structures. Therefore, it results in an increase in the giant Zeeman shift, since the Zeeman shift increases with increasing Mn content below 10%.

A double-quantum dot (DQD) system, one of DMS and the other nonmagnetic, was fabricated from a double-quantum well (DQW) to study the spin manipulation, such as spin injection. By the lithography technique, DQD with a lateral diameter of 30 nm was successfully fabricated from DQW composed of $Zn_{0.70}Cd_{0.22}Mn_{0.08}Se$-MW (5 nm-thick) and $Zn_{0.76}Cd_{0.24}Se$-NW (10 nm thick). These quantum wells were separated by a 5-nm-thick ZnSe barrier. The structure of DQD is schematically illustrated in Figure 16.13(a). The exciton energy of the DMS-QD was designed to be 100 meV higher than that of the nonmagnetic QD, by adjusting the Cd content and the well thickness. In the DQD system, the measured exciton lifetime in the DMS-QD was as short as 20 ps, suggesting that the excitons tunnel from the DMS-QD to the nonmagnetic QD. Also, the intensity of the exciton PL from DMS-QD decreases down to 1/100 upon decrease of the barrier thickness from 15 to 5 nm. By application of magnetic fields, photo-excited carriers and hence excitons in the DMS-QD are highly spin-polarized by the Mn spins, which is directly confirmed by the high value of the circular polarization degree $P = 0.80$ for the exciton PL. Therefore, the spin-polarized excitons can be efficiently injected into the nonmagnetic QD by quantum tunneling through the potential barrier. P values up to 0.20 are observed for exciton PL from the nonmagnetic QD, as shown in Figure 16.13(b). This is experimental evidence for the spin injection in the present DQD system. Such a P value is higher than 0.11, which is found to be in the initial QW before the dot fabrication. The enhanced P value can be attributed to an increase in the number of spin-polarized carriers injected from DMS-QDs into nonmagnetic QDs. As the DMS-QD is next to the ZnSe buffer-layer, a large number of excitons excited in the ZnSe buffer-layer can migrate into DMS-QD and become subsequently spin polarized. Therefore, the number of spin-polarized excitons is significantly increased in this quantum-dot structure, since the area of the dot is markedly smaller than the area of the exciton diffusion, as discussed before. As a result, the larger number of spin-polarized carriers is injected into the nonmagnetic QD, which gives the higher P value for the exciton PL in the nonmagnetic quantum dot.

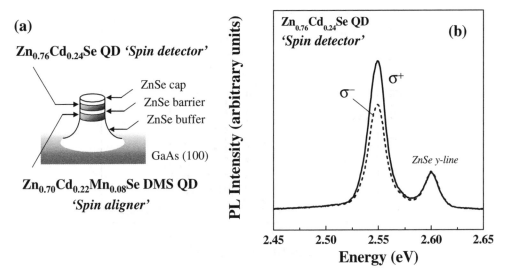

Figure 16.13 (a) A double-quantum dot (DQD), one a magnetic DQD and the other nonmagnetic, structure for the study of spin-injection. (b) Excitonic PL spectra with σ^+ (a solid line) and σ^- (a broken line) polarizations from the nonmagnetic QD of $Zn_{0.76}Cd_{0.24}Se$ in the DQD at a magnetic field of 3 T. A PL peak due to the defect-related ZnSe y-line is also seen. DQD was excited by linearly polarized light [Reprinted from A. Murayama et al., *J. Lumin.* (ICL2005 Conference Proceedings), in press, with permission from Elsevier]

16.4 SELF-ASSEMBLED QUANTUM DOTS

Self-assembled quantum dots (SAQDs) are also very attractive for optical applications because of the discrete density-of-states of photo-excited carriers due to strong quantum-confinement effects. In addition, the spin-relaxation time of an electron has been reported to be as long as several ns in SAQDs [16.46, 16.47]. Such a long time constant of the spin relaxation is an advantage for the spin manipulation in semiconductor materials. Figure 16.14 shows the micro-PL spectra of SAQDs of CdSe, where the plane QD layer was fabricated into a nanopillar shape using electron-beam lithography. The SAQD layer was grown by molecular beam epitaxy with the proper growth conditions to realize the Stranski–Krastanov (SK) growth mode [16.31, 16.32]. As can be seen in Figure 16.14, sharp PL peaks appear with a spectral width of 0.4 meV, which is the instrumental width, indicating excitonic PL from individual SAQDs. The dot diameter and the density were deduced as 3.5 ± 0.2 nm and 5000 μm^{-2}, respectively, from the PL energy and the number of PL peaks in a pillar sample with the specific diameter.

Next, a coupled SAQD system with a DMS layer was studied [16.48]. The magnetic-field dependences of the Zeeman shift of the exciton are shown in Figure 16.15 for SAQDs of CdSe coupled with a DMS-cap layer of $Zn_{0.80}Mn_{0.20}Se$. Typical giant Zeeman shifts of excitons are observed, which can be well fitted by a Brillouin function. The Zeeman shift in SAQDs reaches 20 meV at a magnetic field of 5 T. The exciton lifetime is also affected by the Mn ions, due to the magnetic-field-induced suppression of the energy-transfer process from the exciton to the internal d–d transitions via Auger processes [16.42]. The lifetime of

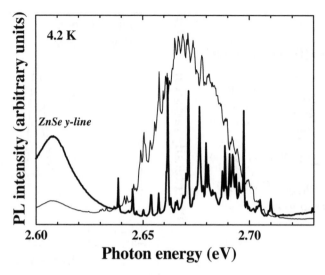

Figure 16.14 Micro-PL spectra of self-assembled quantum dots (SAQDs) of CdSe in the nanopillar sample (a thick solid line) and in the sample of a planar QD layer before the nanopillar fabrication (a thin solid line). The sharp PL peaks of the SAQDs originate from excitons in individual dots [Reproduced from I. Souma et al., *J. Supercond.*, **18**, 219 (2005). With kind permission of Springer Science and Business Media]

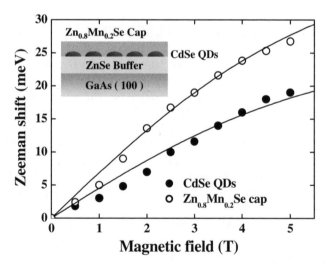

Figure 16.15 Giant Zeeman shifts of excitons in the self-assembled CdSe QDs (closed circles) and in the magnetic cap layer of $Zn_{0.80}Mn_{0.20}Se$ (open circles). Solid lines are fitted by calculations using Brillouin functions [Reproduced from I. Souma et al., *J. Supercond.*, **18**, 219 (2005). With kind permission of Springer Science and Business Media]

the exciton is 40 ps, which is markedly shorter than the 230 ps for the CdSe dots without DMS layer. These properties can be explained by spatial overlapping of wave functions of excitons in QDs with those of Mn ions in the DMS layer, and also due to some diffusions of Mn ions into QDs.

A further study of the coupled SAQDs has been made to demonstrate spin manipulation. Spin injection into nonmagnetic SAQDs is an important subject to investigate, because it opens the way to realize semiconductor spintronic devices and quantum computing based on such nanostructures. The direction of the spin can be controlled by the direction of the external magnetic field in the DMS, due to the giant Zeeman effects of carriers. Therefore, the carrier spins are polarized in SAQDs, if the carriers are injected from spin-conserving states. Thus, spins can be manipulated in QDs by the use of the long relaxation time. A coupled SAQD structure of CdSe with a DMS-QW is schematically illustrated in Figure 16.16(a). In this case, a barrier layer of ZnSe was inserted to separate the SAQDs and DMS-QW. The circularly polarized PL spectra from such a nanostructure are shown in Figure 16.16(b) at a magnetic field of 5 T. A broad emission band with σ^- polarization corresponds to the excitonic PL from the QD ensemble, which is caused by the dot-size distribution. The peak energy of the QD emission band can be controlled by changing the average dot-size. The resultant energy difference between the excitonic PL peak in the QD band and DMS-QW is 50 meV at zero magnetic field. The P value of the exciton PL in the CdSe SAQDs without the DMS-QW is about −0.10 in magnetic fields; the negative P value means that the intensity of the σ^--polarized PL is stronger than that of the σ^+ polarization. The negative P value originates from the negative g-values of excitons in the SAQDs of CdSe. The exciton PL peak from the DMS-QW is indicated by the E^{MW} arrow in Figure 16.16(b). The magnetic-field dependence of the E^{MW} peak shows a giant Zeeman shift and therefore the spectrum appears only in the σ^+-polarized component in nonzero magnetic fields. More-over, as can be seen in Figure 16.16(b), an additional PL peak, indicated by E_A, appears with σ^+ polarization. The energy of the additional PL peak depends on the magnetic field

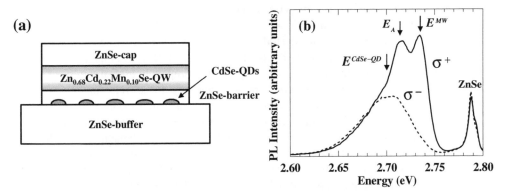

Figure 16.16 (a) A coupled SAQD structure of CdSe with a DMS-QW of $Zn_{0.68}Cd_{0.22}Mn_{0.10}Se$ (MW) and a 5-nm-thick tunneling barrier of ZnSe. (b) Circularly polarized excitonic PL spectra in the coupled SAQDs at a magnetic field of 5 T. The σ^+-polarized PL spectrum (a solid line) is composed of 3 components (see text), while the σ^--polarized spectrum (a broken line) shows the exciton PL from the CdSe-QD ensemble only [Reprinted with permission from A. Murayama et al., *Appl. Phys. Lett.*, **88**, 023114, 1–3. Copyright (2006), the American Institute of Physics]

and agrees with that calculated from a Brillouin function with the same parameters as those for the DMS well. The energy of the additional PL peak is 25 meV, which is lower than that of the exciton energy E^{MW} in DMS-QW. This energy difference is almost equal to the LO-phonon energy of 27 meV for in a single CdSe QD layer. To interpret this additional PL peak, the electron and *hh* levels are calculated for the coupled structure in nonzero magnetic fields. As a result, the type-II band alignment can be assumed for a down-spin electron in the QDs and down-spin *hh* in the DMS-QW in high magnetic fields. The time-resolved circularly polarized PL shows that the lifetime of such additional PL is as long as 3.5 ns. The exciton PL lifetime in the DMS well and CdSe QDs are 60 ps and 240 ps, respectively, which are nearly the same as those obtained from their corresponding single layers. Therefore, we conclude that the additional PL in the coupled structure originates from the type-II transition between electrons injected from the DMS well and *hh*s remaining in the DMS part. The photoexcited down-spin electrons in DMS-QW can efficiently penetrate into SAQDs via LO-phonon scattering, although down-spin *hh*s stay inside the DMS well because of the type-II band alignment. The down-spin *hh* level in DMS-QW is calculated to be 6 meV lower than that in the dot at a magnetic field of 5 T, where the electron energy in this dot is assumed to be 1-LO-phonon energy lower than that in DMS-QW. This scenario is appropriate for realizing the electron-spin injection into SAQDs. The electron-spin injection takes place efficiently into the SAQDs via LO-phonon-assisted resonant tunneling, since the energy level in the QD is completely discrete.

16.5 HYBRID NANOSTRUCTURES WITH FERROMAGNETIC MATERIALS

The characteristic optical properties such as, the giant Zeeman effects of excitons and giant Faraday rotation in II–VI DMS, are attractive for applications in magneto-optical devices operating in the visible-light region, and an optical isolator operating at room temperature has already been developed. In addition, spin-polarized carriers can be generated in DMS due to the giant Zeeman splitting of their electronic states, which is also used as a spin aligner in spintronic devices. In these studies, applications of external fields to the II–VI DMS are necessary for the alignment of paramagnetic spins in DMS. For applying efficient magnetic fields to DMS nanostructures, it is important to prepare hybrid structures of DMS with ferromagnetic materials. It offers a new technology for developing microscopic magneto-optical and spin-related devices without significant energy consumption. Therefore, several types of hybrid structures with DMS-QWs have been proposed [16.38, 16.49]. We have fabricated hybrid nanostructures of DMS-QW with ferromagnetic cobalt (Co) wires [16.39–16.41]. The DMS-QW was made into wires of width down to 100 nm and sandwiched between the Co wires. The resonant spin-flip light scattering was observed at room temperature in the hybrid nanostructure, indicating the Zeeman shift of paramagnetic spins of Mn ions and hence the spin alignment of Mn ions in DMS. Moreover, we have studied giant Zeeman effects of excitons at 4.2 K in the same hybrid nanostructure, where the Co magnetization in the wires is aligned normal to the wire surface by applying the external magnetic field. The advantage of this hybrid nanostructure is that nearly perpendicular magnetic fields can be produced between the Co wires, originating from overlapping magnetic fields generated from the neighboring Co wires. Therefore, the whole DMS-QW

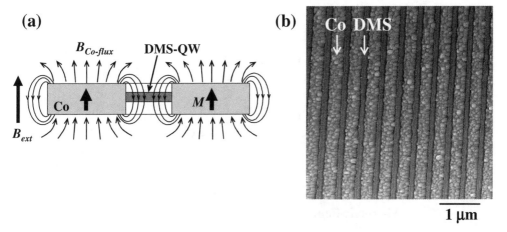

Figure 16.17   (a) A schematic diagram of the hybrid nanostructure of DMS-QW with ferromagnetic Co wires. The magnetic-field distribution generated by the Co magnetization is also illustrated, where the Co magnetization is aligned by applying an external magnetic field. (b) An AFM image of the surface of hybrid nanostructure

part in the hybrid structure can feel the perpendicular fields applied from Co wires, which provides the Faraday configuration for optical measurements. As a result, the magneto-optical properties, such as the giant Zeeman effects of excitons, are expected to be affected by the magnetic flux from Co wires.

The hybrid nanostructure fabricated for this study is schematically illustrated in Figure 16.17(a). Wires of a $Zn_{0.69}Cd_{0.23}Mn_{0.08}Se$-QW of width 200 nm are sandwiched between Co wires of width 400 nm and thickness 50 nm. A calculation of the magnetic-field distribution around the gap of the Co-wire system shows that a field of intensity up to 0.5 T and nearly perpendicular to the well plane can be applied at the position of DMS-QW, when the Co magnetization is perfectly aligned normal to the wire plane. (Note: The magnetic-field distribution is calculated in a direction nearly perpendicular to the QW plane, when the Co magnetization is perfectly aligned perpendicular to the Co-wire plane, that is, the QW plane, by applying the external magnetic field.) The intensity of the lateral component of the magnetic field is less than 10% of that of the perpendicular component. To realize this situation, external magnetic fields higher than an in-plane magnetic shape-anisotropy field of Co $(4\pi M_s = 1.8\,T)$ are applied normal to the wire surface. When the Co magnetization orients perpendicular to the film surface, a magnetic flux, $B_{Co\text{-}flux}$ from the Co magnetization can be applied in opposite (anti-parallel) direction of the external field. Therefore, an effective perpendicular magnetic field, $B^{\perp}_{eff} = B_{ext} - B^{\perp}_{Co\text{-}flux}$, is finally applied to the DMS-QW. The perpendicular magnetic field, $B^{\perp}_{eff}$ induces giant Zeeman effects of excitons in the DMS-QW. The effect of $B^{\perp}_{Co\text{-}flux}$ on the giant Zeeman shift of excitonic PL can be distinguished from the PL spectral shape, since the strength of $B^{\perp}_{Co\text{-}flux}$ is inhomogeneous in the DMS-QW part and thus can cause the broadening in the PL spectrum, while that of B_{ext} is made perfectly uniform in our experimental set-up by using a micro-PL system. An atomic-force microscope (AFM) image of the surface of the hybrid nanostructure is shown in the Figure 16.17(b).

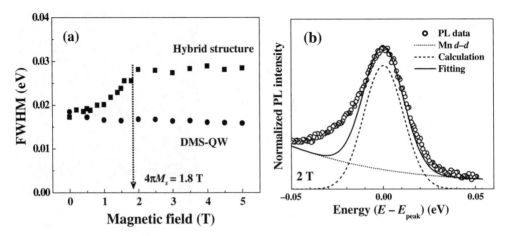

Figure 16.18 (a) FWHM of the exciton PL spectrum as a function of external field in the hybrid nanostructure (closed squares) and in the planar QW before the fabrication of hybrid nanostructure (closed circles). The in-plane shape-anisotropy field of the Co wire is 1.8 T, at which the Co magnetization aligns perpendicularly to the wire surface. (b) PL spectrum of the hybrid nanostructure at 2 T (open circles). A calculated PL spectral shape (a broken line) and a broad spectrum tail due to d–d internal transitions of Mn ions in the DMS-QW (a dotted line). Superposition of both the spectra (a solid line) is compared with the PL spectrum. On the horizontal axis, $E - E_{peak}$ corresponds to the photon-energy difference from the PL peak energy E_{peak} [Reprinted with permission from M. Sakuma et al., *Appl. Phys. Lett.*, **85**, 6203. Copyright (2004) the American Institute of Physics]

The full width at half maximum (FWHM) of the exciton PL spectrum for the hybrid nanostructure increases significantly with increasing external field, as shown in Figure 16.18(a). It is attributed to the giant Zeeman effects with various energy shifts due to the inhomogeneous magnetic flux applied from the Co wires. At a zero external magnetic field, the Co magnetization lies completely in-plane and no magnetic flux in the Faraday configuration is produced from the Co wire system to the DMS-QW. As the external field increases, the Co magnetization rotates gradually toward the perpendicular direction of the film surface and $B^{\perp}_{Co\text{-}flux}$ starts to act on excitons in the DMS-QW, as in the Faraday configuration. Therefore, the FWHM of the exciton PL spectrum in the hybrid nanostructure increases gradually with increasing external magnetic field and saturates at 1.9 T, which is almost equal to the in-plane shape-anisotropy field (1.8 T) of the Co film. In order to confirm the giant Zeeman effects due to the application of $B^{\perp}_{Co\text{-}flux}$, a quantitative analysis of the spectral shape was made based on the field distribution of $B^{\perp}_{Co\text{-}flux}$ in the position of the DMS-QW. The result is compared with the observed PL spectrum at a magnetic field of 2 T, as shown in the Figure 16.18(b). The calculated result for the spectral shape shows a good agreement with the spectrum observed, except for the tail on the lower-energy side. Therefore, it is also an evidence for indicating that the $B^{\perp}_{Co\text{-}flux}$ induces the giant Zeeman effects of excitons in all of the DMS-QW wires. The emission in the lower-energy tail can be attributed to that from local areas with relatively low $B^{\perp}_{Co\text{-}flux}$, possibly due to irregularly formed wider gaps between the Co wires. Further study is necessary using ferromagnetic thin films with the perpendicular residual magnetization at zero external magnetic field.

16.6 CONCLUSIONS

Dynamical spin-injection processes of excitons have been studied in a DQW composed of MW and NW. In the magnetic $Zn_{0.96}Mn_{0.04}Se/Zn_{0.76}Cd_{0.24}Se$ DQWs with 4 nm- and 8 nm-thick barriers, the spin injection takes place by both fast- and slow-injection processes, with time constants of 5 ps and 1 ns, respectively. Excitons in the MW recombine with a faster decay time of 30 ps than the slow spin-injection process. Therefore, the slow spin-injection from the MW is quantitatively explained by individual tunneling of electrons and heavy holes. In the magnetic $Zn_{0.97}Mn_{0.03}Te/ZnTe$ DQW, we have shown the magnetic-field-induced switching of the spin configuration and of the direction of spin injection. The spin-injection time increases significantly toward the level-crossing field of 2.5 T. The obtained results are useful for designing efficient spin injectors and magneto-optical switching devices.

Dynamical processes of optically created e–h plasma and spin-polarized excitons in the DMS-MQW of $Cd_{0.95}Mn_{0.05}Te$ have been studied by using transient pump-probe spectroscopy. The BGR induced by the pump-created e–h plasma is observed in the time region of 0–3 ps. After the e–h plasma decays, the bleached peak of exciton absorption rises and shows spin-dependent transient behavior in magnetic fields. The electron-spin-relaxation time was determined as a function of the magnetic field. The electron-spin-relaxation time is found to be affected by the magnetic field through the s–d exchange mechanism with the local magnetic spins of Mn ions.

Nanoscale dots and wires of $Zn_{1-x-y}Cd_xMn_ySe$-based DMS have been fabricated by means of electron-beam lithography. The blue-shift of the excitonic PL is observed with decreasing dot diameters, which indicates the lateral confinement of the exciton. The exciton diffusion is significantly suppressed in these nanostructures, which also results in the enhancement of the giant Zeeman effects of excitons. Coupled QDs composed of the DMS-MW and NW have been successfully fabricated. Excitons tunnel from the MW into the NW, and the circular polarization degree of 20% is obtained for the exciton PL in the NW due to the spin injection from the MW to the NW. In addition, SAQDs of CdSe coupled with a DMS-QW have been examined. Spin injection via LO-phonon-assisted resonant tunneling is observed. It is expected to open a new technology for spin manipulation in SAQDs.

Hybrid nanostructures of DMS with ferromagnetic materials have been fabricated. Magneto-optical properties of the hybrid nanostructures are studied by spin-flip light-scattering of Mn ions and excitonic PL for the $Zn_{1-x-y}Cd_xMn_ySe$ wires in a hybrid structure. Magnetic fields from the Co wires are confirmed to be efficiently applied to the DMS nanostructure and the field-induced giant Zeeman effects of excitons are observed.

ACKNOWLEDGEMENTS

The authors are indebted to K. Nishibayashi, Z.H. Chen, K. Hyomi, I. Souma, K. Kayanuma, S. Shirotori, H. Sakurai, K. Seo, H. Ikada, A. Uetake, T. Tomita, T. Asahina, and M. Sakuma for their helpful collaborations. This work is supported by the Nanotechnology Project of NEDO and also by the Ministry of Education, Science and Culture, Japan.

REFERENCES

[16.1] *Diluted Magnetic Semiconductors, Semiconductors and Semimetals*, edited by J.K. Furdyna and J. Kossut (Academic Press, New York, 1988), Vol. 25.

[16.2] D.D. Awschalom and N. Samarth, *J. Magn. Magn. Mater.*, **200**, 130 (1999).

[16.3] *Semiconductor Spintronics and Quantum Computation*, edited by D.D. Awschalom, D. Loss, and N. Samarth (Springer, Berlin, 2002).

[16.4] K. Onodera, T. Masumoto, and M. Kimura, *Electron. Lett.*, **30**, 1954 (1994).

[16.5] J.H. Harris and A.V. Nurmikko, *Phys. Rev. Lett.*, **51**, 1472 (1983).

[16.6] D.R. Yakovlev, G. Mackh, B. Kuhn-Heinrich, W. Ossau, A. Waag, G. Landwehr, R. Hellmann, and E.O. Göbel, *Phys. Rev. B*, **52**, 12033 (1995).

[16.7] M. Nogaku, J.X. Shen, R. Pittini, T. Sato, and Y. Oka, *Phys. Rev. B*, **63**, 153314 (2001); M. Nogaku, R. Pittni, T. Sato, J.X. Shen, and Y. Oka, *J. Appl. Phys.*, **89**, 7287 (2001).

[16.8] Y. Oka, K. Kayanuma, S. Shirotori, A. Murayama, I. Souma, and Z.H. Chen, *J. Lumin.*, **100**, 175 (2002).

[16.9] Y. Oka, K. Kayanuma, E. Nakayama, S. Shiratori, I. Souma, T. Tomita, Z.H. Chen, and A. Murayama, *Nonlinear Opt.*, **29**, 491 (2002).

[16.10] R. Fiederling, M. Keim, G. Reuscher, W. Ossau, G. Schmidt, A. Waag, and L.W. Molenkamp, *Nature*, **402**, 787 (1999).

[16.11] M. Oestreich, J. Huebner, D. Haegele, P.J. Klar, W. Heimbrodt, W.W. Ruehle, D.E. Ashenford, and B. Lunn, *Appl. Phys. Lett.*, **74**, 1251 (1999).

[16.12] Y. Ohno, D.K. Young, B. Benschoten, F. Matsukura, H. Ohno, and D.D. Awschalom, *Nature*, **402**, 790 (1999).

[16.13] W.M. Chen, I.A. Buyanova, G.Yu. Rudko, A.G. Mal'shukov, K.A. Chao, A.A. Toropov, Y. Terent'ev, S.V. Sorokin, A.V. Lebedev, S.V. Ivanov, and P.S. Kop'ev, *Phys. Rev. B*, **67**, 125313 (2003).

[16.14] K. Kayanuma, E. Shirado, M.C. Debnath, I. Souma, Z.H. Chen, and Y. Oka, *J. Appl. Phys.*, **89**, 7278 (2001).

[16.15] K. Kayamuma, E. Shirado, M.C. Debnath, I. Souma, S. Permogorov, and Y. Oka, *Physica E (Amsterdam)*, **10**, 295 (2001).

[16.16] K. Kayanuma, S. Shirotori, Z.H. Chen, T. Tomita, A. Murayama, and Y. Oka, *Physica B (Amsterdam)*, **340–342**, 882 (2003).

[16.17] W.M. Chen, I.A. Buyanova, K. Kayanuma, K. Nishibayashi, K. Seo, A. Murayama, Y. Oka, A.A. Toropov, A.V. Lebedev, S.V. Sorokin, and S.V. Ivanov, *Phys. Rev. B*, **72**, 073206 (2005).

[16.18] B.T. Jonker, L.P. Fu, W.Y. Yu, C. Chou, A. Petrou, and J. Warnock, *J. Appl. Phys.*, **73**, 6015 (1993).

[16.19] J.F. Smyth, D.A. Tulchinsky, D.D. Awschalom, N. Samarth, H. Luo, and J.K. Furdyna, *Phys. Rev. Lett.*, **71**, 601 (1993).

[16.20] M.C. Debnath, J.X. Shen, I. Souma, T. Sato, R. Pittini, Z.H. Chen, and Y. Oka, *Physica E (Amsterdam)*, **10**, 310 (2001).

[16.21] S. Shirotori, K. Kayanuma, T. Tomita, Z.H. Chen, A. Murayama, and Y. Oka, *J. Supercond.*, **16**, 457 (2003).

[16.22] S. Shirotori, K. Kayanuma, I. Souma, T. Tomita, A. Murayama, and Y. Oka, *Phys. Status Solidi C*, **1**, 945 (2004).

[16.23] G. Trankle, H. Leier, A. Forchel, H. Haug, C. Ell, and G. Weiman, *Phys. Rev. Lett.*, **58**, 419 (1987).

[16.24] J.J. Baumberg, D.D. Awschalom, N. Samarth, H. Luo, and J.K. Furdyna, *Phys. Rev. Lett.*, **72**, 717 (1994).

[16.25] S.A. Crooker, D.D. Awschalom, J.J. Baumberg, F. Flack, and N. Samarth, *Phys. Rev. B*, **56**, 7574 (1997).

[16.26] Z.H. Chen, H. Sakurai, K. Seo, K. Kayanuma, T. Tomita, A. Murayama, and Y. Oka, *Physica B (Amsterdam)*, **340–342**, 890 (2003).

[16.27] H. Sakurai, K. Seo, Z.H. Chen, K. Kayanuma, T. Tomita, A. Murayama, and Y. Oka, *Phys. Status Solidi C*, **1**, 981 (2004).

[16.28] Z.H. Chen, H. Sakurai, T. Tomita, K. Kayanuma, A. Murayama, and Y. Oka, *Physica E (Amsterdam)*, **21**, 1022 (2004).

[16.29] M. Illing, G. Bacher, T. Kummell, A. Forchel, D. Hommel, B. Jobst, and G. Landwehr, *J. Vac. Sci. Technol. B*, **13**, 2792 (1995).

[16.30] O. Ray, A.A. Sirenko, J.J. Berry, N. Samarth, J.A. Gupta, I. Malajovich, and D.D. Awschalom, *Appl. Phys. Lett.*, **76**, 1167 (2000).

[16.31] K. Shibata, E. Nakayama, I. Souma, A. Murayama, and Y. Oka, *Phys. Status Solidi B*, **229**, 473 (2002).

[16.32] T. Kuroda, N. Hasegawa, F. Minami, Y. Terai, S. Kuroda, and K. Takita, *J. Lumin.*, **83/84**, 321 (1999).

[16.33] A.K. Bhattacharjee and C. Benoit a la Guillaume, *Phys. Rev. B*, **55**, 10613 (1997).

[16.34] N. Takahashi, K. Takabayashi, I. Souma, J.X. Shen, and Y. Oka, *J. Appl. Phys.*, **87**, 6469 (2000).

[16.35] Z.H. Chen, M.C. Debnath, K. Shibata, T. Saitou, T. Sato, and Y. Oka, *J. Appl. Phys.*, **89**, 6701 (2001).

[16.36] H. Ikada, T. Saito, N, Takahashi, K. Shibata, T. Sato, Z.H. Chen, I. Souma, and Y. Oka, *Physica E (Amsterdam)*, **10**, 373 (2001).

[16.37] A. Uetake, H. Ikada, T. Asahina, M. Sakuma, K. Hyomi, T. Tomita, A. Murayama, and Y. Oka, *Phys. Status Solidi C*, **1**, 941 (2004).

[16.38] J. Kossut, I. Yamakawa, A. Nakamura, G. Cywiński, K. Fronc, M. Czeczott, J. Wróbel, F. Kyrychenko, T. Wojtowicz, and S. Takeyama, *Appl. Phys. Lett.*, **79**, 1789 (2001).

[16.39] M. Sakuma, K. Hyomi, I. Souma, A. Murayama, and Y. Oka, *J. Appl. Phys.*, **94**, 6423 (2003).

[16.40] M. Sakuma, K. Hyomi, I. Souma, A. Murayama, and Y. Oka, *Phys. Status Solidi B*, **241**, 664 (2004).

[16.41] M. Sakuma, K. Hyomi, I. Souma, A. Murayama, and Y. Oka, *Appl. Phys. Lett.*, **85**, 6203 (2004).

[16.42] M. Nawrocki, Yu.G. Rubo, J.P. Lascaray, and D. Coquillat, *Phys. Rev. B*, **52**, R2241 (1995).

[16.43] B.E. Larson, K.C. Hass, H. Ehrenreich, and A.E. Carlsson, *Phys. Rev. B*, **37**, 4137 (1988).

[16.44] R. Meyer, M. Dahl, G. Schaack, A. Waag, and R. Boehler, *J. Cryst. Growth*, **159**, 997 (1996).

[16.45] I.A. Merkulov, D.R. Yakovlev, A. Keller, W. Ossau, J. Geurts, A. Waag, G. Landwehr, G. Karczewski, T. Wojtowicz, and J. Kossut, *Phys. Rev. Lett.*, **83**, 1431 (1999).

[16.46] M. Paillard, X. Marie, P. Renucci, T. Amand, A. Jbeli, and J.M. Gérard, *Phys. Rev. Lett.*, **86**, 1634 (2001).

[16.47] S. Mackowski, T.A. Nguyen, H.E. Jackson, L.M. Smith, J. Kossut, and G. Karczewski, *Appl. Phys. Lett.*, **83**, 5524 (2003).

[16.48] I. Souma, K. Kayanuma, K. Hyomi, K. Nishibayashi, A. Murayama, and Y. Oka, *J. Supercond.*, **18**, 219 (2005).

[16.49] H. Schömig, A. Forchel, S. Halm, G. Bacher, J. Puls, and F. Henneberger, *Appl. Phys. Lett.*, **84**, 2826 (2004).

[16.50] A. Murayama, H. Hirano, K. Hyomi, M. Sakuma, I. Souma, and Y. Oka, *J. Lumin.* (ICL2005 conference proceedings), in press.

[16.51] A. Murayama, T. Asahina, K. Nishibayashi, I. Souma, and Y. Oka, *Appl. Phys. Lett.*, **88**, 023114, 1–3 (2006).

Index

Plain type locators refer to text treatments: page references in **bold** or *italic* type indicate reference to a table or figure only. Greek letters and numerals are alphabetized as though spelled out.